AF531176

ROLES OF ORGANIC MATTER IN SEDIMENT DIAGENESIS

Based on a Symposium
Sponsored by the Society of
Economic Paleontologists and Mineralogists

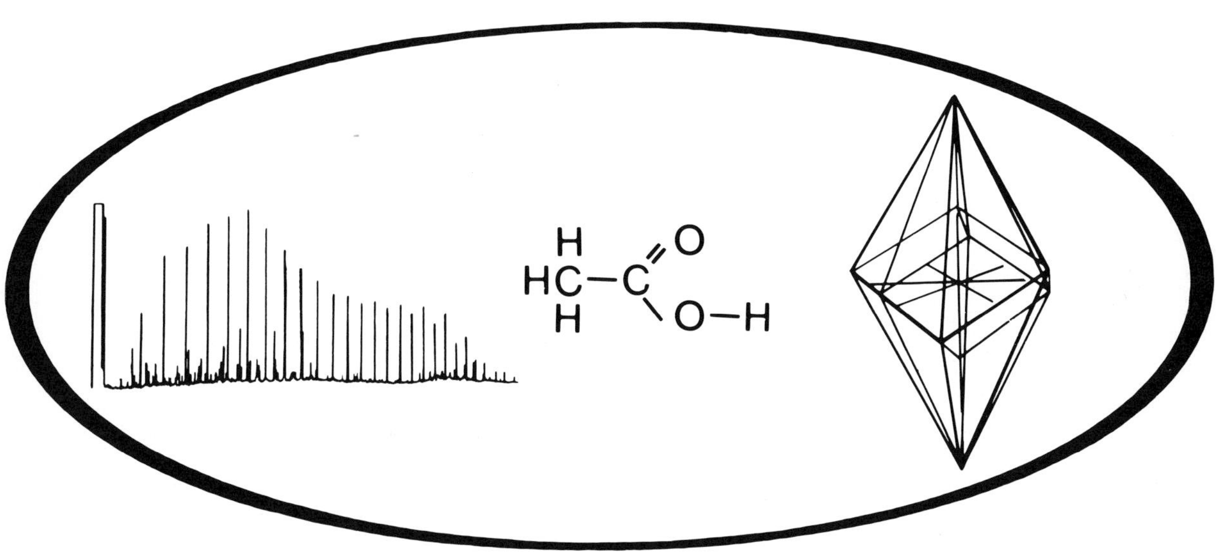

Edited by
Donald L. Gautier
U.S. Geological Survey, Denver, Colorado

SOCIETY OF ECONOMIC PALEONTOLOGISTS AND MINERALOGISTS
Special Publication No. 38

Tulsa, Oklahoma U.S.A. ISBN: 0-918985-59-5 *June, 1986*

A Publication of
The Society of Economic Paleontologists and Mineralogists
a division of
The American Association of Petroleum Geologists

Preface to

Society of Economic Paleontologists and Mineralogists Special Publication
Roles of Organic Matter in Sediment Diagenesis

Donald L. Gautier

This volume is a direct result of a symposium entitled "Relationship of Organic Matter and Mineral Diagenesis" convened as part of the 1984 mid-year meeting of the Society of Economic Paleontologists and Mineralogists (SEPM) in San Jose, California. That symposium, in turn, was the result of a SEPM Research Conference held in October 1983 at Lost Valley Ranch, Colorado. The publication contains, with a few additions, papers presented at the symposium. The goal of the volume "Roles of Organic Matter in Sediment Diagenesis" is to bring to the attention of the sedimentological community the importance of interaction of organic compounds with the inorganic sedimentary system and the degree to which organic compounds drive diagenetic systems.

This volume comprises 16 reports illustrative of the scope and direction of current research in sedimentological and geochemical studies of organic/inorganic interaction. In the first paper, Walter demonstrates that calcite precipitation is inhibited by sulfate, but that aragonite precipitation is inhibited by orthophosphate. Orthophosphate also inhibited dissolution of calcite, Mg calcite, and aragonite, whereas sulfate had little effect on dissolution rate.

John Kocurko, in the second paper, shows that oxidation of organic compounds is the controlling mechanism in the dissolution and precipitation of early diagenetic cements in recent sediments of south Louisiana.

Charles Curtis and Max Coleman carry the interpretation of early diagenetic cements to the ancient sedimentary record of the Carboniferous Coal Measures by means of isotopic and elemental geochemical data.

Laura Crossey and her coauthors discuss use of pyrolytic techniques in the evaluation of the quality, type, and maturity of organic matter in various organic-rich rocks.

In their analysis of diagenesis in Neogene Tertiary rocks of Japan, Taguchi and his coauthors present one of the most comprehensive studies of organic and inorganic diagenesis available. They conclude that transformations of organic compounds do not always drive mineral reactions, but that the thermocatalytic effects of clay minerals may bring about significant changes in lipids, nitrogen fixation, and carbon isotopic trends.

The relationships among transformations of clay minerals, organic compounds, and the thermal evolution of sedimentary basins are discussed by Naeser and by Pollastro and Barker in their studies of the Green River Basin of Wyoming. Naeser, using fission-track dating in formerly annealed apatite grains, provides evidence for the significant recent cooling of the basin. Pollastro and Barker use a combination of clay mineralogy, vitrinite reflectance analysis, and fluid-inclusion homogenization temperatures to arrive at a similar conclusion that the Green River Basin was 30–50° C hotter during the Neogene than at present. The cooling is explained by uplift and erosion of approximately 1,700 m of sedimentary section.

The Green River Basin is also the subject of Edman and Surdam, who conclude that porosity enhancement in the Upper Cretaceous Ericson Sandstone was controlled by position of the sandstone relative to a paleo-oil water contact. The enhanced porosity is inferred to result from organic acid generation and associated aluminum complexing. Other papers also focus on acids derived from organic matter. Kharaka, LeRoy, Carothers, and Goerlitz illustrate the degree to which aqueous organic species, particularly short-chain aliphatic acid anions, control pH and Eh of subsurface waters in sedimentary basins. Indu Meshri compares the stoichiometric and thermodynamic reactivities of carbonic acid and organic acids. Her work shows that organic acids have a hydrogen donor capacity that is 6 to 350 times greater than carbonic acid. Lundegard and Land provide evidence that, from a mass balance standpoint, decarboxylation of organic matter is insufficient to account for all the secondary porosity in sandstones of the Texas Gulf Coast. They suggest that porosity may be greatly enhanced locally by organic acids and carbon dioxide and that hydrous pyrolysis reactions between organic carbon and kerogen may provide a significant source of acids for dissolution reactions in the subsurface.

Crossey, Surdam, and Lahann argue that reduction of mineral oxidants and consequent oxidation of organic matter may be a viable mechanism for releasing additional carboxylic acids from kerogen, thereby increasing porosity by the dissolution of carbonate cements and aluminosilicates.

Knut Bjørlykke provides evidence that meteroic waters, not organic-derived species, are responsible for the development of secondary porosity in the sandstones of the Brent Group (Jurassic, North Sea). In this case, the secondary porosity is not volumetrically important.

Jim Wood's paper presents a theoretical framework for the development of secondary porosity due to decarboxylation of organic matter. Calculated thermal mass transfer coefficients for calcite and quartz show that low temperatures and high P_{CO_2} are most effective in mobilizing calcite, but that high temperatures are more effective in the mobilization of quartz. Another paper by Wood and Hewett shows that, from a theoretical perspective, forced-fluid flow may be significant in controlling the distribution of authigenic mineral phases.

For the last paper, Charles Barker and Robert Halley present a case study of the application of fluid-inclusion analysis, vitrinite reflectance measurement, and stable isotopic studies to the interpretation of the thermal and diagenetic evolution of the Bone Spring Limestone. Barker and

Halley show that the history of cementation, fluid migration, and petroleum migration are clearly recorded during diagenesis.

Thanks are due to all the authors for their patience and diligence and to the Society of Economic Paleontologists and Mineralogists Publications Committee and Council for authorization. I am grateful to Special Publications Editor Barbara Lidz and to the SEPM Research Committee, who recognized the significance of the interactions of organic and inorganic constituents of sedimentary basins.

The U.S. Geological Survey was most generous in permitting me the time and technical support to put the symposium and volume together. Finally, my profound thanks to M. P. Krupa for her technical assistance, organizational skill, and forbearance.

CONTENTS

RELATIVE EFFICIENCY OF CARBONATE DISSOLUTION AND PRECIPITATION DURING DIAGENESIS: A PROGRESS REPORT ON THE ROLE OF SOLUTION CHEMISTRY

LYNN M. WALTER
Department of Earth and Planetary Sciences
Washington University
St. Louis, Missouri 63130

Abstract: The textural, physical and chemical properties of limestones are often controlled by diagenetic reactions involving dissolution and precipitation of carbonate minerals. To constrain the rates and paths of these reactions, chemically controlled dissolution and precipitation experiments were performed using various carbonate solids in a number of solution types at 25°C anda P_{CO_2} of $10^{-2.5}$ atm. The solution chemical variables included Mg/Ca ratio, salinity and dissolved SO_4^{-2} and PO_4^{-3}. Dissolution rates of calcite, aragonite and Mg-calcite as a function of saturation state were similar regardless of changes in Mg/Ca ratio (over the range 1–5), salinity (0–35‰) and SO_4^{-2} (0–0.028 m). The addition of dissolved orthophosphate inhibited dissolution of all three carbonate minerals to similar degrees.

In contrast, calcite precipitation rates in seeded growth experiments responded strongly to changes in Mg/Ca ratio and dissolved SO_4^{-2} concentration. Changes in salinity had no influence on calcite growth rates. The relative precipitation rates of calcite versus aragonite were strongly affected by the presence of SO_4^{-2} and PO_4^{-3}. Calcite precipitation was selectively inhibited by SO_4^{-2}, whereas aragonite precipitation was selectively inhibited by PO_4^{-3}. Thus, not only absolute but also relative growth rates of carbonate cements may be influenced by the nature of the geochemical environment. The results suggest that sedimentary microevironments, where PO_4^{-3} is released from organic matter degradation and SO_4^{-2} is depleted via bacterial SO_4^{-2} reduction, may be favorable sites for calcite or Mg-calcite precipitation.

The data also suggest that carbonate precipitation reactions are much more readily inhibited by various chemical species than are dissolution reactions. This gives rise to a large difference in relative rates of dissolution and precipitation in certain geochemical environments, which generally favors the dissolution side of the reaction.

INTRODUCTION

From the time of their initial deposition continuing through deep burial, carbonate rocks are prone to chemical alteration. Both precipitation and dissolution reactions play important roles in carbonate diagenesis. The relative efficiency of the two competing reactions controls the ultimate porosity, texture and chemistry of the carbonate rock produced.

One approach to understanding the rates and products of carbonate diagenetic reactions is the petrographic and geochemical study of the rocks themselves. Even more can be learned from the study of the chemistry of coexisting carbonate rock-pore fluid systems. In both cases, however, there is always some question concerning which fluid was responsible for producing a given rock texture or chemical signature. The multiple fluids potentially involved in forming a limestone make unraveling chemical cause and effect a formidable and often impossible task.

The alternate approach to understanding relative rates of diagenetic reactions is direct rate determination from laboratory experiments. Necessary prerequisites for obtaining geologically useful results from experimental studies are the selection of reasonable chemical parameters and the rigorous investigation of the influence of each parameter. This paper presents the results of initial investigations of the effects of some of the major chemical variables in natural diagenetic environments on rates of carbonate dissolution and precipitation. The main variables explored here are solution Mg/Ca ratio, salinity, and dissolved SO_4^{-2} and PO_4^{-3} concentrations. The influence of each variable was investigated as a function of solution saturation state with respect to the dissolving or precipitating carbonate phase. Previous papers (Walter and Morse, 1984b; Walter, 1985; Walter and Morse, 1985) have dealt with the effect of other important variables such as biogenic grain microstructure and mineralogic composition on experimentally determined rates of dissolution. The influence of biogenic carbonate microstructure and mineralogy on cement growth rates from seawater is described by Walter and Millero (in prep.). The role of differences in solution chemistry on carbonate reaction kinetics is the main focus of this paper.

Other workers have dealt with the effects of some of the chemical variables investigated here. For basic reviews on mechanisms of carbonate dissolution and precipitation, the reader is referred to Plummer et al. (1979) and Morse (1983). The effect of solution Mg/Ca ratio on the growth rate and composition of calcite has been addressed (e.g., Folk, 1974; Berner, 1975; Mucci and Morse, 1983), but the effect on dissolution rates of different carbonate minerals has not been explored. Dissolved orthophosphate has long been known to inhibit carbonate mineral dissolution and precipitation (e.g., Raistrick, 1949; Simkiss, 1964; Berner and Morse, 1974), but its relative influence on carbonate precipitation versus dissolution rates has not been specifically addressed.

The role of dissolved sulfate in carbonate diagenetic reactions have commanded attention recently, specifically as an inhibitor of dolomitization of calcite and aragonite (Baker and Kastner, 1981). The work of Bischoff and Fyfe (1968) on the inhibition of the aragonite-calcite transformation by dissolved SO_4^{-2} indicates that it may be an important inhibitor of the precipitation of carbonate minerals other than dolomite. Although Sjöberg (1978) reported that SO_4^{-2} inhibits calcite dissolution rates, Kushnir (1983) observed that SO_4^{-2} had no effect on the dissolution rates of dolomite, calcite or aragonite. The present study differs from those mentioned above because it quantifies the influence of dissolved SO_4^{-2} on the seeded precipitation of both calcite and aragonite under chemically controlled conditions.

Increased ionic strength over the range of 0.002 to 0.200 was shown to have no effect on calcite growth rates from

$CaCl_2$ solutions (Kazmierczak et al., 1982); however, the effect of increased salinity (0–35‰) on carbonate reaction kinetics has not been investigated in laboratory experiments. Based solely on geologic observations, it has been suggested that dissolved Na^+ may act as an inhibitor of calcite precipitation (Folk, 1974).

One aspect of carbonate reaction rates that has not been adequately examined is the relative rates of dissolution versus precipitation as a function of departure from equilibrium. The saturation state with respect to the dissolving or precipitating carbonate mineral is a fundamental control on reaction rate. At saturation (equilibrium) the net reaction rate is equal to zero. As the degree of undersaturation or supersaturation is increased, the reaction rate will increase. It is not clear from present date, however, if reaction rates are "symmetrical" about the equilibrium point. For example, would a solution which is 5 times supersaturated with respect to calcite precipitate calcite at the same rate as a solution 5 times undersaturated with respect to calcite would dissolve calcite? Also, certain dissolved species may have selective effects on carbonate dissolution or precipitation reaction rates leading to potential asymmetries in relative rates of dissolution versus precipitation.

This paper presents results based on controlled experimentation concerning how carbonate dissolution versus precipitation rates at known saturation states are affected by changes in Mg/Ca ratio, salinity, PO_4^{-3} and SO_4^{-2}. Specific questions addressed are:

1. Is the relative efficiency of carbonate dissolution versus precipitation reactions as a function of saturation state affected by differences in solution chemistry?
2. What predictions can be made concerning the products of carbonate diagenetic reactions in different geochemical environments (meteoric, marine, hypersaline), and which dissolved ion variations are most significant for understanding carbonate reaction rates in nature?
3. Can changes in solution chemistry promote selective effects on the precipitation rates of aragonite versus calcite (or Mg-calcite)?

METHODS

Solids and Solutions

The carbonate solids used in the dissolution and precipitation experiments are listed in Table 1. The three biogenic solids (coral, *Halimeda* and echinoid) were crushed to the 33–125 μm size range. The samples were cleaned ultrasonically to remove submicron particles and annealed (170° C for 6 hours) to eliminate crystal strain potentially induced by grinding samples. The importance of careful sample pretreatment for obtaining accurate kinetic data has been discussed at length by Walter and Morse (1984a). Synthetic calcite (Mallinkrodt "Analytical Reagent" grade) and synthetic aragonite (synthesized using the method of Wray and Daniels, 1957) were also used. Synthetic solids lack the complicated microstructure of biogenic grains, and their total crystalline surface area generally is available for reaction (Reddy et al., 1981; Walter and Morse, 1984b). The specific surface areas of both synthetic carbonates were determined by the BET krypton gas adsorption method (Brunauer et al., 1938), as modified by deKanel and Morse (1979). The synthetic calcite had a specific surface area of $0.45\ m^2g^{-1}$, synthetic aragonite had a specific surface area of $1.40\ m^2g^{-1}$. Precipitation rates obtained using these solids as seeds were normalized to specific surface area to permit comparison of calcite and aragonite precipitation rates at equal seed surface areas.

TABLE 1.—CARBONATES USED IN THE EXPERIMENTS

Dissolution	Precipitation
Synthetic calcite	Synthetic calcite
Aragonite: coral (*Fungia*)	Synthetic aragonite
green alga (*Halimeda*)	
Mg-calcite: echinoid (*Clypeaster*); (12 mole % $MgCO_3$)	

TABLE 2.—SOLUTIONS USED IN CARBONATE DISSOLUTION AND PRECIPITATION EXPERIMENTS

Dissolution	Precipitation
A. Mg/Ca = 5: seawater (S = 35‰) Mg = 0.05m, Ca = 0.01m (I = 0.7)	A. Mg/Ca = 5: seawater (S = 35‰) Mg = 0.05m, Ca – 0.01m (I = 0.7)
B. Mg/Ca = 5: $MgCl_2$ (0.05m) + $CaCl_2$ (0.01m) (I = 0.18)	B. Mg/Ca = 5: $MgCl_2$ (0.05m) + $CaCl_2$ (0.01m) (I = 0.18)
C. Mg/Ca = 1: $CaCl_2$ (0.03m) + $MgCl_2$ (0.03m) (I = 0.18)	C. Mg/Ca = 1: $CaCl_2$ (0.01m) + $MgCl_2$ (0.01m)
	1. +0.08m NaCl; (I = 0.18)
	2. +0.03m Na_2SO_4; (I = 0.12)
	3. +0.70m NaCl; (I = 0.73)
	D. $CaCl_2$ (0.01m); (I = 0.032)

Note: I denotes ionic strength. $I = 1/2\ \Sigma\ m_iz_i^2$ where m_i = molality of ion i and z_i = charge on ion.

The solution types investigated in the study are listed in Table 2. The first question addressed in the experiments was whether reaction rates in natural seawater (filtered Gulf Stream seawater, S = 35‰) having high NaCl content and dissolved SO_4^{-2} (0.028 m) would be the same as those at equivalent saturation states in low ionic strength $CaCl_2$ + $MgCl_2$ media lacking NaCl and SO_4^{-2}. If the rates in the two endmember solutions were not the same, the various dissolved components which differed had to be investigated individually to detect which of them (NaCl, Mg/Ca, SO_4^{-2}) might be responsible for differences in reaction rates. This approach required investigation of a rather large number of solution types. Some of the dissolution and precipitation experiments also were spiked with dissolved orthophosphate (PO_4^{-3}) to observe its effectiveness as an inhibitor of carbonate precipitation versus dissolution reactions.

EXPERIMENTAL PROCEDURE

All experiments were performed at 25° C and a P_{CO_2} of $10^{-2.5}$ atm using the pH-stat method (Morse, 1974). This method permits precise precipitation and dissolution rate measurements (± 10%) at a known and constant saturation state in solution. To maintain constant saturation state, the Ca^{+2} and CO_3^{-2} ion activity product must remain constant. Detailed descriptions of procedures for pH-stat dissolution and pH-stat precipitation are given in Walter (1983) and Mucci (1981), respectively. Briefly, pH was held at a fixed value during each experiment by automatic addition of acid during dissolution or base during precipitation. Reaction rates were calculated from the addition rate and concentration of the titrant. Dissolved Ca^{+2} was fixed (within 2%) during the dissolution reaction because addition of acid compensated via dilution effects for the amount of Ca^{+2} added. For precipitation experiments, however, a separate Ca^{+2} titrant was co-injected with the basic titrant to replace the Ca^{+2} removed by precipitation.

Once the desired test solution was equilibrated with the CO_2 gas and the pH had stabilized, the weighed carbonate sample was introduced to the solution and maintained in suspension by an overhead stirrer to avoid grinding. Both precipitation and dissolution rates were measured following 30 minutes of reaction. After this time interval, rates were constant for many hours of reaction (as many as 7 hours). Following the precipitation experiments, samples of the solids were retained for X-ray diffraction analysis and examination using the scanning electron microscope. The precipitation is believed to have occurred only on seed surfaces because no precipitates were observed on the walls of the glass reaction vessel. Furthermore, reaction rates were always directly proportional to the mass of seed used to induce precipitation, which would not be expected if spontaneous precipitation were occurring.

Calculation of Saturation State

One of the main factors controlling the rate of reaction is the saturation state with respect to the dissolving or precipitating carbonate mineral phase. The equilibrium constants used to evaluate carbonate mineral saturation state in seawater (35‰ at 25°C) are given in Table 3. Those for solutions other than normal seawater at 25° C are given in Table 4. The saturation state, Ω, is expressed for aragonite and calcite in seawater as:

$$\Omega_{\substack{\text{aragonite}\\ \text{or calcite}}} = \frac{m_{Ca^{+2}} \times m_{CO_3^{-2}}}{K'_{\substack{\text{aragonite}\\ \text{or calcite}}}} \tag{1}$$

where m_i = the total concentration of ion i in moles kg^{-1} seawater and K' is the mineral stoichiometric solubility constant in seawater.

The saturation state with respect to aragonite or calcite in solutions other than seawater is expressed as:

$$\Omega_{\substack{\text{aragonite}\\ \text{or calcite}}} = \frac{a_{Ca^{+2}} \times a_{CO_3^{-2}}}{K_{\substack{\text{aragonite}\\ \text{or calcite}}}} \tag{2}$$

where a_i = thermodynamic activity of $\text{ion}_i = \gamma_i m_i$, and γ_i = free ion activity coefficient and m_i = molality of free ion i in moles kg^{-1} H_2O.

The saturation state for magnesian calcite containing 12 mole % $MgCO_3$ was evaluated similarly in both seawater and other media because no stoichiometric solubility constant has been determined for Mg-calcite in seawater. The fractional exponent form for the expression of magnesian calcite saturation state (Thorstensen and Plummer, 1977) has been used here:

$$\Omega_{\substack{\text{Mg-calcite}\\ \text{12 mole \% MgCO}_3}} = \frac{a_{Ca^{+2}}^{0.88}\, a_{Mg^{+2}}^{0.12}\, a_{CO_3^{-2}}}{K_{\substack{\text{Mg-calcite}\\ \text{12 mole \% MgCO}_3}}} \tag{3}$$

TABLE 3.—CONSTANTS USED TO EVALUATE CARBONATE MINERAL SATURATION STATE IN SEAWATER (S = 35‰) AT 25° C

Constant	*Value*	*Source*
$K'_{calcite}$	4.39×10^{-7}	Morse et al. (1980)
$K'_{aragonite}$	6.65×10^{-7}	Morse et al. (1980)
α_0	2.84×10^{-2}	Weiss (1974)
K'_1	1.00×10^{-6}	Mehrbach et al. (1973)
K'_2	0.767×10^{-9}	Mehrbach et al. (1973)
$K^*_{12\text{ mole \% Mg-calcite}}$	5.25×10^{-9}	Walter and Morse (1984_a)

*The stoichiometric saturation constant for 12 mole % Mg-calcite was adapted for seawater calculations using the total ion activity coefficient values for Ca^{+2}, Mg^{+2} and CO_3^{-2} of Millero and Schreiber (1982).

The activity coefficients, γ_i, are functions of ionic strength and were calculated from the data of Millero and Schreiber (1982) in each solution type. The degree of ion pairing was calculated using an iterative computer program. The total dissolved Ca^{+2} and Mg^{+2} concentrations in each solution were known within 1% from titration analyses. The carbonate ion term was calculated from pH and P_{CO_2} as well as from carbonate alkalinity and pH. The two methods of calculation yielded values for the CO_3^{-2} term which agreed to within 2%.

The relations of carbonate dissolution and precipitation rates to saturation state are given below in equations 4 and 5, respectively. Most workers have observed that these empirical kinetic equations accurately fit the experimental data (e.g., Ohara and Reid, 1973; Morse, 1983; Mucci and Morse, 1983; Walter and Morse, 1985):

$$R = k\,(1 - \Omega)^n \tag{4}$$

and,

$$R = k\,(\Omega - 1)^n \tag{5}$$

where r = dissolution or precipitation rate, k = rate constant, Ω = saturation state with respect to the dissolving or precipitating phases and n = order of reaction. Thus, plots of log rate (dissolution or precipitation) versus the term log (1 − Ω) (dissolution) or log (Ω − 1) (precipitation) provide a simple linear representation of carbonate reaction rates.

The saturation states investigated generally fell in the near equilibrium range. Undersaturations with respect to dis-

TABLE 4.—CONSTANTS USED TO EVALUATE CARBONATE MINERAL SATURATION STATE IN SOLUTIONS OTHER THAN SEAWATER

Constant	Value	Source
K_0	$10^{-1.467}$	Weiss (1974)
K_1	$10^{-6.352}$	Harned and Davis (1943)
K_2	$10^{-10.33}$	Harned and Scholes (1941)
$K_{calcite}$	$10^{-8.481}$	Sass et al. (1982)
$K_{aragonite}$	$10^{-8.284}$	Sass et al. (1982)

solving phases ranged from about 10–80% saturation values. Supersaturations with respect to precipitating phases ranged from about 2–15 times saturation values.

RESULTS

Carbonate Dissolution Rates in Various Solutions

Carbonate dissolution rates in natural seawater (35‰ salinity) versus low ionic strength $CaCl_2$ + $MgCl_2$ solutions at two Mg/Ca ratios (1 and 5) are presented in Figures 1 through 4. The dissolution rates of synthetic calcite as a function of calcite saturation state (Fig. 1) do not differ markedly among the three solution types. The rates observed in seawater (containing NaCl and SO_4^{-2}) are the slowest of all. The maximum difference, however, is a factor of three closer to equilibrium with respect to calcite (Ω = 0.8). At greater undersaturations, reaction rates in the three solutions become more similar. Note also the similarity in calcite dissolution rates in the low ionic strength $CaCl_2$ + $MgCl_2$ solutions of different Mg/Ca ratio. Thus, it appears that large variations in Mg/Ca ratio, dissolved NaCl and SO_4^{-2} do not markedly affect calcite dissolution rates.

Dissolution rates for two biogenic aragonites (coral and *Halimeda*) as a function of aragonite saturation state are presented in Figures 2 and 3, respectively. As with calcite dissolution, the rates of aragonite dissolution for both solids are not strikingly different in the different solution types. Aragonite dissolution behavior in the three solutions differs from that of calcite, however, in that the maximum difference in dissolution rate occurs far from equilibrium and becomes insignificant as equilibrium with respect to aragonite is approached. Also, aragonite dissolution rates are most rapid in seawater. The maximum difference in aragonite dissolution rates between seawater and the $CaCl_2$ + $MgCl_2$ solution having a Mg/Ca ratio of 1 is a factor of 3 at low saturation states with respect to aragonite. The difference in aragonite dissolution rates in the three media is less than

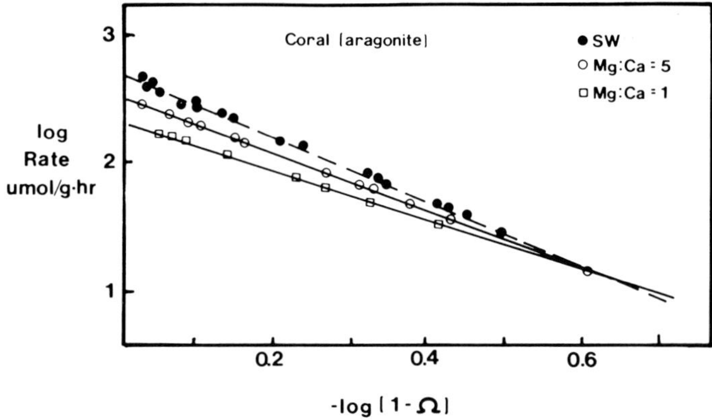

FIG. 2.—Dissolution rates of coral (aragonite) (μmoles $g^{-1}hr^{-1}$) as a function of saturation state with respect to aragonite in seawater (Mg/Ca = 5) and two low ionic strength solutions (Mg/Ca = 1, 5) at 25° C and a P_{CO_2} of $10^{-2.5}$ atm.

a factor of 2 at saturation states (Ω) greater than 0.5 ($-\log\{1 - \Omega\}$ greater than 0.3). Apparently, variations in solution Mg/Ca ratio, dissolved SO_4^{-2}, and NaCl content also do not have large effects on aragonite dissolution rates.

The dissolution rates of a biogenic Mg-calcite (containing 12 mole % $MgCO_3$) in the three solution types as a function of saturation state with respect to Mg-calcite are shown in Figure 4. The dissolution rates in seawater are essentially the same as those in the $CaCl_2$ + $MgCl_2$ solution having a Mg/Ca ratio of 1. The rates measured in the chloride medium having a Mg/Ca ratio of 5 are only about 40% faster than those measured in the two other solutions. Given the ± 10% error involved in determining reaction rates and calculating saturation states, the differences in Mg-calcite dissolution rates among the three solution types are virtually insignificant.

Dissolution rates of calcite, Mg-calcite, and aragonite were all significantly reduced in solutions spiked with small amounts of dissolved orthophosphate (50 μmoles l^{-1}). The

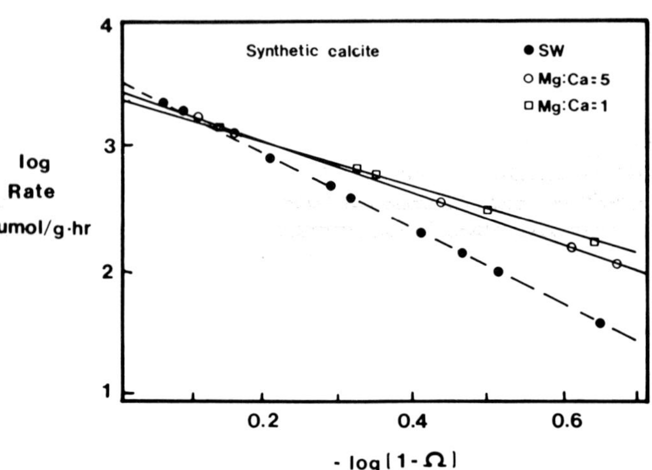

FIG. 1.—Dissolution rates of synthetic calcite (μmoles $g^{-1}hr^{-1}$) as a function of saturation state with respect to calcite in seawater (Mg/Ca = 5) and two low ionic strength solutions (Mg/Ca = 1, 5) at 25° C and a P_{CO_2} of $10^{-2.5}$ atm.

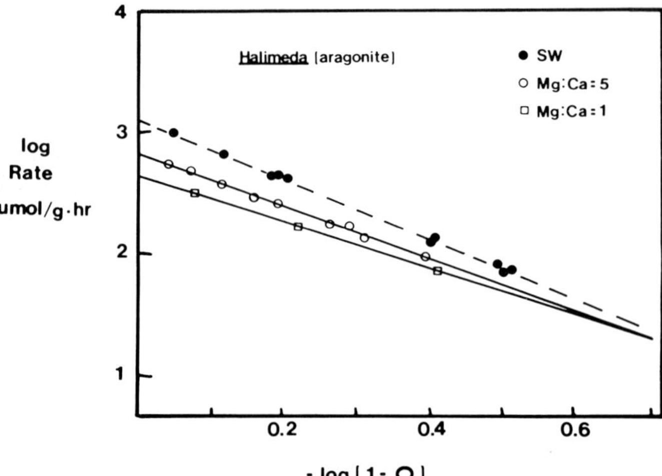

FIG. 3.—Dissolution rates of *Halimeda* (aragonite) (μmoles $g^{-1}hr^{-1}$) as a function of saturation state with respect to aragonite in seawater (Mg/Ca = 5) and two low ionic strength solutions (Mg/Ca = 1, 5) at 25° C and a P_{CO_2} of $10^{-2.5}$ atm.

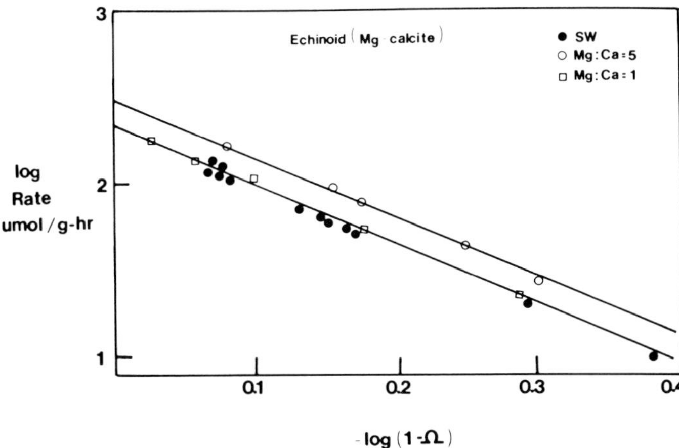

FIG. 4.—Dissolution rates of echinoid (Mg-calcite, 12 mole % $MgCO_3$) (μmoles $g^{-1}hr^{-1}$) as a function of saturation state with respect to Mg-calcite in seawater (Mg/Ca = 5) and two low ionic strength solutions (Mg/Ca = 1, 5) at 25° C and a P_{CO_2} of $10^{-2.5}$ atm.

influence of orthophosphate on carbonate dissolution rates is described in detail by Walter (in prep.), so only the main findings will be summarized here. Table 5 presents data on dissolution rate reductions observed in seawater and a low ionic strength $CaCl_2$ + $MgCl_2$ solution (Mg/Ca = 1) spiked with phosphate. The rate reductions observed for all three carbonate minerals were similar (to within 20%). In general, the dissolution rate reductions increased as equilibrium with respect to the dissolving phase was approached. The effects of a given phosphate concentration level on dissolution rates were greater in the low ionic strength solution than in seawater. For example, at a saturation state (Ω) value of 0.8, dissolution rates were reduced by a factor of 5 in seawater but were reduced by a factor of 20 in the lower ionic strength solution. The dissolution rate reductions reported here are significantly smaller than those of Berner and Morse (1974) at similar phosphate concentrations.

The experimental results show that the dissolution rates of calcite, Mg-calcite and aragonite are not strongly affected by large changes in dissolved NaCl, SO_4^{-2} or Mg/Ca ratio. The maximum difference observed was a factor of 3, but more commonly the difference was less than a factor of 2. Micromolar concentrations of dissolved phosphate have a much greater influence on dissolution rates than do major changes in dissolved species such as Na^+, Mg^{+2} and SO_4^{-2}. The presence of orthophosphate in diagenetic systems, however, has no selective effects on the relative dissolution rates of different carbonate minerals. Thus, carbonate dissolution rates in low-phosphate marine and low ionic strength meteoric solutions should be similar at similar saturation states with respect to calcite, Mg-calcite and aragonite.

TABLE 5.—REDUCTIONS IN CARBONATE MINERAL DISSOLUTION RATES, R/R_o, IN SOLUTIONS SPIKED WITH 50 μMOLES l^{-1} PO_4^{-3}

Solution	Ω	R/R_o*
Seawater	0.2	0.60
(I = 0.7)	0.8	0.20
$CaCl_2$ + $MgCl_2$:Mg/Ca= 1	0.2	0.15
(I = 0.18)	0.8	0.04

*R/R_o = rate in phosphate-spiked experiments/rate in phosphate-free experiments.

Calcite Precipitation Rates in Various Solutions

The problem.—

Based on previously published data, it is clear that calcite precipitation rates decrease with increasing Mg/Ca ratio, but the influences of other solution variables on calcite precipitation rates are unclear. Some pertinent data are summarized in Figure 5. The kinetic plots for precipitation are based on the empirical equation (5) for precipitation given earlier.

The calcite precipitation rates have been normalized to BET-measured calcite seed surface areas. Mucci and Morse (1983) performed seeded calcite precipitation experiments using a pH-stat at 25° C and a P_{CO_2} of $10^{-2.5}$ atm, the same conditions used in the present study. Their data from normal seawater (35‰ salinity-Mg/Ca = 5) and "pseudo-seawater" (Mg/Ca = 1) are presented in Figure 5. The "pseudo-seawater" differed from normal seawater only in its Mg/Ca ratio; its ionic strength (NaCl content), SO_4^{-2} and other major constituents were identical to seawater of 35‰ salinity. At the same saturation state with respect to calcite, calcite growth rates are about 16 times slower in normal seawater (Mg/Ca = 5) than they are in "pseudo-seawater" having a lower Mg/Ca ratio of 1. Because the only difference in the two solutions is Mg/Ca ratio, this variable must be solely responsible for the large difference in calcite precipitation rates.

Calcites precipitated from solutions of elevated Mg/Ca ratio have a higher mole % $MgCO_3$. In Figure 5, all calcite (or Mg-calcite) growth rates have been plotted as a function of saturation state with respect to low-Mg calcite. Berner (1975) suggested that the slower calcite growth rates result from precipitation of a more soluble (less supersaturated) phase on the calcite seed surface. Mucci and Morse (1984) demonstrated, however, that the solubility of calcites (or

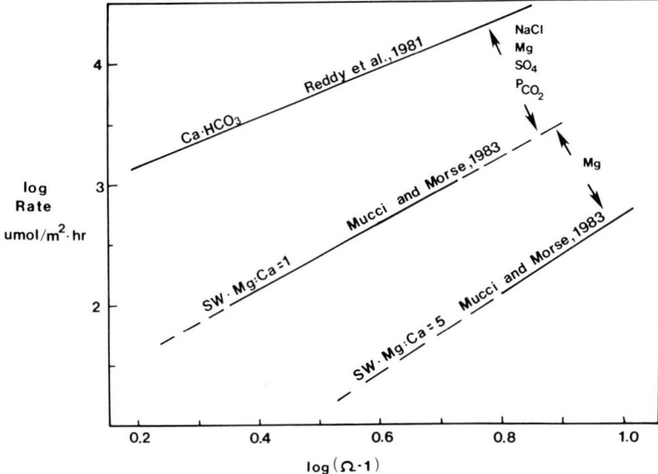

FIG. 5.—Calcite precipitation rates on calcite seeds as a function of calcite saturation state normalized to crystal seed surface area (μmoles $m^{-2}hr^{-1}$) obtained by previous workers in various solution types at 25° C.

Mg-calcites) precipitated from solutions of different Mg/Ca ratios do not differ significantly. They attribute slower calcite growth rates to growth inhibition by adsorbed Mg^{+2} on the calcite surface. In contrast, Walter and Morse (1984a) observed that Mg-calcite solubilities obtained from dissolution experiments increased sharply with their $MgCO_3$ content in agreement with basic trends reported by other workers (Chave et al., 1962; Plummer and Mackenzie, 1974). The reader is referred to a recent review by Mackenzie et al. (1983) for a more detailed discussion of Mg-calcite stability during dissolution and precipitation. For the purpose of this paper, it suffices to state that by some mechanism, either increased solubility or Mg^{+2} adsorption and growth inhibition, calcites precipitate more slowly at elevated Mg/Ca ratios.

Figure 5 also depicts the calcite precipitation rate data of Reddy et al. (1981) as a function of saturation state with respect to calcite. Their experiments were performed using the free-drift method, where rates are determined from pH changes per unit time. The solution used was a low ionic strength $Ca\text{-}HCO_3$ medium at a P_{CO_2} of $10^{-1.5}$ atm. The reported precipitation rates are 16 to 20 times more rapid than those of Mucci and Morse (1983) in low-Mg seawater (Mg/Ca = 1). Four chemical variables differ between the two studies. Reddy et al. (1981) used a medium lacking NaCl, dissolved Mg^{+2} or SO_4^{-2}. Also, they worked at a higher P_{CO_2} value than did Mucci and Morse (1983). Comparison of the results from the two studies suggests that some other chemical variable besides solution Mg/Ca ratio strongly influences calcite precipitation rates. The following series of experiments test each of the variables (NaCl, SO_4^{-2}, Mg^{+2}) independently to evaluate the relative importance of each for calcite precipitation rates. Preliminary data on the effect of some dissolved components on aragonite versus calcite precipitation rates are also presented.

Effect of Mg/Ca ratio.—

In order to verify that calcite growth rates do not depend solely on solution Mg/Ca ratio, seeded calcite precipitation experiments were performed in low ionic strength solutions at Mg/Ca ratios of 1 and 5. Unlike seawater, these solutions were essentially sodium- and sulfate-free. Figure 6 shows the results of calcite precipitation experiments in a low ionic strength $CaCl_2$ + $MgCl_2$ medium having a Mg/Ca ratio of 1. The precipitates were analyzed by X-ray diffraction and SEM and were found to be calcite. This was the case for all other precipitation experiments where calcite seeds were used. The calcite precipitation rates are normalized to the calcite seed surface area and plotted relative to the saturation state with respect to low-Mg calcite. The calcite precipitation rates are about 10 times more rapid in the SO_4^{-2}-free and low-NaCl solution than they are in "pseudo-seawater" at the same saturation state and Mg/Ca ratio (1). The previous results of Mucci and Morse (1983) and Reddy et al. (1981) are plotted for comparison. The experiments of Mucci and Morse (1983) employed the same calcite seeds and were performed under the same P_{CO_2} and experimental conditions as those used in this study and, thus, are directly comparable. The difference in the data supports the contention that some other dissolved component besides

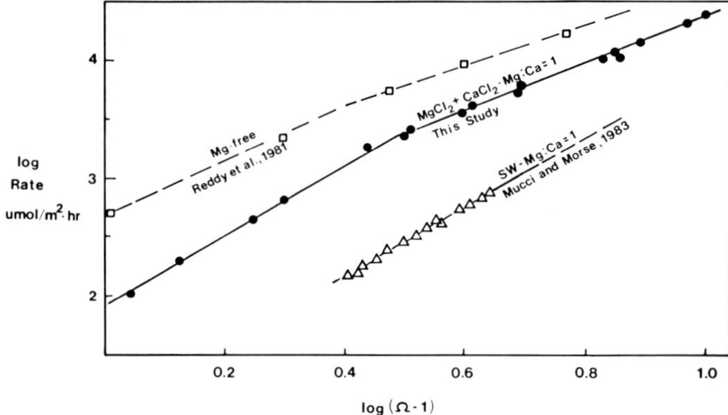

FIG. 6.—Calcite precipitation rates on calcite seeds at 25° C and a P_{CO_2} of $10^{-2.5}$ atm as a function of calcite saturation state from a low ionic strength $CaCl_2$ + $MgCl_2$ solution (Mg/Ca = 1), compared to the results of Mucci and Morse (1983) from a seawater-like medium of low Mg/Ca ratio (1) and those of Reddy et al. (1981) from a Mg-free, $Ca\text{-}HCO_3$ solution. All precipitation rates are normalized to calcite seed surface area (μmoles $m^{-2}hr^{-1}$). Note that precipitation rates are slower at the same Mg/Ca ratio in the solution of higher ionic strength.

Mg/Ca ratio in seawater is responsible for the rate differences.

The calcite growth rates measured by Reddy et al. (1981) are 2 to 5 times more rapid than those measured in the low ionic strength $CaCl_2$ + $MgCl_2$ solution. There are a number of possible reasons for this. Reddy et al. (1981) used a completely Mg-free medium, worked at a higher P_{CO_2} value ($\sim 10^{-1.5}$ atm) and used a free-drift method to obtain their calcite growth rate data. It is not possible to decide without further experimentation which of these variables is responsible for the difference in the rates. It is known, however, that higher P_{CO_2} values generally lead to more rapid calcite precipitation at equivalent saturation states (Reddy et al., 1981).

A similar experimental test was performed at a higher Mg/Ca ratio of 5. Calcite precipitation experiments were performed in natural seawater having a Mg/Ca ratio of 5 and also in a $CaCl_2$ + $MgCl_2$ solution of the same Mg/Ca ratio, but lacking SO_4^{-2} and NaCl. The experimental results are depicted in Figure 7. As before, calcite precipitation rates at equivalent saturation states with respect to calcite were 5 to 7 times more rapid in solutions lacking SO_4^{-2} and NaCl than they were in seawater at the same Mg/Ca ratio and dissolved Ca^{+2} and Mg^{+2} molalities. The calcite growth rates obtained in normal seawater agreed well with those of Mucci and Morse (1983) from natural (Mg/Ca = 5) seawater.

Effect of dissolved NaCl.—

These experimental results demonstrate that there must be a dissolved component or components in seawater other than Mg^{+2} that reduces calcite growth rates. One obvious dissolved component differing between seawater and the $CaCl_2$ + $MgCl_2$ media investigated is NaCl. A series of calcite precipitation experiments was performed in the $CaCl_2$ + $MgCl_2$ medium (Mg/Ca = 1) to which 0.7m NaCl was added, making the solution ionic strength near that of seawater. The precipitation rates of calcite on calcite seeds in

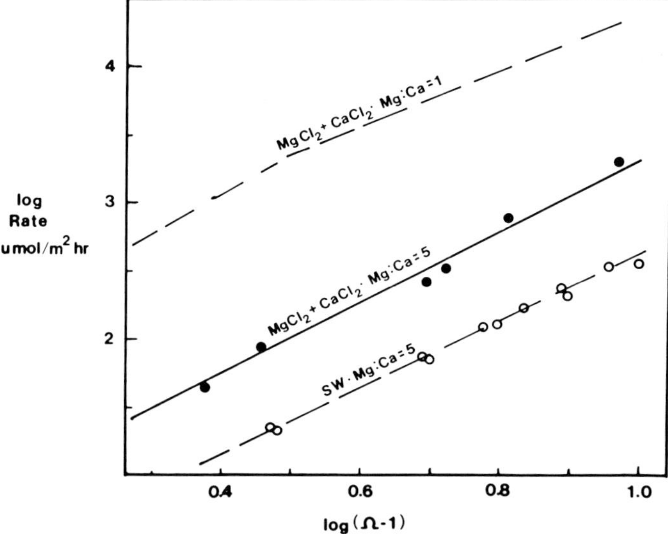

FIG. 7.—Calcite precipitation rates on calcite seeds at 25° C and a P_{CO_2} of $10^{-2.5}$ atm as a function of calcite saturation state from a low ionic strength $CaCl_2$ + $MgCl_2$ solution (Mg/Ca = 5) and normal seawater (Mg/Ca = 5). Results from the Mg/Ca = 1, low ionic strength medium are included for comparison. Rates are expressed in μmoles $m^{-2}hr^{-1}$. Note that precipitation rates are slower in seawater than they are in the low ionic strength chloride medium at the same Mg/Ca ratio.

the altered medium are presented in Figure 8. The experimental data obtained from the $CaCl_2$ + $MgCl_2$ medium prior to the addition of NaCl are plotted for comparison. Calcite growth rates do not change significantly when large amounts of NaCl are added to the medium. Thus, the large discrepancy between calcite precipitation rates in $CaCl_2$ + $MgCl_2$ media and seawater of the same Mg/Ca ratio cannot be attributed to the influence of dissolved NaCl.

It was also of interest to observe how calcite precipitation rates would respond to the complete removal of Mg^{+2} from the medium. A series of calcite precipitation experiments were performed in a low ionic strength $CaCl_2$ solution (0.01m, I = 0.03). The results are presented in Figure 8. The calcite precipitation rates in the Mg-free medium do not differ significantly from those obtained at a Mg/Ca ratio of 1. This result indicates that the discrepancy between the calcite precipitation rates of Reddy et al. (1981) in a Ca-HCO_3 solution and those determined here at a Mg/Ca ratio of 1 (shown in Fig. 6) are probably caused by differences in experimental methods or in the experimental P_{CO_2}. Thus, a further decrease in the Mg/Ca ratio below value of 1 has little or no influence on calcite growth rates.

Effect of dissolved SO_4^{-2}.—

The other main dissolved component in seawater which could influence calcite precipitation rates is SO_4^{-2}. The final series of experiments involved the addition of 0.028m Na_2SO_4 to the $CaCl_2$ + $MgCl_2$ medium (Mg/Ca = 1) to bring the solution up to the dissolved SO_4^{-2} level of natural seawater. The experimental results obtained in the SO_4^{-2}-enriched medium are depicted in Figure 9. The calcite precipitation rates obtained from the SO_4^{-2}-free $CaCl_2$ + $MgCl_2$ solution and those obtained by Mucci and Morse (1983) for seawater at a Mg/Ca ratio of 1 are included for comparison. The addition of SO_4^{-2} to the medium reduces calcite growth rates by a factor of 18. In fact, the calcite precipitation rates are reduced by a greater amount than is required to reproduce the calcite growth rates obtained from seawater at the same Mg/Ca ratio. The stronger inhibition of calcite growth rates in the $CaCl_2$ + $MgCl_2$ solution probably arises from the higher dissolved activity ($a_i = \gamma_i m_i$, where γ_i = activity

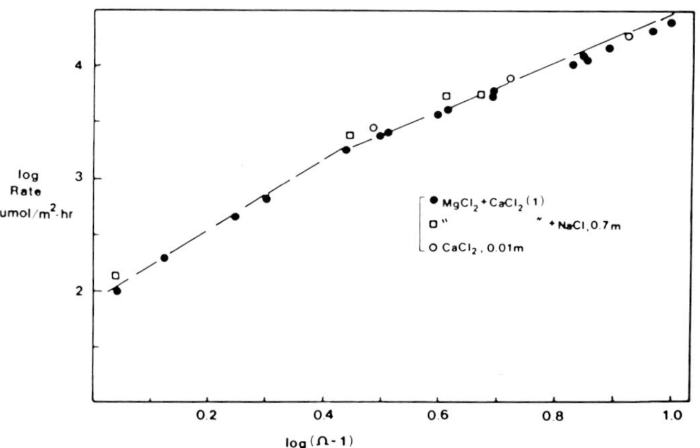

FIG. 8.—Calcite precipitation rates at 25° C and a P_{CO_2} of $10^{-2.5}$ atm on calcite seeds as a function of calcite saturation state. The results from the $MgCl_2$ + $CaCl_2$ (Mg/Ca = 1) experiments presented in Figure 7 are replotted to permit comparison with rates obtained in two altered solution types. Rates obtained from this solution after the addition of 0.7 m NaCl are not significantly different from the low sodium solution. Rates measured in Mg-free $CaCl_2$ also are not significantly different from solutions containing 0.01 m $MgCl_2$ (Mg/Ca = 1). All rates expressed in μmoles $m^{-2}hr^{-1}$.

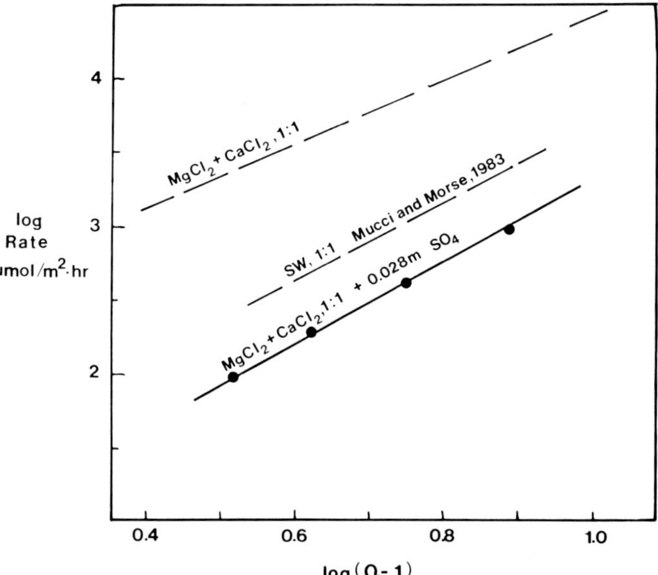

FIG. 9.—Calcite precipitation rates at 25° C at a P_{CO_2} of $10^{-2.5}$ atm on calcite seeds as a function of calcite saturation state. The addition of 0.028 m Na_2SO_4 to the $CaCl_2$ + $MgCl_2$ medium (Mg/Ca = 1) significantly decreases calcite growth rates from their original values in the SO_4-free medium. Note that rates are fairly close to those measured by Mucci and Morse (1983) in seawater (containing 0.028m SO_4) at the same Mg/Ca ratio (1). All rates expressed in μmoles $m^{-2}hr^{-1}$.

coefficient of ion i and m_i = free molality of ion i) of SO_4^{-2} at the lower ionic strength of this medium. Further experimentation regarding SO_4^{-2} effects on carbonate precipitation is in progress, but the results at this point indicate that SO_4^{-2} is the only dissolved component capable of explaining why calcite precipitation rates are slower in seawater than they are in lower ionic strength media at the same Mg/Ca ratio. Because changes in ionic strength (NaCl content) have no apparent influence on calcite precipitation rates, the rate reduction observed following the addition of Na_2SO_4 is attributable solely to the inhibitory effect of dissolved SO_4^{-2}.

Relative Rates of Aragonite and Calcite Precipitation

A set of experiments was conducted using synthetic aragonite seeds to induce precipitation from the $CaCl_2$ + $MgCl_2$ solution having a Mg/Ca ratio of 1. X-ray diffraction and SEM analyses confirmed that the precipitate which nucleated on the seed surfaces was aragonite. Thus, even in a meteoric-type solution, the precipitation of aragonite occurred readily on aragonite seeds. The precipitate mineralogy was controlled by that of the seed in all experiments with calcite precipitating on calcite seeds and aragonite precipitating on aragonite seeds. The effects of substrate mineralogy was so strong that in experiments where both aragonite and calcite seeds were used, aragonite and calcite co-precipitated from solution and did so selectively on their respective seeds. This phenomenon permitted the investigation of the relative growth rates of the two mineral phases (normalized to equivalent seed surface areas) and how these relative rates could be influenced by changes in solution chemistry.

Figure 10 presents data for the precipitation rates of calcite and aragonite normalized to the BET-measured surface area of their respective seed types. The rates have been plotted versus the ion activity product of dissolved calcium and carbonate (IAP) to permit comparison of the relative precipitation rates of the two phases from the same solution chemistry. At a given IAP or pH value in solution, calcite precipitates more rapidly than does aragonite, generally by about a factor of 2. This is expected because at a given IAP in solution, calcite is about 1.5 times more supersaturated than aragonite, the more soluble phase, and should precipitate more rapidly.

Calcite and aragonite precipitation rates were also determined in the same $CaCl_2$ + $MgCl_2$ solution in the presence of 0.028m SO_4^{-2}. The relative precipitation rates are plotted as a function of solution IAP in Figure 11. The first point to note is that both calcite and aragonite precipitate at much slower rates in the presence of SO_4^{-2}; however, calcite precipitation is more effectively inhibited by SO_4^{-2} than is aragonite precipitation. This leads to a reversal in the relative precipitation rates of aragonite and calcite from that observed in the SO_4^{-2}-free solution. Aragonite precipitates about 4 times more rapidly than does calcite from the SO_4^{-2}-enriched solution, whereas without SO_4^{-2}, it precipitated only half as rapidly as calcite. Thus, SO_4^{-2} appears to be selective in inhibiting the precipitation of calcite.

The influence of dissolved orthophosphate (PO_4^{-3}) on cal-

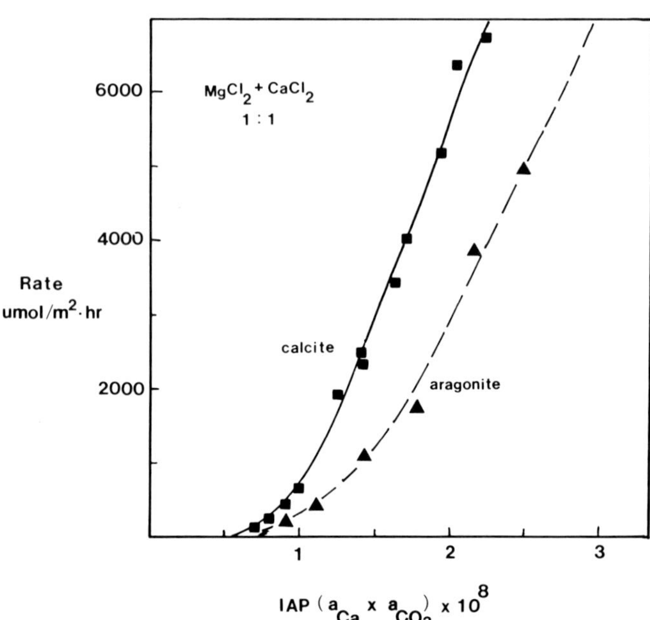

FIG. 10.—Relative precipitation rates of calcite and aragonite as a function of ion activity product ($a_{Ca^{+2}} \times a_{CO_3^{-2}}$) in a $CaCl_2$ + $MgCl_2$ solution (Mg/Ca = 1) at 25° C and a P_{CO_2} of $10^{-2.5}$ atm. Rates are normalized to calcite and aragonite seed surface areas (μmoles $m^{-2}hr^{-1}$). Note that calcite precipitates more rapidly on calcite seeds than aragonite does on aragonite seeds at a given IAP value.

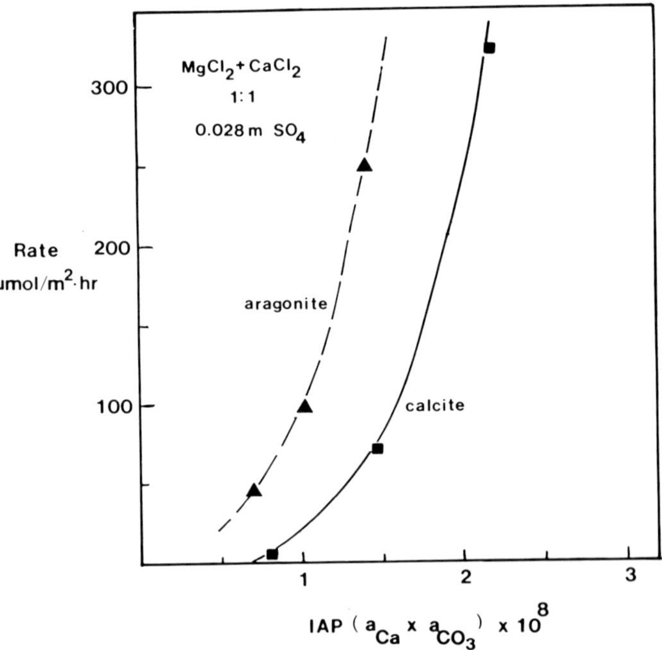

FIG. 11.—Relative precipitation rates of calcite and aragonite as a function of ion activity product ($a_{Ca^{+2}} \times a_{CO_3^{-2}}$) in a $CaCl_2$ + $MgCl_2$ solution (Mg/Ca = 1) spiked with 0.028 m Na_2SO_4 at 25° C and a P_{CO_2} of $10^{-2.5}$ atm. Rates are normalized to calcite and aragonite seed surface areas (μmoles $m^{-2}hr^{-1}$). Note that in the presence of 0.028 m SO_4^{-2}, aragonite now precipitates more rapidly than does calcite at a given IAP value.

TABLE 6.—REDUCTIONS IN CARBONATE MINERAL PRECIPITATION RATES, R/R_0, IN SOLUTIONS SPIKED WITH 10 μMOLES l^{-1} PO_4^{-3}.

Solution	Sample	Ω	R/R_0*
Seawater	synthetic calcite	8	0.30
	synthetic aragonite	7	0.06
	coral (aragonite)	7	0.05
Mg/Ca = 1, $CaCl_2$ + $MgCl_2$	synthetic calcite	7	0.05
	synthetic aragonite	7	0.008
	coral (aragonite)	7	0.007

*R/R_0 = rate in phosphate-spiked experiments/rate in phosphate-free experiments.

cite and aragonite precipitation rates was also investigated at a dissolved concentration of phosphate of 10 μmoles l^{-1}. Precipitation experiments in the presence of phosphate were conducted in normal seawater (Mg/Ca = 5) and the low ionic strength $CaCl_2$ + $MgCl_2$ solution (Mg/Ca = 1) at saturation states near 8 times supersaturation with respect to calcite or aragonite. Both aragonite and calcite (Mg-calcite in seawater) precipitation rates are markedly reduced by only a small amount of orthophosphate (Table 6). The amount of rate reduction is 6 to 7 times greater in the low ionic strength solution than it is in seawater for both calcite and aragonite precipitation. Carbonate dissolution rate reductions were also considerably greater (about a factor of 5–8) in the low ionic strength solution (see Table 5). The interesting difference between phosphate effects on precipitation versus dissolution is that a *smaller* phosphate concentration produces a *larger* degree of rate reduction for precipitation than for dissolution. Thus, as with the other dissolved species, phosphate appears to influence precipitation rates of carbonate minerals more strongly than dissolution rates.

More importantly, phosphate appears to be selective in inhibiting aragonite precipitation. Precipitation rates of aragonite on coral and synthetic aragonite seeds are both reduced nearly 6 time more strongly in seawater and low ionic strength solutions than are calcite (or Mg-calcite) precipitation rates on calcite seeds. The results show that dissolved SO_4^{-2} and PO_4^{-3} both can have selective effects on carbonate precipitation rates, with SO_4^{-2} selectively inhibiting calcite precipitation and PO_4^{-3} selectively inhibiting aragonite precipitation. The preliminary findings are exciting because the co-variation of dissolved SO_4^{-2} and PO_4^{-3} in natural environments is conducive to simple modeling of aragonite and calcite (or Mg-calcite) cement occurrences.

DISCUSSION AND CONCLUSIONS

Controls on Carbonate Cement Mineralogy

The data obtained from this study regarding the relative efficiency of calcite dissolution versus precipitation and the relative precipitation rates of aragonite and calcite (or Mg-calcite) have interesting geologic applications. The results indicate that aragonite precipitation may be favored in high-sulfate, low-phosphate solutions while calcite (or Mg-calcite) precipitation may be favored in low-sulfate, high-phosphate solutions. In normal seawater and marine sediments typified by rapid circulation of overlying oxygenated seawater, the dissolved phosphate concentration is generally low because of rapid nutrient utilization by organisms. Also, in well-oxygenated environments, the dissolved sulfate concentration remains high because bacterial sulfate reduction does not begin until oxygen is depleted in the sediment-pore water system. In well flushed and ventilated shallow marine carbonates, the anticipated precipitate would be aragonite.

In more restricted sedimentary microenvironments, where exchange with overlying seawater is less vigorous, the oxidative decomposition of organic matter releases dissolved phosphate which accumulates in the pore water system. If the environment becomes functionally anoxic, organic matter decomposition via sulfate reduction will occur. Under these conditions, phosphate continues to be released while sulfate is consumed by bacterially-driven sulfate reduction. In microenvironments where organic matter decomposition products can accumulate, Mg-calcite precipitation may be favored over aragonite. Such changes in marine carbonate pore water chemistries have been observed by Berner (1966) and Rosenfeld (1979). In fact, Rosenfeld (1979) reports dissolved phosphate concentrations up to 30 μmoles l^{-1} in sediment pore waters of Florida Bay, levels considerably higher than those investigated in the present study. Also, many surficial crust cements in the shallow marine environment are aragonitic, whereas reef cements are commonly Mg-calcite (e.g., Bathurst, 1975). While it is recognized that controls on marine carbonate cement mineralogy in nature may be complex, it is possible that the occurrence and mineralogy of marine carbonate cements may be partially influenced by dissolved sulfate and phosphate in the diagenetic fluids. The results presented here also show that the mineralogy of the substrate is an important control.

In addition, understanding the cause of secular trends in the mineralogy of marine carbonate precipitates such as ooids and cements (Sandberg, 1983 and references therein) requires knowledge of chemical controls over aragonite versus calcite precipitation from seawater. The main controlling variable favored at present is fluctuation in atmospheric P_{CO_2} (Sandberg, 1983). It is possible that other chemical factors such as dissolved SO_4^{-2} and PO_4^{-3} may be involved in controlling the mineralogy of carbonate cements besides P_{CO_2}. More work is needed, both in the laboratory and in recent carbonate environments of active cementation, to correlate pore water chemistry with cement growth rates and compositions. Continuing studies in both the laboratory and field are being conducted to define these critical variables more rigorously.

Relative Efficiency of Carbonate Dissolution and Precipitation Reactions

Dissolution rates of aragonite, calcite and Mg-calcite at a given saturation state are relatively insensitive to changes in solution Mg/Ca ratio, NaCl content and dissolved SO_4^{-2}. In contrast, the precipitation rates of calcite (and Mg-calcite) as a function of saturation state are severely affected by changes in dissolved SO_4^{-2} and by changes in Mg/Ca ratio above a Mg/Ca value of 1. Calcite precipitation rates do not appear to be sensitive to changes in NaCl or Mg/Ca ratio changes below a value of 1. Because of these influences, calcite precipitation rates become strongly

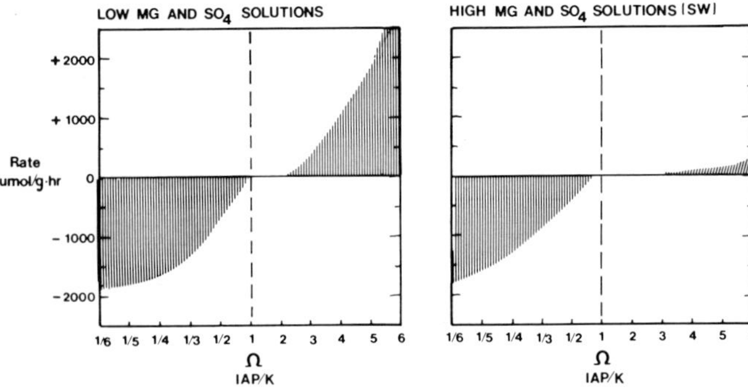

FIG. 12.—Plot of the symmetry of precipitation versus dissolution rates for calcite (μmoles $g^{-1}hr^{-1}$) at 25° C and a P_{CO_2} of $10^{-2.5}$ atm as a function of a 6-fold departure either side of equilibrium with respect to calcite. Note the excellent symmetry of reaction rates in low Mg^{+2} and low SO_4^{-2} solutions. As Mg^{+2} and SO_4^{-2} increase, precipitation rates are reduced dramatically, while dissolution rates are less significantly affected, leading to marked asymmetry for the two reactions at equivalent departures from equilibrium.

asymmetric as a function of the degree of departure from calcite equilibrium in solutions of high Mg/Ca ratio and dissolved SO_4^{-2}. This behavior is illustrated in Figure 12 which plots the integrated rates of both calcite dissolution and precipitation over equivalent ranges of saturation states, $\Omega = 1 - 6$ (supersaturation) and $\Omega = 1/6 - 1$ (undersaturation) in different solution types. In low-Mg^{+2} and SO_4^{-2} solutions, calcite growth and dissolution rates are almost perfectly symmetrical as a function of departure from equilibrium, although the near equilibrium ($\Omega < 2$) growth rates are lower than dissolution rates at an equivalent distance from equilibrium. In contrast, relative rates of calcite precipitation versus dissolution become sharply asymmetric in solutions of high Mg/Ca ratio and dissolved SO_4^{-2}

The presence of orthophosphate in diagenetic systems will increase the relative rate asymmetry with saturation state even further because phosphate inhibits calcite precipitation rates to a much greater degree than those of dissolution at a given concentration level. Thus, it appears that there is greater potential for carbonate porosity enhancement by diagenetic solutions characterized by high Mg/Ca, SO_4^{-2} and PO_4^{-3} because each of these dissolved species inhibits precipitation to a much greater extent than dissolution.

During early meteoric diagenesis, metastable carbonate phases (aragonite and Mg-calcite) are progressively converted via dissolution/precipitation reactions to low-Mg calcite. Given their greater chemical instability relative to low-Mg calcite, the asymmetry in aragonite or Mg calcite dissolution rates versus the reprecipitation of low-Mg calcite would be even more pronounced than those shown in Figure 12 for low-Mg calcite dissolution in high Mg/Ca and SO_4^{-2} solutions. The important question of how directly these results may be applied to the study of early- and late-stage carbonate diagenesis awaits further study. Kinetic experiments performed in brines of various compositions at elevated temperatures would be helpful in constraining relative rates and products of deep burial diagenetic reactions. Further study of the actual variation in bonate mineral saturation states in surficial carbonate diagenetic environments is also critical because few data are available for these systems. Berner (1966) observed saturation states near equilibrium with respect to low-Mg calcite in sediment pore waters in contact with shallow marine carbonates (undersaturation for aragonite and high-Mg calcite). Plummer et al. (1976) report calcite saturation states (Ω values) ranging from 0.5 to 2.0 in Bermudan meteoric diagenetic systems.

The results presented here stress that controlled kinetic experiments can play an important role in constraining models for carbonate diagenesis. While some of the results are preliminary, they indicate directions for future research and provide models which should be testable through the study of diagenesis in recent carbonate sediments as well as ancient limestones.

ACKNOWLEDGMENTS

This research was supported by American Chemical Society Petroleum Research Fund Grant 14509-AC2 and National Science Foundation Grant OCE79-19242. I thank John Morse and Frank Millero for numerous helpful discussions during the course of this research. In particular, I acknowledge the assistance of Matt Tagett, whose cheerful help enabled me to complete the large number of precipitation experiments. Others thanked for providing sounding boards at different times during the study are Liz Burton, Hal Wanless, and Charles Bukowski. Freda Sofian, the dominatrix of the word processor, is thanked for her expert typing and patience during the preparation of this manuscript.

REFERENCES

BAKER, P. A., AND KASTNER, M., 1981, Constraints on the formation of sedimentary dolomite: Science, v. 213, p. 214–216.

BATHURST, R., 1975, Carbonate Sediments and Their Diagenesis: New York, Elsevier, 658 pp.

BERNER, R. A., 1966, Chemical diagenesis of some modern carbonate sediments: American Journal of Science, v. 264, p. 1–36.

———, 1975, The role of magnesium in the crystal growth of calcite and aragonite from seawater: Geochimica et Cosmochimica Acta, v. 39, p. 489–504.

———, AND MORSE, J. W., 1974, Dissolution kinetics of calcium carbonate in seawater. IV. Theory of calcite dissolution: American Journal of Science, v. 274, p. 108–134.

BISCHOFF, J. L., AND FYFE, W. S., 1968, Catalysis, inhibition, and the calcite-aragonite problem. I. The aragonite-calcite transformation: American Journal of Science, v. 266, p. 65–79.

BRUNAUER, S., EMMET, P. H., AND TELLER, E., 1938, Adsorption of gases in multimolecular layers: Journal of the American Chemical Society, v. 60, p. 309–319.

CHAVE, K. E., DEFFEYES, K. S., WEYL, P. K., GARRELS, R. M., AND THOMPSON, M. E., 1962, Observations on the solubility of skeletal carbonates in aqueous solutions: Science v. 137, p. 33–34.

DE KANEL, J., AND MORSE, J. W., 1979, A simple technique for surface area determination: Journal of Physics and Earth Science Instrumentation, v. 12, p. 272–273.

FOLK, R. L., 1974, The natural history of crystalline calcium carbonate: Effect of magnesium content and salinity: Journal of Sedimentary Petrology, v. 44, p. 40–53.

HARNED, H. S., AND DAVIS, R. D., 1943, The ionization constant of carbonic acid in water and the solubility of carbon dioxide in water and aqueous salt solutions from 0 to 50° C: Journal of the American Chemical Society, v. 65, p. 2030–2037.

———, AND SCHOLES, S. R., 1941, The ionization constant of HCO_3^- from O to 50° C: Journal of the American Chemical Society, v. 63, p. 1706–1709.

KAZMIERCZAK, T. F., TOMSON, M. B., AND NANCOLLAS, G. H., 1982, Crystal growth of calcium carbonate. A controlled composition study: Journal of Physical Chemistry, v. 86, p. 103–107.

KUSHNIR, J., 1983, The rate of dissolution of carbonate minerals in the presence of sulfate ions: Geological Society America Abstracts with Programs, v. 15, p. 620.

MACKENZIE, F. T., BISCHOFF, W. D., BISHOP, F. C., LOIJENS, M., SCHOONMAKER, J., AND WOLLAST, R., 1983, Low temperature occurrence, solubility, and solid solution behavior, *in* Reeder, R. J., ed., Carbonates: Mineralogy and Chemistry: Mineralogical Society of America, Reviews in Mineralogy, v. 11, p. 97–144.

MEHRBACH, C., CULBERSON, C. H., HAWLEY, J. E., AND PYTKOWICZ, R. M., 1973, Measurements of the apparent dissociation constants of carbonic acid in seawater at atmospheric pressure: Limnology and Oceanography, v. 18, p. 897–907.

MILLERO, F. J., AND SCHREIBER, D. R., 1982, Use of the ion pairing model to estimate activity coefficients of the ionic medium of natural waters: American Journal of Science, v. 282, p. 1508–1541.

MORSE, J. W., 1974, Dissolution kinetics of calcium carbonate in seawater. III. A new method for the study of carbonate reaction kinetics: American Journal of Science, v. 274, p. 97–107.

———, 1983, The kinetics of calcium carbonate dissolution and precipitation, *in* Reeder, R. J., ed., Carbonates: Mineralogy and Chemistry: Mineralogical Society of America, Reviews in Mineralogy, v. 11, p. 227–264.

———, MUCCI A., AND MILLERO, F. J., 1980, The solubility of calcite and aragonite in seawater of 35‰ salinity at 25° C and atmospheric pressure: Geochimica et Cosmochimica Acta, v. 44, p. 85–94.

MUCCI, A., 1981, The solubility of calcite and aragonite and the composition of calcite overgrowths in seawater and related solutions: Ph.D. Thesis, Miami, Florida, University of Miami, 237 p.

———, AND MORSE, J. W., 1983, The incorporation of Mg^{+2} and Sr^{+2} into calcite overgrowths: Influences of growth rate and solution composition: Geochimica et Cosmochimica Acta, v. 47, p. 217–233.

——— AND ———, 1984, The solubility of calcite in seawater solutions of various magnesium concentrations, I_t = 0.697m at 25° C and one atmosphere total pressure: Geochimica et Cosmochimica Acta, v. 48, p. 815–823.

OHARA, M., AND REID, R. C., 1973, Modeling Crystal Growth Rates From Solution: New Jersey, Prentice-Hall, Inc., 272 p.

PLUMMER, L. N., AND MACKENZIE, F. T., 1974, Predicting mineral solubility from rate data. Application to the dissolution of magnesium calcites: American Journal of Science, v. 274, p. 61–83.

———, PARKHURST, D. L., AND WIGLEY, T. M. L., 1979, Critical review of the kinetics of calcite dissolution and precipitation, *in* Jenne, E. A., ed., Chemical Modeling in Aqueous Systems: American Chemical Society Symposium Series, v. 93, p. 537–573.

———, VACHER, H. L., MACKENZIE, F. T., BRICKER, O. P., AND LAND, L. S., 1976, Hydrogeochemistry of Bermuda: A case history of groundwater diagenesis of biocalcarenites: Geological Society of America Bulletin, v. 87, p. 1301–1316.

RAISTRICK, B., 1949, The influence of foreign ions in crystal growth from solution. I. The stabilization of supersaturation of calcium carbonate solutions by anions possessing O-P-O-P-O chains: Discussions of the Faraday Society, v. 5, p. 234–237.

REDDY, M. M., PLUMMER, L. N., AND BUSENBERG, E., 1981, Crystal growth of calcite from calcium bicarbonate solutions at constant P_{CO} and 25° C: A test of the calcite dissolution model: Geochimica et Cosmochimica Acta, v. 45, p. 1281–1291.

ROSENFELD, J. K., 1979, Interstitial water and sediment chemistry of two cores from Florida Bay: Journal of Sedimentary Petrology, v. 49, p. 989–994.

SANDBERG, P. A., 1983, An oscillating trend in Phanerozoic non-skeletal carbonate mineralogy: Nature, v. 305, p. 19–22.

SASS, E., MORSE, J. W. AND MILLERO, F. J., 1982, Dependence of the values of calcite and aragonite thermodynamic solubility products on ionic models: American Journal of Science, v. 283, p. 218–229.

SIMKISS, K., 1964, The inhibitory effects of some metabolites on the precipitation of calcium carbonate from artificial and natural seawater: Journal Consiel International Exploration Mer, v. 29, p. 6–18.

SJÖBERG, E. L., 1978, Kinetics and mechanism of calcite dissolution in aqueous solutions at low temperatures: Stockholm Contributions to Geology, v. 32, p. 1–96.

THORSTENSEN, D. C., AND PLUMMER, L. N., 1977, Equilibrium criteria for two component solids reacting with fixed composition in an aqueous phase-example the magnesian calcites: American Journal of Science, v. 277, p. 1220–1223.

WALTER, L. M., 1983, The dissolution kinetics of shallow water carbonate grain types: Effects of mineralogy, microstructure, and solution chemistry: Ph.D. Thesis, Miami, Florida, University of Miami, 318 pp.

———, 1985, Relative reactivity of skeletal carbonates during dissolution: Implications for diagenesis, *in* Schneidermann, N. and Harris, P. M., eds., Carbonate Cements, Society of Economic Paleontologists and Mineralogists Special Publication. 36, p. 3–16.

———, and MORSE, J. W., 1984a, Magnesian calcite stability: A reevaluation: Geochimica et Cosmochimica Acta, v. 48, p. 1059–1070.

——— and ———, 1984b, Reactive surface area of skeletal carbonates during dissolution: Effect of grain size: Journal of Sedimentary Petrology, v. 54, p. 1081–1090.

WALTER, L. M., AND MORSE, J. W., 1985, The dissolution kinetics of shallow marine carbonates in seawater: A laboratory study: Geochimica et Cosmochimica Acta, v. 49, p. 1503–1513.

WEISS, R. F., 1974, Carbon dioxide in water and seawater: The solubility of a non-ideal gas: Marine Chemistry, v. 2, p. 203–215.

WRAY, G. L., AND DANIELS, F., 1957, Precipitation of calcite and aragonite: Journal of the American Chemical Society, v. 79, p. 2031–2034.

INTERACTION OF ORGANIC MATTER AND CRYSTALLIZATION OF HIGH MAGNESIUM CALCITE, SOUTH LOUISIANA

M. JOHN KOCURKO
Midwestern State University, Wichita Falls, Texas 76308

ABSTRACT: Rim cements of high magnesium calcite, with exceptionally high magnesium carbonate content, are presently lithifying quartz sands at the surface along the South Louisiana Gulf Coast. Cements accumulate at the margin of a marsh environment on the leeside of a coastal beach complex. Based on X-ray diffraction analyses, surface cements range from 35 to 50 mole % magnesium carbonate. Cements show a well defined calcite lattice structure with no indication of dolomite ordering. Surface cements are associated with evaporites and blue-green algae but are not restricted to areas of algal activity.

Surface crusts are periodically buried by storm action and preserved in the subsurface. X-ray diffraction analyses of buried samples indicate a magnesium carbonate content of 17 to 33 mole %. Scanning electron microscopy indicates that cements are "fiber-bundles" with individual fibers constructed of crystallites with an average diameter of 0.3 μm.

Sandstone clasts, found along beaches of South Louisiana, are interpreted to be ripped-up fragments of older surface crusts. Clasts are cemented by high magnesium calcite with 8 to 17 mole % magnesium carbonate. Cements appear as blades with well defined trigonal terminations. These blades are also constructed of crystallites, averaging 0.3 μm in diameter. The surface crusts, buried crusts and rip-up clasts can be categorized into three cement types based on composition, crystal morphology and environment of occurrence.

The three cement types reported here are considered to be representative of diagenetic stages involving dissolution and precipitation. After burial, magnesium carbonate content decreases while the δC^{13} values become more negative. Isotope analyses yield average δC^{13} values of −1.0‰, −4.0‰, and −37.4‰ PDB for surface, buried, and rip-up-clast cements, respectively. Oxidation of organics is probably the controlling mechanism of the dissolution/precipitation process. Extreme negative δC^{13} values of the rip-up-clast cements and a progressive decrease in values with time and depth of burial support this interpretation.

INTRODUCTION

Carbonate deposits are currently being studied in selected areas along the Louisiana coast (Fig. 1). The primary study area is Elmer's Island, west of Grand Isle and located in Lafourche Parish approximately 80 km south of New Orleans. It is but one of many barrier island complexes that separate marshlands of southeast Louisiana from the Gulf of Mexico.

Intergranular cements of high magnesium calcite are presently precipitating in coastal marsh and barrier island complexes of the Gulf Coast of Louisiana. The cements locally form sandstone crusts at the sediment surface. These surface crusts consist of cemented terrigenous clastics and organic debris. Individual crusts average 5 mm in thickness. Accumulations of superimposed layers have been noted up to 30 mm thick. The crusts form in a supratidal environment between a lagoonal marsh and the leeside of a coastal beach complex (Fig. 2). Submergence coupled with tidal and eolian forces are currently causing a migration of beach sediments across the marsh environment. This is evidenced by the presence of marsh sediments with grass and mangrove remains exposed in the beach swash zone. Similar submergence and beach migration have been reported by Cuomo et al. (1984) in the Ship Shoal and Trinity Shoal areas of Louisiana.

The environment of deposition is characterized by seasonally warm surface temperatures (up to 50° C), organic activity and periodic hypersaline conditions. Figure 3 shows a typical supratidal flat from which crusts have been collected on Elmer's Island. Intermittently these carbonate-cemented sandstone crusts are covered by wind-blown sands, washover deposits from storms, or backwash deposits from high lagoonal water. This results in the accumulation of numerous layers of carbonate-cemented material that can be found in the subsurface. Rate of burial of these layers varies with the frequency of storm activity. For example, a local surface crust was covered by approximately 100 cm of sediment during the passage of Hurricanes Bob and Frederick in 1979. The crust was subsequently excavated and found to be intact. A new surface crust formed within a two-month period and presently covers the area.

Sedimentary structures and inclusions are also commonly preserved during the lithification process. Both surface and buried crusts exhibit small-scale cross laminae, planar laminae, root casts, burrows and incorporated wood fragments. Flat-pebble conglomerates form by disruption of crusts during storm action. The broken fragments may be subsequently reincorporated in new surface crusts.

Sandstone and mudstone clasts (or nodules) (Fig. 4), exhibiting similarities to present-day surface and buried crusts, can be found along numerous Gulf Coast beaches. Samples have been collected from South Louisiana as well as from Mustang Island and Padre Island, Texas. The sandstone clasts are interpreted to be ripped-up fragments of older surface crusts. Rip-up clasts range in size from pebbles to small boulders. The clasts are generally flattened (parallel to bedding) with well rounded edges. Even though a transport history is evident, the size and intricate shapes indicate a relatively short time or distance of movement. The rip-up clasts, like the surface and buried crusts, are characterized by small-scale cross laminations, burrows, root casts and inclusions of wood fragments. Flat-pebble conglomerates have also been noted. Roberts and Whelan (1975) and Morgan and Treadwell (1954) reported similar rip-up clasts from the Chandeleur Islands, Louisiana. It is probable that ancient buried crusts are subjected to erosional processes (i.e., wave action) and may be redeposited on present-day beaches. Clasts have been collected *in situ* near Port Fourchon, Louisiana, from marsh sediments exposed in the beach swash zone. The ancient crusts are fragmented but can be traced along bedding planes. The largest single fragment is approximately 1 cm thick and 30 cm by 50 cm across its maximum diameters. The fragments show no indication of

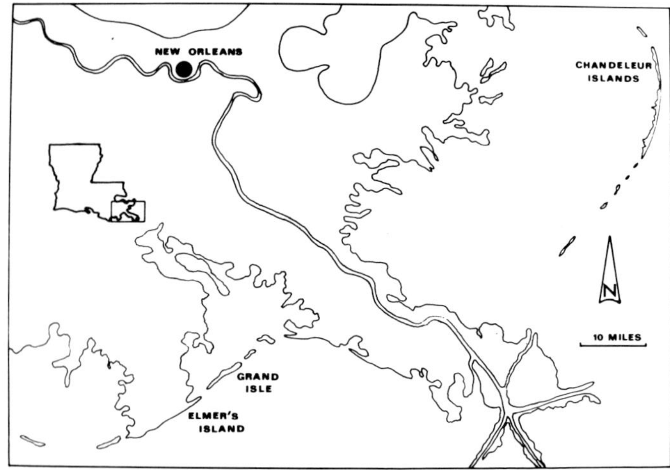

FIG. 1.—General reference map of the South Louisiana area.

transport and the lithified layers are encased in marsh muds. Laminae of mud and crust are conformable.

In general, three distinct cement types can be found in the South Louisiana area. They show differences in composition, crystal morphology, and environment of occurrence. These groups are interpreted to represent progressive diagenetic stages. They will be referred to as cements of "surface crusts," "buried crusts," and "rip-up clasts."

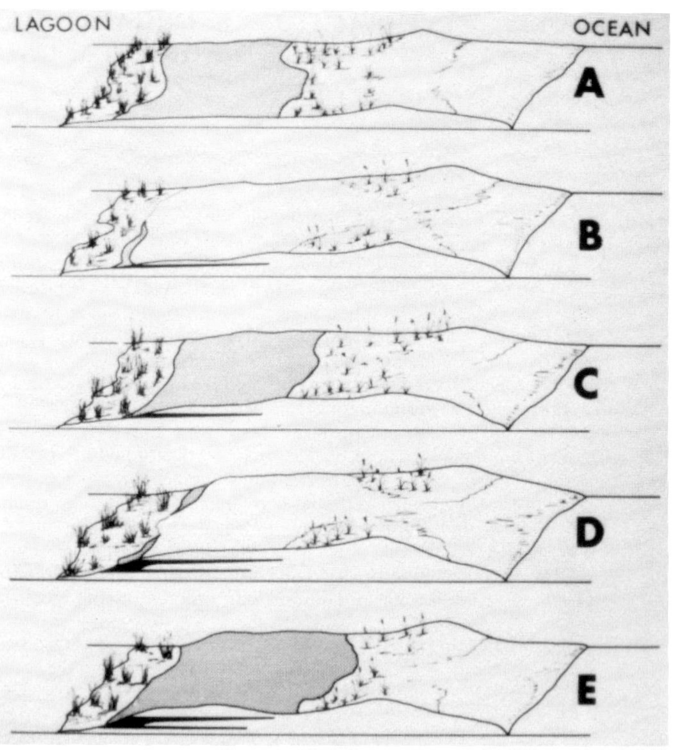

FIG. 2.—The sequence of block diagrams (A through E) represents periodic formation of surface crusts (A, C, and E) and subsequent burial by washover deposits (B & D) in a barrier island complex.

CEMENT COMPOSITION

X-ray diffraction analysis indicates a wide range of carbonate mineralogy in all three cements. Samples of surface crusts collected from the area depicted in Figure 3 range from a combination of aragonite, dolomite and high magnesium calcite, to samples cemented entirely by high magnesium calcite. This probably indicates periodic changes in the precipitating environment.

The analyses also indicate the presence of three distinct cement types based on the magnesium carbonate content

FIG. 3.—Area of crust development on Elmer's Island, Louisiana. Lagoonal marsh is to the left.

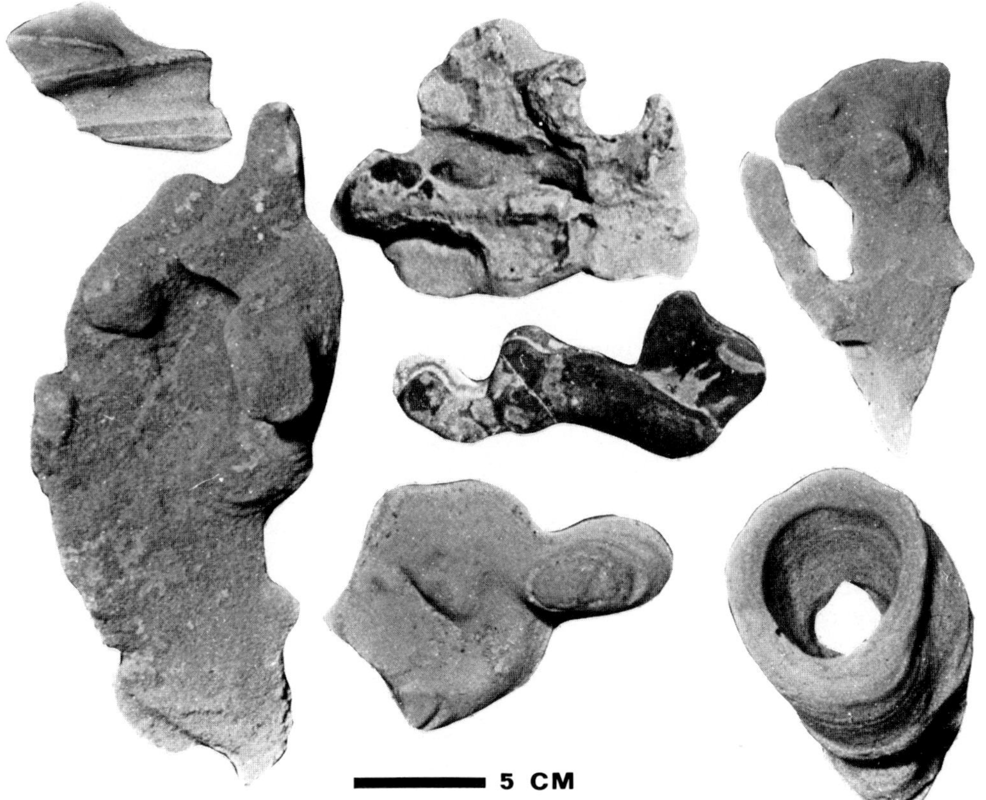

FIG. 4.—Typical rip-up clasts collected from beaches along the South Louisiana coast.

(Fig. 5). Surface crust cements generally range from 35 to 50 mole % magnesium carbonate. These cements are typically of calcite structure and show no evidence of dolomite ordering. In addition to the high magnesium calcite, surface crusts from the area of investigation commonly contain aragonite and ordered dolomite. The occurrence of unusually high mole % magnesium carbonate is rare, but several localities have been cited in the literature. A few examples are Jørgensen (1976, 1979); Brown and Farrow (1978); Atwood and Bubb (1970); Shinn et al. (1965) and Shinn (1964). With the exception of Jørgensen (1976, 1979), the areas reported are dominated by carbonate sediments rather than clastics.

The second group of high magnesium calcite cements is found in the buried crusts. The magnesium carbonate content of these cements ranges from 17 to 33 mole %. Buried crust cements are typically void of aragonite but may contain ordered dolomite.

The rip-up clast cements represent the third group of high magnesium calcites. Generally the magnesium carbonate content ranges from 8 to 17 mole %. Rip-up clasts contain no aragonite but may be composed entirely of ordered dolomite.

The three groups of cements can be categorized on the basis of their average mole % magnesium carbonate. Surface crust cements have an average of 41 mole %, buried crust cements = 29 mole % and rip-up-clast cements = 10 mole %.

Cement types can also be distinguished on the basis of carbon and oxygen isotope ratios. Carbon isotope values are calculated by the following formula and standard:

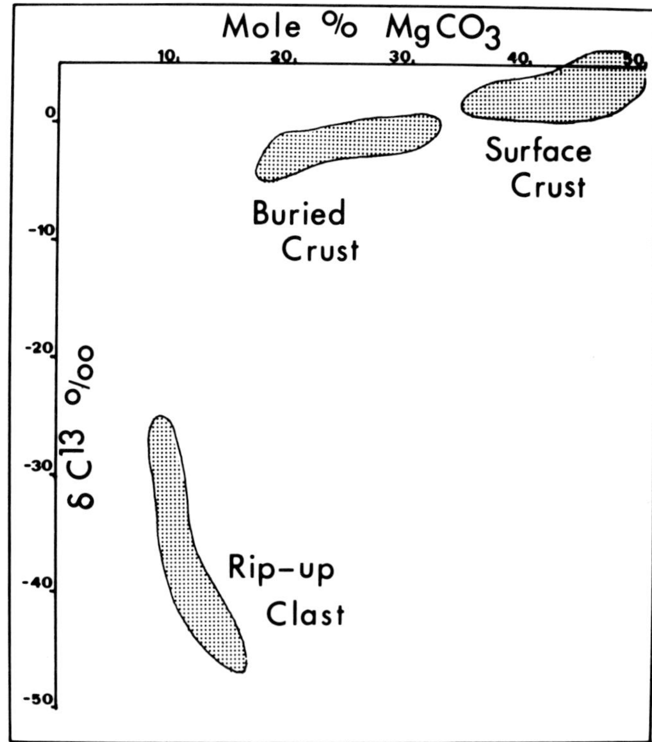

FIG. 5.—Distribution of mole % magnesium carbonate relative to δC^{13} PDB in high magnesium calcite cements.

$$\delta C^{13}‰ \text{ PDB} = \left(\frac{C^{13}/C^{12} \text{ sample}}{0.011237} - 1\right) \times 1000.$$

Surface crusts have an average δC^{13} value of −1.0‰ PDB and range from +6.2‰ to −3.9‰. Average δO^{18} is +0.8‰ PDB. As surface crusts are progressively buried, the δC^{13} value for buried crusts is −4.0‰ PDB, with a range of −1.9‰ to −7.3‰. Average δO^{18} is −0.6‰ PDB. The rate of accumulation of the lighter carbon isotope is uncertain. Buried crusts are estimated, at the most, to be only a few tens of years old. It is evident that with sufficient time, progressive cementation and persistent organic activity, the cements will develop more negative δC^{13} values. This is supported by isotopic values derived from the analysis of rip-up clasts. Rip-up clasts from the Elmer's Island area have been dated at approximately 1500 yrs BP by standard C^{14} methods. These clasts correspond to an ancient barrier island complex described by Shea Penland (Louisiana Geological Survey, pers. commun., 1983). The rip-up-clast cements have an average δC^{13} value of −37.4‰ PDB, with a range of −25.9‰ to −48.1‰. The average δO^{18} value is −0.6‰ PDB. Similar findings have been reported by Roberts and Wheland (1975), Hathaway and Degens (1969), Jørgensen (1976, 1979), van Straaten (1957) and Allen et al. (1969). The uniformity of oxygen isotope ratios indicates little or no variation in this aspect of the pore fluids. The primary variables seem to be methane availability and rate of crystal growth (i.e., Mg^{++} activity).

CEMENT TEXTURE AND CRYSTAL MORPHOLOGY

Samples were thin-sectioned and analyzed petrographically. The cement is generally of such fine texture that only gross distribution within the sample can be observed by this method. Petrographically, the cement is classified as a combination of micrite and microspar. In many of the rip-up clasts the cement is of a coarser texture and a distinct fiber-radiate structure is visible (Fig. 6).

Crystal morphology is similar in all three cement categories. There is, however, a distinct difference in the apparent degree of crystallinity (refinement of geometry). Figure 7 depicts the general crystal morphology of the three cement types. In each example the cement "crystals" are found to be an accumulation of small, subhedral crystallites that are aligned in near perfect crystallographic registry. The crystallites average 0.3 μm in diameter.

Surface crust cements form uniform rims around preexisting detrital grains (Fig. 7B). The rim cements average 20 μm in thickness and are composed of a mosaic of subhedral and anhedral crystals. Even though the geometry of the mosaic crystal is poorly developed, there is a distinct uniformity of crystallographic angles (Fig. 7A).

After burial and continued precipitation, the cement crystals maintain crystallographic geometry (Fig. 7C). The alignment of crystallites in the structure gives the overall crystal a fiber-bundle appearance (Fig. 6B). Similar bundles have been reported by Devery and Ehlmann (1981) and van Straaten (1957).

Rip-up-clast cements are composed of blade-like cystals (Figs. 6C and 7F) that tend to reduce porosity. Similar to the previously noted cements, the blades are constructed of aligned crystallites with an average diameter of 0.3 μm. The blade-like crystals have well defined geometry and trigonal terminations. Crystallographic angles are the same as those found in surface and buried crust cements.

The degree of geometric refinement in the crystal shape is probably controlled by growth rates. Surface cements form rapidly (within a month or two), whereas buried cements and rip-up clasts are progressively slower to accumulate.

ENVIRONMENTAL CHARACTERISTICS

Initial precipitation of surface crust cement is apparently controlled by two factors: evaporation rates and biologic activity. Many areas of crust development are characterized by the formation of evaporitic minerals, primarily halite. Halite formation is observable in the field and documented by X-ray diffraction analysis. In addition to hypersaline conditions, algal growth is usually associated with areas of crust development. The occurrence of two distinct types of algal growth can be documented. First, the formation of typical stromatolitic mats (Fig. 8) has been noted throughout the area of investigation. In areas of substantial mat development, surface crusts of sandstone seldom form but buried crusts may be present. The underside of the algal mat is typically calcareous. X-ray diffraction analyses of the mat material indicate the presence of ordered dolomite and high magnesium calcite. The second situation in which algal growth has been noted is in areas of extensive surface crust development. Surface crusts typically have a slight green discoloration in the upper 2 to 4 mm of the laminae. It should be noted that, even though the green color is attributed to algal growth, there is no mat development or green color at the sediment surface. The presence or absence of stromatolitic mats is probably related to variations in marine inundation. The formation of carbonate cement at the surface is probably favored by evaporation, coupled with algal activity. Similar carbonate cements have been reported by Gebelein and Hoffman (1973). The magnesium carbonate content of their samples, however, was only in the range of 17 to 20 mole %.

In order to check the importance of algal activity in cement formation, a field sample of algal mat was cultured in the laboratory. Algal material was allowed to grow into and through filter paper. The algae were then removed from the sediment substrate, washed and isolated for future growth. A culture chamber (Fig. 9) was prepared using sterilized, dry, quartzose beach sand. The sand was previously boiled in hydrochloric acid to remove all carbonate material and wet-sieved to remove clay-size material. The clean sand was then used as a substrate for algal growth. The filter paper with algal material was placed on the sediment surface and covered with a thin layer of additional sand. The entire chamber was then flooded with filtered seawater. After flooding, the seawater was allowed to drain through the sediment column. As algal growth progressed, additional layers of sediment were introduced into the chamber. A moist substrate and hypersaline conditions were generated by periodic seawater flooding. The chamber maintained algal growth for approximately five years. Examination of

FIG. 6.—The SEM photomicrographs depict the three types of carbonate cements found along the South Louisiana coast. (A) Surface crust cement, bar scale = 5 μm. (B) Buried crust cement, bar scale = 10 μm. (C) Rip-up-clast cement, bar scale = 10 μm.

the culture revealed a dominance of algal filaments (ca. *Schizothrix*) and diatoms (*Encyonema*).

X-ray diffraction analyses of the substrate indicate the development of ordered dolomite, which was not present in the orginal sediment sample. Halite was also present in minor quantities. Isotope analyses of the bulk sediment yield a δC^{13} value of $-2.5‰$ PDB and δO^{18} of $-1.2‰$ PDB. These values correspond with similar data from naturally occurring surface crusts. Based on the experiment, organic activity at the sediment surface probably has a significant effect on the generation of carbonate cements found in the surface crusts, particularly the dolomite phase. Algal activity in the surface crust is probably responsible for the accumulation of relatively large quantities of Mg^{++}, as noted by Gebelein and Hoffman (1973), and thereby the generation of dolomite. Excess Mg^{++} is probably incorporated into penecontemporaneous calcite cements during evaporation processes. This gives rise to a sandstone crust cemented by both dolomite and high magnesium calcite. By periodic fluctuation of the Mg^{++} source (i.e., decrease in algal activity or dilution of solutions), cement precipitation may change from high magnesium calcite to aragonite.

After burial, organics derived from algae and diatoms (as well as higher plants) decompose. In most areas of this field

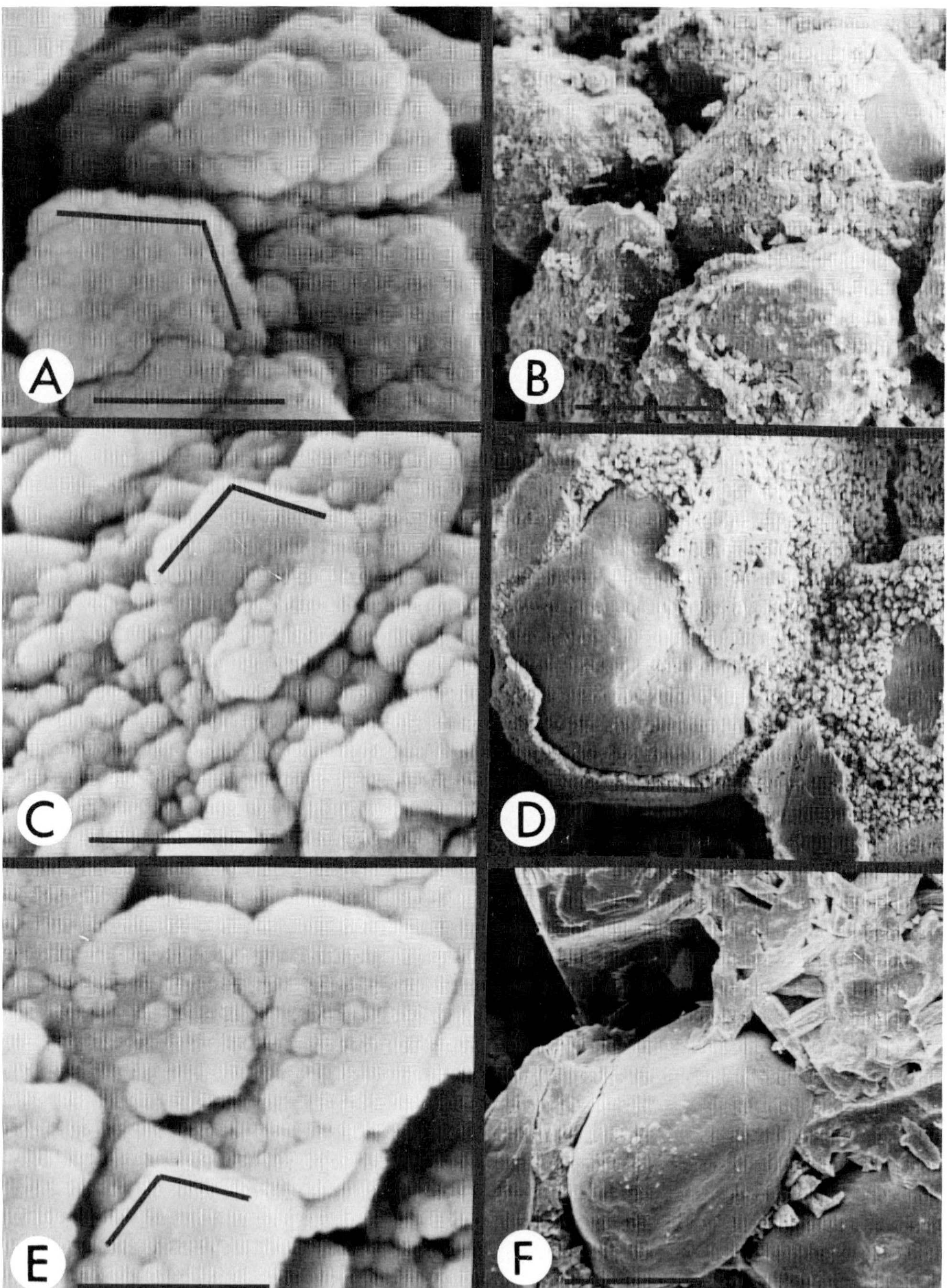

FIG. 7.—Crystal morphology and distribution of cements in surface crusts (A and B), buried crusts (C and D) and rip-up clasts (E and F). In SEM photomicrographs A, C and E the bar scales represent 100 μm. Note consistency of crystallographic angles and the refinement of crystal geometry.

study, organics in the upper 25 cm of the sediment column tend to be oxidized. The oxidation/reduction boundary is controlled by the depth of the water table. It is probable that in this oxidation/reduction environment, sulfate-reducing bacterial action is dominant and organic decomposition progresses. The interaction of decomposed organics and groundwater produces another environment for the generation of carbonate cements. As noted above, buried crust cements are typically lower in magnesium carbonate content and have slightly more negative δC^{13} values. This change in cement composition probably occurs on a laminae-by-laminae basis during shallow burial and is controlled by microenvironments. Analyses of a shallow surface core (Fig. 10) support this interpretation. Note in Figure 10 that with depth there is a distinct decrease in δC^{13} values from −0.9‰ to −5.9‰ PDB. This progressive change correlates with a decrease in the magnesium carbonate content from 42.3 mole % at the surface to 25.6 mole % at a depth of 80 mm. There is also a loss of aragonite with depth. This variation in cement composition can be explained by groundwater interaction during the burial process. Groundwater analyses from the study area during January 1984 indicate an average salinity of 35.8% at 12.6°C, with a pH of 6.7. Pore fluids are practically normal seawater with the exception of a

FIG. 8.—Typical stromatolitic mat development. Lagoonal marsh is to the right.

slightly lower pH. The decreased pH is probably a result of organic decay. The slightly acidic conditions in the subsurface may account for the absence of aragonite in buried crust and rip-up-clast samples (i.e., dissolution).

Seasonal fluctuations produce changes in groundwater chemistry. For comparison, groundwater analyses performed during August 1983 in the same study area indicate an average salinity of 45.2% at 30.1°C, with a pH of 8.2. Clearly, this is a solution with potential for carbonate precipitation rather than dissolution.

With seasonal variation in pore fluid chemistry, a simple model for alternating dissolution/precipitation can be applied. This model can explain the existence of nonlithified (or poorly lithified) layers of sand between buried crusts, the lack of aragonite in buried crusts, and the decrease in magnesium carbonate content downward through the sediments. If the diagenetic progression is projected through time (i.e., 1500 yrs) and depth of burial (i.e., to 5 m), cements similar to those found in rip-up clasts can be expected. As organics are decomposed and methane generated, carbon dioxide may be generated in the zone of oxidation by the reaction

$$CH_4 + 2O_2 = CO_2 + 2H_2O.$$

Carbon dioxide dissolved in water can then form a bicarbonate ion which can be incorporated into the carbonate cements. In the reducing zone, sulfate-reducing bacterial action may cause the development of carbonate cements by

$$CH_4 + SO_4^{=} = CO_3^{=} + H_2S + H_2O,$$

then

$$CO_3^{=} + H_2O + CO_2 = 2HCO_3^{-}.$$

With increasing depth of burial and time, the pore fluids stabilize to near normal seawater conditions with respect to salinity and oxygen isotope ratios but are charged with methane. This situation give rise to a carbonate cement with

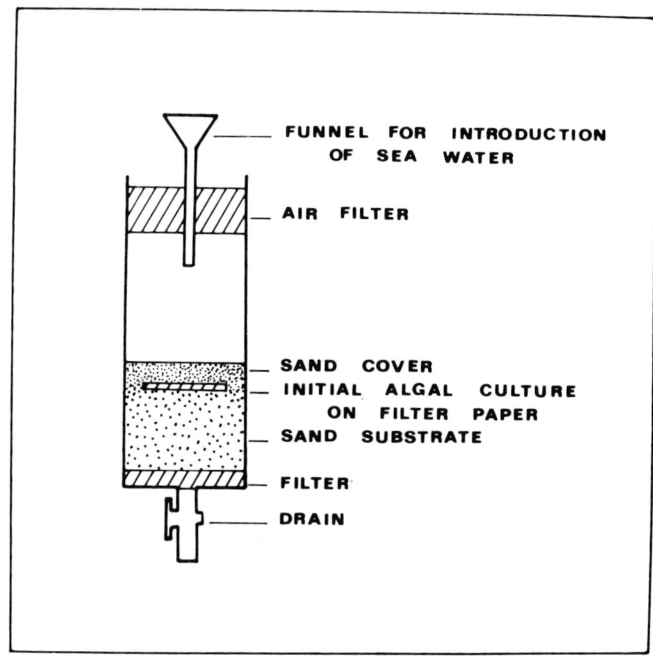

FIG. 9.—Laboratory culture chamber for the growth of algal material.

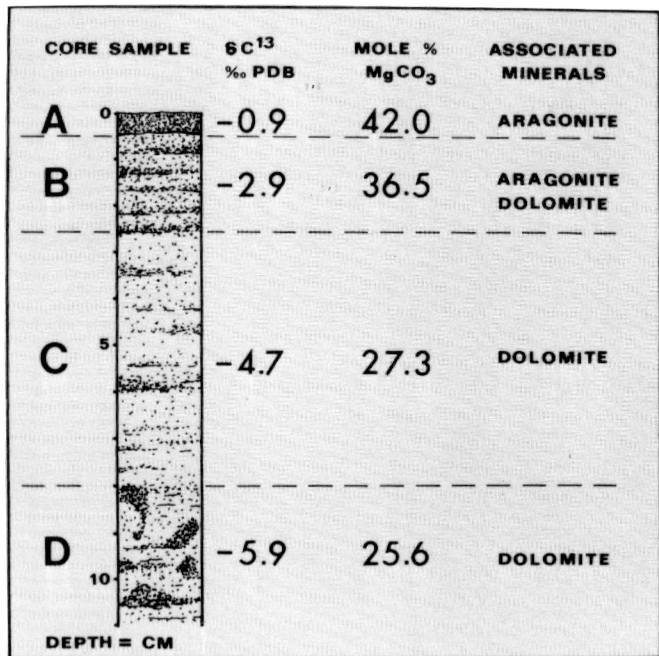

FIG. 10.—Analyses of a shallow surface core. Note decrease in δC^{13} and mole % magnesium calcite with increasing depth. Unit A = surface crust, Unit B = buried crusts, Unit C = buried crust with laminae of nonlithified or poorly lithified sediment, Unit D = buried crusts with mottled organic accumulations.

extremely negative δC^{13} values but a magnesium carbonate content that is stable with respect to seawater (i.e., 8 to 17 mole %).

DISCUSSION AND CONCLUSIONS

The coast of South Louisiana provides a unique opportunity to observe the formation of unusual carbonate cements. The extremely high magnesium carbonate content of the calcite (i.e., 35 to 50 mole %) is rare and probably among the highest reported in the literature to date. In addition, cements can be found with extremely negative δC^{13} values (i.e., −48.1‰ PDB). Data indicate that precipitation of these cements is directly affected by biochemical activity and organic decomposition.

Surface crusts, buried crusts and rip-up clasts represent three successive stages of carbonate cement development. These stages reflect changes in the environment of precipitation that cause both chemical and morphological variations in cement structure.

Surface crust cement is composed of a combination of aragonite, dolomite and high magnesium calcite with 35 to 50 mole % magnesium carbonate. Precipitation of these rapidly forming cements is apparently controlled by evaporation rates and organic activity (i.e., algal growth). Dolomite and, in part, high magnesium calcite seem to be affected most by Mg^{++} availability due to selective up-take by blue-green algae. Precipitation of aragonite or high magnesium calcite is regulated by seasonal evaporation rates. Even though there is penecontemporaneous formation of dolomite and aragonite (or high magnesium calcite), the two microenvironments of precipitation are distinctively different (i.e., biochemical vs. physicochemical environments). The only apparent interaction is in transfer of Mg^{++} from one system into the other. Carbon and oxygen isotopes indicate surface crust cements form from solutions which are isotopically very much like normal seawater.

Buried crust cements form in a subsurface environment where oxidation/reduction reactions and sulfate-reducing bacterial actions cause decomposition of organics. The primary organic source is algal material from buried surface crusts. With burial and decomposition, Mg^{++} is no longer accumulated by the algal population. As a result, with depth, there is a decrease in available Mg^{++}, and thus a decrease in the mole % magnesium carbonate of the newly precipitated cements. Seasonal variation causes fluctuations in the groundwater chemistry which, in turn, controls dissolution and precipitation of the carbonate cements. The result is periodic dissolution and removal of aragonite, and probably to some extent, the more unstable phases of high magnesium calcite (i.e., those of exceptionally high $MgCO_3$ content). When precipitation occurs, high magnesium calcites of 15 to 30 mole % magnesium carbonate form. During this stage of cement formation, there is interaction between the pore fluids and the decomposing organics (i.e., methane). Isotopically heavier CO_2 is mixed with methane generated CO_2 and the overall δC^{13} values tend to decrease. With increasing depth of burial and time, methane-generated CO_2 becomes dominant and δC^{13} values are progressively more negative. Cement generation probably occurs on a laminae-by-laminae basis with burial. Buried crust cements seem to accumulate at a slower rate than cements found at the surface and, therefore, show greater refinement of crystallographic geometry.

Rip-up-clast cements represent a continuation of burial and progressive interaction with methane-generated CO_2. The pore fluids are probably near normal seawater charged with isotopically lighter CO_2. High magnesium calcites in the range of 25 to 50 mole % magnesium carbonate may be unstable with respect to these "normal" fluids and dissolution probably occurs. Reprecipitation of carbonates then follows with a reduced magnesium carbonate content (i.e., a more stable phase) but with extremely negative δC^{13} values. Slow crystallization should lead toward greater textural simplicity and equilibrium, and therefore, crystal geometry is more distinct than in buried crusts. In all three cement categories, ordered dolomite is the most stable carbonate phase. It is persistent from the surface to buried samples to rip-up clasts.

ACKNOWLEDGMENTS

Appreciation is extended to Steve Laney, Pat Welch, Chris Salmon and Sidney Rasbury for their assistance in the field and laboratory. Manuscript preparation was aided by comments and suggestions from Jay Murray, Rick Williams and Walt Dalquest. Special gratitude goes to Jack Watkins, who has been an important influence since the beginning. For all the hours of typing, deadlines, etc., appreciation is given to Marilyn.

REFERENCES

ALLEN, R. C., GAVISH, ELIEZER, FRIEDMAN, G. M., AND SANDERS, J. E., 1969, Aragonite-cemented sandstone from outer continental shelf off Delaware Bay: Submarine lithification mechanism yields product resembling beachrock: Journal of Sedimentary Petrology, v. 39, p. 136–149.

ATWOOD, D. K., AND BUBB, J. N., 1970, Distribution of dolomite in a tidal flat environment, Sugarloaf Key, Florida: Journal of Geology, v. 78, p. 499–505.

BROWN, B. J., AND FARROW, G. E., 1978, Recent dolomite concretions of crustacean burrow origin from Lock Sunary, west coast of Scotland: Journal of Sedimentary Petrology, v. 48, p. 825–834.

CUOMO, R. F., KOFRON, F. M., AND NUMMEDAL, DAG, 1984, Shelf morphodynamics of drowned barriers on Louisiana shelf: American Association of Petroleum Geologists Bulletin, v. 68, p. 468.

DEVERY, D. M., AND EHLMANN, A. J., 1981, Morphological changes in a series of synthetic Mg-calcites: American Mineralogist, v. 66, p. 592–595.

GEBELEIN, C. D., AND HOFFMAN, PAUL, 1973, Algal origin of dolomite lamination in stromatolitic limestone: Journal of Sedimentary Petrology, v. 43, p. 603–613.

HATHAWAY, J. C., AND DEGENS, E. T., 1969, Methane-derived marine carbonates of Pleistocene age: Science, v. 165, p. 690–692.

JØRGENSEN, N. O., 1976, Recent high magnesium calcite/aragonite cementation of beach and submarine sediments from Denmark: Journal of Sedimentary Petrology, v. 46, p. 940–951.

———, 1979, Magnesium incorporation in recent marine calcite cement from Denmark: Journal of Sedimentary Petrology, v. 49, p. 945–950.

MORGAN, J. P., AND TREADWELL, R. C., 1954, Cemented sandstone slabs of the Chandeleur Islands, Louisiana: Journal of Sedimentary Petrology, v. 24, p. 71–75.

ROBERTS, H. H., AND WHELAN, THOMAS, 1975, Methane-derived carbonate cements in barrier and beach sands of a subtropical delta complex: Geochimica et Cosmochimica Acta, v. 39, p. 1085–1089.

SHINN, E. A., 1964, Recent dolomite, Sugarloaf Key, *in* Geological Society of America Guidebook, Field Trip No. 1, South Florida Carbonate Sediments, p. 62–67.

———, GINSBURG, R. M., AND LLOYD, R. M., 1965, Recent supratidal dolomite from Andros Island, Bahamas: Society of Economic Paleontologists and Mineralogists Special Publication 13, p. 112–123.

VAN STRAATEN, L. M. J. V., 1957, Recent sandstones on the coasts of the Netherlands and of the Rhone Delta: Geologie in Mijnbouw, v. 19, p. 196–213.

CONTROLS ON THE PRECIPITATION OF EARLY DIAGENETIC CALCITE, DOLOMITE AND SIDERITE CONCRETIONS IN COMPLEX DEPOSITIONAL SEQUENCES

C. D. CURTIS
University of Sheffield, England
AND
M. L. COLEMAN
British Petroleum Research Centre, Sunbury, England

ABSTRACT: Carbonate concretions occur commonly in organic-rich clastic sediments of ancient coastal floodplains and deltas. They consist of calcite, dolomite, or siderite or combinations of these. The influence of marine, mixed or non-marine depositional waters is seen in early cements; solute sources from decomposition of unstable detrital minerals become more important in later diagenesis.

The mineralogy and chemistry of early cements are dependent on dissolved sulphate concentrations and processes which affect them (notably depositional water composition, microbial sulphate reduction and mixing with waters of different origin). Calcite precipitates from marine-influenced pore waters rich in dissolved sulphate, dolomite from similar waters depleted in sulphate and siderite from meteorically-derived pore waters.

Systematic compositional variations with depth of precipitation are preserved both within and between siderite concretions. Earliest cements have higher Mn/Fe, Fe/Mg and $\delta^{13}C$ values: all decline in later cements. These trends are linked with solute supply for cations and microbial metabolism for bicarbonate.

INTRODUCTION

Precipitation of Diagenetic 'Cements'

Authigenic mineral segregations of one type or another, "-though a minor feature of most sedimentary rocks, are very common" (Pettijohn, 1957). Another significant point is made subsequently: "The mineral constituents of concretions are those which normally play the role of cement in the rocks in which these structures occur." Authigenic mineral phases thus occur widely as *disseminated* precipitates, less commonly (but more obviously) as *segregations* caused by local chemical perturbations of the post-depositional environment. Obviously, the cements are derived from pore water solutes and their composition should reflect pore water composition.

Many concretions are obviously 'early,' having precipitated and displaced pore water when the host sediment had experienced negligible compaction (authigenic cements making up 80% or more of the concretion volume). Very obviously, so much mineral matter could not have been present in solution within the pore water originally trapped: solutes must have been subsequently introduced to that volume of initial sediment to become a concretion. Any explanation for concretion growth, therefore, must not only comment on probable pore water composition but also on the source and transport mechanism because they control composition throughout precipitation.

Mineral segregations may represent post-depositional redistribution of solutes within a sedimentary unit or they may be the result of a specific chemical 'sink,' fixing solutes derived from 'outside' and not precipitating elsewhere in the unit. This possibility cautions against too literal an interpretation of the second quotation above.

Common Authigenic Precipitate Minerals

Carbonates are probably the most common concretion body minerals. Pyrite, various phosphates and silica also occur frequently. Iron oxides and hydrated iron oxides are most probably secondary oxidation products when seen in ancient concretions.

Calcite is found in ferroan and magnesian varieties. Dolomite shows a wide range of iron substitution, with ankerite now recognized as a very common late diagenetic cement in some sandstones (Boles, 1978). Siderite also occurs very widely as a concretionary 'cement,' and, as is perhaps less well known, exhibits a considerable range of composition with major substition of Mg^{2+}, Mn^{2+} and Ca^{2+} (Curtis et al., 1975). Matsumoto and Iijima (1981) have even recorded ferroan magnesites in freshwater coal measures from Japan.

Although several different iron sulphides may be present in Recent unconsolidated sediments, pyrite is the most common in ancient sedimentary rocks. Phosphates (apatite, francolite), like pyrite, form segregations, often in direct association with carbonate minerals. Concretions are most often described from clastic sediment sequences, particularly mudrocks. Their occurrence in carbonate sediments, however, is by no means rare (Sass and Kolodny, 1972; Dickson and Barber, 1976).

Solute Sources

Some very obvious statements can be made about mineral precipitation reactions. In bicarbonate-rich pore water systems, predominant availability of Fe^{2+} should ensure siderite precipitation; however, the precise Fe/(Ca + Mg) ratio and other possible controls are not known. In the absence of high concentrations of dissolved iron, presence of Ca^{2+} and Mg^{2+} would favor calcite or dolomite precipitation, depending on relative availability. High partial pressures of oxygen or abundant H_2S would precipitate Fe_2O_3 or FeS_2, respectively, in both cases depleting available Fe^{2+} (therefore iron in carbonates) to very low levels. Phosphate availability would favor carbonate-apatite or francolite relative to carbonate, especially under more acid conditions (Nathan and Sass, 1981). Less obvious but of prime interest here are the probable sources of these solutes.

In marine systems, one obvious source of bicarbonate is detrital 'primary' carbonates of varying stability and composition, i.e., high- and low-magnesian calcite and aragonite. On dissolution/replacement, the carbon would retain a $^{13}C/^{12}C$ ratio characteristic of the initial carbonate minerals ($\delta^{13}C \sim$ zero on PDB scale). Organic matter oxidized by aerobic or sulphate reducing bacteria would introduce bicarbonate much depleted in ^{13}C. In contrast, 'heavy' bicarbonate ($\delta^{13} \sim +15$) results from microbial methanogenesis, although subsequent oxidation of the methane so produced could yield extremely 'light' bicarbonate. At considerable burial depths, where temperatures in excess of about 50° C are encountered, thermal maturation of organic matter also produces ^{13}C-depleted HCO_3. The two fundamental sources are thus seen to be 'primary' skeletal calcium carbonate and organic matter, but derivation may be via one of several mechanisms. Each mechanism tends to dominate specific burial depth zones within unconsolidated sediments (Claypool and Kaplan, 1974; Curtis, 1977). Sources of solutes and bicarbonate production mechanisms are summarized in Figure 1.

Degradation of organic matter liberates ammonia and is an important source of phosphate. Skeletal phosphates may be included with the detrital sediment constituents. Sulphate reduction (mentioned above) is the principal mechanism whereby sulphide (as H_2S or HS^-) is introduced to diagenetic environments. The important source of dissolved SO_4^{2-} in pore water is overlying depositional waters or, in evaporitic sequences, calcium sulphate minerals.

Calcium and magnesium, like bicarbonate, may be incorporated as detrital carbonates. Both are present at relatively low concentrations in depositional and pore waters and also in exchange sites on clay minerals. Irwin (1980) discusses these sources in the context of diagenetic carbonate precipitation.

Iron and manganese are present in abundance in the hydrated oxide/hydroxide component of soils, in clay interlayer hydroxide sheets and in various ferromagnesian silicate minerals. Clastic sediments, especially muds derived from lateritic weathering profiles, are potentially rich in both.

EARLY DIAGENETIC CARBONATES IN MARINE SEDIMENTS

Limestones

Although here we are not primarily concerned with limestones, it is important to remember that the calcium carbonates (calcite varieties and aragonite) are the only carbonates of significance to precipitate from 'normal' marine waters. Ancient shelf limestones, furthermore, are now predominantly cemented by calcite sparites, whether as disseminated cements or as segregations (Bathurst, 1975; Campos and Hallam, 1979). The *norm* situation in marine sediments without significant terrigenous or organic carbon input is thus calcium carbonates as primary detrital grains with calcium carbonate cements dominating diagenetic modification. The source of solutes is almost invariably limited to decomposition of unstable minerals (aragonite, high magnesian calcite) or localized dissolution (resulting in stylolites).

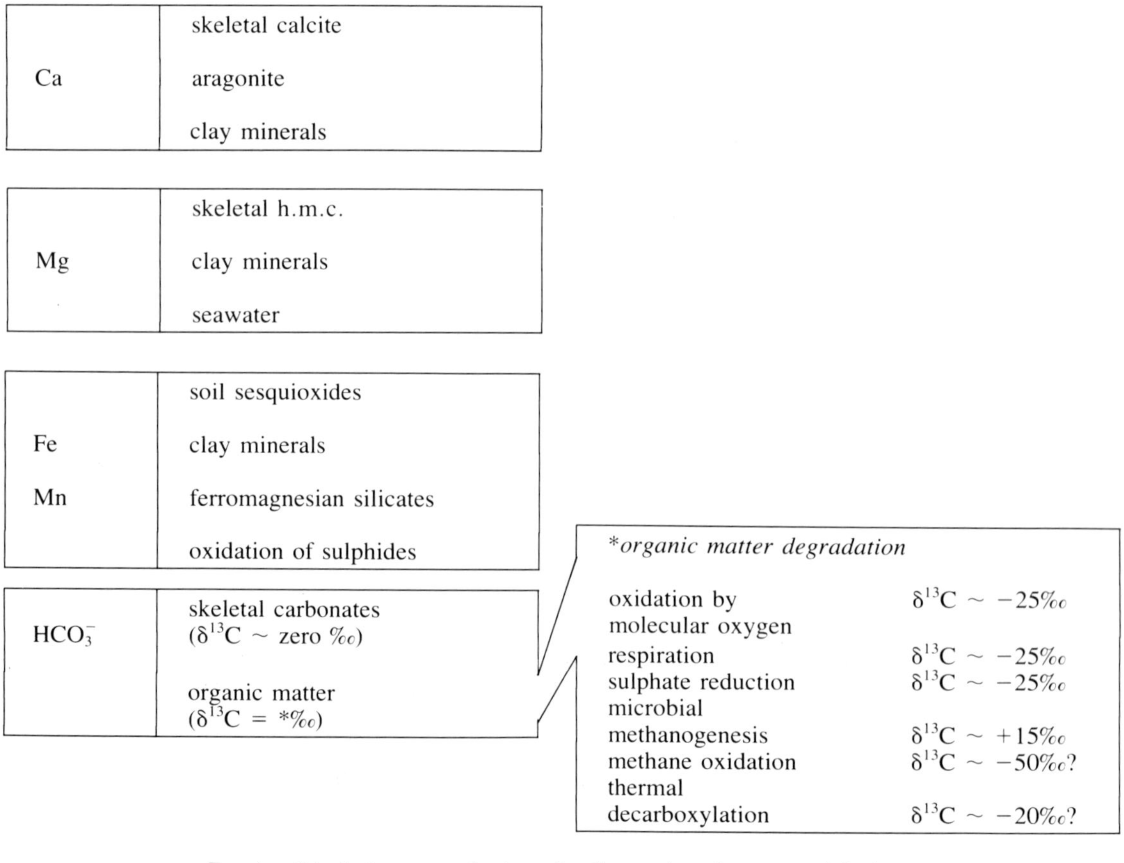

Ca	skeletal calcite aragonite clay minerals
Mg	skeletal h.m.c. clay minerals seawater
Fe Mn	soil sesquioxides clay minerals ferromagnesian silicates oxidation of sulphides
HCO_3^-	skeletal carbonates ($\delta^{13}C \sim$ zero ‰) organic matter ($\delta^{13}C$ = *‰)

*organic matter degradation	
oxidation by molecular oxygen	$\delta^{13}C \sim -25‰$
respiration	$\delta^{13}C \sim -25‰$
sulphate reduction	$\delta^{13}C \sim -25‰$
microbial methanogenesis	$\delta^{13}C \sim +15‰$
methane oxidation	$\delta^{13}C \sim -50‰$?
thermal decarboxylation	$\delta^{13}C \sim -20‰$?

FIG. 1.—Principal sources of solutes for diagenetic carbonate precipitation.

Dolomite in Marine Sediments

Studies of deep-sea sediments in recent years have identified an important environment of authigenic dolomite precipitation: organic-rich fine-grained sediments (Pisciotto and Mahoney, 1981; Matsumoto et al., 1978). There are plenty of examples of similar occurrences in rapidly deposited, organic-rich sediments in continental margin situations (Irwin et al., 1977; Pisciotto, 1982). Dolomite and its ferroan variety (termed ankerite if more than 20% of the magnesium sites are occupied by iron or manganese; Deer et al., 1966) both precipitate in pore space and replace primary carbonate (Irwin, 1980). From this brief survey, it is quite clear that dolomite must also be considered a 'normal' early diagenetic carbonate in organic-rich marine mudrocks. The necessary condition is simply that accumulation rate and composition are such that sulphate is rapidly depleted prior to significant compaction. Unstable primary carbonates and pore water solutes (notably Mg^{2+}) then will react with organic degradation products (mainly microbial methanogenesis and thermal decarboxylation) to produce authigenic and replacement dolomite. Authigenic carbonates other than calcite and dolomite are generally rarer in 'fully' marine successions (siderite is found in minette-type iron ores).

The Importance of Dissolved Sulphate Levels

'Primary' carbonate precipitation in organic-rich sediments ('coal balls') has been known and recognized as an early post-burial phenomenon for well over a century (Hooker and Binner, 1854). Both calcite and dolomite are found, usually in association with marine 'roof measures' and abundant pyrite (see Teichmuller and Teichmuller, 1975). Association of authigenic pyrite with carbonates is thus widespread but intricacies of the interaction have only emerged recently. Apparently, sulphate reduction and pyrite precipitation increase saturation with respect to iron-poor carbonates by generating both bicarbonate and hydroxyls, yet maintaining dissolved iron activity at low levels:

$$15CH_2O + 2Fe_2O_3 + 8SO_4^{2-} \rightarrow 4FeS_2 + 7H_2O + 15HCO_3^- + OH^-. \quad (1)$$

This process, operating in the presence of marine pore waters, should cause supersaturation with respect to dolomite rather than calcite because of the high molar Mg^{2+}/Ca^{2+} ratio in seawater (5.3). Calcite, however, as has already been mentioned, is the normal precipitate from marine *depositional* waters. This indicates that supersaturation and high Mg/Ca ratio are not the only controls on dolomite precipitation.

This fundamental problem seems to have been resolved by systematic experimental work on inhibition of dolomite precipitation in seawater (Kastner and Baker, 1983; Baker and Kastner, 1981). They have shown that sulphate itself is the most important inhibitor: dolomitization of calcite is inhibited at concentrations of around 2 mM, or aragonite at levels approaching 50% of the seawater value (25 mM). Dolomitization rates are decreased at concentrations lower than these threshold levels.

Precipitation of authigenic dolomite (and replacement of aragonite and calcite) is thus favored by any process which reduces sulphate concentration. Dilution by fresh water of meteoric origin is one possibility, although this must also dilute the other important solutes (Ca^{2+}, Mg^{2+}, HCO_3^-). Microbial sulphate reduction, however, can reduce sulphate levels effectively to zero without significant reduction in overall ionic strength.

Sulphate reduction has been an extremely significant process throughout most of geological time and, indeed, has been a major influence on atmospheric oxygen and seawater sulphate levels (Berner, 1984). If sulphate is available, it will be reduced wherever and whenever sufficient organic matter is present to eliminate molecular oxygen and act as substrate for the microbes. Numerous studies of shallow marine, organic-rich sediment cores have demonstrated virtually complete elimination of sulphate from pore waters within tens of centimeters of the sediment/water interface (e.g. Nissenbaum et al., 1972). Dolomite, ankerite or calcite (including ferroan calcite) can then form, depending on cation availability and rate of bicarbonate generation by one or more of several possible organic matter degradation reactions. Supersaturation is also caused by any mechanism which adds hydroxyls or removes hydrogen ions:

$$2Fe_2O_3 + CH_2O + 3H_2O \rightarrow 4Fe^{2+} + HCO_3^- + 7OH^- \quad (2)$$

(reduction of iron oxides; Curtis, 1977), or

$$SO_4^{2-} + 2CH_2O \rightarrow HS^- + 2HCO_3^- + H^+ \quad (3)$$
$$(H^+ + HS^- \rightarrow H_2S)$$

(sulphate reduction *followed by* loss of H_2S; Coleman and Raiswell, in prep.).

EARLY DIAGENETIC CARBONATES IN NON-MARINE SEDIMENTS

Depositional Environments

There is an enormous variety of non-marine depositional environments. Considerable numbers of the possible permutations of water depth and salinity are found today and probably are represented in the geological column. The term 'salinity' as used above conceals further complexity in the sense that solutes present will differ markedly in concentration from those in seawater, reflecting the provenance of material and style of weathering in part'cular drainage basins. In the discussion that follows, the non-marine depositional environments considered will be those of humid climatic environments where waters are fresh. For the main part this means sediments of coastal plains, estuary systems and deltas.

Diagenetic Carbonate Minerals

In Recent freshwater swamp environments, siderite occurs commonly both as concretions and disseminated cements. Calcite is also found although, in general, it seems to be much less widespread (Ho and Coleman, 1969; Whelan and Roberts, 1973). Matsumoto and Iijima (1981) studied authigenic carbonates in Pleistocene, Paleogene and Upper Triassic coal fields of Japan. They found that the earliest cements strongly reflected depositional environment: low Ca-Mg-siderite in freshwater environments, calcite and dolomite in brackish to marine sediments. Siderite is the common authigenic carbonate in freshwater sediments of the Carboniferous Coal Measures of England (Curtis, 1967), iron being present mostly as pyrite in the marine sediments.

The problem with all these reports is that more or less continuous freshwater sediment sequences in continental margin situations almost invariably exhibit intercalated marine sediments representing periods of transgression. Diagenetic history is thus complicated by trapped pore waters of very variable composition and occasional input of detrital marine carbonate. The sequences of diagenetic cements seen in such rocks will thus be influenced by pore water mixing and chemical evolution, thereby obscuring the more distinct character of the earliest cements. This point was made by Matsumoto and Iijima (1981).

CYCLOTHEMIC COAL MEASURES SEQUENCES

Patterns of Sedimentation

The Carboniferous coal deposits of Europe and eastern North America accumulated in the equatorial zone. This has been established by paleomagnetics and also by the virtual absence of growth rings, in sharp contrast to high latitude Permian Gondwana coal deposits (Hallam, 1981). The coal seams themselves often occur at the top of coarsening-upward deltaic cycles, when emergence (or near emergence) allowed plants to colonize.

Teichmuller and Teichmuller (1975) present a helpful review of coal-accumulating environments which draws important analogies with present-day peat swamps. For significant accumulation, subsidence must be at a rate which can be matched by plant growth, the swamp must be protected from the sea (beaches, spits) and there must be low relief in the hinterland in order to restrict (or channel) supply of detrital sediment. Failure of any of these conditions results in inundation followed by marine or limnic sediment accumulation. Marine transgressions cause coal swamps to move inland and carbonate sediment may be deposited. Eventually, conditions favorable to peat accumulation may be reestablished after overbank flooding or by minor crevasse delta deposition.

Pore Water Environments and Composition

Seat earth/underclay.—

Emergent conditions prevail at times within deltaic complexes. Soils form as a consequence of net downward movement of CO_2, oxygen and organic acid-charged meteoric waters (Fig. 2). Lateritic soils are enriched in Fe, Al and Mn as other elements (including Si) are leached. Groundwaters in the phreatic recharge zone are thus somewhat enriched in Ca, Mg, Na, K and Si, the common solutes of river and stream waters (Livingstone, 1963). Carbonates do not precipitate within humid climate soil profiles,

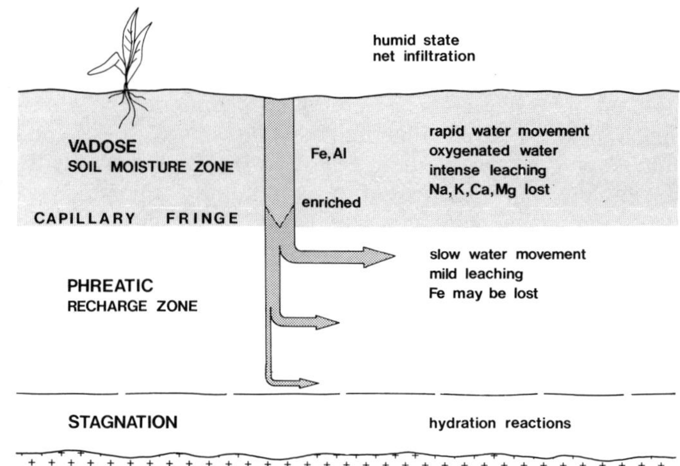

FIG. 2.—Water movement within a soil profile: humid climate.

only where net evapotranspiration is comparable with, or exceeds, precipitation. Pore waters are oxygenated and have low to very low solute concentrations.

Peat swamp 1.—

The sediments beneath a peat swamp are likely to be either stagnant or support minor phreatic recharge from above. In either case, pore waters would be anoxic, of low ionic strength and probably somewhat acid. Some leaching might continue and, under these anoxic conditions, Fe would be much more mobile. These points are illustrated in Figure 3. The extent of lateral iron mobilization would depend very much on local hydrological factors and might be significant, especially when sediments beneath the peat were lat-

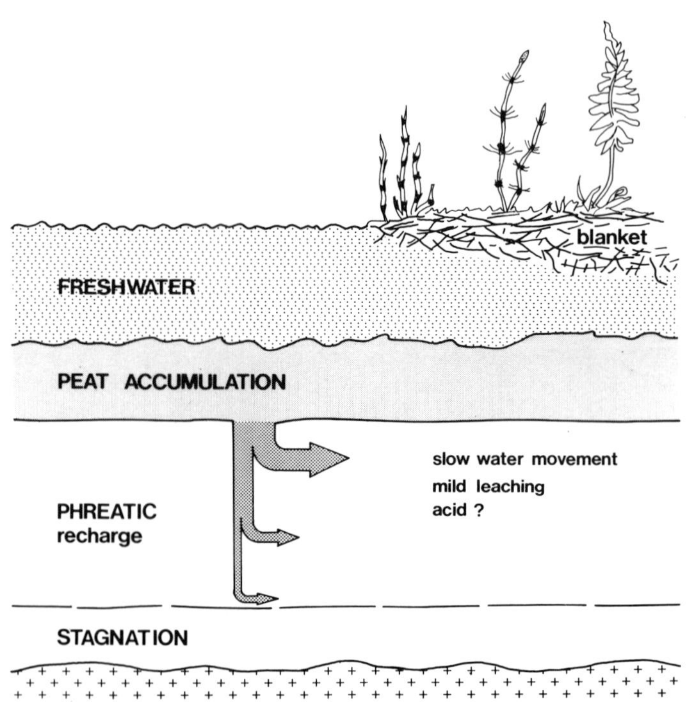

FIG. 3.—Possible water movement within a freshwater swamp.

eritic. The same would be true of manganese. Lateral migration and water mixing (especially with oxygenated waters) can lead to iron segregation as hydrated oxides, as in the high latitude bog iron ores which are forming now. Precipitation as siderite is more likely in tropical swamps where a combination of iron reduction (equation 2) and microbial organic degradation (especially microbial methanogenesis) would cause supersaturation. Molecular oxygen, however, would have to be totally excluded.

Peat swamp 2 (following subsidence/transgression).—

Eustatic sea-level rise (major transgression), delta lobe abandonment or storm breaching of sand spits or beaches would cause immediate replacement of fresh depositional waters by saline. In the former case, detrital sediment supply would be made more remote and significant water deepening would be expected as a result of imbalance between subsidence and sedimentation. Fine clastics would accumulate on top of the peat, possibly with some 'primary' detrital calcium carbonate.

Inundation of the swamp would place high ionic strength, sulphate-rich waters immediately above low ionic strength pore waters below. The concentration gradient so established would promote downward diffusion of marine solutes, notably SO_4^{2-}, Mg^{2+} and Ca^{2+}. Sulphate reduction would start at once (ample organic substrate, formerly no sulphate), and any iron present within the peat pore waters would precipitate as pyrite.

Pore water composition would thus be intermediate between marine and freshwater, with Mg^{2+}, Ca^{2+} and HCO_3^- (from sulphate reduction) plentifully available. It seems likely that sulphate reduction would outpace supply of iron, H_2S would be lost to depositional waters and alkalinity would rise in consequence. Under these conditions either calcite or dolomite, depending upon the extent of sulphate depletion, would precipitate. Early diagenesis in the upper part of coal seams beneath marine sediments is thus anticipated to be dominated by association of pyrite with iron-poor carbonates. Whether these carbonates are calcite or dolomite precipitates would depend on the effective dilution of sulphate by microbial reduction to sulphide. Low sedimentation rate with concomitant aerobic or suboxic degradation of organic matter would favour calcite because sulphate would remain unreduced.

Marine mud.—

Clastic sediment accumulating after marine incursion would trap seawater (80 to 90 % by volume in muds). Sulphate reduction would be expected, again with calcite as the most likely carbonate to precipitate in response to alkalinity generation.

A complication must be anticipated here. As soon as marine sediments start to accumulate above the buried peat, compaction will start and the hydrodynamic flow regime must reverse, as shown in Figure 4. Sediment pore waters would be diluted by sulphate-free pore waters from below, again contributing to sulphate depletion. Dolomite precipitation is perhaps more probable than in 'fully marine' depositional sequences. Absence of sulphate means absence of sulphide in these situations. It is thus not unreasonable to anticipate that if sulphate concentrations are reduced to very low levels, finite dissolved iron concentrations would develop such that dolomite, if formed, should be ferroan.

The marine shales in Coal Measures sequences are often laminated as well as organic-rich. This implies development of euxinic conditions. Sulphate reduction in the depositional water column would lower the concentration of sulphate in trapped pore waters, hence further increasing the chances of significant or even total depletion. Much, of course, would still depend on sedimentation rate with more rapid accumulation favoring dolomite.

Non-marine muds.—

The same kind of analysis can be extended to the brackish-to-freshwater sequences above marine 'bands.' Marine pore waters (sulphate-free by this time) would tend to be expressed upward by compaction into terrestrially-derived sediment deposited with low ionic strength pore water. The most important consequence for carbonate precipitation would be the high Mg/Ca ratio. Carbonates unusually rich in magnesium might document this influence.

Upper Carboniferous Coal Measures, Northern England

The Alton (Gastrioceras listeri) Marine Band.—

We have recently reported detailed descriptions of diagenetic sulphides (Love et al., 1983) and carbonates (Curtis et al., in prep.) extensively developed in marine and non-marine sediments of Early Westphalian age cropping out near Penistone, Yorkshire. The sequence consists of a coal (the Halifax Hard Bed, approximately 1 m thick) resting on a seat earth which is a quartz arenite underlain by a siltstone. Fissile, laminated black shales immediately overlie the coal (the Alton Marine Band) and contain the abundant remains of marine organisms (goniatites and marine bivalves). Above the marine band rests a rather monotonous sequence of mudrocks notable only, perhaps, for the abundant siderite concretions and thin concretionary beds. Non-marine bivalves attest to a dominantly brackish-to-freshwater origin for these sediments with the exception of a second, poorly developed marine horizon (the Parkhouse Marine Band) some 7 m above the coal. The *Gastrioceras listeri* horizon has been traced throughout most of northwest Europe and clearly represents a major marine incursion (Calver, 1968).

Early diagenesis within the coal peat.—

The coal contains classic 'coal balls' which preserve uncompacted plant material in calcite mixed with pyrite (see Love et al., 1983, Fig. 4). The calcite is relatively iron-poor and magnesium-rich:

$$(Ca_{0.89}\,Mg_{0.09}\,Fe_{0.01})\,CO_3.$$

The stable carbon and oxygen isotope ratios are, respectively, $\delta^{13}C = -21.49‰$ and $\delta^{18}O = -8.91‰$, both relative to PDB.

Although these sediments accumulated in truly freshwater swamps, the earliest stages of diagenesis, as seen in

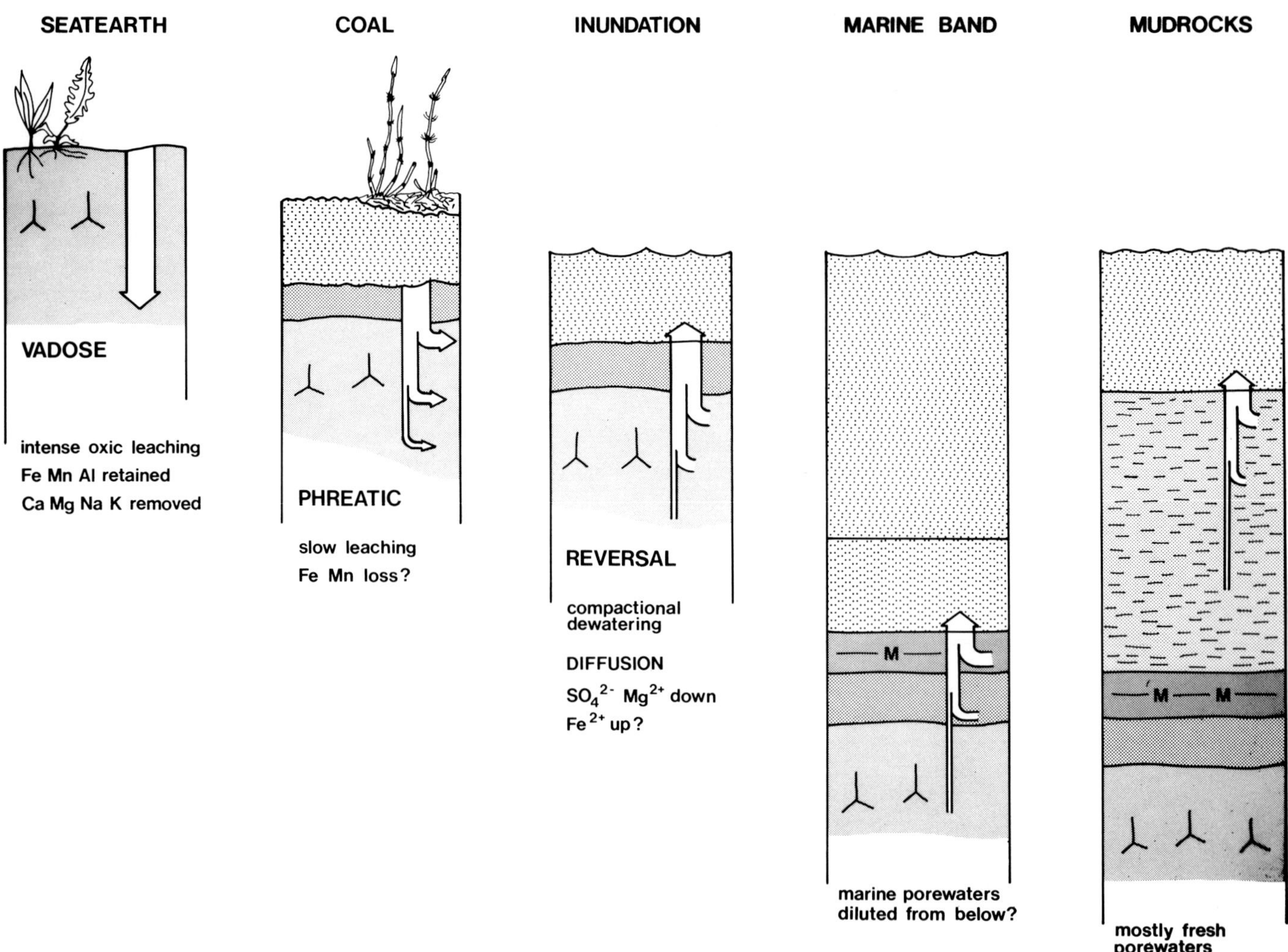

FIG. 4.—Hydrological aspects of a possible cyclothemic sequence: soil profile, peat swamp, marine inundation, deposition of marine muds, establishment of freshwater, limnic deposition.

carbonate and sulphide precipitation, were dominated by the ready availability of SO_4^{2-}, Ca^{2+} and Mg^{2+} in overlying waters. Sulphate reduction occurred extensively (isotopically 'light' bicarbonate, extensive pyrite precipitation), yet was not depleted sufficiently to allow dolomite to form; magnesium was plentiful. This interpretation is supported by the low level of iron in the calcite: H_2S was available in excess of iron (all precipitated as pyrite); therefore, sulphate must also have been available.

The only geochemical (as opposed to the obvious paleontological) indication that fresh, meteoric waters were also present in the system is obvious ^{18}O depletion in the calcite.

A brief survey of the literature suggests that 'coal balls' are almost invariably restricted to seams with marine roof measures and are usually calcitic. Occasionally, the remains of marine organisms are found mixed with plant remains when concretions occur at the very top of a seam (Holmes and Scott, 1981). Involvement of marine pore waters, presumably via bioturbation or storm activity, is then beyond question. In other cases (Teichmuller and Teichmuller, 1975) the carbonate phase is dolomite, presumably indicating much diminished sulphate availability relative to calcium and magnesium.

Marine band concretions.—

The laminated marine shales also contain abundant carbonate concretions. These enclose uncrushed goniatites and bivalves in sharp contrast to crushed, decalcified and pyritized remains in the shales. Here, too, concretions grew very early. Each concretion is enclosed within a perfectly continuous rim of pyrite, some 0.5 to 1.0 cm thick and more pyrite is present, mostly as framboids, throughout the Marine Band. The carbonate phase, however, in spite of such ample evidence of former presence of sulphate, invariably turns out to be dolomite or ankerite, ranging in composition between

$$(Ca_{0.54}\ Mg_{0.43}\ Fe_{0.02})\ CO_3 \qquad \text{(sample B11 0A)}$$

and

$$(Ca_{0.51}\ Mg_{0.34}\ Fe_{0.14})\ CO_3. \qquad \text{(sample IS1 4A)}$$

Accepting the generality of dolomite inhibition by sulphate (Kastner and Baker, 1983), it is difficult to see why SO_4^{2-} depletion should be more complete in marine sediments than in essentially freshwater ones below. Three factors, however, could have been influential; two have been mentioned previously:

(a) compaction would tend to introduce sulphate-depleted waters from below;
(b) euxinic conditions certainly prevailed—bottom waters may have been significantly depleted;
(c) pyrite-rimmed concretions are sites where microbial sulphate reduction was formerly localized (Raiswell, 1976; Coleman and Raiswell, in prep.).

It seems likely that pyrite and carbonate precipitated simultaneously within a spatially zoned system. Since sulphate had to be supplied from 'outside,' sulphate concentrations within the concretion would have been depleted.

Apparently anomalous dolomite precipitation can thus be at least rationalized in terms of the accumulative effect of three processes, all of which would have caused sulphate depletion, thereby favoring dolomite rather than calcite. It should be noted, however, that the concretions described by Raiswell (1976) occur in marine depositional sequences and are calcitic. Factor (c) above is not sufficient in itself to reduce sulphate concentrations below the critical level.

Freshwater siderites.—

An altogether simpler picture was obtained from the non-marine sediments lying between the marine horizons. Carbonate segregations are numerous and, without exception, sideritic. Seven concretions or concretionary horizons were sampled and each subsampled vertically. Precise chemical analyses were obtained for the carbonate phase, identified by X-ray diffraction analysis, and stable oxygen and carbon isotope ratios determined. The range in composition, as documented by cation mole fraction, was considerable:

Fe : 0.735 → 0.898
Mg : 0.011 → 0.184
Ca : 0.000 → 0.076
Mn : 0.010 → 0.143

(The lower Ca limit is probably spurious due to over correction for Ca present as apatite, calculated from the phosphate analyses.)

Considerable variation was also documented for isotope ratios:

$$\delta^{13}C(‰\ PDB) : -3.07 \rightarrow +10.35$$
$$\delta^{18}O(‰\ PDB) : -8.94 \rightarrow -5.35.$$

The significance of this variation is thrown into remarkably sharp focus, first by consideration of center-to-edge trends (Fig. 5) and then by analysis of within- and between-con-

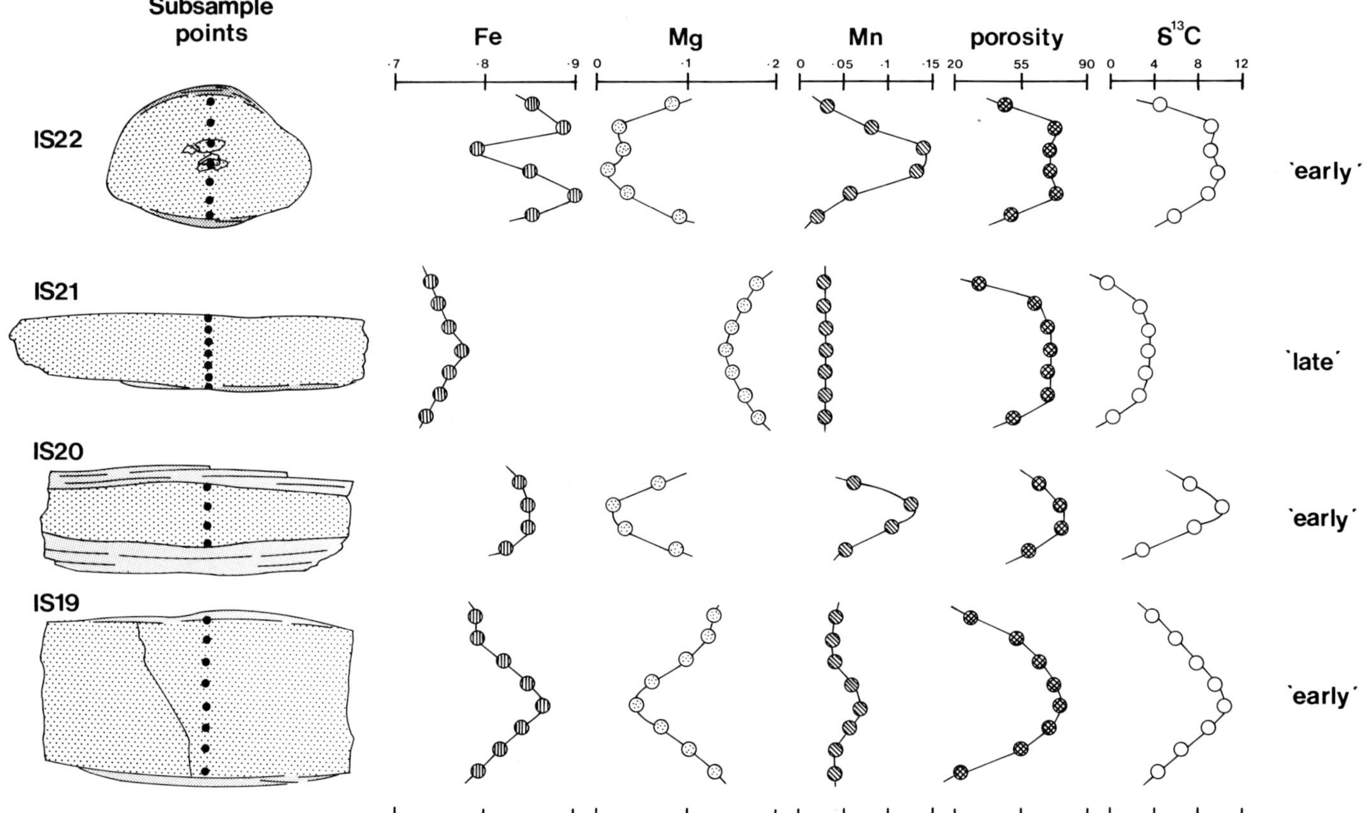

FIG. 5.—Compositional variation within some siderite concretions in the non-marine measures above the *Gastrioceras listeri* horizon, Westphalian A of Yorkshire (after Curtis et al., in prep.).

cretion covariation of molar Mg/Fe ratio with $\delta^{13}C$ (Fig. 6).

Oxygen isotope compositions generally showed a trend to more negative values at the edges. This suggests sequential growth from the center during burial (temperature increase). Absence of satisfactory calibration of the siderite-water oxygen isotope fractionation at low temperatures, however, precludes further interpretation. The source area for these fine-grained detrital sediments would have been mantled by iron-rich lateritic soils. Muddy sediments, so derived, were equally iron-rich and, when buried with organic detritus, supported methanogenic bacteria. The combined effects of iron reduction and organic degradation by microbial methanogenesis would cause massive supersaturation with respect to ^{13}C-rich siderite (Curtis, 1983).

$$7CH_2O + 2Fe_2O_3 \rightarrow 3CH_4 + 4FeCO_3 + H_2O \quad [4]$$

Fine-grained, poorly crystallized iron(III)-hydrated oxides (limonite, goethite) are unstable under reducing conditions and therefore would have readily supplied the cations for early carbonate precipitates. With progressive burial and reaction, this source of iron eventually would have become depleted and less reactive sources more important. Figure 6 demonstrates that Mg became increasingly important as diagenesis proceeded; the later siderites in peripheral subsamples were very significantly magnesian. The most likely source in these freshwater sediments would have been Mg present in interlayer hydroxide sheets (soil 'clay vermiculites') or in exchange sites. Mg supply would have to be maintained during burial, but it is not possible to estimate if its absolute concentration increased. It is important to note, however, that increase in Mg/Fe ratio in solution would occur by precipitation of siderite in a closed system.

'Porosity' values in Figure 5 are simply the volume percentages of carbonate present in each subsample. The assumption here is that carbonate passively displaced pore water when it precipitated, thereby documenting porosity (or degree of compaction) experienced at the time of precipitation. The central subsamples of concretions IS19, IS20, and IS22 obviously precipitated very early, as primary porosities in excess of 80% were estimated (Curtis et al., in prep). The very strong influence of ^{13}C-rich methanogenic bacterial bicarbonate is seen in these earliest precipitates with a smooth trend (covariation with porosity, Fig. 5) to lighter values in later, deeper cements. This trend presumably records the increasing importance of bicarbonate generated by thermal decarboxylation (see also Irwin et al., 1977).

Perhaps most interesting of all in Figure 6 is the observation that the same trend persists *between* as well as *within concretions*. It would appear the concretions IS19, IS20 and IS22 started to grow at shallower burial depths than concretions IS18, IS21, IS23 and IS25 (IS25 developed within the marine sediments of the Parkhouse Marine Band and will be considered in the next section). Siderite spans different burial depth ranges in different concretions.

In Figure 7, the Mn/Fe ratio is plotted against $\delta^{13}C$. This plot confirms the suggestion in Figure 5 that the very earliest (and only the very earliest) precipitated subsamples are strongly enriched in Mn. 'Later' concretions, which started to form at significant burial depths, exhibit remarkably constant Mn/Fe ratios. Pore waters close to the sediment/water interface were thus significantly enriched in Mn^{2+}. This is entirely to be expected in light of the differential redox behavior and mobility of these two metals (Lynn and Bonatti, 1965). Mn(IV) oxides concentrate at the redoxcline and Mn^{2+} concentrations are enhanced just below.

The story of the siderites is thus a simple one: sources of both cations and anions can be identified and systematic changes with depth mostly reflect depletion of highly re-

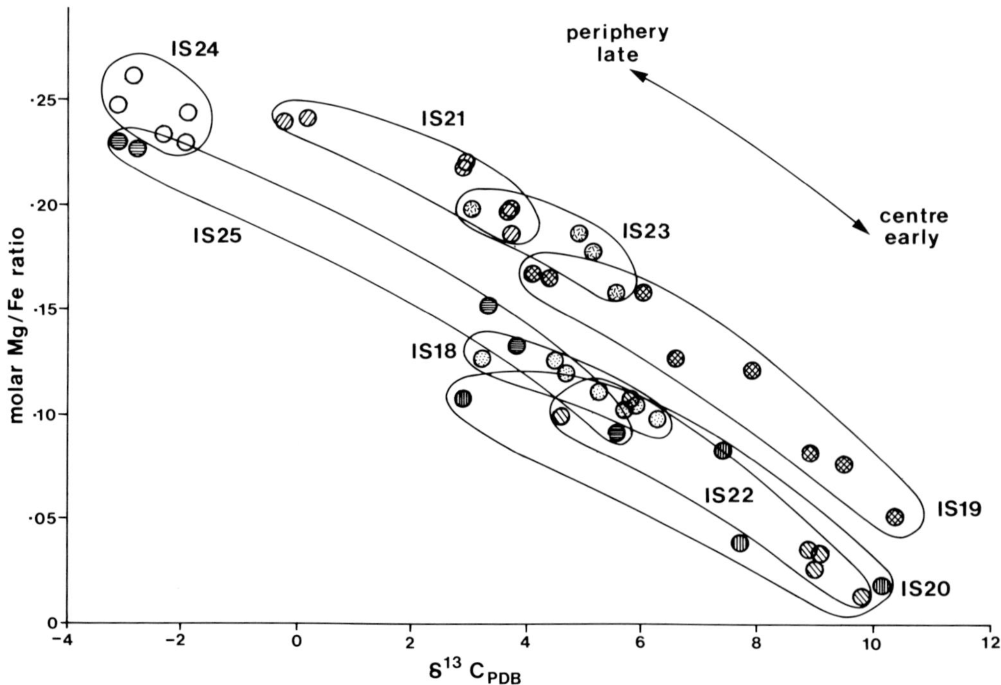

FIG. 6.—Covariation of molar Mg/Fe ratios with $\delta^{13}C$: siderite concretions (individually enclosed within sauciform fields) from Yorkshire (after Curtis et al., in prep.).

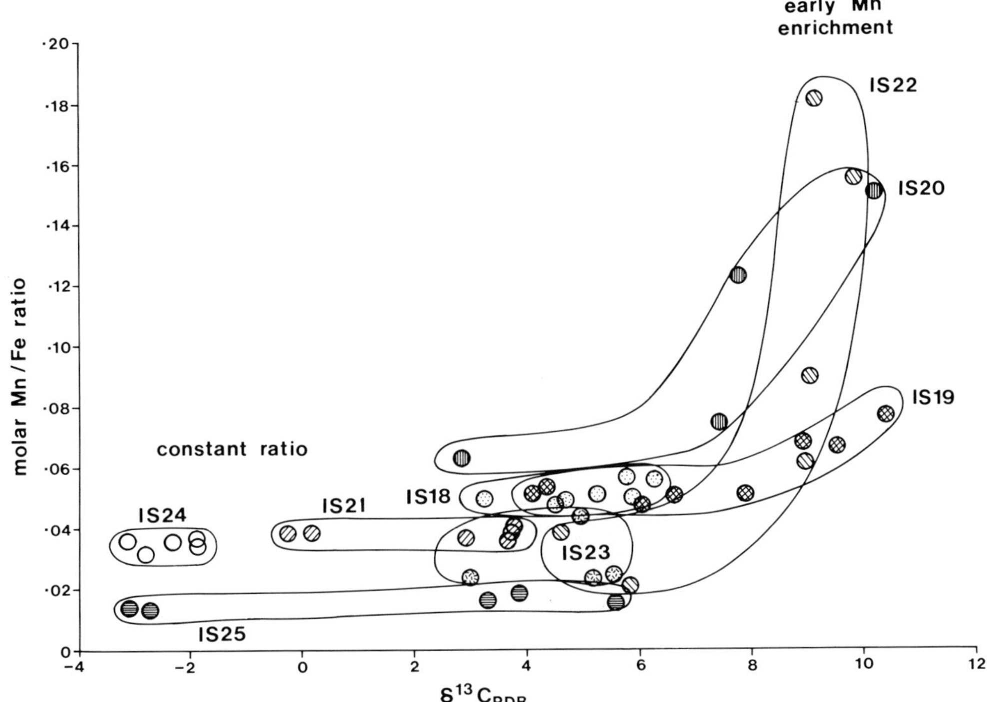

Fig. 7.—Plot of molar Mn/Fe ratio against $\delta^{13}C$ for Yorkshire siderite concretions (after Curtis et al., in prep.). Note that only the very 'earliest' (^{13}C-rich) subsamples have high Mn/Fe ratios.

active oxides of Fe(III) and Mn(IV). With increasing burial, the principal source of bicarbonate switched from microbial methanogenesis to thermal decarboxylation, although the latter was never dominant. The implied increase in temperature is confirmed qualitatively by the $\delta^{18}O$ data.

Siderite concretion IS24, from the Parkhouse marine band.—

This concretion occurs within fine-grained pyritic shale containing siliceous foraminifera and designated the Parkhouse Marine Band by Calver (1968). This represents a minor incursion compared with the Alton horizon below, but saline waters were certainly present and significant sulphate reduction took place. Generation of H_2S, however, was not sufficient overall to maintain very low solution Fe^{2+} concentrations. Siderite precipitated slightly below the sediment/water interface and the influence of marine depositional waters is seen in two ways, the isotopic and the trace element analyses. The *earliest* subsamples have negative $\delta^{13}C$ values, in sharp contrast to all other concretions (Fig. 7), and Mg/Fe ratios are all very high (Mg^{2+} readily available from the inception of precipitation as it would be in seawater). The negative $\delta^{13}C$ values record availability of bicarbonate generated by sulphate reduction.

Triassic, Paleogene and Pleistocene Coal Fields of Japan

Finally, reference must be made to the comprehensive study of diagenesis in younger coal fields by Matsumoto and Iijima (1981). They also found low Mg-Ca siderites characteristic of diagenesis in freshwater sediments and magnesian calcites and dolomites characteristic of marine-influenced diagenesis. Intermediate stages of diagenesis in siderites followed precisely the same chemical trends described here. It is remarkable that such systemtic yet complex patterns are found in sediments of widely differing ages. Perhaps the most significant conclusion to be drawn is that microbiological processes are extremely important in sediment diagenesis. Evolution, which has made the task of paleoecologists interested in higher life forms so difficult, has mercifully left these simpler organisms alone.

There is one point of disagreement. Matsumoto and Iijima (1981) attribute the trend of increasing Mg/Fe ratio with depth in siderites and the fact that later diagenetic events become homogenized (between different depositional environments) to general mixing of pore waters at depth, thus making Mg^{2+} available from seawater. Undoubtedly, this also was the case in the Carboniferous Coal Measures, but we are inclined to think it less important than detrital source control, because marine sediments make up such a small proportion of the total succession.

SUMMARY AND DISCUSSION

Post-depositional chemical alteration in Coal Measures is extensive and varied. Since depositional water composition is the only significant source of sulphur, it imposes one very important control on diagenetic mineralogy. Under conditions of slow detrital sediment accumulation, sulphate and H_2S are plentifully available (active sulphate reduction), iron is quantitatively precipitated as pyrite and isotopically light (^{13}C depleted HCO_3^- from organic matter), and iron-poor carbonates precipitate. Mg^{2+} and Ca^{2+} are also plentifully available; high magnesian calcite is the normal precipitate.

In freshwater sediments, sulphate is scarce and pore water solutes are present at lower concentrations. The major control on diagenetic modification is again solute availability, but this time from within the sediment itself by breakdown or degradation of detrital constituents. The most unstable react first, essentially kinetic control. Hydrated iron oxides are reduced, raising Fe^{2+} concentrations, Fe/Ca ratio and alkalinity. In the absence of sulphate, the principal organic degradation pathway is microbial methanogenesis. Early carbonates are thus Fe (and Mn)-rich siderites.

Rapid changes in depositional environment, particularly in the composition of depositional and trapped pore water, overprint the simple patterns seen above. Sediment pore waters become mixed by downward diffusion of solutes (marine over freshwater) or upward compactional flow. This effect is most clearly seen after marine transgressions, when early diagenesis within freshwater coal peats is entirely marine in character (calcite plus pyrite). Where marine and freshwaters mix, and particularly when sulphate reduction is active, the principal control on carbonate precipitation would appear to be pore water sulphate concentrations: high magnesian calcite precipitates at higher levels, dolomite at lower levels.

Later diagenetic cements (seen here only in the non-marine depositional sequence) seem to reflect only changing availability of solutes from within the sediments themselves. The more reactive iron oxides are reduced and Fe^{2+} no longer 'swamps' supply. As microbial decomposition rates decline (presumably substrate quality control), thermal sources become more important.

ACKNOWLEDGMENTS

One of us (CDC) acknowledges the support of the University of Sheffield Foreign Travel Fund, the Society of Economic Paleontologists and Mineralogists and the Marathon Oil Company for financial assistance toward travel (both this meeting and the Society of Economic Paleontologists and Mineralogists Research Conference, 1983). Don Gautier is thanked for his interest and for bringing together and organizing such a stimulating group at the 1983 Lost Valley Ranch 'Conference.' We are grateful to Pat Mellor (University of Sheffield) and the young ladies of the EPD typing office (British Petroleum), who prepared successive versions of the manuscript under unreasonable conditions of haste. Lynn Walter is thanked for her comprehensive and helpful review.

REFERENCES

BAKER, P. A., AND KASTNER, MIRIAM, 1981, Constraints on the formation of sedimentary dolomite: Science, v. 213, p. 214–216.

BATHURST, R. G. C., 1975, Carbonate Sediments and their Diagenesis, 2nd ed.: Elsevier, Amsterdam, 620 p.

BERNER, R. A., 1984, Sedimentary pyrite: an update: Geochimica et Cosmochimica Acta, v. 448, p. 605–615.

BOLES, J. R., 1978, Active ankerite cementation in the subsurface Eocene of southwest Texas: Contributions to Mineralogy and Petrology, v. 68, p. 13–22.

CALVER, M. A., 1968, Distribution of Westphalian marine faunas in northern England and adjoining areas: Proceedings of the Yorkshire Geological Society, v. 37, p. 1–72.

CAMPOS, H. S., AND HALLAM, ANTHONY, 1979, Diagenesis of English Lower Jurassic limestones as inferred from oxygen and carbon isotope analysis: Earth and Planetary Science Letters, v. 45, p. 23–31.

CLAYPOOL, G. E., AND KAPLAN, I. R., 1974, The origin and distribution of methane in marine sediments, *in* Kaplan, I. R., ed., Natural Gases in Marine Sediments: Plenum Press, p. 99–139.

CURTIS, C. D., 1967, Diagenetic iron minerals in some British Carboniferous sediments: Geochimica et Cosmochimica Acta, v. 31, p. 2109–2123.

———, 1977, Sedimentary geochemistry: Environments and processes dominated by involvement of an aqueous phase: Philosophical Transactions of the Royal Society of London, v. 286A, p. 353–372.

———, 1983, Geochemistry of porosity enhancement and reduction in clastic sediments, *in* Brooks, Jim, ed., Petroleum Geochemistry and Exploration of Europe: London, Blackwell, The Geological Society, p. 113–125.

———, PEARSON, M. J., AND SOMOGYI, W., 1975, Mineralogy, chemistry and origin of a concretionary siderite sheet (clay-ironstone band) in the Westphalian of Yorkshire: Mineralogical Magazine, v. 40, p. 385–393.

DEER, W. A., HOWIE, R. A. AND ZUSSMAN, J., 1966, An Introduction to the Rock-Forming Minerals: Longmans, Green and Company, Ltd., London. 528 p.

DICKSON, J. A. D., AND BARBER, C., 1976, Petrography, chemistry and origin of early diagnetic concretions in the Lower Carboniferous of the Isle of Man: Sedimentology, v. 23, p. 189–211.

HALLAM, ANTHONY, 1981, Facies Interpretation and the Stratigraphic Record: Freeman, Oxford and San Francisco, 291 p.

HO, C., AND COLEMAN, J. M., 1969, Consolidation and cementation of recent sediments in the Atchafalaya Basin: Geological Society of America Bulletin, v. 80. p. 183–192.

HOLMES, J., AND SCOTT, A. C., 1981, A note on the occurrence of marine animal remains in a Lancashire coal ball (Westphalian A): Geological Magazine, v. 118, p. 307–308.

HOOKER, J. D., AND BINNER, E. W., 1854, On the structure of certain limestone nodules in seams of bituminous coal with a description of some Trigonocarpons contained in them: Philosophical Transactions of the Royal Society of London, v. B145, p. 145–156.

IRWIN, HILARY, 1980, Early diagenetic carbonate precipitation and pore fluid migration in the Kimmeridge Clay of Dorset, England: Sedimentology, v. 27, p. 577–591.

———, COLEMAN, M. L., AND CURTIS, C. D., 1977, Isotopic evidence for source of diagenetic carbonates formed during burial of organic-rich sediments: Nature, v. 269, p. 209–213.

KASTNER, MIRIAM, AND BAKER, P. A., 1983, A new insight into the origin of sedimentary dolomites: 1982/1983 Yearbook of Science and Technology, McGraw Hill, p. 406–408.

LIVINGSTONE, D. A., 1963, Chemical composition of rivers and lakes, *in* Fleischer, M., ed., Data of Geochemistry, 6th ed.: U.S. Geological Survey Professional Paper 440-G.

LOVE, L. G., COLEMAN, M. L., AND CURTIS, C. D., 1983, Diagenetic pyrite formation and sulphur isotope fractionation associated with a Westphalian marine incursion, northern England: Transactions of the Royal Society of Edinburgh: Earth Sciences, v. 74, p. 165–182.

LYNN, D. C., AND BONATTI, E., 1965, Mobility of manganese in diagenesis of deep-sea sediments: Marine Geology v. 3, p. 457–474.

MATSUMOTO, R., AND IIJIMA, A., 1981, Origin and diagenetic evolution of Ca-Mg-Fe carbonates in some coalfields of Japan: Sedimentology, v. 28, p. 239–259.

———, UTADA, M., AND KAGAMI, H., 1978, Sedimentary petrology of DSDP cores from sites 362 and 363, the Walvis Ridge and site 364 the Angola Basin, drilled on Leg 40: Initial Reports of the Deep Sea Drilling Project, v. XL, Washington (U.S. Government Printing Office), p. 469–485.

NATHAN, Y., AND SASS, E., 1981, Stability relations of apatites and calcium carbonates: Chemical Geology, v. 34, p. 103–111.

NISSENBAUM, ARIE, PRESLEY, B. J., AND KAPLAN, I. R., 1972, Early diagenesis in a reducing fjord, Saanich Inlet, British Columbia—I. Chemical and isotopic changes in major components of interstitial water: Geochimica et Cosmochimica Acta, v. 36, p. 1007–1027.

PETTIJOHN, F. J., 1957, Sedimentary Rocks, 2nd ed.: Harper and Brothers, New York, 718 p.

PISCIOTTO, K. A., 1982, Review of secondary carbonates in the Monterey Formation, California, *in* The Monterey Formation and Related Siliceous Rocks of California: Society of Economic Paleontologists and Mineralogists Pacific Section Special Publication, p. 273–283.

———, AND MAHONEY, J. J., 1981, Isotopic survey of diagenetic carbonates, Deep Sea Drilling Project Leg 63: Initial Reports of the Deep

Sea Drilling Project, v. LXIII, Washington (U.S. Government Printing Office), p. 595–609.

RAISWELL, ROB., 1976, The microbiological formation of carbonate concretions in the Upper Lias of N.E. England: Chemical Geology, v. 18, p. 227–244.

SASS, EITAN, AND KOLODNY, YESHOU, 1972, Stable isotopes, chemistry and petrology of carbonate concretions (Mishash Formation, Israel): Chemical Geology, v. 10, p. 261–286.

TEICHMULLER, M., AND TEICHMULLER, R., 1975, The geological basis of coal formation, *in* Stach, E., ed., Coal Petrology, 2nd ed.: Berlin, Gebruder Borntraeger, 428 p.

WHELAN, T., AND ROBERTS, H. H., 1973, Carbon isotope compostion of diagenetic carbonate nodules from freshwater swamp sediments: Journal of Sedimentary Petrology, v. 43, p. 54–58.

CORRELATION OF ORGANIC PARAMETERS DERIVED FROM ELEMENTAL ANALYSIS AND PROGRAMMED PYROLYSIS OF KEROGEN

LAURA J. CROSSEY, E. SVEN HAGEN, RONALD C. SURDAM
Department of Geology/Geophysics, University of Wyoming, Laramie, Wyoming 82071
AND
THOMAS W. LAPOINT
Department of Zoology & Physiology, University of Wyoming, Laramie, Wyoming 82071

ABSTRACT: The quality, quantity, type, and level of hydrocarbon maturation of various organic-rich rock samples were evaluated by two different techniques: elemental analysis and programmed pyrolysis. Eighty samples varying in both total organic carbon (0.5–75 wt% OC) and kerogen type were analyzed. Effects of the mineral matrix on measured geochemical parameters were examined by analysis of synthetic mixtures of separated kerogen (Types I, II, and III) and mineral matter (bentonite and illite at various percentages).

Statistical analyses were performed on bulk rock samples, separated kerogen, and bitumen-extracted kerogen. Analyses include population descriptions, correlations, and stepwise multiple regression. Data parameters include: bulk rock Hydrogen Index (HI), separated kerogen HI, bitumen-extracted kerogen HI, atomic H/C and O/C (bitumen-extracted kerogen), and % OC in bulk, separated, and bitumen-extracted samples.

Results from a wide variety of hydrocarbon source rocks indicate that programmed pyrolysis can be used to predict atomic *H/C*. Although the variance of both methods is similar, the *HI* values derived from bulk rock samples consistently underestimate those derived from bitumen-extracted samples. Both measures are highly correlated ($r = 0.87$, $P = 0.001$). Guidelines for the practical application of bulk rock pyrolysis data to source rock evaluation studies are presented.

In summary, standard pyrolysis techniques yield information which can be used for rapid, accurate appraisal of the quality, maturity, and type of organic material contained within a wide variety of organic-rich rocks.

INTRODUCTION

In the past decade programmed-temperature pyrolysis has become a widely accepted analytical technique for hydrocarbon source rock evaluation, including the quality, type, and stage of thermal maturation of kerogen. Pyrolysis allows rapid and relatively inexpensive appraisal of source rock potential for a large number of samples. Numerous publications attest to the acceptance (Claypool and Reed, 1976; Espitalié et al., 1977; Monin et al., 1980; Orr, 1983; Peters et al., 1983; Dembicki et al., 1983; Verheyen et al., 1984; Tissot, 1984) and limitations (Clementz, 1979; Espitalié et al., 1980; Horsfield and Douglas, 1980; Katz, 1981) of programmed pyrolysis. The process of programmed pyrolysis consists of heating a sample aliquot (either whole rock or isolated kerogen samples) in an inert atmosphere along a linear thermal ramp causing an increasing degree of thermal degradation with time. Pyrolytic yields or evolved hydrocarbon-like products are then measured (for more details, see methods section below). All quantitative indices (hydrogen index, production index, genetic potential) are based on the calibrated peak areas of S_1 or S_2 (see page 776 of Orr, 1983). Hydrogen Index (HI) is defined as the quantity of hydrocarbon compounds (calculated from S_2) normalized to the total organic carbon of the sample. Production Index (PI) refers to the ratio of the quantity of hydrocarbons measured as S_1 to the total quantity of hydrocarbons generated by the sample ($S_1 + S_2$). In general, there is a good correlation between pyrolytic yield and elemental analysis of isolated kerogen. This relationship is also valid for whole-rock results in many cases (Orr, 1983). Orr has shown that pyrolysis data (i.e., HI) of isolated kerogen agrees with elemental *H/C* and *O/C* by ±5 to 10%. Whole rock pyrolysis may be affected by as much as a factor of 2 when compared to pyrolysis of isolated kerogen (Tissot, 1984). Whole rock pyrolysis may yield significantly different results due to the retention effect of mineral matrix (Espitalié et al., 1980), contribution from heavy hydrocarbons (Clementz, 1979), and variations in type and quantity of organic matter in the rock (Orr, 1983).

Although most of the data presented by previous workers verifies the uses and limitations of programmed pyrolysis, data groups are typically small and the relationship among variables is treated in a qualitative manner. There are four primary goals of this study:

1) to rigorously examine the relationship between rapidly derived bulk rock organic parameters and those derived from isolated kerogen;
2) to develop a better understanding of the effects of kerogen type, quality, and bulk total organic carbon on organic parameters, based on statistical treatment of data from a wide variety of hydrocarbon source rocks;
3) to define the effects of mineral matrix on measured geochemical parameters using synthetic mixtures which represent normal to low (0.5 to 5 wt %) total organic carbon ranges for a variety of kerogen types; and
4) to give guidelines for utilization of pyrolysis data in basin-wide studies.

A comprehensive data set of potential source rocks was developed which represents the complete range of natural kerogen (Types I, II, III), total organic carbon, thermal maturity, as well as available published data. In this way we have developed a data set that may be representative of a typical hydrocarbon source rock study within a basin. Other analyses include: S_1, S_2, HI, elemental H/C and O/C, and PI. Statistical procedures using least squares regression allowed us to determine the form and significance of functional relationships between two or more variables. In addition, multiple linear regression quantitatively analyzes the interrelation among variables such as elemental H/C, O/C, and HI to explain the observed variations better.

In a basin source rock study, samples with a wide range in TOC, maturity, and kerogen type may be encountered. An optimum analytic scheme would include kerogen iso-

TABLE 1.—Summary of samples and organic parameters

Source	TOC	Hydrogen Index	Maturity
Monterey Formation Miocene, California	4.59	623	Immature
Green River Formation Eocene, Wyoming	7.15	861	Immature
Woodford Shale Devonian, Oklahoma	4.71	609	Immature
Cody, Frontier, Mowry, Thermopolis Cretaceous, Wyoming	0.74–5.86	51–552	Immature—post mature
Stevens Sandstone Miocene, California	0.10–7.27	58–382	Marginally mature
Australian Source Mesozoic-Paleozoic	15.6–61.0	91–722	Immature—mature

lation (elemental analysis) and *GC* characterization of extracted hydrocarbons and pyrolysate from numerous samples. In reality, time and cost often limit the extent of geochemical analysis, and therefore the extent of understanding of the source rock characteristics on a basin-wide scale. By rigorously evaluating the relationships among organic parameters such as bulk-rock HI, kerogen-extract HI, and atomic H/C, we are able to make generalizations regarding these relationships. When the mechanisms responsible for the behavior of bulk rock samples during pyrolysis are understood and accounted for quantitatively, predictions from simple bulk rock pyrolysis become more reliable.

METHODS

Sample Selection

The sample suite of potential hydrocarbon source rocks represents a wide range in total organic carbon (TOC), kerogen type, thermal maturity, and geologic age (see Table 1). Samples were selected on the basis of hydrogen index derived from bulk rock pyrolysis to obtain a broad spectrum of kerogen types (I, II, and III; Hunt, 1979). It was assumed that this choice of algal, marine, and terrestrial kerogen would also yield a similar range in elemental H/C and O/C. In addition, four kerogens were selected for use in synthetic mixtures of kerogen and clay (illite and bentonite) at low TOC values. These experiments are described in a later section.

Analytical Methods

A coulometric system was used to determine inorganic carbon; CO_2 was measured after acid treatment of the sample. Total carbon was measured on a Carlo Erba 1106 Elemental Analyzer; organic carbon was determined by difference. Hydrogen, nitrogen, and oxygen were also measured on the Carlo Erba. All samples were analyzed in duplicate.

Programmed pyrolysis was performed on a modified Varian 3760 Gas Chromatograph and Chemical Data Systems pyroprobe coil controlled externally with a Hewlett Packard power supply, multimeter, and microcomputer. Samples were weighed into quartz tubes and packed with quartz wool on both sides to prevent migration of the sample during analysis. This enabled the samples to be maintained consistently in the center of the pyroprobe coil, resulting in uniform heating of the samples and symmetrical S_2 peaks. Samples were pyrolyzed in a helium atmosphere for 15 minutes over a linear thermal ramp from 300 to 600°C. All volatile reduced-carbon species are carried to a flame ionization detector and measured as a detector response to carbon. A liquid hydrocarbon standard was used (Mobil's synthetic Delvac oil). S_1 and S_2 are calibrated to milligrams hydrocarbon per gram sample (mg HC/g spl). For this study, S_3 (CO_2) was not measured by pyrolysis. Again, all samples were analyzed in duplicate. Hydrogen Index and Production Index were calculated as follows:

$$\text{HI (mgHC/gOC)} = (S_2/\%\text{OC}) \times 100; \text{PI} = S_1/(S_1 + S_2).$$

Sample Preparation and Analytic Scheme

Analytic flow scheme.—

Figure 1 schematically shows the analytic flow scheme utilized in this study. Sample selection and preparation procedures are described below.

Bulk rock samples.—

Bulk rock samples consist of original sampled material (washed core and cuttings) ground to a powder in a Spex shatterbox. Inorganic carbon, total carbon, and S_1 and S_2 (pyrolysis) were determined for each sample.

Kerogen separates.—

By definition, kerogen separates consist of organic material which has been isolated from the inorganic constituents of the rock sample. Kerogen separates were obtained by treating bulk material with 6N HCl to remove carbonate minerals, and 52% HF to remove silicates. During the HF treatment the samples were heated for two hours at 50°C, then allowed to stand for 24 hours. Residues of the acid treatments were then mixed with a solution of $ZnBr_2$ (specific gravity = 2.1) and centrifuged to float off the organic

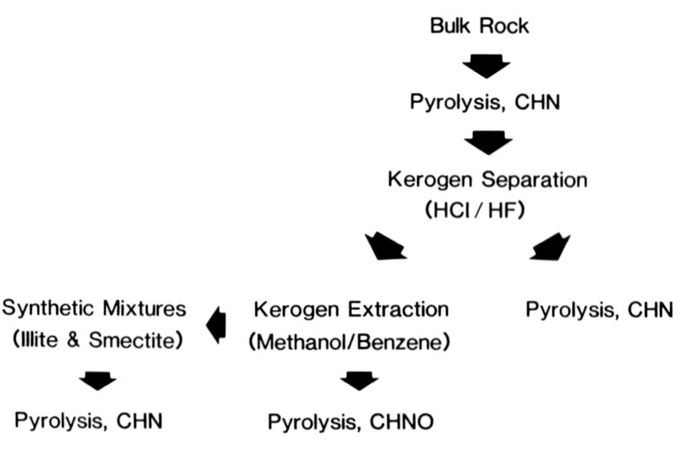

FIG. 1.—Schematic analytical flow scheme. Sample selection, preparation, and analytical methods are described in the text.

material. The float was filtered, washed, and vacuum-dried at 50°C for 10 hours. Total carbon, hydrogen, nitrogen, and S_1 and S_2 (pyrolysis) were measured on the separates.

Kerogen extracts.—

By definition, a kerogen extract or bitumen-extracted kerogen consists of a kerogen separate that has been treated to remove all soluble hydrocarbon phases. Kerogen extracts were prepared by extracting the kerogen separate material with a Soxhlet Extraction apparatus using a benzene/methanol solvent. The samples were extracted for 15 hours (more if necessary to obtain colorless solvent). The extracts were then vacuum-dried for 10 hours at 50°C. Total carbon, nitrogen, hydrogen, oxygen, and S_1 and S_2 (pyrolysis) were determined.

Synthetic mixtures.—

Synthetic mixtures of four extracted kerogens and two clay standards were prepared by weighing the kerogen extract and an appropriate aliquot of the clay to attain a target TOC value. Clays were selected to demonstrate maximum effect on pyrolytic yield (from information in Orr, 1983). The TOC of each mixture was less than 5 wt %. Clay standards used were illite (Fithian, Illinois) and bentonite (Polkville): both are API clay standards. Samples were shaken in a ball mill for 5 minutes, and actual TOC values determined by the Carlo Erba Elemental Analyzer. In addition, programmed pyrolysis was performed as described above.

Statistical techniques.—

Least squares regression analyses were performed on data collected from the literature and on data collected in our laboratory. For proper application of linear regression techniques, collected data must meet certain parametric assumptions. These include normality (the data are drawn from a normal distribution), homogenous variances (sample variances are approximately equivalent), and independence of samples. All variables in this study were checked for departures from parametric assumptions. Statistical characterizations of the data include:

r = correlation coefficient: reflects strength of correlation between the variables;

r^2 = coefficient of determination: reflects the goodness of fit of the data to the regression line;

P = probability: for convenience, the significance of a correlation will be reported according to the following convention: * = 0.05–0.10 probability, ** = 0.01–0.001 probability, and *** = <0.001 probability that the data are not correlated (r = 0).

TABLE 2.—Linear regression data for relationships between measured organic parameters

Relationships	Samples Considered	n	r	r^2	P
Atomic H/C with $HI^1_{extract}$	all samples	70	0.88	0.78	<.001
Atomic H/C with $HI^2_{separate}$	Stevens and Mowry	44	0.91	0.82	<.001
$HI_{extract}$ with HI^3_{bulk}	all samples	70	0.87	0.75	<.001
Atomic H/C with HI^4_{bulk}	all samples	73	0.75	0.56	<.001
Atomic H/C with HI^5_{bulk}	<1% TOC	22	0.64	0.41	<.001
Atomic H/C with HI^6_{bulk}	1–5% TOC	33	0.62	0.39	<.01
Atomic H/C with HI^7_{bulk}	TOC > 5%	18	0.96	0.92	<.001
Atomic H/C with HI^8_{bulk}	Mowry	31	0.81	0.66	<.001

EQUATIONS

1 Atomic H/C = 0.00079 ($HI_{extract}$) + 0.612
2 Atomic H/C = 0.00097 ($HI_{separate}$) + 0.618
3 $HI_{extract}$ = 1.077 (HI_{bulk}) + 76.95
4 Atomic H/C = 0.00084 (HI_{bulk}) + 0.676
5 Atomic H/C = 0.00180 (HI_{bulk}) + 0.526
6 Atomic H/C = 0.00076 (HI_{bulk}) + 0.724
7 Atomic H/C = 0.00091 (HI_{bulk}) + 0.596
8 Atomic H/C = 0.00066 (HI_{bulk}) + 0.625

These characterizations are compiled in Table 2.

The regression and parametric statistical programs used are described in Nie et al. (1975). A forward stepwise multiple linear regression was used to derive an empirical formula for the relationship of HI to atomic H/C and O/C. The pertinent multiple regression equations, with their significance levels, are listed in Table 3.

Interpretation of regression results yields three distinct types of information: 1) prediction, 2) explanation of variance, and 3) substitution of variables. The proportion of the variation in one variable explained by the variation in the second (the coefficient of determination or r^2) is a measure of the degree of covariation between the variables. If there is a strong correlation between two variables (for example, atomic H/C and extract HI) then pyrolysis-derived HI can be used to 'predict' H/C. Regression programs allow the subdivision of samples into specific subgroups (i.e., TOC or kerogen type) to facilitate interpretation by isolating factors contributing to variances in a measured parameter. If variables are sufficiently well correlated and reasons for deviations understood (i.e., low TOC), then the statistically derived empirical expressions may be applied to new data. An optimum goal would be the substitution of an 'inexpensive' (bulk rock pyrolysis) analytical technique for an 'expensive' (elemental analysis) technique.

RESULTS

The data set used in this study encompasses a wide range of natural kerogen compositions, as well as some synthetic

TABLE 3.—Multiple regression equations derived from measured organic parameters

Regression equation	Samples Considered	n	1st Variable Entered: (r^2)	2nd Variable Entered: (r^2)
$HI_{extract} = 694(H/C - 0.29) - 800(O/C)$	Orr (1983)	15	H/C: 0.76	O/C: 0.92
$HI_{extract} = 993(H/C - 0.44) - 839(O/C)$	This study	70	H/C: 0.78	O/C: 0.81
$S_{2\ extract} = 1.12(S_{2\ separate}) - 215(PI) + 19$	Stevens & Mowry	48	S_2 separate: 0.89	PI: 0.96

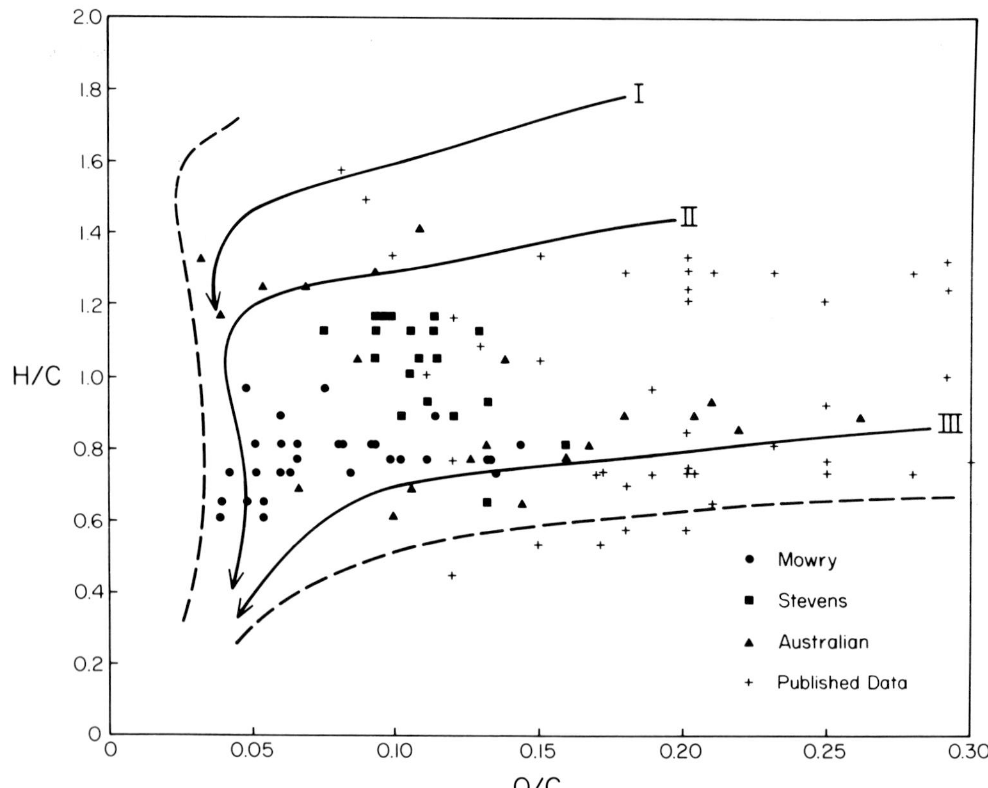

FIG. 2.—A van Krevelen plot showing the distribution of samples from this study (Mowry, Stevens, and Australian) as well as published data (see text for details). Arrows represent maturation pathways for Type I, II, and III kerogen. Note that the samples cover a complete range of kerogen types and maturation. For more discussion of van Krevelen diagrams, see Tissot and Welte (1978).

and natural humic acids presented by Orr (1983) (see Fig. 2). The data set differs from previous studies (compiled by Orr, 1983) in several respects: 1) large sample size (n = 77; Appendix), 2) wide range in TOC, 3) wide range in H/C and O/C, and 4) both multiple samplings of one formation and several different formations (~18) are represented. Table 1 itemizes TOC, maturity, HI, and source of the samples. In the following discussion, the data are grouped in several different ways. Based on Figure 2, overlap between previously published data (Orr, 1983) and samples examined in this study occurs only along the trend for immature Type III kerogens (see Tissot and Welte, 1978, for a discussion of maturity trends and van Krevelen plots). In general, our samples represent higher degrees of thermal maturity, whereas most published data sets consist of immature samples. In many cases, data from the literature are not included in the statistical analysis. Within our data set, groupings by source (Mowry, Stevens, etc.) and TOC of the original sample are frequently employed.

For all sources, kerogen types, and TOC values, the correlation of atomic H/C ratio and the pyrolysis yield (HI) of the kerogen extract is excellent (see Fig. 3). The correlation coefficient (r) is 0.88 for the data set (excluding literature values) and the slope is highly significant ($P \leq .001$). The H/C ratio also correlated well with the HI determined from the kerogen separate (r = 0.91***) despite a range in PI of 0.07 to 0.76. The presence of *in situ* hydrocarbons may affect S_2 as well as S_1 (Clementz, 1979). Note that the regression lines for bulk and extracted kerogen versus atomic H/C are nearly parallel (Fig. 3, Table

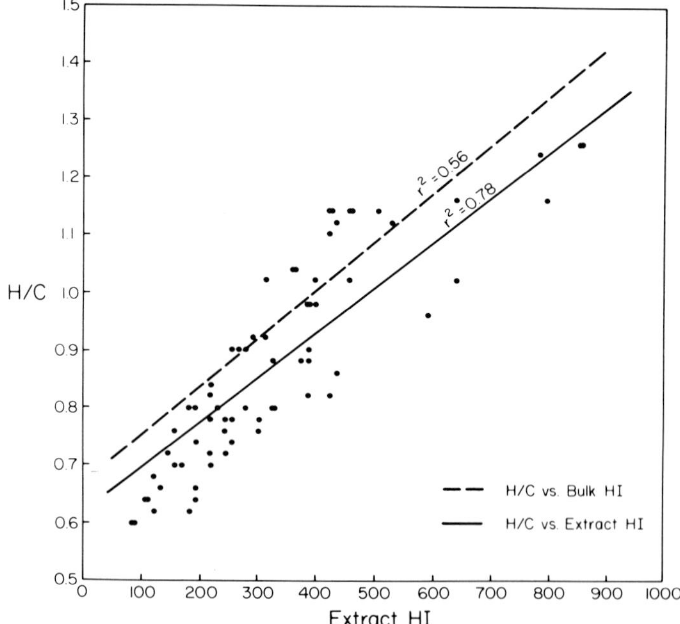

FIG. 3.—Measured HI from kerogen extracts is plotted against measured atomic H/C of the kerogen. The solid line represents the best-fit regression through the data. The dashed line represents the regression obtained from bulk rock samples (see Fig. 4). Note that the lines are approximately parallel and that, for a given atomic H/C value, the bulk-derived HI underestimates the extract-derived HI.

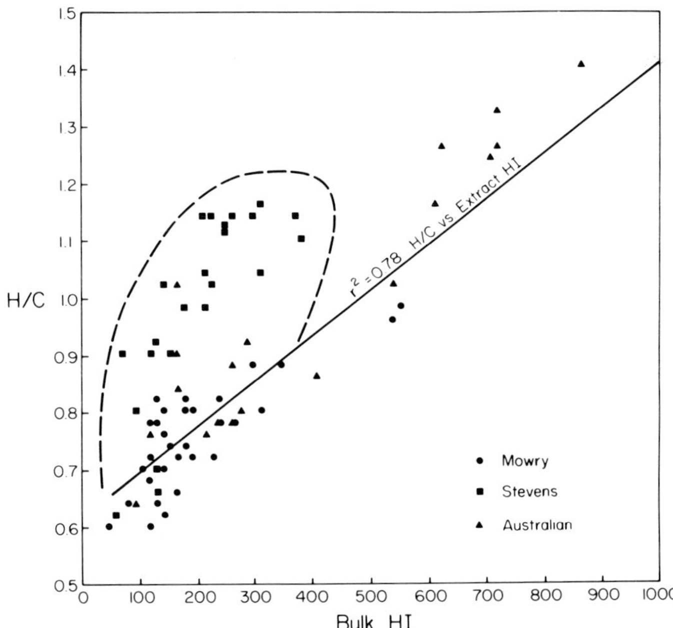

FIG. 4.—Bulk rock HI is plotted against atomic H/C of the kerogen. The line obtained from extracted kerogens is plotted for reference (see Fig. 3). Samples are keyed according to source. Note that the region of data outlined with the dashed line contains samples from the Stevens data set; but several Australian samples are also included. Samples in this area of the plot have anomalously low HI values for their atomic H/C.

2). The position of these lines suggests that bulk HI typically underestimates extract HI over the entire range of HI and H/C (Fig. 3). The HI values derived from whole rock (bulk) pyrolysis and atomic H/C ratio correlate well (r = 0.75***; see Fig. 4). Note that, compared to the regression line derived from extracted kerogen HI values, most of the deviations are associated with one source (Stevens, Fig. 4) and low TOC (Fig. 6). The correlation coefficient improves substantially when the samples are grouped on the basis of bulk TOC. Samples of low (<1), intermediate (1–5), and high (>5) % TOC represent all kerogen types (Fig. 6, also Figs. 4 and 5). Thus, grouping on the basis of TOC does not bias the data with respect to kerogen type. Samples containing less than 1% TOC are poorly correlated (r = 0.64**). Samples having TOC values between 1 and 5% are also relatively poorly correlated (r = 0.62**). Samples having >5% TOC, however, have strongly correlated bulk HI and atomic H/C values (r = 0.96***). A breakdown of the various slopes for regressions on the TOC subgroups is given in Figure 6. Note that the position of the >5 wt % regression line falls in the approximate position of the regression line for the extract HI versus H/C plot (Figs. 3 and 6).

A multiple regression of extract HI with atomic H/C and O/C shows a significant improvement in r compared to the correlation of extract HI and atomic H/C alone (see Tables 2 and 3). The empirical equation relating H/C and O/C to extract-derived HI is given in Table 3 and differs slightly from the equation derived by Orr (1983). These differences may be due to the size of the respective data sets (n = 15,

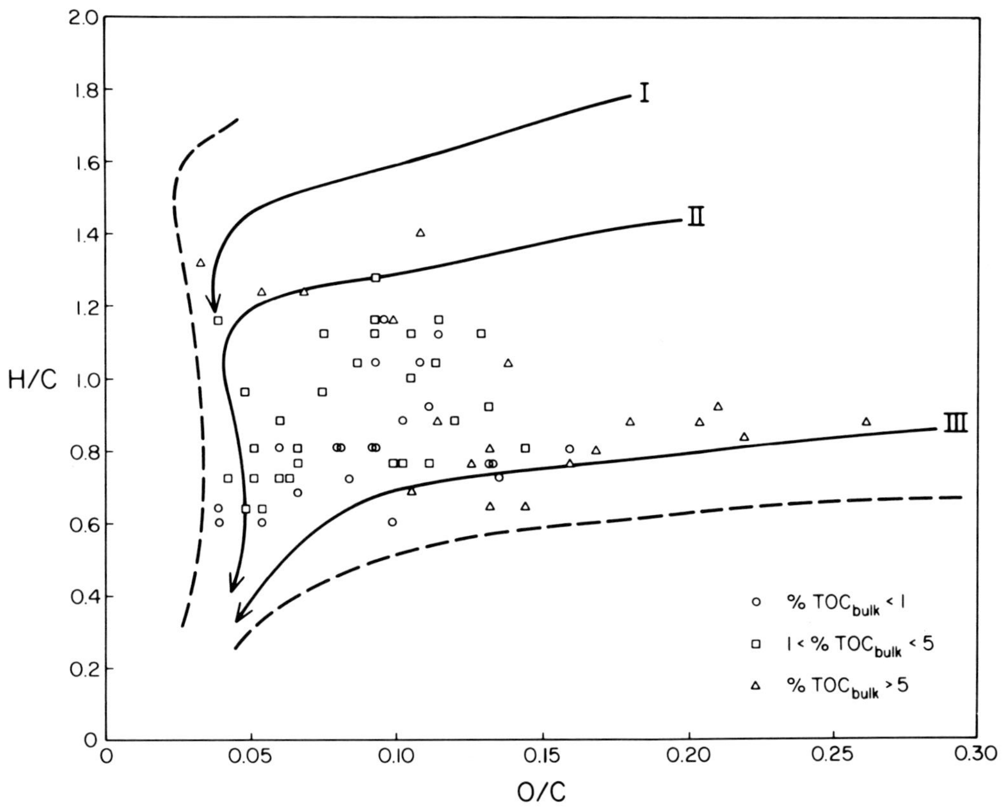

FIG. 5.—A van Krevelen plot showing the complete data set (Fig. 2) divided into subgroups based on TOC. Note that the TOC subgroups are spread across the diagram; no subgroup is biased toward any specific kerogen type.

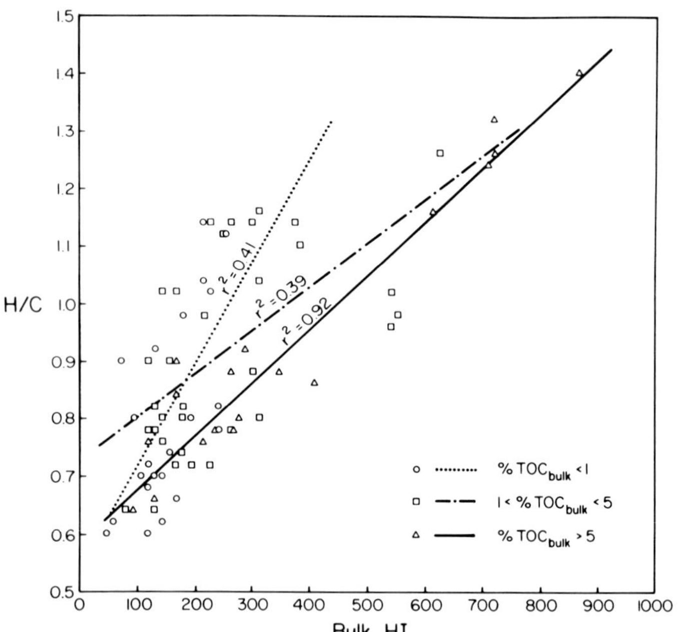

FIG. 6.—Bulk rock-derived HI is plotted against atomic H/C of the kerogen. Samples are divided into subgroups based on TOC. A regression line for each subgroup is also plotted. Note that as % TOC increases the regression line shifts towards the regression line derived from kerogen extracts (see Fig. 4 for comparison) at TOC >5%. There is an excellent correlation for samples containing >5% TOC: r^2 = 0.92.

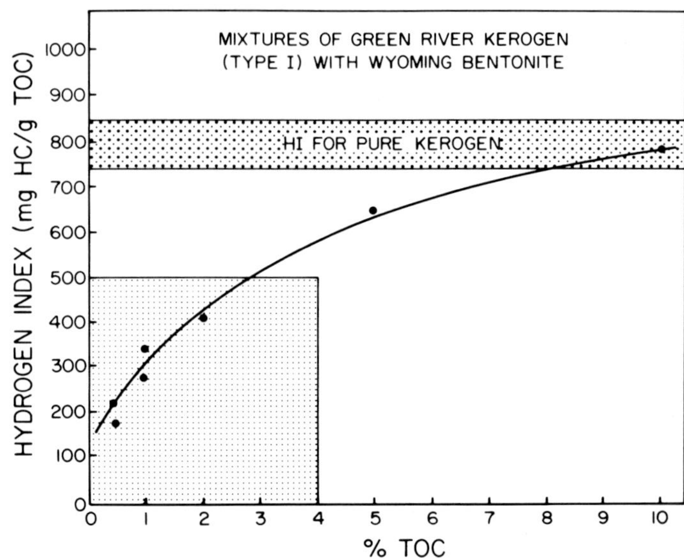

FIG. 7.—Hydrogen Index values for synthetic mixtures of Green River kerogen are plotted against % TOC (from Orr, 1983).

Orr; n = 70, this study), variations in analytical technique (Rock-Eval, Orr; CDS and microcomputer-controlled pyrolysis, this study), and finally, the exclusion of more mature samples may bias the Hydrogen Index-atomic H/C relationship.

Multiple regression of PI and pyrolytic yield (S_2) of the kerogen separate on the S_2 derived from the kerogen extract were satisfactory (Table 3). Separate-derived S_2 explained 95% of the variance in extract S_2, and, including production index, 98% of the variance was explained. Clementz (1979) has shown that large quantities of heavy bitumen or migrated hydrocarbon can affect the size and maximum temperature of the S_2 peak during pyrolysis. Our multiple regression analysis suggests that for the data presented in this study, migrated hydrocarbons do not significantly affect S_2.

Previous studies of synthetic mixtures of kerogen and mineral matrix incorporate mixtures with a broad range of total organic carbon content. The trend obtained is shown in Figure 7 (from Orr, 1983, Fig. 7; Espitalié et al., 1980). In general, the measured bulk HI approaches the HI for pure kerogen over the 1 to 4% TOC range. This investigation examines this lower TOC range for synthetic kerogen extract-clay mixtures and shows the effect of decreasing pyrolysis-derived HI with decreasing TOC. Samples having a low HI (extract determined) were most affected. Figures 8 and 9 illustrate the 'matrix' effect for each kerogen type and clay. Green River shale samples have the steepest slope on an HI versus TOC plot. This sample had the highest HI (1000 mgHC/gOC). The Woodford shale has the next highest extract HI (800) and yields a less steep slope. The Mowry shale has a moderate HI (370) and shows a gradual slope. Finally, the Stevens material has a very low HI (150) and the measured HI values from the mixtures consistently underestimate the actual HI (as determined from the extracts). In general, the illite suppresses the HI value to a greater extent than the bentonite as shown by the shift in X-axis intercept from approximately 0 with bentonite to almost 1.0 wt % TOC with illite. This work is in agreement with previous investigations (Espitalié et al., 1980). Fig-

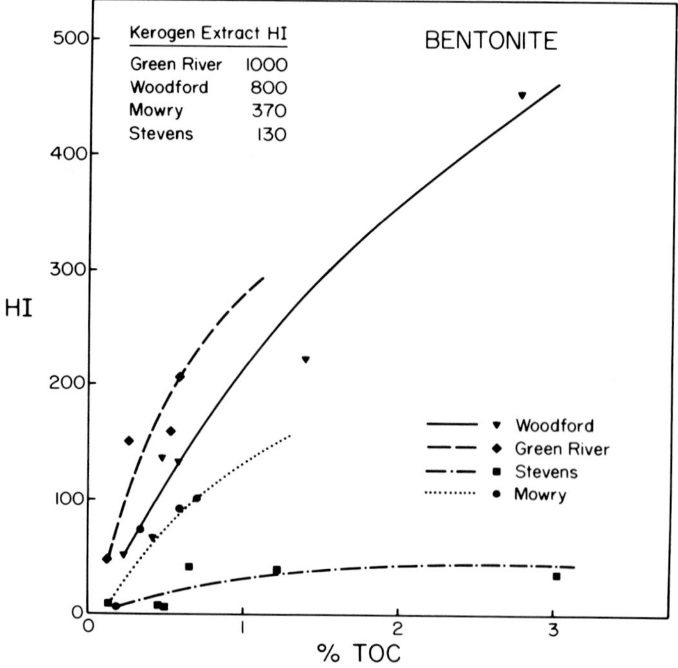

FIG. 8.—Hydrogen Index values from mixtures of kerogen and bentonite are plotted against % TOC of the mixture. Curves are drawn through the data to indicate trends.

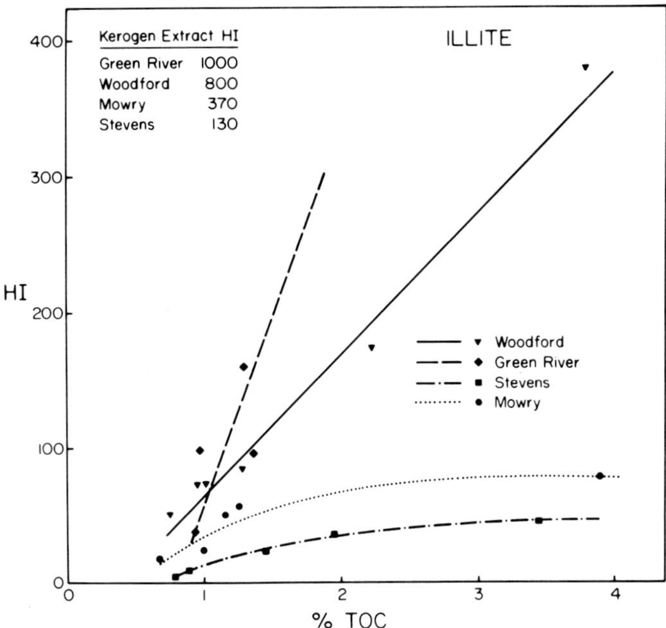

FIG. 9.—Same as Figure 8, but the kerogen has been mixed with illite. Note scale changes. In general, illite suppresses the pyrolytic yield to a greater extent than bentonite.

ures 10 and 11 suggest that one mineral matrix may drastically suppress a specific kerogen type (illite, Mowry), whereas another may not (bentonite, Mowry). In any case, Type III kerogen shows the lowest recovery of HI with increasing TOC (Figs. 10 and 11).

DISCUSSION

Synthetic Mixtures

Experimental results indicate that the source rock mineral matrix is responsible for discrepancies between bulk rock pyrolysis and kerogen analysis. A model for the retention effect of the mineral matrix has been proposed by Espitalié et al. (1980). Heavy and light hydrocarbons are released by the thermal degradation of kerogen disseminated throughout the source rock. Light hydrocarbons are able to migrate out of the source rock, but heavy hydrocarbons are trapped on 'active sites' of the clay mineral matrix. Further heating of the source rock may in turn crack these retained heavy hydrocarbons, allowing a second phase of light hydrocarbon release; but a carbon residue remains on the matrix and is not measured as pyrolytic yield. Thus, the effect of an 'active' mineral matrix is to underestimate the pyrolytic yield.

Clay type.—
Pyrolysis of synthetic mixtures of kerogen and clay isolates the effects of three factors on bulk pyrolysis: mineral matrix composition, actual H/C of the kerogen, and % TOC of the mixture. Some minerals have relatively little effect on the bulk rock determinations of HI (quartz and kaolinite, for example: Espitalié et al., 1980), whereas others suppress the HI by as much as 90% at 1% TOC (see Fig. 11). In many shales (the Mowry shale, for example), mixed-layer clays (smectite and illite, in particular) constitute the bulk of the fine clay fraction. Our evaluation of bentonite and illite suggests that illite consistently suppresses the HI of the synthetic mixture to a greater extent than bentonite (Figs. 8, 9, 10, and 11).

Kerogen type.—
As work by Espitalié et al. (1980) indicates, Type III kerogen is most affected by the mineral matrix (see Figs. 10 and 11). At 3% TOC, samples with the lowest extract HI (Stevens and Mowry samples) have only attained 20–30% of their actual HI, while Type I (Green River) and II (Woodford) kerogen have attained 40 and 90%, respectively. In terms of absolute HI units, however, bulk rock HI values from Type I kerogen will most dramatically underestimate the extract-derived HI because these kerogens have the highest absolute amount of HI to recover as TOC increases.

The synthetic mixtures of low % TOC illustrate the importance of considering the 'matrix effect' on bulk rock pyrolysis. Although the magnitude of the effect is determined by the above two factors (matrix composition and atomic H/C of the kerogen), the importance of the actual ratio of kerogenous material to matrix material is evident from the similar shapes of all curves (Figs. 8, 9, 10, and 11). The effect of TOC is seen in our sample data as well.

Sample Data

Figure 12 shows the relationship between bulk HI and extract HI. A significant portion of the samples falls in a

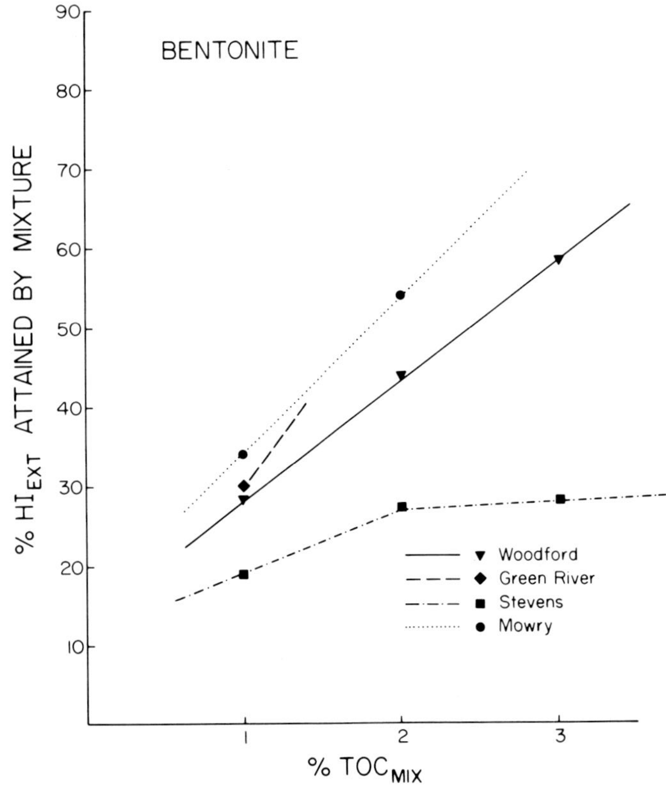

FIG. 10.—The fraction of the kerogen extract HI attained by the mixture shown in Figure 8 (bentonite) plotted against % TOC of the mixture.

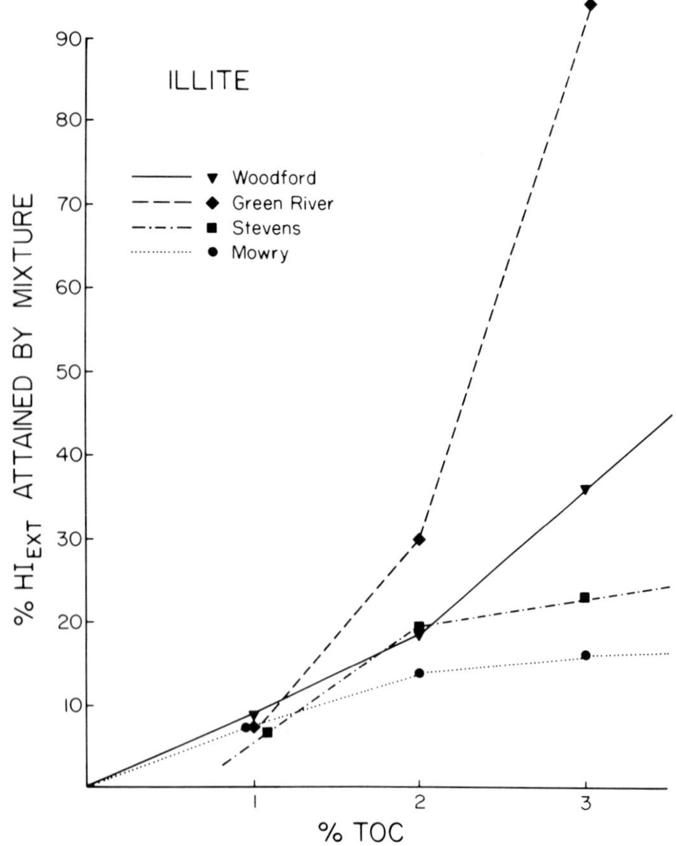

FIG. 11.—Same as Figure 10, but mixtures of kerogen and illite are plotted. Note that Type III kerogen (Stevens and Mowry, Table 1) is most affected by the mineral matrix. The Type I kerogen (Green River Shale) has almost completely recovered its extract HI at 3% TOC.

region where the bulk HI is lower than would be expected on the basis of both the 1:1 line and the best-fit regression line. Figure 4 also highlights these 'anomalous' data points, too numerous actually to be considered anomalous. This is precisely the shift in bulk HI one would predict on the basis of low TOC. Figure 6 isolates sample groups on the basis of % TOC; samples within the 'anomalous' region consist primarily of samples containing <5% TOC. In general, bulk pyrolysis data may be viewed as a collection of samples exhibiting varying degrees of the matrix effect. The parallel nature of the H/C versus bulk HI and H/C versus extract HI correlations (Fig. 2, Table 2) indicate that the effect is relatively constant over the entire H/C range. A 'quick' correction for all bulk rock pyrolysis data is the addition of 75 HI units to the bulk HI (see Fig. 12). Figure 6 indicates that at >5 wt % TOC this correction is unnecessary; the regression line for the sample subgroup containing >5 % TOC virtually overlies the H/C-extract HI regression line.

Guidelines for Application

Information derived from the sample data set, combined with the results of the synthetic kerogen-clay mixtures, allows us to suggest guidelines for the constructive application of pyrolysis data in a hydrocarbon exploration context.

Our 31 Mowry samples serve as an example of a single source rock evaluation. This type of evaluation typifies the study of a particular formation known to source hydrocarbon in an oil province (for example, the Kimmeridgian of the North Sea). Because one specific source rock is being evaluated, the best approach is to separate the kerogen and do a comparative study of the bulk rock and extract behavior of the source rock, following a similar analytic and

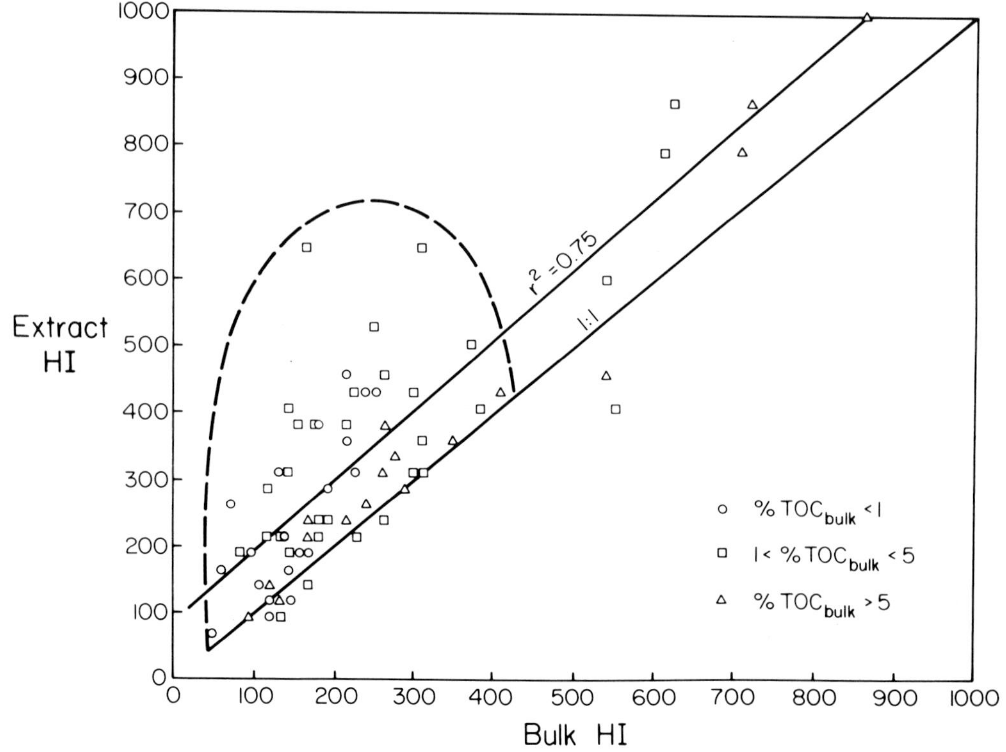

FIG. 12.—Bulk rock-derived HI values are plotted against kerogen extract-derived HI values. The 1:1 correlation line is shown for reference. Samples are grouped on the basis of % TOC. Note that the lines are approximately parallel and offset by about 75 HI units. The region where bulk rock HI underestimates extract HI is outlined with a dashed line and is represented by samples from the <1% and 1–5% TOC subgroups.

TABLE 4.—Single source evaluation (Mowry samples)

(mgHC/ g OC) HI_{bulk}	(mgHC/ g OC) HI_{ext}	Measured H/C	Calculated* H/C	% DEV	Calculated** H/C	% DEV
137	106	0.59	0.79	+34	0.72	+22
115	87	0.56	0.77	+38	0.70	+25
145	166	0.66	0.80	+21	0.72	+9
259	244	0.73	0.89	+22	0.80	+10
304	323	0.81	0.93	+15	0.83	+2
195	279	0.75	0.84	+12	0.75	0
126	119	0.63	0.78	+24	0.71	+13
109	154	0.65	0.72	+18	0.70	+8
172	187	0.61	0.82	+34	0.74	+21
51	80	0.57	0.72	+26	0.66	+16
552	397	0.91	1.14	+25	0.99	+9
133	—	0.73	0.79	+8	0.71	−3
178	226	0.74	0.83	+12	0.74	0
133	219	0.77	0.79	+3	0.71	−8
140	301	0.71	0.79	+11	0.72	+1
145	182	0.75	0.80	+7	0.72	−4
121	—	0.68	0.78	+15	0.70	+3
229	216	0.67	0.87	+30	0.78	+16
86	196	0.60	0.75	+25	0.68	+13
158	195	0.70	0.81	+16	0.73	+4
192	242	0.68	0.84	+24	0.75	+10
318	321	0.74	0.94	+27	0.83	+12
118	222	0.72	0.78	+8	0.70	−3
348	370	0.82	0.97	+18	0.85	+4
173	148	0.68	0.82	+21	0.74	+9
179	248	0.68	0.83	+22	0.74	+9
142	120	0.58	0.80	+38	0.72	+24
237	—	0.73	0.88	+21	0.78	+7
180	384	0.76	0.83	+9	0.74	−3
545	589	0.90	1.13	+26	0.98	+9
244	422	0.76	0.88	+16	0.79	+4
				AVE DEV = +20%		AVE DEV = 8%

*H/C = (0.00084 × HI_{bulk}) + 0.676 Equation #4, Table II
**H/C = (0.00066 × HI_{bulk}) + 0.625 Equation #8, Table II

statistical flow scheme to that presented here. According to Sokal and Rohlf (1981), a sample size of approximately 25 will sufficiently approach a normal distribution. Table 4 shows the difference between using the general bulk HI-atomic H/C relationship and the regression determined from the Mowry samples alone. When the general relationship is used, the average error of the calculated H/C is +20%. When the equation derived from the Mowry samples alone is used, the average error of the calculated H/C is <10%.

TABLE 5.—Multiple source rock evaluation

	(mgHC/ g OC) HI_{bulk}	(mgHC/ g OC) HI_{ext}	Measured H/C	Calculated* H/C	% DEV
Woodford	609	797	1.08	1.16	+7
Monterey	623	856	1.18	1.17	−1
Green River	861	1000	1.31	1.36	+4
Aus. Spls.	259	302	0.73	0.88	+21
"	91	105	0.59	0.75	+27
"	265	383	0.82	0.89	+9
"	171	641	0.96	0.81	−16
"	221	238	0.70	0.85	+21
"	122	153	0.71	0.75	+6
"	288	287	0.86	0.91	+6
"	170	215	0.79	0.81	+3
"	718	857	1.17	1.25	+7
"	243	253	0.72	0.87	+21
"	413	426	0.81	1.01	+25
"	275	329	0.75	0.90	+20
"	722	1000	1.24	1.25	+1
"	708	780	1.16	1.24	+7
"	539	453	0.96	1.11	+16
"	174	251	0.84	0.82	−2
				AVE % DEV =	+9.6%

*H/C = (0.00084 × HI_{BULK}) + 0.676 Equation #4, Table II

A study encompassing the entire spectrum of organic-rich potential source rocks within the stratigraphic column presents a different problem. Our suite of Australian samples serve as an example of this type of source rock analysis. This suite represents 13 different formations of various kerogen types. For this survey, the general bulk rock HI-atomic H/C relationship yields excellent results. The average error of the calculated H/C is <10% (Table 5). In both of the exploration scenarios outlined above, the information presented in this study is utilized to yield adequate information with minimal time and cost. As different pyrolysis techniques may yield slightly different sensitivities, a check of the actual equipment used with kerogen extraction and bulk rock samples is recommended.

CONCLUSIONS

We have presented organic data from numerous samples covering a wide range of kerogen type, maturity, and organic richness. The sample data have been rigorously treated to determine relationships between the rapidly der[illegible]d bulk rock organic parameters and those derived from [illegible]ated kerogen. Experimental work on synthetic clay and kerogen mixtures has aided in understanding the magnitude of deviations of bulk rock pyrolysis data from isolated kerogen measurements. In general, samples with >5% TOC do not exhibit significant matrix effects. Information from the experiments combined with sample data analysis led to the formulation of specific guidelines for the practical utilization of pyrolysis data in actual exploration applications: the evaluation of potential hydrocarbon source rocks of varying kerogen type containing <5% TOC, and encompassing a broad range of thermal maturity.

ACKNOWLEDGMENTS

The authors express their appreciation to Dr. Steven Boese for valuable discussion and analytical assistance. Connie Boese and Pegui Yin provided analytical assistance throughout the study. Al Deiss and Bonnie Boland assisted with the graphics.

REFERENCES

Claypool, G. E., and Reed, P. R., 1976, Thermal-analysis technique for source-rock evaluation: Quantitative estimate of organic richness and effects of lithologic variation: American Association of Petroleum Geologists Bulletin, v. 60, no. 4, p. 608–612.

Clementz, D. M., 1979, Effect of oil and bitumen saturation on source-rock pyrolysis: American Association of Petroleum Geologists Bulletin, v. 63, p. 2227–2232.

Dembicki, Jr., H., Horsfield, B., and Ho, T. T. Y., 1983, Source rock evaluation by pyrolysis-gas chromatography: American Association of Petroleum Geologists Bulletin, v. 67, no. 7, p. 1094–1103.

Espitalié, J., Laporte, J. L., Madec, M., Marquis, F., Leplat, P., Paulet, J., and Boutefeu, A., 1977, Méthode rapíde de caractérisation des roches méres, de leur potentiel pétrolier et de leur degré d'evolution: Revue de l'Institut Francais du Pétrole, v. 32, p. 32–42. English edition available. Ref. 26 447, Institut Francais du Pétrole (1978).

———, Madec, M., and Tissot, B., 1980, Role of mineral matrix in kerogen pyrolysis: Influence on petroleum generation and migration: American Association of Petroleum Geologists Bulletin, v. 64, no. 1, p. 59–66.

Horsfield, B., and Douglas, A. G., 1980, The influence of minerals on the pyrolysis of kerogens: Geochimica et Cosmochimica Acta, v. 44, p. 1119–1131.

Hunt, J. M., 1979, Petroleum Geochemistry and Geology: San Francisco, W. H. Freeman and Co., 617 p.

Katz, B. J., 1981, Limitations of Rock-Eval pyrolysis for typing organic matter: American Association of Petroleum Geologists Book of Abstracts, San Francisco, May 31–June 3.

Monin, J. C., Durand, B., Vandenbroucke, M., and Huc, A. Y., 1980, Experimental simulation of the natural transformation of kerogen, *in* Douglas, A. C. and Maxwell, J. R., eds., Advances in Organic Geochemistry 1979: Physics and Chemistry of the Earth, Oxford, England, Pergamon Press, v. 12, p. 517–530.

Nie, N. H., Hull, C. H., Jenkins, J. G., Steinbrenner, K., and Bent, D. H., 1975, Statistical Package for the Social Sciences, 2nd ed.: New York, McGraw-Hill, 675 p.

Orr, W. L., 1983, Comments on pyrolytic hydrocarbon yields in source-rock evaluation, *in* Bjorøy, M., Albrecht, C., Corning, C., de Groot, K., Eglinton, G., and Speers, G., eds., Advances in Organic Geochemistry 1981: Chichester, John Wiley and Sons, Ltd., p. 775–787.

Peters, K. E., Whelan, J. K., Hunt, J. M., and Tarafa, M. E., 1983, Programmed pyrolysis of organic matter from thermally altered Cretaceous black shales: American Association of Petroleum Geologists Bulletin, v. 67, no. 11, p. 2137–2146.

Sokal, R. R., and Rohlf, F. J., 1981, Biometry: San Francisco, W. H. Freeman and Co., 859 p.

Tissot, B. P., 1984, Recent advances in petroleum geochemistry applied to hydrocarbon exploration: American Association of Petroleum Geologists Bulletin, v. 68, no. 5, p. 545–563.

———, and Welte, D. H., 1978, Petroleum Formation and Occurrence: New York, Springer-Verlag, 538 p.

Verheyen, T. V., Johns, R. B., and Espitalié, J., 1984, An evaluation of Rock-Eval pyrolysis for the study of Australian coals including their kerogen and humic acid fractions: Geochimica et Cosmochimica Acta, v. 48, p. 63–70.

Appendix: ORGANIC DATA

Source	SPL#	%OC	BULK HI	BULK PI	EXT HI	EXT H/C	EXT O/C	SEP HI	SEP PI
Stevens	1	1.54	382	0.16	417	1.11	0.09	479	0.16
″	2	1.62	220	0.30	379	0.99	—	337	0.26
″	3	0.71	135	0.54	—	—	—	319	0.28
″	4	0.47	130	0.53	215	0.70	—	244	0.26
″	5	0.19	337	0.59	—	—	—	218	0.36
″	6	1.43	230	0.28	425	1.15	0.11	469	0.26
″	7	0.10	225	0.76	—	—	—	662	0.35
″	8	0.97	254	0.33	427	1.12	0.11	496	0.29
″	9	1.23	300	0.31	425	1.13	0.08	419	0.28
″	10	0.24	183	0.67	380	0.98	—	352	0.51
″	11	0.99	222	0.39	309	1.02	0.09	339	0.21
″	12	0.15	253	0.57	—	—	—	322	0.34
″	13	0.96	138	0.35	310	0.93	0.11	294	0.25
″	14	0.30	220	0.51	360	1.05	0.11	388	0.36
″	15	0.65	214	0.35	460	1.14	0.10	454	0.26
″	16	1.04	150	0.52	400	1.01	0.10	420	0.29
″	17	1.09	125	0.30	282	0.89	0.12	275	0.23
″	18	1.60	154	0.24	378	0.90	0.13	328	0.18
″	19	2.52	311	0.07	642	1.15	0.09	516	0.10
″	20	0.33	58	0.41	180	0.62	—	205	0.25
″	21	3.38	246	0.14	530	1.12	0.10	412	0.16
″	22	0.54	72	0.60	258	0.89	0.10	303	0.36
″	23	0.67	97	0.32	196	0.79	0.13	227	0.24
″	24	2.24	258	0.21	453	1.15	0.10	464	0.19
″	25	1.81	318	0.10	355	1.03	0.11	435	0.14
″	26	4.40	373	0.07	502	1.14	0.13	426	0.08
″	27	7.27	131	0.15	130	0.65	0.13	149	0.15
Mowry	28	1.31	137	0.44	106	0.64	0.05	—	—
″	29	0.74	115	0.56	87	0.60	0.05	—	—
″	30	0.91	145	0.16	166	0.71	0.14	—	—
″	31	1.23	259	0.11	242	0.78	0.11	—	—
″	32	1.03	304	0.16	323	0.87	0.06	280	0.18
″	33	0.95	195	0.22	279	0.80	0.06	244	0.20
″	34	0.92	126	0.33	119	0.67	0.11	—	—
″	35	0.90	109	0.49	154	0.70	0.06	—	—
″	36	0.96	172	0.29	187	0.66	0.04	—	—
″	37	0.74	51	0.41	80	0.61	0.10	—	—
″	38	3.31	552	0.10	397	0.98	0.07	—	—
″	39	1.02	133	0.11	—	0.78	0.10	—	—
″	40	1.25	178	0.14	226	0.80	0.08	—	—
″	41	1.16	133	0.14	219	0.82	0.09	—	—
″	42	1.28	140	0.14	301	0.77	0.10	—	—
″	43	1.81	145	0.08	182	0.80	0.14	—	—
″	44	0.98	121	0.17	—	0.72	0.09	—	—
″	45	1.29	229	0.23	216	0.72	0.04	—	—
″	46	1.18	86	0.23	196	0.64	0.06	—	—
″	47	1.00	158	0.21	195	0.75	0.13	—	—
″	48	1.11	192	0.24	242	0.72	0.05	—	—
″	49	1.42	318	0.27	321	0.80	0.05	—	—
″	50	1.10	118	0.27	222	0.77	0.07	—	—
″	51	5.86	348	0.05	370	0.88	0.11	—	—
″	52	1.57	173	0.26	148	0.73	0.06	—	—
″	53	1.42	179	0.36	248	0.73	0.06	—	—
″	54	0.92	142	0.57	120	0.62	0.04	—	—
″	55	0.91	237	0.14	—	0.79	0.08	—	—
″	56	1.02	180	0.20	384	0.81	0.07	—	—
″	57	1.92	545	0.12	589	0.96	0.05	—	—
″	58	0.87	244	0.19	422	0.81	0.09	—	—
Woodford	59	4.71	609	0.05	797	1.16	0.04	822	0.07
Monterey	60	4.59	623	0.10	856	1.26	0.09	550	0.11
Green River	61	7.15	861	0.04	1026	1.40	0.11	792	0.05
Australian	62	61.0	259	0.07	302	0.78	0.13	214	0.07
″	63	43.4	91	0.06	105	0.63	0.14	86	0.11
″	64	50.4	265	0.05	383	0.88	0.20	252	0.04
″	65	4.70	171	0.08	641	1.03	0.09	423	0.07
″	66	22.3	221	0.06	238	0.75	0.13	173	0.08

Appendix: ORGANIC DATA

Source	SPL#	%OC	BULK HI	BULK PI	EXT HI	EXT H/C	EXT O/C	SEP HI	SEP PI
"	67	21.4	122	0.13	153	0.76	0.13	100	0.14
"	68	55.0	288	0.05	287	0.92	0.21	260	0.09
"	69	29.3	170	0.06	215	0.84	0.22	204	0.07
"	70	19.5	718	0.07	857	1.26	0.05	686	0.06
"	71	33.0	243	0.06	253	0.77	0.16	226	0.07
"	72	45.3	413	0.07	426	0.87	0.18	331	0.07

Appendix: ORGANIC DATA

Source	SPL#	%OC	BULK HI	BULK PI	EXT HI	EXT H/C	EXT O/C	SEP HI	SEP PI
"	73	48.8	275	0.05	329	0.80	0.17	244	0.06
"	74	15.6	722	0.11	1171	1.32	0.03	816	0.07
"	75	17.6	708	0.06	780	1.24	0.07	635	0.07
"	76	5.60	539	0.05	453	1.02	0.14	362	0.07
"	77	51.0	174	0.05	251	0.90	0.26	163	0.04

RELATIONSHIP OF ORGANIC AND INORGANIC DIAGENESIS OF NEOGENE TERTIARY ROCKS, NORTHEASTERN JAPAN

KAZUO TAGUCHI
Institute of Mineralogy, Petrology and Economic Geology
Faculty of Science, Tohoku University
Sendai, Japan
SUSUMU SHIMODA
Department of Geoscience, University of Tsukuba
Sakura-mura, Ibaraki Prefecture, Japan
YUUKO ITIHARA
Department of Geosciences, Faculty of Science
Osaka City University
Osaka, Japan
NOBUHIRO IMOTO
Department of Earth Sciences, Kyoto University of Education
Kyoto, Japan
RYOSHI ISHIWATARI
Department of Chemistry, Tokyo Metropolitan University
Setagaya-ku, Tokyo, Japan
AKIRA SHIMOYAMA
Department of Chemistry, University of Tsukuba
Sakura-mura, Ibaraki Prefecture, Japan
MASAHIKO AKIYAMA
Department of Geology and Mineralogy, Hokkaido University
Sapporo, Hokkaido, Japan
AND
NORIYUKI SUZUKI
Department of Geology, Shimane University
Matsue, Japan

ABSTRACT: The stratigraphic distribution and diagenetic changes of various organic and inorganic materials present in the Middle Miocene marine shales and mudstones from the Shinjo Oil Field, northeastern Japan, were compared for possible diagenetic relationships. Vitrinite reflectance, bitumen and hydrocarbon ratios, Rock-Eval pyrolysis, elemental analysis of kerogen, CPI values of n-alkanes and kerogen-form fatty acids, $\delta^{13}C$ of bitumen and kerogen, ^1H-NMR T_1 of the dioxane-insoluble organic matter, epimerization of steranes and triterpanes, aromatization of steranes, organic and ammonium nitrogen contents, surface texture of sponge spicules, and the transformation of silica, clay and zeolite minerals were investigated as diagenetic indices and correlation among them was examined. The transformation of smectite to illite/smectite mixed-layer minerals and of organic N to fixed NH_4-N, and the disappearance of sponge spicules by corrosion exhibited a good correlation with major indices of organic diagenesis near the threshold of intense oil generation. It is suggested that the diagenesis of organic matter (particularly changes in the lipid materials, fixation of nitrogen, and carbon isotope fractionation trends) are caused by the thermocatalytic effect of clay minerals.

INTRODUCTION

Studies of diagenetic processes in sedimentary rocks have progressed rapidly during the last decade. Previous studies have focused on inorganic (mineral) diagenesis, prior to the introduction of the concept of organic diagenesis or organic metamorphism by petroleum geochemists. Moreover, most studies on organic and inorganic diagenesis have been done as separate and distinct studies, as pointed out by Taguchi (1978). Thus, the present study was undertaken to examine principally the relationships among the various parameters of inorganic and organic diagenesis. Such a study, which attempts to determine various properties on one sample, requires sizable amounts of material, particularly for the analyses of various organic constituents, that are contained in the sedimentary rocks in an extremely small quantity. Core or cuttings samples usually provide small samples and may be contaminated due to the circulating mud water. Consequently, in this study, surface samples, collected from the Oashizawa Valley, Shinjo Oil Field, Yamagata Prefecture, were used despite some of the problems caused by the use of outcrop samples. Each rock sample collected was divided into a number of small pieces, which were distributed to each co-worker of this research group. Although we attempted to do a comparative study with specific data, such as vitrinite reflectance or hydrocarbon ratio between outcrop and core (or cuttings) samples, we could not obtain suitable cores or cuttings from the study area.

In this paper, the relationship of time and temperature factors to maturation parameters remains mostly unsolved. A significant inconsistency exists between the time-temperature regime of Japanese Tertiary oil genesis and the thermal regimes for oil generation proposed by Connan (1974) and Waples (1980), as pointed out by Taguchi (1983), and Sato and Asakawa (1982), respectively.

Prior to this co-operative research, geochemical studies of metal-porphyrin compounds (Taguchi et al., 1970), amino acids (Sasaki, 1974), vitrinite reflectance (Ikehara et al.,

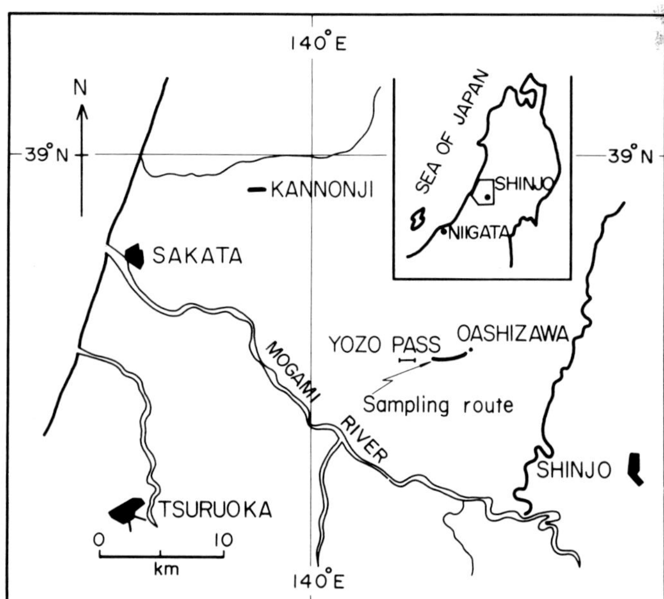

FIG. 1.—Index map of the study area, showing sampling route.

1980), and fatty acids (Suzuki and Taguchi, 1984) on rocks collected from the same study area have been reported elsewhere. In this paper, some parts of the above investigations are redescribed at need.

SAMPLE DESCRIPTION AND METHODS

Thirty-eight rock samples were taken from the Kusanagi, Furukuchi, Noguchi, and Nakawatari Formations of Miocene to Pleistocene age exposed along the Oashizawa Valley, the Shinjo Oil Field, Yamagata Prefecture (Figs. 1 and 2). This area was selected for study because the strata are well exposed along a gorge and because other studies have been done in the area (Taguchi, 1962, 1967; Inomata, 1962; Inoue, 1962). A location map, the general stratigraphic sequence of the Shinjo Oil Field, and the lithologic properties and stratigraphic position of rock samples are shown in Figures 1, 2 and 3. Samples to be analyzed were taken at about 1.5 m below the non-weathered surface exposure so as to avoid surface contamination as much as possible. The analyzed samples are fine-grained hard shale and/or mudstone, except for four samples of siltstone from the Noguchi and Nakawatari Formations. Therefore, no significant contamination is expected in these samples.

The geology of the Shinjo Oil Field is classified into the following stratigraphic units in ascending order: Okawa, Kusanagi, Furukuchi, Noguchi, Nakawatari, Sakekawa, Yamuke and Motoaikai Formations (Fig. 2). Except for a slight and partial disconformity at the basal part of the Sakekawa Formation, all these formations are essentially conformable. The basal part of the Neogene rocks unconformably overlies the Paleozoic complexes and associated granitic rocks which intrude in the east and south sides of the Shinjo basin. The total thickness of the Neogene and Quaternary sediments is about 4,800 m. The study area, corresponding to the west part of the present Shinjo basin, existed as an active sedimentary basin until after the deposition of the Nakawatari Formation. Since then, a mild arching of the study area has divided the basin into a deep Japan Sea side basin and the shallower Shinjo basin. The study area attained maximum burial depth at the time of deposition of the Nakawatari Formation. The thickness of these rocks is about 3,500 m; probably the original thickness was considerably more. The thickness of strata from top of the Nakawatari Formation to the bottom of the Kusanagi Formation is estimated to be about 2,000 m.

Studies of organic diagenesis included the analysis of organic carbon and nitrogen, bitumen, various hydrocarbons, fatty acids and kerogen, and the measurement of vitrinite reflectance, stable carbon isotope ratio, ^{1}H spin-lattice relaxation time (T_1), and pyrolytic characteristics using whole-rock samples (Rock-Eval method; Espitalié et al., 1977). Analyses of rock and kerogen samples for organic carbon, hydrogen, and oxygen were carried out using a Perkin-Elmer CHNO analyzer except for the ammoniacal nitrogen study. The extractable organic matter was obtained by shaking with benzene:methanol (90:10 v:v) at room temperature. Saturated hydrocarbons were isolated from the extracts using column chromatography on silicic acid with elution by *n*-hexane. Normal alkanes were separated from branched and cyclic alkane by urea adduction. Hydrocarbons (*n*-alkanes, steranes and hopanes) and fatty acids (after methyl esterification) were analyzed using a Hitachi

AGE		FORMATION	LITHOLOGY
QUAT.	PLEIS-TOCENE	SHIMOYAMASAKI (TERRACE DEPOSITS)	Dacitic tuff (Sands & Gravels)
TERTIARY	PLIOCEN	YAMAYA	Conglomerate
		FUNAGATA	Siltstone & Sandstone
		SHIBAKURAYAMA	Sandstone
		MOTOAIKAI	Sandstone & Siltstone
		YAMUKE	Sandstone & Siltstone
		SAKEKAWA	Tuffaceous Sandstone
		NAKAWATARI	Siltstone
		NOGUCHI	Siltstone
	?		
	MIOCENE	FURUKUCHI	Black mudstone
		KUSANAGI	Hard banded shale
		ŌKAWA	Pyroclastics and mudstone

FIG. 2.—Generalized stratigraphic column in Shinjo Oil Field.

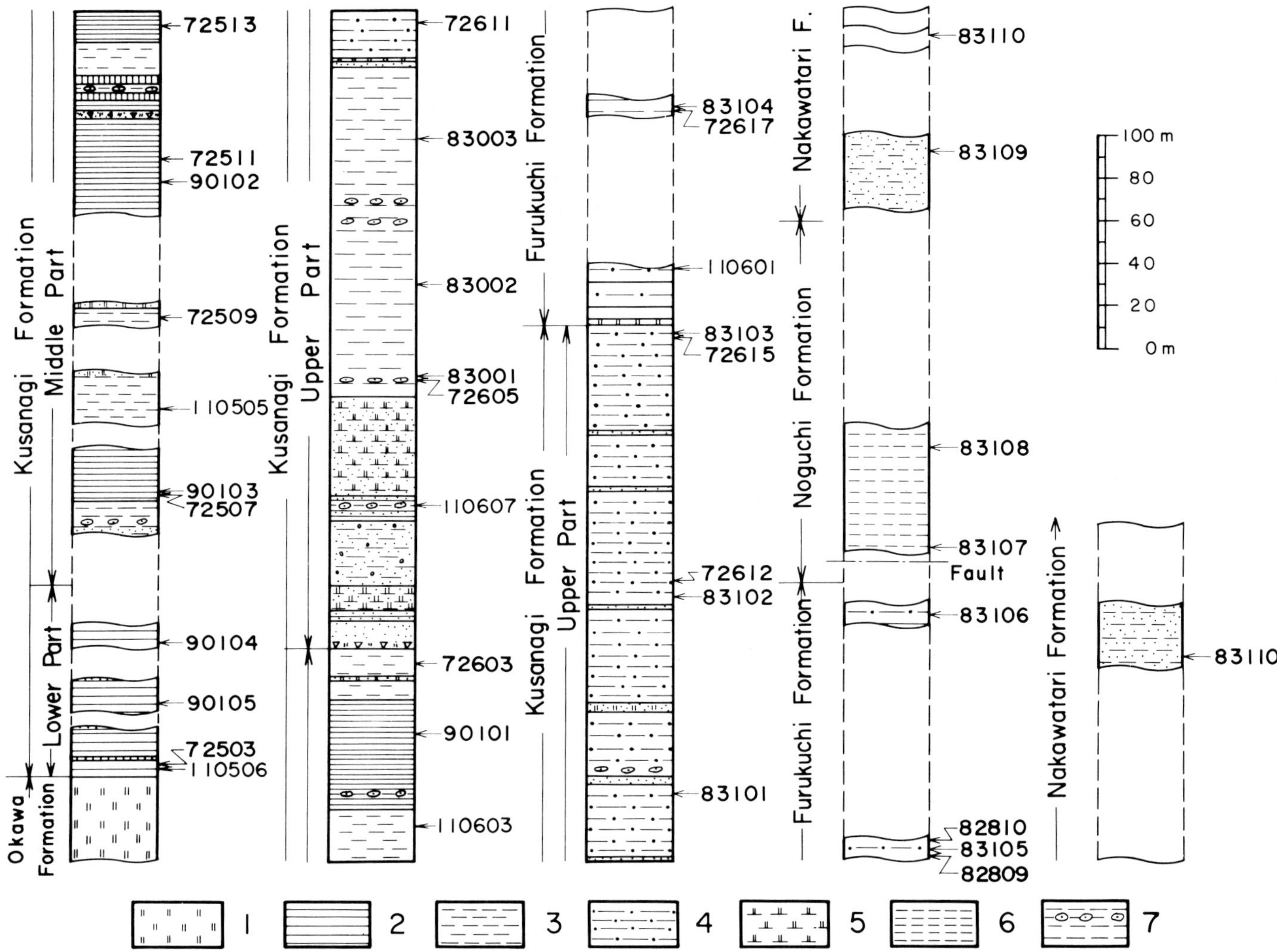

FIG. 3.—Schematic stratigraphic and lithologic sections of the studied area. Numbers in the column indicate sampling points. Major rock facies: 1: Dark brown andesitic tuff breccia; 2: Very hard siliceous shale; 3: Grey hard siltstone; 5: Sandy tuff; 6: Grey sandy siltstone; 7: Calcareous concretion.

Perkin-Elmer F6D gas chromatograph and a JEOL QH-100 gas chromatograph-mass spectrometer with a 2 m/3 mm i.d. SUS column packed with 2% Apiezon L on Chromosorb GAW DMCS (180–100 mesh) or with a 25 m/0.25 mm i.d. FS-WCOT OV-101 capillary column. The mass spectrometer was operated at an electron energy of 70eV, ion source temperature of 200°C and separator temperature of 300°C. The structures of the steranes and the hopanes were assigned by comparison of literature mass spectra and retention times (Mulheirn and Ryback, 1977; Seifert and Moldowan, 1978; Shi-Ji-Yang et al., 1982). The ratios of $20S-/20R-C_{29}$-sterane and $22S-/22S-C_{32}$ hopane were calculated by measuring their peak areas in the mass fragmentograms of m/z 217 and m/z 191, respectively.

The analysis of fatty acids (Suzuki and Taguchi, 1984), based on the incorporation state in the samples, was carried out as follows. Extractable *n*-fatty acids, which are present as unbound phases (i.e., easily extracted with organic solvents), were divided into two categories, free- and bound-form fatty acids. Each category was subdivided as follows: (1) free fatty acids (FFA), extracted with alkaline solution from extractable organic matter (EOM), and (2) esterified *n*-fatty acids (EFA), obtained by saponification of neutral EOM. Bound-form fatty acids present in extracted residues (i.e., kerogen) were also divided into two fractions: (3) secondary extractable *n*-fatty acids (SEFA), recovered by hydrolysis of secondary extractable organic matter (SEOM), obtained after demineralization using HF:HC1 of the primary extracted residues, and (4) kerogen-form *n*-fatty acids (KFA), obtained by saponification of the re-extracted residues. No humic acids were detected in any samples studied.

Carbon stable isotope measurements were made using a Varian MAT 250 Mass Spectrometer. Isotopic compositions were referred to the PDB carbon isotopic standard via reference gases calibrated against the TKL-1 and K-2 isotopic standards. Each CO_2 was obtained by combustion of extractable organic matter and kerogen in the presence of O_2 over CuO at 900°C for 7 minutes. ^{1}H spin-lattice relaxation time (T_1) of the dioxane-insoluble organic matter was measured by Bruker Sxp 4-100 pulse and FT NMR Spectrometer at the reasonance frequency of 90 MHz, according to the method of previous workers (Yokono and Sanada, 1978; Maruyama, 1979; Hirata and Akiyama, 1982). Vitrinite reflectance determinations were made with a Leitz MPV-2 Orthoplane microscope.

Studies for inorganic diagenesis included the analysis of ammonium nitrogen, which was separated into three parts: water soluble NH_4-N, exchangeable NH_4-N with 1N KCl, and fixed NH_4-N (Bremner and Harada, 1959). In this study, to elucidate the origin of ammoniacal nitrogen, the concentration of organic carbon, organic nitrogen, and total nitrogen was determined. Organic carbon content was determined by a wet-oxidation method, and total nitrogen by the sealed tube digestion method proposed by Stevenson (1960). This method involves the digestion of specimens in concentrated H_2SO_4 in sealed tubes at a temperature of 420°C for 2 hrs, after which the ammonia is distilled in a micro-Kieldahl unit and estimated colorimetrically with Nessler's reagent. Organic nitrogen content was determined by difference between total nitrogen and ammoniacal nitrogen contents. The concentration of total and ammoniacal nitrogen was determined by applying the same method as Stevenson (1960) to the material obtained by digesting sedimentary rock samples in HF(1:1) and 1N H_2SO_4 at a temperature of 80°C for 1 hr. The content of fixed NH_4-N was determined by subtracting water soluble and exchangeable NH_4-N contents from total ammoniacal nitrogen content.

Major mineral constituents were examined by X-ray diffractometry, supplemented by chemical, differential thermal (DTA), and infrared (IR) analyses. The clay analysis was performed using the less than 2μm clay fraction obtained by an ordinary sedimentation method. Other minerals were examined using powder samples prior to a hydraulic elutriation. Sponge spicules were separated from the rock samples using a sodium sulphate method (Takayanagi, 1978), then observed with a binocular microscope and scanning electron microscopy (SEM). Porosity measurement was made using a poresizer (Shimazu Poresizer 9300). Further details of these procedures have been described elsewhere by Taguchi et al. (1983), Suzuki and Taguchi (1984), Akiyama and Ujiie (1983), Shioya and Ishiwatari (1983), Suzuki and Shimada (1983), and Shimoyama and Matsubaya (1983) for organic diagenesis, and by Itihara (1983a, 1983b), Imoto and Saito (1983), and Shimoda et al. (1983) for inorganic diagenesis.

RESULTS AND DISCUSSION

Organic Diagenesis

Organic matter content and type.—Total organic carbon values for all samples are shown in Table 1 and Figure 4. The average total organic carbon content is highest in the Kusanagei Formation from sample nos. 110506 to 83103, though the content in individual rock samples shows a wide variation. There is a marked decrease in total organic carbon from the Furukuchi Formation to the upper formations. This vertical distribution pattern is similar to that discussed by Taguchi (1983). Comparing the data presented in this study with those of source rocks from the correlative formations in typical oil-producing fields near the Japan Sea side, the average total organic carbon content in the study area is relatively low.

Rock-Eval pyrolysis (Espitalié et al., 1977) shows that samples from the Kusanagi and the Furukuchi Formations, with the exception of one sample, contain type I or II kerogen. Samples from the Noguchi and the Nakawatari Formations have mostly type III kerogen (Fig. 5 and Table 1). In contrast, results obtained on the basis of elemental analysis of demineralized kerogens from the selected samples plotted on a Van Krevelen diagram (Fig. 6) are slightly

TABLE 1.—ROCK-EVAL PYROLYSIS RESULTS

Sample	Org.C	T	S1	S2	S3	H.I.	O.I.	P.I.	S2/S3
83110	0.29	411	0.065	0.192	0.249	67.4	85.3	0.249	0.8
83109	0.49	406	0.064	0.351	0.247	71.1	50.3	0.154	1.4
83108	0.64	411	0.084	1.023	0.153	161.2	24.0	0.076	6.7
83107	0.46	420	0.040	0.227	0.174	49.1	38.2	0.149	1.3
83106	0.51	419	0.034	0.315	0.159	62.7	31.6	0.096	2.0
83105	0.94	419	0.090	1.569	0.239	166.3	25.3	0.054	6.6
83104	0.88	421	0.104	1.379	0.144	158.0	16.5	0.070	9.6
110601	0.74	427	0.594	1.889	0.140	253.5	18.7	0.239	13.5
83103	0.66	426	0.064	0.799	0.172	121.8	26.2	0.074	4.6
83102	1.12	429	0.228	3.073	0.138	275.4	12.3	0.069	22.3
83101	1.06	426	0.166	2.701	0.168	253.3	15.7	0.058	16.1
83003	1.52	426	0.775	5.021	0.179	331.1	11.9	0.134	27.9
83002	1.47	427	0.763	4.943	0.250	356.5	18.0	0.134	19.8
83001	1.07	428	0.577	3.375	0.179	313.1	16.5	0.146	18.9
110607	0.92	431	0.931	2.985	0.192	321.8	20.7	0.238	15.6
90101	1.23	426	1.514	4.384	0.187	357.8	15.3	0.257	23.4
110603	0.61	424	0.116	1.241	0.158	203.2	26.0	0.085	7.8
90102	1.01	428	1.515	3.022	0.182	297.5	17.9	0.334	16.6
110505	0.79	431	0.852	2.009	0.163	346.3	28.1	0.298	12.3
90103	1.49	431	2.677	4.984	0.148	334.5	9.9	0.349	33.7
90104	1.20	434	1.732	3.833	0.117	319.5	9.7	0.311	32.8
90105	0.62	436	0.310	2.284	0.064	368.4	10.3	0.120	35.9
110506	0.63	437	0.142	1.966	0.049	312.1	7.7	0.067	40.5

Org. C: Organic Carbon [wt%], S1: Hydrocarbons present in the rock [mgHC/g Rock], S2: Hydrocarbons from kerogen cracking [mgHC/g Rock], S3: CO_2 from kerogen cracking [mgCO_2/g Rock], H.I.: Hydrogen Index [mgHC/g Org.C], O.I.: Oxygen Index [mgCO_2/gOrg.C], T: Temperature at the peak of S2[°C], P.I.: Production Index [S1/(S1 + S2)], S2/S3: A gross indicator of kerogen quality.

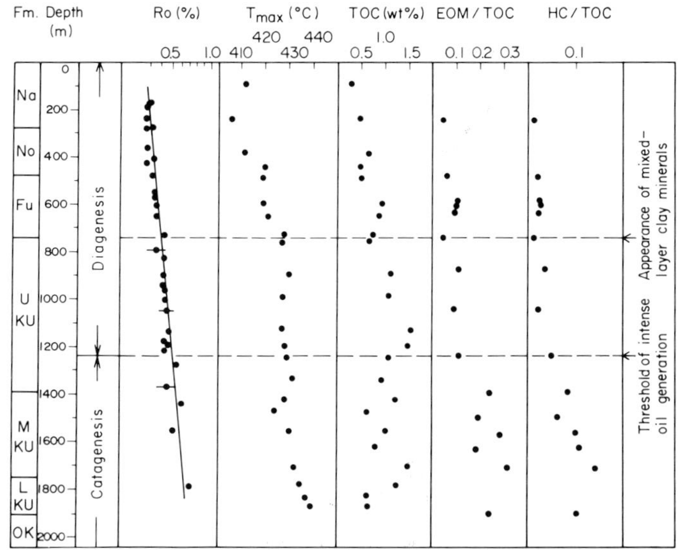

FIG. 4.—Variation with depth of vitrinite reflectance, temperature of maximum pyrolytic yield, total organic carbon, and ratios of extractable organic matter/total organic carbon and hydrocarbons/total organic carbon.

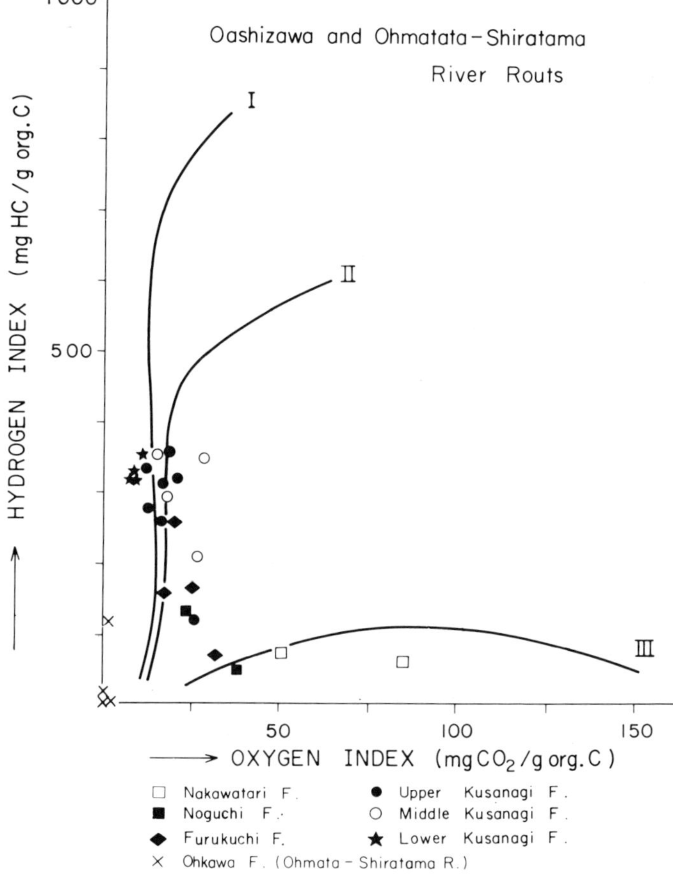

FIG. 5.—Rock-Eval pyrolysis data—Hydrogen Index (H.I.) versus Oxygen Index (O.I.) for kerogens. Samples from the Okawa Formation, the lowermost formation of the section, are plotted for comparison. The samples were collected from outside of the study area, because the Okawa Formation in the study area consists mostly of andesitic tuff breccia (Fig. 2).

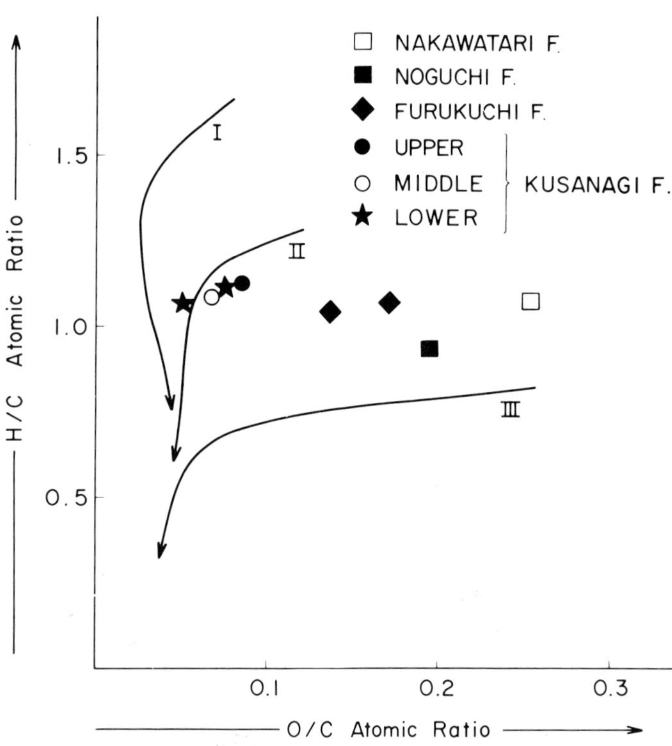

FIG. 6.—Elemental kerogen analysis for selected samples compared with Rock-Eval analysis.

different from Rock-Eval data (Fig. 5). Four samples from the Kusanagi Formation contain type II kerogen. Two samples from the Furukuchi and one sample from each of the Noguchi and Nakawatari Formations contain mixtures of types II and III. The difference between whole-rock pyrolysis and isolated kerogen analysis may be primarily due to the catalytic effect of the mineral matrix (Espitalié et al., 1977; Kruge, 1983), the effect of heavy bitumen amount, and the decomposition of carbonate matrix at relatively low temperature (Taguchi et al., 1983).

Bitumen and hydrocarbon ratios, and vitrinite reflec tance—Determination of the threshold of intense oil generation.—Plots of bitumen and hydrocarbon ratios with burial depth in the study area are shown in Figure 4. Marked increases of bitumen and hydrocarbon ratios occur in the region of sample no. 83001, which is equivalent to the upper part of the Kusanagi Formation. The corresponding vitrinite reflectance value is about 0.50% (R_0). Aiba (1980) and Ikehara et al. (1980) report that Japanese Tertiary source rocks containing type I or II kerogen begin to generate oil at a R_0 value of about 0.50%. T_{max} values (temperature of maximum hydrocarbon generation at S_2; Espitalié et al., 1977) show an increasing trend from about 411°C to 437°C with an increase of burial depth (Table 1 and Fig. 4). For Japanese source rocks, a T_{max} of about 427°C is considered to be equivalent to the threshold of intense oil generation (Connan, 1974; Kaneko et al., 1983). Studied samples no. 110601 and no. 83103, which are equivalent to the upper part of the Kusanagi Formation, show T_{max} values of about 427°C.

Based on these data, it is inferred that the threshold of

TABLE 2.—CARBON PREFERENCE INDICES OF *n*-ALKANES AND *n*-FATTY ACIDS

Sample Number	Total Organic Carbon(%)	EOM. (g/gC)	SEOM. (g/gC)*	H.C. (g/gC)	HC. from SEOM. (g/gC)*	CPI of *n*-Alkane** (C_{24}–C_{34})	CPI of *n*-Alkane*** (C_{24}–C_{35})	CPI of FFA (C_{13}–C_{21})	CPI of EFA (C_{13}–C_{21})	CPI of SEFA (C_{13}–C_{21})	CPI of KFA (C_{13}–C_{21})
82810	0.44	0.105	0.119	0.025	tr	2.66	1.45	4.38	7.08	5.25	7.41
82809	0.54	0.098	0.073	0.026	tr	n.d.	n.d.	n.d.	n.d.	n.d.	n.d.
72617	0.67	0.096	0.090	0.024	tr	2.18	1.50	4.59	6.39	6.26	5.63
72615	0.73	0.047	0.088	0.011	tr	1.83	1.12	high	high	8.94	6.13
72612	1.15	0.108	0.021	0.037	tr	1.59	1.04	high	high	high	4.09
72611	0.45	0.093	0.039	0.020	tr	n.d.	1.23	high	high	4.37	n.d.
72605	1.08	0.158	0.037	0.047	tr	1.14	1.11	high	high	7.40	4.32
72603	1.45	0.241	0.031	0.081	tr	1.23	1.08	high	high	9.07	2.38
72513	1.01	0.198	0.040	0.060	tr	1.33	1.18	high	high	high	1.49
72511	1.03	0.280	0.025	0.098	tr	1.38	1.03	high	high	high	n.d.
72509	1.14	0.185	0.026	0.103	tr	1.08	1.07	high	high	6.86	1.64
72507	1.45	0.262	0.032	0.141	tr	1.16	1.12	high	high	4.83	1.25
72503	0.58	0.241	0.015	0.100	tr	1.16	1.09	high	6.75	8.86	1.92

*: g per g kerogen-form organic carbon; **: *n*-alkanes from primary extractable organic matter (EOM); ***: *n*-alkanes from secondary extractable organic matter (SEOM); tr: trace (lower than 0.002 g/gC; n.d.: not determined; high: higher than 10.0 (After Suzuki and Taguchi, 1984)

intense oil generation occurs in the region of samples no. 83001 and no. 72605 as shown in Figure 4.

Carbon preference indices of n-alkanes, and extractable and bound form fatty acids (Table 2 and Fig. 7).—Analytical results of samples studied are summarized in Table 2. The values of carbon preference indices of *n*-alkanes (CPI_p) and *n*-fatty acids (CPI_{FA}) were obtained by the following formulas, respectively.

$$CPI_p = \frac{1}{2}\left(\frac{C_{25} - C_{33}}{C_{24} - C_{32}} + \frac{C_{25} - C_{33}}{C_{26} - C_{34}}\right)$$

$$CPI_{FA} = \frac{1}{2}\left(\frac{C_{14} - C_{20}}{C_{13} - C_{19}} + \frac{C_{14} - C_{20}}{C_{15} - C_{21}}\right)$$

It is a well known fact that *n*-alkanes in the heavy range, obtained from EOM, and *n*-fatty acids, obtained by hydro-

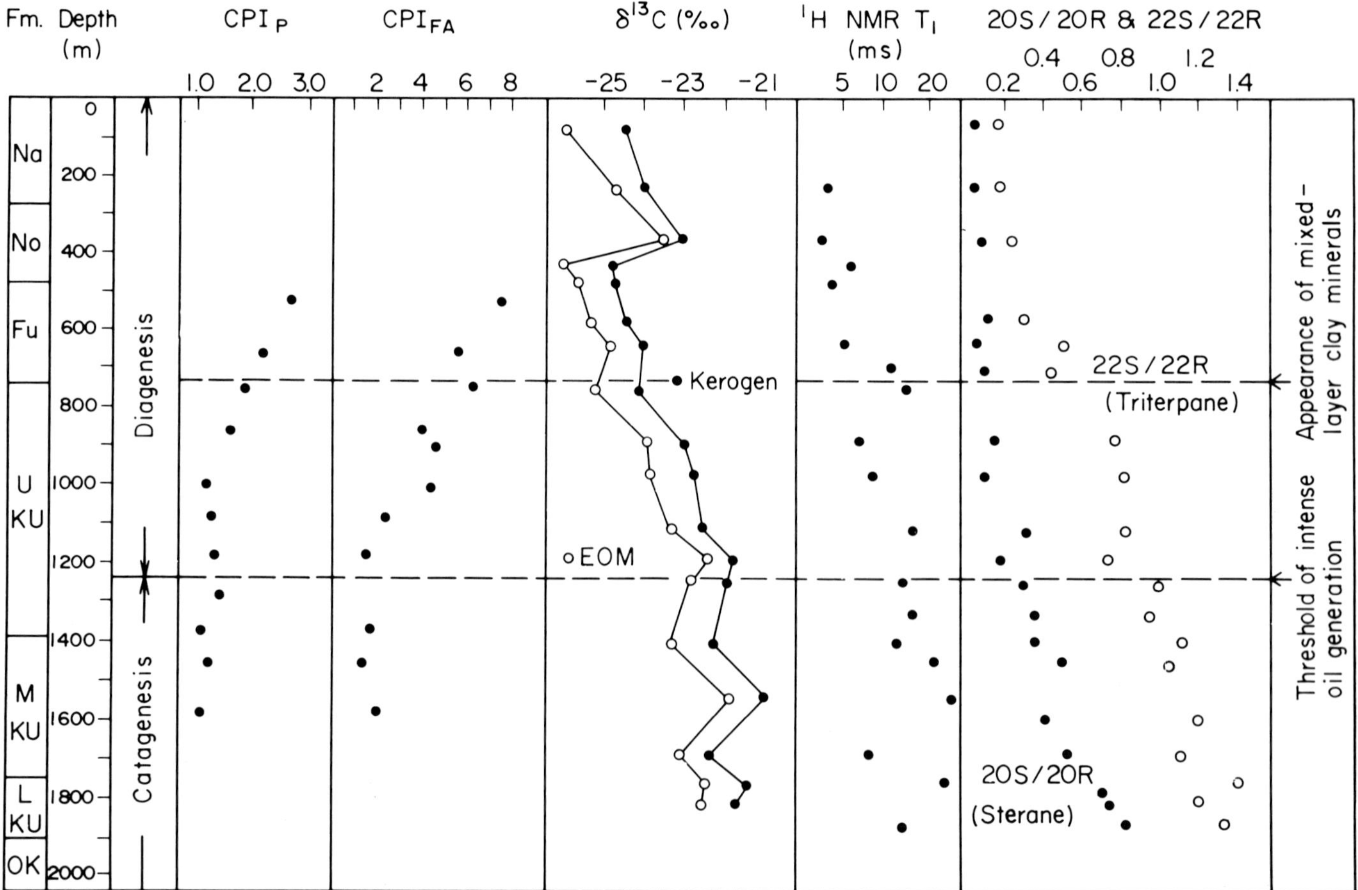

FIG. 7.—Variation with depth of Carbon Preference Indices of *n*-alkanes and *n*-fatty acids, carbon isotope ratio, 1H spin-lattice relaxation time (T_1), and ratios of 20*S*/20*R* (sterane) and 22*S*/22*R* (triterpane).

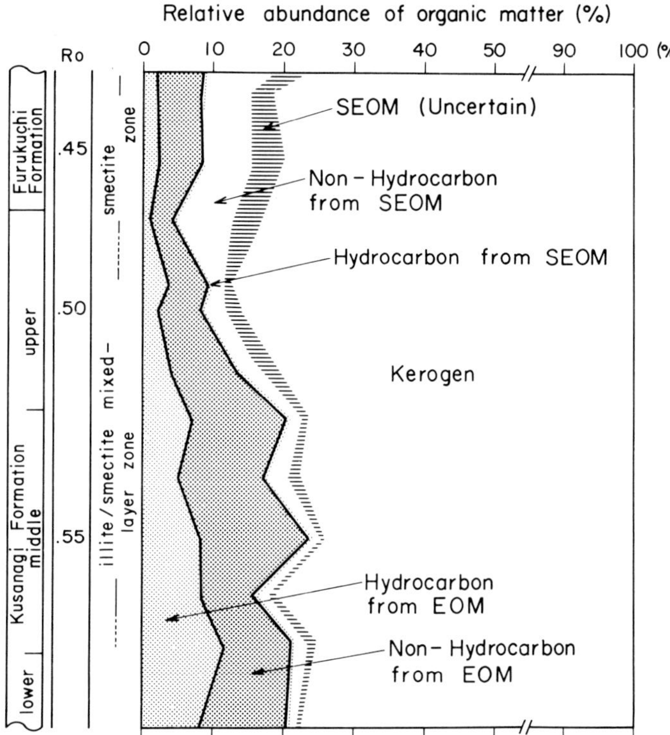

FIG. 8.—Variation in relative abundances of various organic constituents in the studied samples, all as a function of depth. Contents of total organic matter were obtained by using an organic conversion factor of 1.20. The striped pattern area indicates uncertain recovery amounts of secondary extractable organic matter (SEOM), because the recovery was not perfect. The right and left side lines of the striped pattern area represent the presumable case of 75% and 100% in the recovery, respectively. (After Suzuki and Taguchi, 1984)

lysis of the whole-rock sample, provide high values of CPI_p (odd-even ratios) and CPI_{FA} (even-odd ratios) for recent sediments, and low values of CPI_P and CPI_{FA} for ancient sediments, respectively. This study was undertaken to examine the relative molecular distribution pattern of *n*-paraffins and, in particular, *n*-fatty acids existing in various incorporation states in the sediments and their diagenetic change with depth compared with other diagenetic parameters. The *n*-alkanes from EOM show higher values of CPI_P at shallow burial depth and successively lower CPI_P values with increasing burial depth, thereby agreeing with previous studies of the variation of CPI_P values with burial depth. On the other hand, CPI_P values from SEOM show lower values than those from EOM in all the samples and no systematic change with depth (Table 2).

The relative concentrations of FFA and EFA indicate similar distribution patterns throughout the samples, with a relative decrease of the higher molecular weight acids with increasing depth; however, no systematic change of CPI_{FA} values from FFA and EFA with burial depth is recognizable. Values of CPI_{FA} are high compared to those of recent sediments, even at greater depths such as the so-called "principal zone of oil formation" (Vassoevich et al., 1970). The CPI_{FA} values of SEFA are lower than those of FFA and EFA. There is no systematic change with burial depth in CPI_{FA} values of SEFA, as is also noted for FFA and EFA. Kerogen-form fatty acids (KFA) consist predominantly of the low molecular weight acids ($<C_{21}$) in all samples; the CPI_{FA} values of KFA are lower than those of FFA, EFA and SEFA. The CPI_{FA} values of KFA show a systematic change with burial depth, however (Table 2 and Fig. 7).

High CPI_{FA} values of FFA and EFA from samples near the "principal zone of oil formation" (the lower Upper Kusanagi Formation) seem to indicate that these fatty acids have not been affected significantly by usual burial diagenesis. At the present time, the authors consider FFA and EFA fractions to be controlled strongly by sedimentary environments rather than by diagenetic change, as reported previously (Suzuki and Taguchi, 1984).

SEOM has poor hydrocarbon content and consists mainly of asphaltene or NOS compounds with higher polarity, as shown in Figure 8. In addition, both *n*-alkanes and *n*-fatty acids extracted from SEOM contain a greater proportion of higher molecular weight components than those from EOM. These facts indicate that the higher molecular weight fatty acids are closely associated with asphaltene or NOS compounds, and consequently, those acids are incorporated rather tightly in the sedimentary rocks.

CPI_{FA} of KFA and CPI_P of EOM obtained from the same samples show nearly parallel changes with increasing burial (Fig. 7). Of course, this change does not mean a comparison of *n*-alkane and *n*-fatty acid molecules with the same carbon number as seen in Table 2. As *n*-fatty acids of recent sediments and living organisms invariably have high even-odd ratios, the successive change of CPI_{FA} found in KFA can be regarded as due to diagenetic processes.

Carbon isotope ratios for bitumen and kerogen.—Stable carbon isotope ratios for the bitumen and kerogen from collected samples are shown in Table 3 and Figure 7. $\delta^{13}C$ values for bitumen and kerogen show the general trend from light to heavy carbon isotope values (from −26 to −22 ppt and −25 to −21 ppt, respectively) with increasing burial.

Bitumen $\delta^{13}C$ values are generally 0.5 to 1.0 ppt lighter

TABLE 3.—$\delta^{13}C$ OF EXTRACTABLE ORGANIC MATTER (EOM) AND KEROGEN (per mil)

Sample No.	Extractable O.C. first	Extractable O.C. duplicate	Kerogen first	Kerogen duplicate
STD*	−24.5	−24.5	−24.5	−24.4
83110	−25.9		−24.6	−24.5
83109	−24.7		−24.1	
83108	−23.4		−23.1	−23.1
83107	−26.0	−26.1	−24.2	−25.4
83106	−25.6		−24.7	
83105	−25.4		−24.5	
83104	−24.8		−24.0	
83103	−25.2		−24.2	
83102	−23.9		−23.0	
83101	−23.8		−22.7	
83003	−23.3		−22.6	
83002	−22.4		−21.8	−21.6
83001	−22.8	−22.7	−22.1	−21.9
90001	−23.3		−22.3	
90102	−21.7		−20.9	
90103	−23.1	−22.9	−22.5	−22.2
90104	−22.3		−21.2	
90105	−22.4		−21.5	

*Graphite as working standard. All values are the deviation in ppt of the $^{13}C/^{12}C$ ratio of the sample from Peedee Belemnite Standard (PDB).

than the kerogen from which the bitumen was derived. No significant change in the difference between $\delta^{13}C$ values of bitumen and kerogen was noted through the vertical interval studied, even through the zone of the hydrocarbon liquid window.

Previous studies (McKirdy and Powell, 1974; Hoefs and Frey, 1976; Peters et al., 1978; Redding et al., 1979) suggest that $\delta^{13}C$ values of kerogens from field samples which have undergone regional metamorphism show practically no significant fractionation during early catagenesis (corresponding to the threshold of intense oil generation), and only begin to increase at higher levels of catagenesis. Theoretical calculations (Redding et al., 1979) suggest significant fractionation in $\delta^{13}C$. Omokawa (1982) pointed out that variations of $\delta^{13}C$ for kerogens with a maturation level lower than $R_0 = 0.8\%$ are mostly due to kerogen type regardless of the maturation process, based on Japanese Neogene Tertiary source rock studies. Considering these results, carbon isotopic trends for kerogen and bitumen observed in this study are attributed to sources of organic materials (kerogen type). Geologic data (Taguchi, 1962) indicate that the depositional environment in the study area has changed from deep marine to shallow marine with frequent small reversals from the lower to the upper strata.

Values of $\delta^{13}C$ ($-21.7 \sim -23.8$ ppt) of EOM present in samples from depths greater than sample no. 83003 show similar values to those of crude oils (-22 to -24 ppt) from the Neogene Tertiary of Japan (Nakai et al., 1974). This indicates that rocks belonging to the Kusanagi formation are possible source rocks for these crude oils and that the Kusanagi Formation has been at the level of the intense oil generation zone during burial. A possible explanation for the absence of significant carbon isotope fractionation in kerogens from the region of principal oil generation may be found in the work of Redding et al. (1979) and Peters et al. (1981). They suggest that a two-step process would give the low carbon isotope fractionation. That is, most hydrocarbons formed during catagenesis are released as a liquid phase from which methane is formed. On the other hand, Sacket (1978) suggested that the first-formed methane in pyrolysis experiments carried out in the presence of sodium montmorillonite shows little, if any, carbon isotope fractionation. At present, we are unable to resolve this problem. Despite the important role of clay minerals in other organic maturation reactions (discussed below), no significant carbon isotope fractionation observed in this study may be attributed to the presence of clay minerals.

^{1}H-NMR T_1 of the dioxane-insoluble organic matter.—^{1}H-NMR T_1 was compared with established maturation parameters, for instance, vitrinite reflectance.

The ^{1}H spin-lattice relaxation time (T_1) versus burial depth for the dioxane-insoluble organc matter is shown in Table 4 and Figure 7. As is evident from Figure 7, the log T_1 values increase with increasing burial depth (from the upper toward the lower strata). Therefore, the ^{1}H spin-lattice relaxation time T_1 can be regarded as being a possible parameter for organic maturation. Comparing vitrinite reflectance values R_0(%) with the ^{1}H spin-lattice relaxation time (T_1), an R_0 value of 0.5% (which indicates the entrance to the hydrocarbon liquid window for Japanese Neogene Tertiary source rocks) corresponds to a T_1 value of 8 ms in the MITI-Hamayuchi borehole core (Hirata and Akiyama, 1982) and to a value of 10 ms ± in the Neogene Tertiary of the study area, respectively.

TABLE 4.—^{1}H-NMR T_1 OF THE DIOXANE-INSOLUBLE ORGANIC MATTER

Sample no.	Residue(%)	^{1}H-NMR T_1
83109	1.3	3.6
83108	1.4	3.1
83107	1.3	5.2
83106	0.6	3.8
83104	1.0	5.2
110601	1.6	12.3
83103	—	15.9
83102	0.9	6.6
83101	2.6	9.3
83003	4.2	14.4
83001	2.0	14.1
110607	2.6	16.4
90101	—	15.1
110603	1.6	23.3
90102	0.7	28.4
90103	—	9.3
90104	0.5	26.2
90105	—	20.0
110506	0.3	18.7

Epimerization of steranes and triterpanes, and aromatization of steranes.—Analytical results of the steranes and the triterpanes for 19 samples from the study area are shown in Table 5. Both the amount of total steranes ($C_{27} \sim C_{33}$ steranes) and total triterpanes ($C_{27} \sim C_{33}$ triterpanes) show relatively lower values at the upper strata and a general increase toward the lower strata with a marked increase from sample no. 83101 near the threshold of intense oil generation. Their amounts decrease rapidly in the lowest strata, however.

The relative abundance of steranes and triterpanes near the principal zone of oil formation as compared to the upper strata is due to organic maturation. Douglas et al. (1977) reported that the increase in the amount of steranes and triterpanes was caused by heating recent sediments at 250 to ~275°C.

For simplicity and because of the poor resolution of C_{27} and C_{28} steranes, and C_{31} and C_{33} hopanes only the 5α(H), 14α(H), 17α(H) C_{29} steranes and the 17α(H), 21β(H) C_{32} hopanes are discussed in this study. Figures 9 and 10 show mass fragmentograms of m/z 217 and m/z 191 for the steranes and triterpanes from four selected samples, respectively. Each figure shows a relative increase of 20*S* isomer to 20*R* isomer of the C_{29} sterane and 22*S* isomer to 22*R* isomer of the C_{31} and C_{32} hopanes with increased burial depth (from the upper to lower strata).

Plots of the ratio of 20*S* to 20*R* isomer of 5α(H), 14α(H), 17α(H) C_{29} steranes and of 22*S* to 22*R* isomer of 17α(H), 21β(H) C_{32} hopanes against burial depth are shown in Figure 7. Both ratios of 20*S*/20*R* of C_{29} steranes and of 22*S*/22*R* of C_{32} hopanes increase steadily with increasing burial depth. The configurational change at C-20 in the 5α(H), 14α(H), 17α(H) C_{29} steranes showed approximately equal amounts of the *R* and *S* configuration in the deepest samples, which corresponds to the middle zone of the principal phase of oil formation. On the other hand, the epimeriza-

TABLE 5.—ANALYTICAL RESULTS OF STERNANES AND TRITERPANES

Sample No.	Steranes			Triterpanes		
	(μg/g dry sed.)*1	(μg/g O.C.)*2	(20S/20R)*3	(μg/g dry sed.)*4	(μg/g O.C.)*5	(22S/22R)*6
83110	0.178	61.4	0.04	0.468	161	0.17
83109	0.091	18.6	0.04	0.267	54.5	0.15
83108	0.551	86.1	0.09	0.330	51.6	0.22
80105	0.790	84.0	0.11	1.73	184	0.30
83104	0.614	69.8	0.06	1.87	212	0.48
110601	0.690	93.2	0.10	1.86	251	0.42
83102	0.803	71.1	0.13	1.76	157	0.70
83101	1.66	156	0.10	3.38	319	0.80
83003	2.62	172	0.26	4.93	324	0.80
83002	4.19	285	0.19	5.45	370	0.71
83001	1.71	160	0.29	5.22	488	0.96
110607	1.84	200	0.29	4.88	530	0.94
90101	5.48	445	0.35	4.54	369	1.10
110603	1.33	218	0.46	3.94	646	1.06
110505	1.52	192	0.43	4.38	555	1.20
90103	8.45	567	0.56	11.2	755	1.13
90104	7.32	610	0.76	13.9	1160	1.40
90105	1.49	240	0.78	6.55	1060	1.20
110506	1.26	201	0.85	4.01	636	1.38

*1: Total steranes (C_{27}, C_{28} and C_{29} steranes) weight per dry sediment weight, *2: Total sterane weight per total organic carbon weight, *3: Ratio of 20*S*/20*R*-24-ethyl-5α(H), 17α(H)-cholestane, *4: Total triterpanes (C_{27} ~ C_{33} triterpanes) per dry sediment, *5: Total triterpanes per total organic carbon, *6: Ratio of 22*S*/22*R*-17α(H), 21β(H)-bis-homohopane (C_{32}).

tion of the C-22 position of the 17α(H), 21β(H) C_{32} hopanes occurs more rapidly with increasing depth compared to C-20 in C_{29} steranes, as suggested by Mackenzie et al. (1980). Equal amounts of the *R* and *S* configuration of C_{32} hopanes occur near the threshold of intense oil generation.

Figure 11 shows gas chromatograms of the unsaturated and aromatic hydrocarbon fractions. Peaks 17, 19, 21, 22 and 24 have an intense m/z 253 ion on their mass spectra, suggesting they are ring *C*-aromatized steranes (Seifert and Moldowan, 1978; Mackenzie et al., 1981). Peaks 25 through 29 have an intense ion of m/z 231 in their mass spectra, suggesting they are ring *A*, *B*, *C* tri-aromatized steranes. Peaks 14, 15, 16, 18 and 20 have an intense m/z 257 ion on their mass spectra and were tentatively identified as C_{27}, C_{28} and C_{29} steranes (Table 6).

Results of the identification are shown in Table 6. Mackenzie et al. (1981) proposed that tri-aromatized steranes are generated from mono-aromatized steranes by aromatization accompanied by demethylation during diagenesis. As is evident from Figure 11, mono-aromatized steranes (peaks

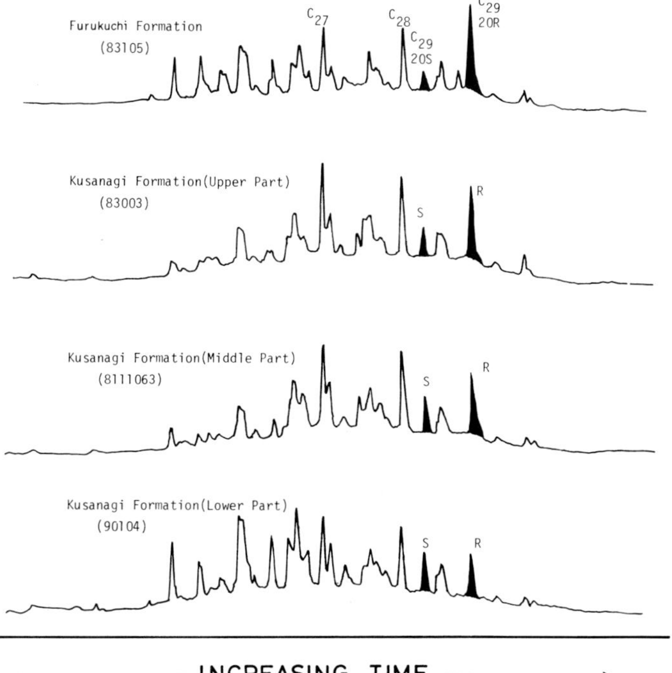

FIG. 9.—Variation with depth of steranes, represented by m/z 217 mass fragmentograms of four selected samples.

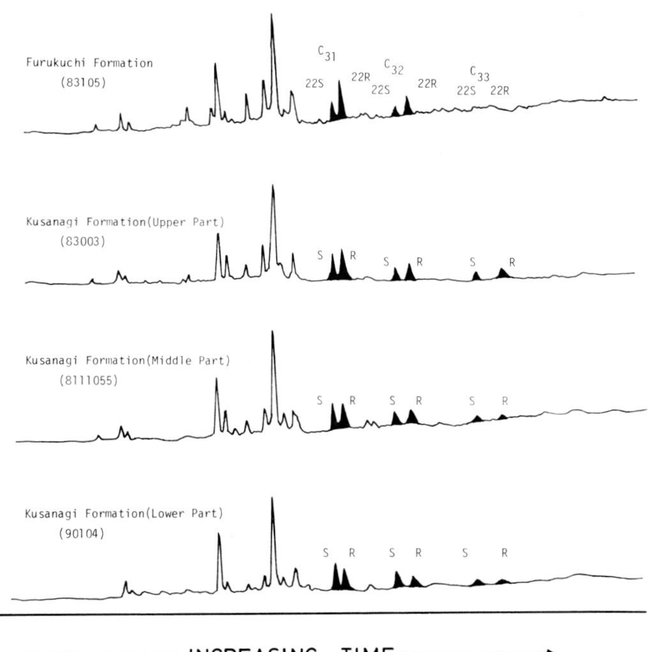

FIG. 10.—Variation with depth of triterpane, represented by m/z 191 mass fragmentograms of four selected samples.

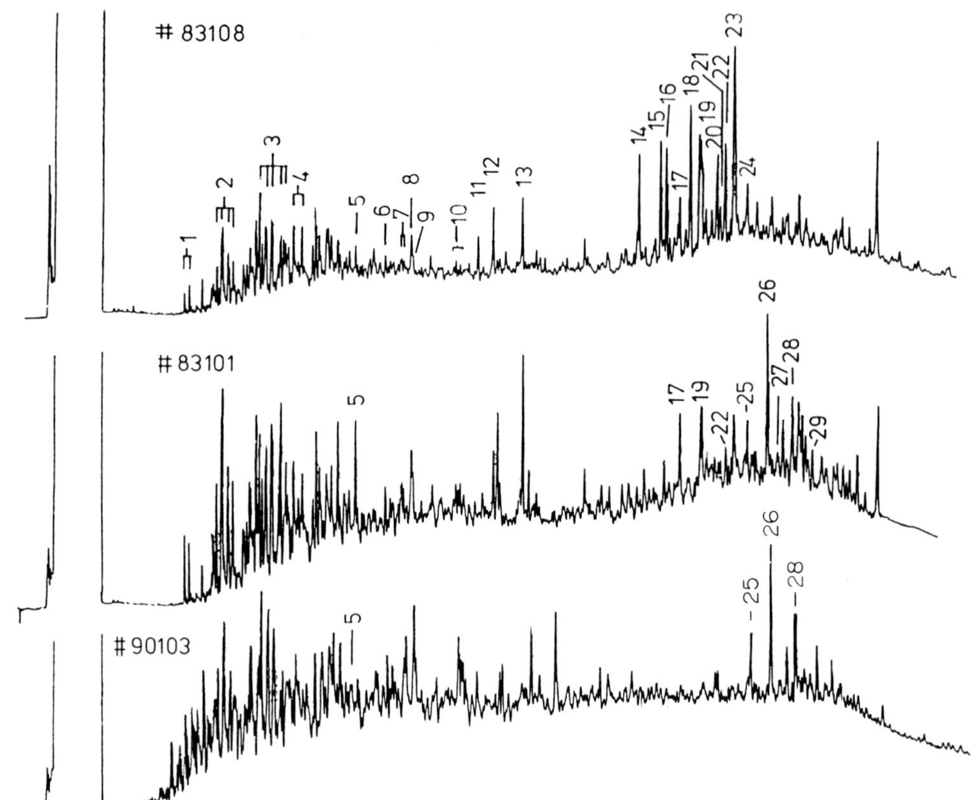

FIG. 11.—Gas chromatograms of sterenes and aromatic hydrocarbons. (See Table 6 for identification of peaks.)

17, 19, 21, 22 and 24) decrease from sample no. 83108 from the upper strata toward sample no. 90103 from the principal zone of oil generation, while the tri-aromatized steranes (peaks 25 through 29) are present in sample nos. 83101 and 90103, and absent in sample no. 83108. These results also suggest that aromatization of steranes occurs in the principal zone of oil generation. Detailed studies along these lines are being carried out by Shioya, a member of this research group.

Inorganic Diagenesis

Organic and ammonium nitrogen.— Figure 12 shows the distribution patterns of organic carbon and nitrogen, total nitrogen content, and organic C/total N ratio with increasing burial depth. Organic C and total N contents show a parallel change with depth, while the depth distribution pattern of organic N is different. This result suggests that part of the organic N in sediments has changed to inorganic N during diagenesis. On the other hand, the organic C/total N ratio shows a nearly constant value (10) except for the samples from the Okawa Formation, the lowermost formation in the study area (C/N ≒ 20).

The Okawa Formation is regarded as a shallow marine deposit, based on the work of Taguchi (1962), and contains terrestrial carbonaceous matter. The high ratio in the Okawa Formation may be attributed to the composition of the higher plant material. The C/N ratio of nearly 10, obtained from samples other than the Okawa, is comparable to values ob-

TABLE 6.—UNSATURATED AND AROMATIC HYDROCARBONS DETERMINED BY GAS CHROMATOGRAPHY AND MASS SPECTROMETRY FOR SELECTED SAMPLES

Peak No.	Compound
1	methylnaphthalenes
2	dimethylnaphthalenes
3	trimethylnaphthalenes
4	tetramethylnaphthalenes
5	phenanthrene
6	methylbenzothiophene
7	3- and 2-methylphenanthrenes
8	9-methylphenanthrene
9	1-methylphenanthrene
10	dimethylphenanthrenes
11	unknown
12	unknown
13	unknown
14	C_{27} sterene
15	C_{27} sterene
16	C_{28} sterene
17	monoaromatized sterane
18	C_{28} sterene
19	monoaromatized sterane
20	C_{29} sterene
21	monoaromatized sterane
22	monoaromatized sterane
23	phthalate ester (contamination)
24	monoaromatized sterane
25	triaromatized sterane
26	triaromatized sterane
27	triaromatized sterane
28	triaromatized sterane
29	triaromatized sterane

Peak numbers refer to gas chromatograms shown in Figure 11.

tained from Miocene and Oligocene sediments by Trask (1932). This implies that the C/N ratio formed in the sediments during early diagenesis remains unchanged, because inorganic N derived from nitrogenous organic compounds has been preserved in the sediments by adsorption on clay minerals during subsequent diagenesis.

Inorganic N in the samples was separated on the basis of forms present into three parts: water soluble NH_4-N, exchangeable NH_4-N with 1N KC1, and fixed NH_4-N. Figure 12 shows the percentage of total N in the samples. The content of soluble NH_4-N is almost negligible in all samples, ranging from < 1% to 2.2%. Exchangeable NH_4-N constitutes less than 10% except in the two samples from the Nakawatari Formation, which contain abundant clinoptilolite and mordenite. These zeolite minerals have extremely high cation exchange capacities. All the samples (except the two samples from the Nakawatari Formation) contain mainly organic N and fixed NH_4-N. The ratio of fixed NH_4-N/total N shows relatively higher values in the middle strata studied, including the principal zone of oil formation. On the other hand, organic N/total N ratio shows relatively higher values in both the upper and lower strata studied. If ammoniacal nitrogen has been derived from organic nitrogen through thermal degradation, the ratio of fixed NH_4-N/total N should show successively higher values in the lower strata. The ratios of fixed NH_4-N/total N and organic N/total N, however, show general decreases and increases, respectively, from the lower Upper Kusanagi toward the Okawa Formation. This observation could not be attributed to a thermal effect with increasing burial depth. It is suggested that the transformation of organic N to fixed NH_4-N has been affected considerably by clay catalytic reactions, as will be discussed below.

Clay minerals.—Table 7 shows the vertical distribution of minerals found in the samples. Figure 13 shows the results schematically: the absolute percentage values are not considered to be significant, but the relative differences are reasonably reliable.

Smectite clays decrease gradually with increasing burial depth and disappear near the upper part of the Kusanagi Formation. Chlorite/smectite mixed-layer clays are found in only two samples from the lower part of the Kusanagi Formation; no other minerals are associated with them. A small amount of chlorite is found only in the samples from the middle part of the Kusanagi Formation. Mica and kaolinite minerals were detected in a small quantity in most samples, but their abundance varies in every formation.

Chemical analyses of smectite (no. 83107), illite/smectite (no. 90103) and chlorite/smectite (no. 90105), and X-ray diffraction patterns of nos. 83107 and 90103 are shown in Table 8, and Figures 14 and 15, respectively. The smectite seems to belong to the montmorillonite-beidellite-like minerals, based on the analytical results. As is evident from Table 8 and Figure 14, the illite/smectite mineral (no. 90103)

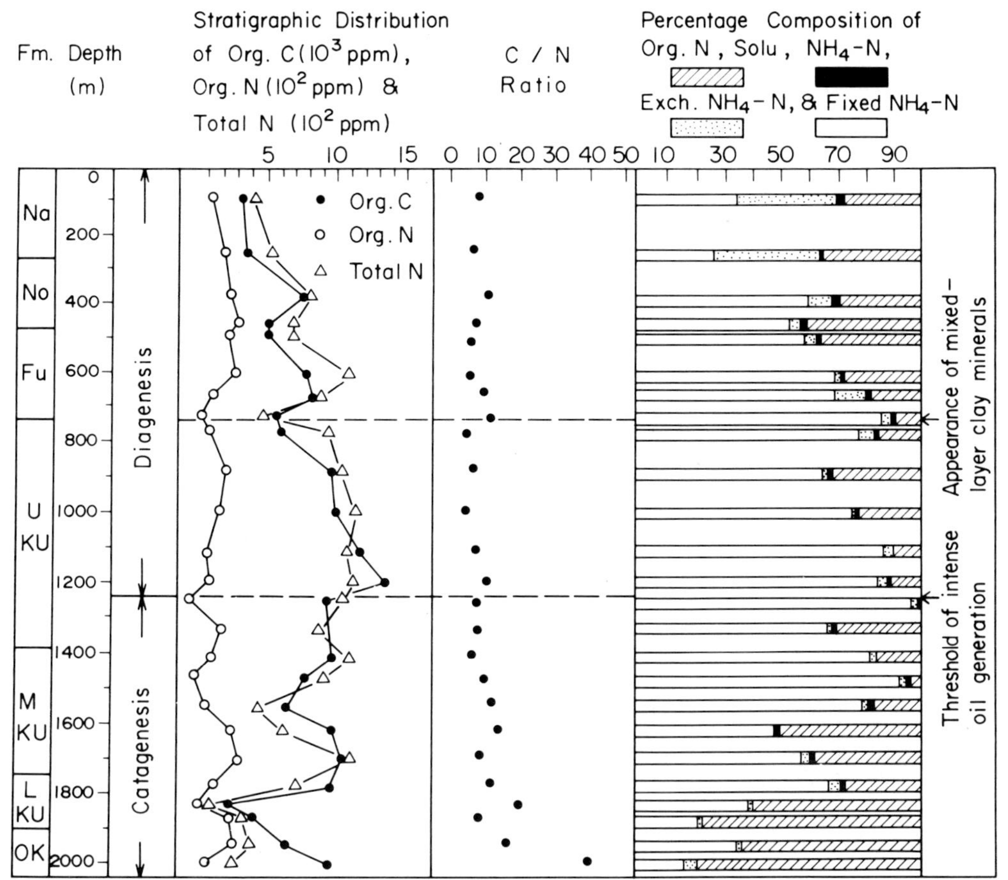

FIG. 12.—Variation with depth of organic carbon and nitrogen, ratio of organic carbon to total nitrogen, and percentage composition of nitrogen in various forms. In the variation of nitrogen composition, samples from the Okawa Formation were analyzed to confirm the decreasing trend with depth of fixed nitrogen.

TABLE 7.—MINERALS FOUND IN THE STUDIED SAMPLES

Sample	Fm.		Smec.	I/S	C/S	C.	Mica	Kaol.	Cp.+Mo.	An.	Qt.	Cr.	Pl.	Kf.	Py.	Cal.
83110	Nakawatari F.		++				−	−?	+		−	+	+			
83109				+			−	−	+		−	+	+			
83108	Noguchi F		++				+	+			+		+	−	−	
83107			++			?	−	+			+		+	−	·	
83106	Furukuchi F.		++				+	+			+		·	−	·	
83105			++			?	+	+		+	+		·	−	·	?
83104			+				−	−		+	++		·	−	·	?
11061				+			−	−		·	++		·	−	·	
83103	Kusanagi F.	U.	+			−?	+	+		·	+		−	−	·	
83102			?	+		?	+	+		·	+		−	−	·	
83101			?	+			+	+			+		−	−	·	
83003				+?			−	−?		+	++		−	−	·	
83002				+			?	−?		+	++		·	−	·	
83001				−?			?	−		+	++		−	−	·	
11067				?		·?	?	·?		−	++		−	−	−	−
90101		M.		+		−	+	−			++		−	−	−	
11063				+		?	−	−			++		−	−	−	−
90102				+		−	+	?			++		−	−	−	
11055				−		−	−	−			++		−	−	−	
90103				−		−	−	−			++		·	−	−	
90104		L.		−		−	−	?			++		·	−	−	
90105					++		·				++		·	−		
11056					++		·?				++		·	−		

F: Formation, Smec: Smectite, I/S: Mixed-layer illite/smectite, C/S: Mixed-layer chlorite/smectite, C: Chlorite, Kaol: Kaolinite, Cp + Mo: Clinoptiolite + mordenite, An: Analcime, Qt: Quartz, Cr: Cristobalite, Pl: Plagioclase, Kf: K-feldspar, Py: Pyrite, Cal: Calcite. Symbols, ++, +, −, ·, and ? indicate major, moderate, minor and trace amounts, and doubtful identification of minerals found in X-ray patterns.

is relatively enriched in MgO, K_2O, and Na_2O and lies in the area of abundant K_2O on the triangular diagram compared with the smectite, though no significant difference is observed in SiO_2, Al_2O_3, and Fe_2O_3. The higher content of Na_2O in the illite/smectite is remarkable. If we consider this result and the decrease of plagioclase content in sample 90103, it is estimated that Na_2O component is mostly contained in the illite/smectite mineral. This illite/smectite may be regarded as rectorite.

Figure 16 shows a plot of smectite and illite/smectite minerals for identification of illite/smectite interstratification. This type of graph was originally proposed by Watanabe (1981). The values of Reicheweite (*S*) (Jagodzinski, 1949; Kakinoki and Komura, 1952) and the ratio of illite layers to smectite layers can be read from the graph, if only the two values ($\Delta 2\theta_1$ and $\Delta 2\theta_2$) of the angular differences between three *001* reflections of interstratified illite/smectite treated with ethylene glycol are plotted on the graph

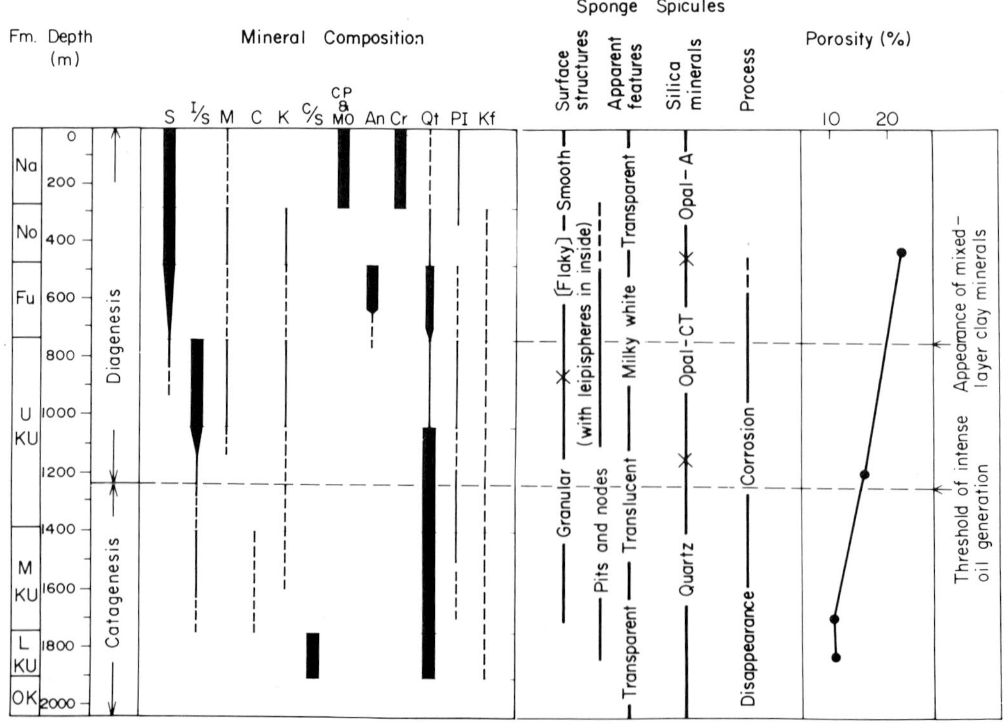

FIG. 13.—Variation with depth of mineral composition, sponge spicule features, and porosity. (See Table 7 for abbreviated words.)

TABLE 8.—CHEMICAL ANALYSIS OF SELECTED SAMPLES

	83107		90103		90105	
SiO_2	57.14	46.75	75.36	44.01	62.90	39.01
TiO_2	0.55	0.68	0.43	0.98	0.32	0.53
Al_2O_3	17.31	21.50	10.12	23.00	8.94	14.71
Fe_2O_3	4.11	5.11	2.48	5.64	6.53	10.75
FeO						
MnO						
MgO	1.58	1.96	1.24	2.82	9.88	16.26
CaO	0.97	1.21	0.26	0.59	0.44	0.72
K_2O	0.90	1.12	0.94	2.14	0.17	0.28
Na_2O	0.21	0.26	0.55	1.25	0.33	0.54
H_2O^+	8.61	10.69	6.74	15.32	7.00	11.52
H_2O^-	8.63	10.72	1.87	4.25	3.45	5.68
Total	100.01%	100.00%*	99.99%	100.00%*	99.96%	100.00%*

*corrected for quartz impurity.

(Watanabe, 1981). As is seen from Figure 16, the samples are divided into two groups; smectite minerals having >70% expandability with $S = 0$, and illite/smectite minerals having 50–70% illite layers with $S = 1$. The samples from the upper strata (from the Nakawatari to Furukuchi Formations) contain smectite minerals (except for one sample from the lowest part of the Furukuchi). On the other hand, all samples from the lower strata (the Kusanagi Formation) contain mixed-layer illite/smectite minerals (except for one sample from the top of the Kusanagi Formation), though the abundance ratio of illite to smectite does not vary systematically with burial depth. If we consider previous studies by Burst (1969), Perry and Hower (1972), and others, and the results

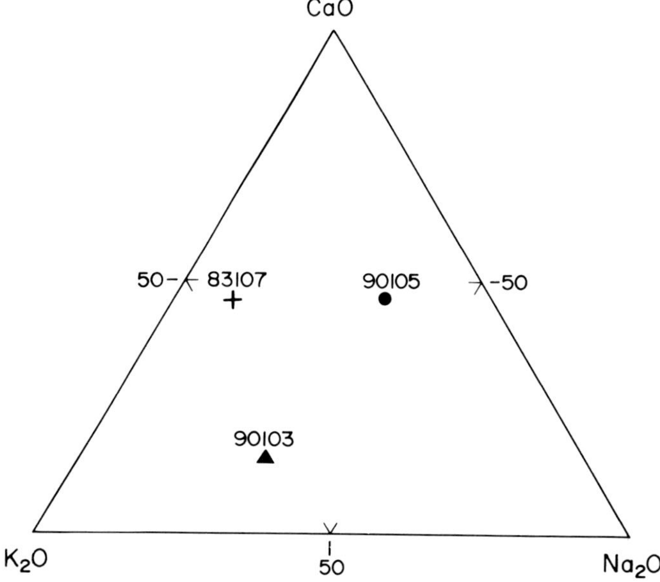

FIG. 14.—CaO − K_2O − Na_2O diagram of smectite and mixed-layer illite/smectite.
(83107): smectite, (90103): mixed-layer illite/smectite
(90105): mixed-layer chlorite/smectite.

mentioned above (Table 8 and Fig. 14), it is apparent that the illite/smectite mineral was formed from the smectite by the addition of alkali cations with increasing temperature. In general, the formation of these clay minerals begins with

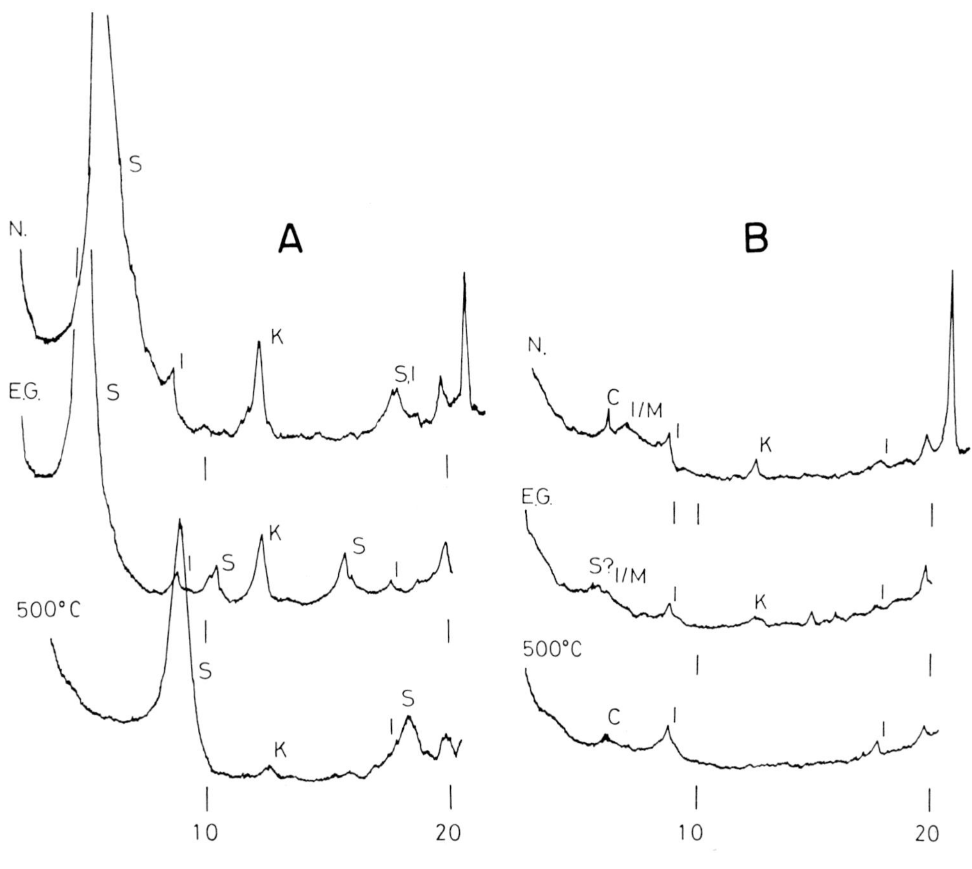

FIG. 15.—X-ray diffraction patterns of selected samples. N: natural state; E.G.: After treatment with ethylene glycol; 500°C; After treatment at 500°C for 2 hrs S; smectite; I: illite; K: kaolinite; I/M: mixed-layer illite/smectite; A: sample no. 83107; B: sample no. 90103.

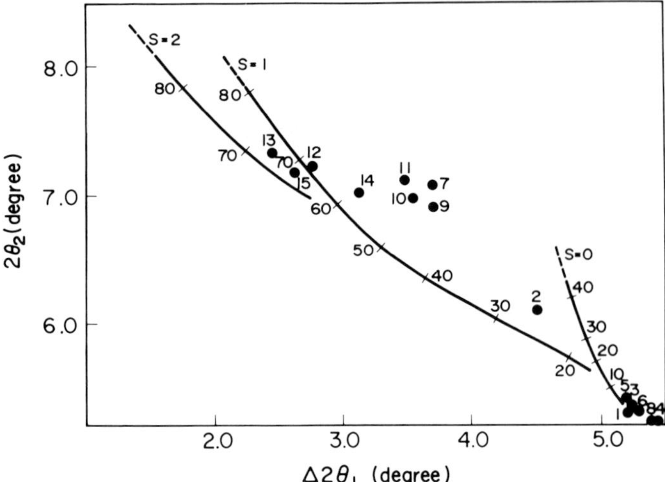

FIG. 16.—Diagram showing illite/smectite interstratifications of the studied samples, expressed by Reichweite (S) and the ratio of illite/smectite layer.
1:83110, 2:83109, 3:83108, 4:83107, 5:83106, 6:83105, 7:83104, 8:83103, 9:83102, 10:83101, 11:83002, 12:90101, 13:90102, 14:90103, 15:90104. S means Reichweite (S), and the numbers along the curves show percentage of illite layers.

smectite layers with Reicheweite $S = 0$ and moves to illite/smectite minerals with $S = 2$ through minerals with $S = 1$. In this study, the minerals with $S = 1$ occur mainly in the upper and middle parts of the Kusanagi Formation.

Chlorite/smectite minerals found in two samples from the lower part of the Kusanagi Formation consist of Fe-Mg chlorite and smectite, probably Fe-saponite based on the chemical analysis (Table 8). If mixed-layer chlorite/smectite clays were formed from smectite and/or mixed-layer illite/smectite, a large addition of Mg and Fe cations to mixed-layer chlorite/smectite must be explained. It is known generally that mixed-layer chlorite/smectite is formed during the alteration of smectite to chlorite. At present, however, the source and period of supply of Mg and Fe cations are not known in this study. As mixed-layer chlorite/smectite is markedly different in chemical composition and crystal structure from the other clay minerals found in the upper strata, the genesis of mixed-layer chlorite/smectite in the study area is not certain.

Zeolites and other minerals (Table 7 and Fig. 13).—Clinoptilolite, mordenite, and cristobalite are found in the samples from the Nakawatari Formation. Analcime occurs between the top of the Furukuchi Formation and the base of the upper part of the Kusanagi Formation, showing variations in relative abundance. Quartz and plagioclase are present in almost all samples. Samples from the lower strata are composed largely of quartz with lesser amounts of plagioclase. K-feldspar is present in all samples (except those from the Nakawatari Formation) and is considered to be authigenic.

Siliceous sponge spicules and silica phase minerals.—Samples from the Nakawatari, Noguchi, Furukuchi, and the upper part of the Kusanagi Formations contain abundant siliceous sponge spicules. Spicules are virtually absent, however, in those from the middle to the lower part of the Kusanagi Formation. Figure 13 and Plate 1 show variation of the characteristic appearances of sponge spicules with increasing burial depth, based on observations by SEM and binocular microscope. The observations reveal a characteristic change with increasing burial depth from milky white spicules having smooth surfaces to translucent or transparent ones having granular structures and finally to spicules with a flaky appearance. Spicules from the lower strata have many pits and/or nodes caused by surface corrosion. As stated above, cristobalite was detected by X-ray analysis in the samples from the Nakawatari Formation. Probably, the silica-phase mineral found on the surface of sponge spicules from the Nakawatari Formation corresponds to opal-CT. Sponge spicules from below the Noguchi Formation consist mostly of quartz.

Porosity.—Figure 13 shows the variation of porosity, which was measured by a poresizer, in selected samples with increasing burial depth. Porosity decreases systematically with burial depth, ranging from 22.8 to 10.8%. Burial depth of the samples having about 10% porosity corresponds to a burial of approximately 2,000 m, based on geological evidence (Taguchi, 1962). A porosity of about 10% is comparable to the early stage of the Late Compaction Stage of Aoyagi (1978) and from the second stage to the middle of the third stage of Burst (1969). The apparent diagenetic processes among constituent minerals and the mineralogical assemblage apparently associated with increasing burial depth are consistent with the data of Aoyagi (1978). Burial depths obtained from this geological study, however, differ significantly from burial depths estimated from porosity and mineral assemblages based on diagenetic study of Neogene argillaceous sediments of Japan (Aoyagi, 1978). A burial depth of 2,000 m is shallower than that reported by Aoyagi (1978). This may be due to the slower sedimentation rate and/or rather higher geothermal gradients in the study area in comparison with the Japan Sea side area.

RELATIONSHIPS BETWEEN ORGANIC AND INORGANIC DIAGENESIS

This investigation suggests a relationship between clay diagenesis and organic maturation at the stage of intense oil generation. As stated above, the threshold of the intense oil formation zone has been placed in the region of sample nos. 72605 and/or 83001, which correspond to the stratigraphic interval of the upper to middle parts of the Kusanagi Formation. On the other hand, the diagenetic transformation of smectite to illite/smectite mixed-layer clay was confirmed in the region of sample nos. 83104 and 83102, which correspond to the stratigraphic boundary between the Kusanagi and Furukuchi Formations. The crystal structure of smectite in the upper part of the section shows a change from highly expandable ($I/S = 0/100\%$) to less expandable ($I/S = 40/60\%$) illite/smectite with Reicheweite $S = 0$, whereas illite/smectite minerals in the Kusanagi Formation showed the transformation of illite/smectite minerals with $S = 1$, ranging about 50/50% to 70/30% I/S (Fig. 16). It is apparent that the conversion reaction of smectite to illite/smectite minerals occurs at a shallower depth than the zone of intense oil generation (compare Fig. 4 with Fig. 13). It is suggested, therefore, that water from clay dehydration

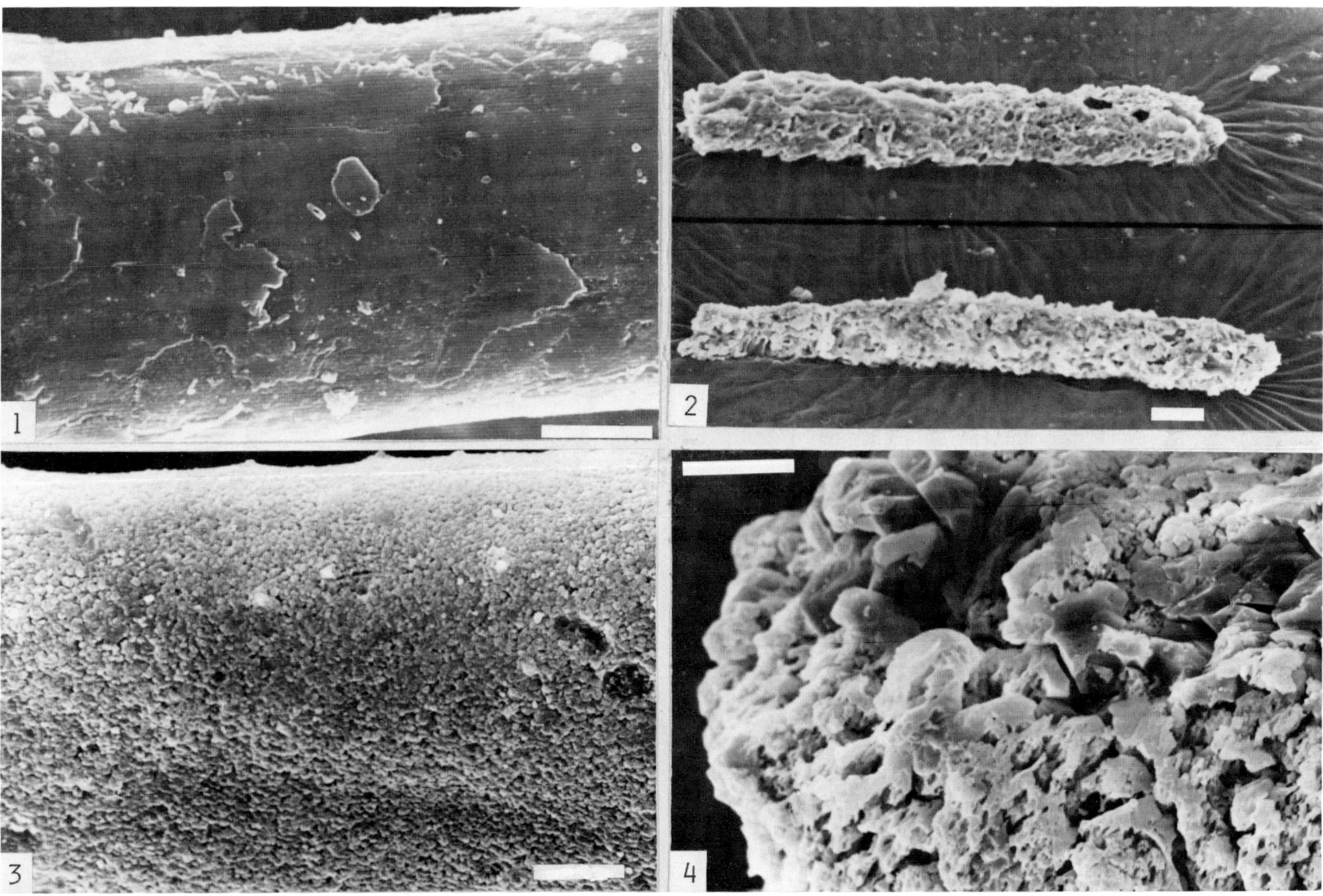

SEM photomicrographs of siliceous sponge spicules.

1: Smooth surface with leaf-like structure. Scale bar 10 μm. (Sample no. 83103).
2: Surface with granular structure and corrosion. Scale bar 20 μm. (Sample no. 11505).
3: Fine granular structure. Scale bar 10 μm. (Sample no. 101803).
4: Coarse granular structure. Scale bar 10 μm. (Sample no. 102003; Okawa formation).

reactions is not of much importance as a hydraulic drive for hydrocarbon migration in the study area. The major zone of intense oil generation, however, occurs at almost the same depth interval as the zone of decreasing expandability of illite/smectite clays from 50/50% to 70/30% (*I/S*).

As stated above, the ratio of fixed NH_4-N/total N shows relatively lower values in the middle to lower part of the Kusanagi Formation and is contrary to our initial expectation that the ratio of fixed NH_4-N/total N should increase toward the lower part of the strata if ammoniacal nitrogen is derived from nitrogenous organic matter only by thermal degradation of organic matter. According to Itihara and Honma (1979), the NH_4 ion is formed as a result of decomposition of organic matter in sediments and is adsorbed and fixed by clay minerals. Consequently, metasedimentary rocks contain significant amounts of ammoniacal nitrogen in biotite minerals inherited from the original organic material in sedimentary rocks. In this study, however, the ratio of fixed NH_4-N/total N shows a distinct decrease in the older rocks, though the treated sediments are not metasedimentary rocks but sediment as young as middle Miocene. At present, we do not have any direct evidence for the conversion of organic N to ammoniacal N. We consider, however, that the high ratio of fixed NH_4-N/total N in the upper to middle part of the Kusanagi Formation is attributable to the generation of NH_4 from nitrogenous compounds associated with resins and other high molecular weight compounds through the thermocatalytic effect of clay minerals. The region that displays the highest ratio of fixed NH_4-N/total N corresponds to a "chemically active zone," as indicated by the coexistence of the principal phase of oil formation and the transformation of illite/smectite minerals. This region still contains nitrogen-bearing compounds such as asphaltene, NOS compounds, and even various amino acid compounds (Sasaki, 1974).

Suzuki and Taguchi (1984) report that secondarily extractable organic matter (SEOM) obtained after demineralization of kerogen consists mainly of non-hydrocarbon constituents corresponding to asphaltene or NOS compounds with higher polarity (Fig. 8). On the other hand, Theng (1979) reported, based on experimental work, that organic compounds with high polarity can be adsorbed easily on clay minerals. Accordingly, it is suggested that SEOM is incorporated in the sedimentary rocks in close association with clay minerals. A part of SEOM may have been derived from the kerogen, because the kerogen has an adsorptive affinity for other organic compounds (Ishiwatari et al., 1981). These observations seem consistent with the interpretation proposed for the genesis of ammoniacal N, i.e., derivation from nitrogen-bearing compounds through the thermocatalytic reaction of clay minerals.

As stated above, the CPI_{FA} values of SEFA are lower than those of FFA and EFA, and KFA consists predominantly of the lower molecular weight acids ($<C_{21}$) in all samples. The CPI_{FA} values of KFA are lower than those of FFA, EFA and SEFA. On the other hand, kerogens from the Kusanagi Formation consist of type I or type II, which are regarded as mostly amorphous, and KFA recovered from the amorphous kerogen consists mainly of lower molecular weight fatty acids (Suzuki, 1984). Moreover, amorphous kerogen is commonly present in the rocks in close association with clay minerals (Suzuki and Taguchi, 1983). Therefore, relatively lower CPI_{FA} values of SEFA and KFA, particularly the decrease of CPI_{FA} from KFA with increasing depth, and abundance of the lower molecular weight *n*-fatty acids ($<C_{21}$) in KFA are attributable to the thermocatalytic effect of clay minerals on amorphous kerogen.

The relative quantity of each *n*-fatty acid fraction is uncertain, as quantitative analysis of each *n*-fatty acid fraction was not carried out in this study. The recovered KFA is estimated to be much more abundant, however, than other *n*-fatty acid fractions based on visual observation. Probably, *n*-fatty acids obtained as KFA have been incorporated predominantly over other *n*-fatty acid fractions to amorphous kerogen associated with clay minerals during the advancing stage of organic maturation.

On the other hand, the quantity of SEOM decreases rapidly with increasing depth below the transition zone of smectite to mixed-layer illite/smectite. If most SEOM is present in the smectite interlayer, it is suggested that SEOM may be expelled from the interlayer with the transformation of smectite to mixed-layer clay minerals. Consequently, the expelled SEOM will be analyzed as EOM. This transformation from SEOM to EOM to be one of the important causes for increase in the relative abundance of EOM with increasing maturity within a limited maturation zone. Of course, we do not deny the role of kerogen as a precurser of EOM. SEOM, as stated previously, is derived primarily from the clay interlayer and partly from the kerogen. Some SEOM is considered to be derived from the solvent-soluble organic matter present in the boundary region between clay mineral particles.

Kvenvolden (1966) has pointed out that the total *n*-fatty acids from limestones show higher even-odd ratios when compared to those of shales. On the other hand, it has been reported that limestones generally have high bitumen and hydrocarbon ratios and low kerogen content compared to shales (Hunt, 1961; Taguchi, 1982). Consequently, it is suggested that limestones have a relatively high EFA content and lower SEFA and KFA content compared to shales. Probably, the higher even-odd ratios in limestones are due to the low clay content, which minimizes the potential for thermocatalytic reaction and the abundance of extractable fatty acids.

As pointed out above, practically no carbon isotope fractionation occurs in the zone of intense oil formation. if, however, we consider 1) the experimental study by Sackett (1978), in which the first methane formed in pyrolysis experiments carried out in the presence of a clay mineral showed little, if any, carbon isotope fractionation; and 2) the two-step process proposed by Peters et al. (1981); then the above observations are reasonably explained as the result of clay mineral reactions.

ACKNOWLEDGMENTS

This work was largely supported by a research grant (Grant-in-Aid for Cooperative Research A) from the Ministry of Education, Science and Culture of Japan. We are grateful to the members of the Research Party of "Diagenesis in Sedimentary Rocks," A. Aihara, S. Honda, O. Matsubaya, Y. Saito, I. Shimada and K. Tomita, for their valuable help in this study. We also thank M. Shioya, K. Baba, Y. Ikehara, K. Mori and M. Kaneko for assistance in field sampling and in the laboratory. The senior author is grateful to T. Asakawa and N. Takeda of the Technology Research Center, Japan Petroleum Development Corporation, Tokyo, for performing Rock-Eval analysis, and to D. L. Gautier of the U.S. Geological Survey, for the encouragement to publish this paper and his critical comments on the manuscript.

REFERENCES

AIBA, J., 1980, Timing of hydrocarbon migration in the main oil and gas fields in the Northeast Japan: Journal of Japanese Association of Petroleum Technologist, v. 45, p. 373–382 (in Japanese with English abstract).

AKIYAMA, M. AND UJIIE, Y., 1983, ^{1}H-NMR T_1 of the dioxane-insoluble organic matter of the Miocene argillaceous rock developed around Shinjo Bain, Yamagata Prefecture: Journal of the Sedimentological Society of Japan, Nos. 17/18/19, p. 28–32 (in Japanese with English abstract).

AOYAGI, K., 1978, Diagenesis of marine argillaceous sediments, *in* Taguchi, K., Iijima, A., and Aoyagi, K., eds., Diagenesis: The Memoirs of the Geological Society of Japan, No. 15, p. 3–14 (in Japanese with English abstract).

BREMNER, J. M., AND HARADA, T., 1959, Release of ammonium and organic matter from soil by hydrofluoric acid and effect of hydrofluoric treatment on extraction of soil organic matter by neutral and alkaline reagents: Journal of Agricultural Science, v. 52, p. 137–146.

BURST, J. F., 1969, Diagenesis of Gulf Coast clayey sediments and its possible relation to petroleum migration: American Association of Petroleum Geologists Bulletin, v. 53, p. 73–93.

CONNAN, J., 1974, Time-temperature relation in oil genesis: American Association of Petroleum Geologists Bulletin, v. 58, p. 2516–2521.

DOUGLAS, A. G., COATES, R. C., BOWLER, B. F., AND HALL, K., 1977, Alkanes from the pyrolysis of Recent sediments, *in* Campos, R., and Goni, J., eds., Advances in Organic Geochemistry 1975: Madrid, Enadimsa, p. 357–374.

ESPITALIÉ, J., LAPORTE, J. L., MADEC, M., MARQUIS, F., LEPALAT, P., PAULT, J., AND BOUTEFEU, A., 1977, Méthode rapide de charactérisation des roches méres de leur potential pétrolier et de leur degré d'évolution: Revue de l'Institute Français du Pétrole, v. 32, p. 23–42.

HIRATA, S. AND AKIYAMA, M., 1982, ^{1}H-NMR T_1 as a possible parameter for diagenesis: Geochemical Journal, v. 16, p. 36–38.

HOEFS, J., AND FREY, M., 1976, The isotopic composition of carbonaceous matter in a metamorphic profile from the Swiss Alps: Geochimica et Cosmochimica Acta, v. 40, p. 945–951.

HUNT J. M., 1961, Distribution of hydrocarbons in sedimentary rocks: Geochimica et Cosmochimica Acta, v. 22, p. 37–49.

IKEHARA, Y., TAGUCHI, K., HIRAI, A., AND SEKIGUCHI, Y., 1980, The relationship between the vitrinite reflectance and "principal zone of oil formation" in northeastern Japan: Chikyukagaku (Geochemistry), v. 14, p. 77–87 (in Japanese with English abstract).

IMOTO, N., AND SAITO, Y., 1983, Diagenetic alteration of siliceous sponge spicules in the Neogene mudstone: Journal of the Sedimentological Society of Japan, Nos. 17/18/19, p. 81–85 (in Japanese with English abstract).

INOMATA, T., 1962, Biofacies research on the Upper Miocene in the northeastern part of Yamagata Prefecture: Journal of Japanese Association of Petroleum Technologists, v. 27, p. 465–502 (in Japanese with English abstract).

INOUE, H., 1962, Neogene Paleogeography of the surrounding Dewa Hilly Lands: Journal of Japanese Association of Petroleum Technologists, v. 27, p. 445–463 (in Japanese with English abstract).

ISHIWATARI, R., YAMAMOTO, S., AND FUKUSHIMA, K., 1981, Possible interactions of insoluble organic geopolymers (kerogen) with organic compounds in the aquatic environment. I. Interaction with fatty acid methyl esters, *in* Douglas, A. D., and Maxwell, J. R., eds., Advances in Organic Geochemistry 1979: Oxford, Pergamon Press. p. 495–504.

ITIHARA, Y., 1983a. Studies on the estimation of fixed NH_4−N in sedimentary rocks: Journal of the Sedimentological Society of Japan, Nos. 17/18/19, p. 69–73 (in Japanese with English abstract).

———, 1983b, The relation between organic and ammonium nitrogen in the Shinjo mudstone samples: Journal of the Sedimentological Society of Japan, Nos. 17/18/19, p. 75–80 (in Japanese with English abstract).

———, AND HONMA, H., 1979, Ammonium in the biotite from metamorphic and granitic rocks of Japan: Geochimica et Cosmochimica Acta, v. 43, p. 75–84.

JAGODZINSKI, H., 1949, Eindimensionale Fehlordnung in Kristallen und ihr Einfluss auf die Röntogeninterferenzen. I. Berechnung des Fehlordnugsgrades aus den Röntogenintensitäten:" Acta Crystallographica, v. 2, p. 201–207.

KAKINOKI, J., AND KOMURA, Y., 1952, Intensity of X-ray diffraction by a one-dimensionally disordered crystal: (1) General derivation in cases of the "Reichweite" S = 0 and 1: Journal of the Physical Society of Japan, v. 7, p. 30–35.

KANEKO, M., TAGUCHI, K., AND ASAKAWA, T., 1983, Comments on pyrolytic analysis using ROCK-EVAL and its study case in Niigata Oil Field: Journal of Japanese Association of Petroleum Technologists, v. 48, p. 319 (in Japanese).

KRUGE, M. A., 1983, Diagenesis of Miocene biogenic sediments in Lost Hills Oil Field, San Joaquin basin, California, *in* Isaacs, C. M., Garrison, R. E., Graham, S. A., and Jensky II, W. A., eds., Petroleum Generation and Occurrence in the Miocene Monterey Formation, California: Los Angeles, The Pacific Section, Society of Economic Paleontologists and Mineralogists, p. 39–51.

KVENVOLDEN, K. A., 1966, Molecular distribution of normal fatty acids and paraffins in some Lower Cretaceous sediments: Nature, v. 209, p. 573–577.

MACKENZIE, A. S., HOFFMANN, C. F., AND MAXWELL, J. R., 1981, Molecular parameters of maturation in the Toarcian shales, Paris Basin, France. III. Changes in aromatic steroid hydrocarbons: Geochimica et Cosmochimica Acta, v. 45, p. 1345–1355.

———, PATIENCE, R. L., MAXWELL, V. M., AND DURAND, B., 1980, Molecular parameters of maturation in the Toarcian shales, Paris Basin, France. I. Changes in the configurations of cyclic isoprenoid alkanes, steranes, and triterpanes: Geochimica et Cosmochimica Acta, v. 44, p. 1709–1712.

MARUYAMA, T., 1979, Coalification in the Ishikari coal field, Hokkaido. Report of Hokkaido Industrial Research Institute, v. 278, p. 187–209.

MCKIRDY, D. M., AND POWELL, T. G., 1974, Metamorphic alteration of carbon isotopic composition in ancient sedimentary organic matter: New evidence from Australia and South Africa: Geology, v. 2, p. 591–595.

MULHEIRN, L. J., AND RYBACK, G., 1977, Isolation and structure analysis of steranes from geological sources, *in* Campos, R., and Goni, J., eds., Advances in Organic Geochemistry 1975: Madrid, Enadimsa, p. 173–191.

NAKAI, N., YOSHIDA, Y., AND ANDO, N., 1974, Isotopic studies on oil and natural gas fields in Japan: Chikyukagaku (Geochemistry), v. 7/8, p. 87–98 (in Japanese with English abstract).

OMOKAWA, M., 1982, Carbon isotope variations in kerogen types: Researches in Organic Geochemistry, v. 3, p. 39–45 (in Japanese).

PERRY, E. A., AND HOWER, J., 1972, Late-stage dehydration in deeply buried pelitic sediments: American Association of Petroleum Geologists Bulletin, v. 56, p. 2013–2021.

PETERS, K. E., ROHRBACK, B. G., AND KAPLAN, I. R., 1978, Vitrinite reflectance-temperature determinations for intruded Cretaceous black shale in the eastern Atlantic, *in* Oltz, D. F., ed., Symposium on Geochemistry, Low Temperature Metamorphism of Kerogen and Clay Minerals: Pacific Section, Society of Economic Paleontologists and Mineralogists, p. 53–58.

———, ———, AND ———, 1981, Carbon and hydrogen stable isotope variations in kerogen during laboratory-simulated thermal maturation: American Association of Petroleum Geologists Bulletin, v. 65, p. 501–508.

REDDING, C. E., SCHOELL, M., MONIN, J. C., AND DURAND, B., 1979, Hydrogen and carbon isotopic composition of coals and kerogens, *in* Douglas, A. G., and Maxwell, J. R., eds., Advances in Organic Geochemistry 1979: Oxford, Pergamon Press, p. 711–724.

SACKETT, W. M., 1978, Carbon and hydrogen isotope effects during the thermo-catalytic production of hydrocarbons in laboratory simulation experiments: Geochimica et Cosmochimica Acta, v. 42, p. 571–580.

SASAKI, K., 1974, Stratigraphic distribution of amino acids assemblage in the Kusanagi Formation of Oashizawa, Yamagata Prefecture: Journal of the Japanese Association of Petroleum Technologists, v. 39, p. 44–52 (in Japanese with English abstract).

SATO, S., AND ASAKAWA, T., 1982, Estimation of organic maturity in Niigata Plain: Researches in Organic Geochemistry, v. 3, p. 27–28 (in Japanese).

SEIFERT, W. K., AND MOLDOWAN, J. M., 1978, Applications of steranes, triterpanes and monoaromatics to the maturation, migration and source of crude oils: Geochimica et Cosmochimica Acta, v. 42, p. 77–95.

SHI-JI-YANG, MACKENZIE, A. S., ALEXANDER, R., EGLINTON, G., GOWAR, A. P., WOLFF, G. A., AND MAXWELL, J. R., 1982, A biological marker investigation of petroleums and shales from the Shengli oil field, The people's Republic of China: Chemical Geology, v. 35, p. 1–31.

SHIMODA, S., TOMITA, K., AND MATSUKURA, Y., 1983, Diagenetic alteration of clay minerals—An example at Oashizawa in the Shinjo basin: Journal of Sedimentological Society of Japan, Nos. 17/18/19, p. 89–98 (in Japanese with English abstract).

SHIMOYAMA, A., AND MATSUBAYA, O., 1983, Stable carbon isotope study on organic matter in the Neogene sediments from Shinjo Basin: Journal of Sedimentological Society of Japan, Nos. 17/18/19, p. 57–63 (in Japanese with English abstract).

SHIOYA, M., AND ISHIWATARI, R., 1983, Diagenetic alteration of organic matter (hydrocarbons) in Shinjo sedimentary rocks of Miocene age: Journal of Sedimentological Society of Japan, Nos. 17/18/19, p. 33–39 (In Japanese with English abstract).

STEVENSON, F. J., 1960, Microdetermination of nitrogen in rocks and silicate minerals by sealed tube digestion: Analytical Chemistry, v. 32, p. 1704–1706.

SUZUKI, N., 1984, Characteristics of amorphous kerogens fractionated from terrigenous sedimentary rocks: Geochimica et Cosmochimica Acta, v. 48, p. 243–249.

———, AND SHIMADA, I., 1983, Considerations in epimerization of sterane and triterpane as indicators of thermal history of sedimentary rocks: Journal of Sedimentological Society of Japan, Nos. 17/18/19, p. 47–55 (in Japanese with English abstract).

———, AND TAGUCHI, K., 1983, Characteristics and diagenesis of kerogens associated with clay fractions of mudstone, *in* Bijorøy, M. et al., eds., Advances in Organic Geochemistry 1981: Chichester, John Wiley, p. 607–612.

———, AND ———, 1984, Diagenesis of extractable and bound fatty acids in possible source rocks in Japan: Organic Geochemistry, v. 6, p. 125–133.

TAGUCHI, K., 1962, Basin architecture and its relation to the petroleum source rocks development in the region bordering Akita and Yamagata Prefectures and the adjoining areas, with the special reference to the depositional environment of petroleum source rocks in Japan: Science Report of Tohoku University, Series III, v. 7, p. 293–342.

———, 1967, Geological map of Osawa and its explanatory text: Publication of Prefectural Government of Yamagata, Yamagata, Japan. p. 24 (in Japanese).

———, 1978, Organic maturation—A review, with the special reference to its significance in the inorganic process of diagenesis and incipient metamorphism, *in* Taguchi, K., Iijima, A., and Aoyagi, K., eds., Diagenesis: The Memoirs of the Geological Society of Japan, No. 15, p. 165–190 (in Japanese with English abstract).

———, 1982, On the characteristics of carbonate sediments as petroleum source rocks—A review: Science Report of Tohoku University, Series III, v. 15, p. 149–161.

———, 1983, Petroleum geochemistry in the Neogene sedimentary basin of Japan, with special reference to primary migration: Journal of Geological Society, London, v. 140, p. 475–487.

———, Ikehara, Y., Kano, T., Mori, K., and Kaneko, M., 1983, Organic maturation of Neogene Tertiary sediments from Oashizawa, Ohmata-Shiratama and Oguni routes in the Shinjo Oil Field, northeastern Japan: Journal of Sedimentological Society of Japan, Nos. 17/18/19, p. 12–28 (in Japanese with English abstract).

———, Sasaki, K., and Ushijima, N., 1970, Porphyrin pigments in the Neogene Tertiary rocks of the Shinjo oil field, Yamagata Prefecture—Sedimentological and stratigraphical study of the porphyrin pigments in Neogene Tertiary rocks. III. Journal of Geological Society of Japan, v. 76, p. 559–566 (in Japanese with English abstract).

Takayanagi, Y. (ed.), 1978, Manual of Microfossil Research: Tokyo, Asakura-Shoten, p. 161 (in Japanese).

Theng, B. K., 1979, Formation and properties of clay-polymer complexes: Amsterdam, Elsevier, p. 283–326.

Trask, P. D., 1932, Origin and environment of source sediments of petroleum: Gulf Publishing Company, Houston, p. 284.

Vassoevich, N. B., Korchagina, Y. I., Lopatin, N. V., and Chernyshv, V. V., 1970, Principal phase of oil formation: Moskow, University Vestink, No. 6, p. 3–27: English translation, International Geology Review, v. 12, p. 1276–1296.

Waples, D. W., 1980, Time and temperature in petroleum formation: Application of Lopatin's method to petroleum exploration: American Association of Petroleum Geologists Bulletin, v. 64, p. 916–926.

Watanabe, T., 1981, Identification of illite/montmorillonite interstratifications by X-ray powder diffraction: Journal of the Mineralogical Society of Japan, v. 15, p. 32–41 (in Japanese with English abstract).

Yokono, T., and Sanada, Y., 1978, Nuclear magnetic relaxation studies of coal: Fuel, v. 57, p. 334–336.

NEOGENE THERMAL HISTORY OF THE NORTHERN GREEN RIVER BASIN, WYOMING—EVIDENCE FROM FISSION-TRACK DATING

NANCY D. NAESER
U.S. Geological Survey, Box 25046, Mail Stop 424, Denver Federal Center, Denver, Colorado 80225

ABSTRACT: Fission-track dating is a valuable method for studying the thermal history of sedimentary basins because apatite and zircon fission-track annealing temperatures span the main temperature range for oil generation, and both minerals are present in the heavy-mineral suites of many sedimentary rocks. Fission tracks in apatite are totally annealed at temperatures of about 105°C (221°F) to 150°C (302°F) over periods of heating of 10^8 to 10^5 yr duration, respectively. The temperatures at which fission tracks in zircon anneal are not as well known but are probably in the range of 200°C ± 30°C (392°F ± 54°F).

Fission-track ages of detrital apatite separated from Upper Cretaceous and lower Tertiary sedimentary rocks in the El Paso Natural Gas Wagon Wheel no. 1 well on the Pinedale anticline, northern Green River basin, Wyoming, indicate that there has been significant cooling in the rocks in this part of the basin. The data suggest that at least the latest phase of cooling began only 2 to 4 m.y. ago and involved a relatively rapid temperature decrease of 20°C (36°F) or more. The magnitude of cooling is supported by several other paleotemperature indicators in the drill hole.

INTRODUCTION

Fission-Track Dating and the Thermal History of Sedimentary Basins

Temperature is one of the most important parameters involved in understanding the history of sedimentary basins, diagenesis, and hydrocarbon generation. Fission-track dating has the potential for wide application in studying the thermal history of sedimentary basins because the fission-track annealing temperatures of two of the minerals most commonly used in fission-track studies, apatite and zircon, span the main temperature range of oil generation, and both of these minerals are found in the detrital grain suites of many sedimentary rocks. Fission-track dating also has the advantage, not found in many other indicators of thermal maturity, of providing data on the timing of events in a basin.

Fission tracks are zones of intense damage formed when fission fragments pass through a solid. Several naturally occurring isotopes undergo spontaneous fission, but only ^{238}U, which is present in trace amounts in many minerals and glasses, has a fission half-life that is sufficiently short to produce a significant number of tracks through geologic time. Because the spontaneous fission of ^{238}U takes place at a constant rate, the age of a mineral or glass can be calculated by determining its spontaneous track density and the amount of uranium which it contains. Details of the fission-track method are given in Naeser (1976, 1979a).

Annealing of fission tracks.—

Once fission tracks are formed, they are stable in most common minerals at temperatures below about 80°C (176°F). If a mineral containing spontaneous fission tracks is heated to a sufficiently high temperature, however, the tracks fade and disappear, resulting in an anomalously young fission-track age. The temperature at which this "annealing" occurs depends on (1) the mineral involved—different minerals anneal at different temperatures, and (2) the duration of heating—the longer a mineral is heated, the lower the temperature that is required to anneal its tracks completely.

Annealing conditions have been determined for many minerals by short-term laboratory experiments. Temperature-time relationships more directly applicable to heating durations of geologic interest, however, have been obtained geologically by determining the temperature at which minerals are totally annealed (i.e., yield a zero age) in deep drill holes in several areas where the approximate duration of heating is known.

The most complete drill-hole data available are for apatite. They indicate that total annealing of tracks in apatite occurs at temperatures that range from ca. 105°C (221°F) (for 10^8 yr heating duration) to ca. 150°C (302°F) (10^5 yr). The drill-hole data and several examples of the laboratory data are shown in Figure 1. The extrapolated laboratory data predict slightly lower temperatures than indicated by the geologic data. This may be at least in part because the laboratory data do not take into account the significant number of new tracks that are formed during long-term geologic annealing (see discussion in Sanford, 1981).

The annealing temperatures of zircon are not as well known, but limited data suggest they are probably in the range of 200°C ± 30°C (392°F ± 54°F) for geologic heating. (For example, see data in Harrison et al., 1979 and Hurford, 1984.)

Application to Sedimentary Basins.—

Fission-track annealing offers a powerful tool for studies of thermal, tectonic, and burial history in both basement and sedimentary rocks. Application of fission-track dating to sedimentary rocks, however, poses special problems because, unlike basement rocks, the detrital grains that make up the sediments can be of widely divergent ages. They may be equal in age to the stratigraphic age of the rock, as in the case of grains derived from contemporaneous volcanism, but much more commonly they will be older.

The theoretical basis for the application of fission-track dating to sedimentary basins and the expected distribution of fission-track ages with depth in sedimentary rocks were first described by Naeser (1979b). The expected distribution of apparent apatite ages in sedimentary rocks that are at their maximum burial temperature is as follows (Fig. 2). Apatite grains from shallow samples at relatively low temperatures will not have been annealed during burial of the sediments, and their ages will still reflect the age(s) of the

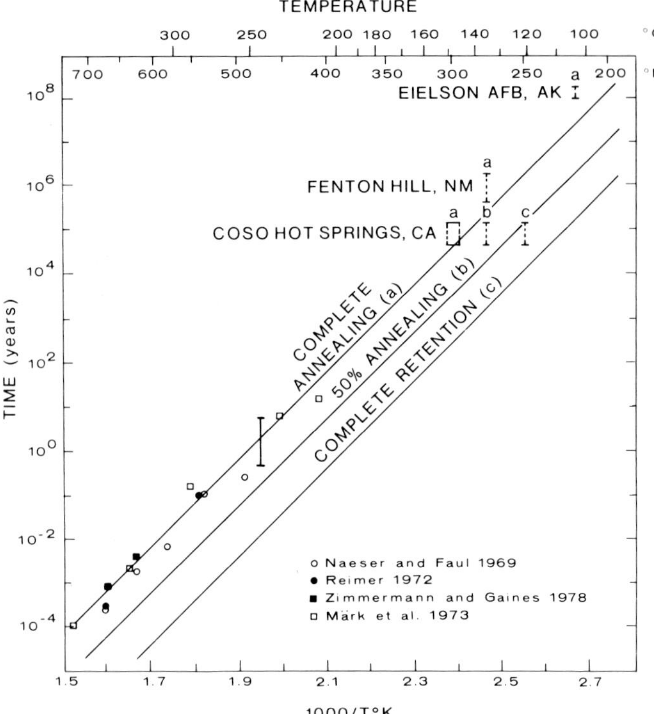

FIG. 1.—Temperatures required to anneal fission tracks in apatite for heating durations ranging from short-term laboratory experiments to 10^8-yr geologic annealing (figure modified from Sanford, 1981, Fig. 2). Temperatures for geologic annealing (dotted vertical bars) are from Naeser (1981) and were determined by dating apatite in drill holes in three areas where the approximate duration of heating is known. Laboratory data are shown by circles and squares. Extrapolation of the laboratory data (solid lines) is based on the rate constant for fission-track annealing determined from a least squares fit of the laboratory data by Zimmermann and Gaines (1978). a, b, and c on the solid lines and dotted vertical bars identify data for complete annealing (a), 50% annealing (b), and complete retention (c) of fission tracks. Gleadow and Duddy (1981) report a temperature of 125°C (257°F) for total annealing of apatite in drill holes in Australia, where the duration of heating is thought to be between 1×10^7 and 4×10^7 yr. Apatite in the Cerro Prieto geothermal field, Mexico, is totally annealed at a temperature between 160°C and 180°C (320°F–356°F) in 10^3–10^4 yr (Sanford, 1981).

source rock(s) for the detrital grains ("zone of no annealing"). In sediments subjected to progressively higher temperatures, apatite will undergo partial annealing and give progressively younger apparent apatite ages ("zone of partial annealing"). Within this zone, the apparent apatite ages become younger than the stratigraphic age of the sediments, and finally decrease to zero at the depth where the temperature for total annealing is attained for sufficient exposure time (see Fig. 1, "zone of total annealing"). Naeser's (1981; Fig. 1) data suggest that, for geologic rates of annealing, the zone of partial annealing (from total track retention to total track loss) is 20°C to 30°C (36°F to 54°F) wide. Duddy and Gleadow (1982) show a width of 45°C to 50°C (81°F to 90°F).

When sediments cool below their maximum paleotemperature, apatite from the zone of total annealing once again begins to accumulate tracks and to record a fission-track age (Fig. 3). The recorded apatite age, the slope of the depth (or temperature) versus age plot, and the thickness of the zone of cooling ages give information on the time, rate, and amount of cooling, respectively.

Zircon fission-track ages should generate similar curves, but the annealing will occur at higher temperatures than in apatite.

Most of the information about the thermal history of sedimentary basins that is obtained by dating detrital grains is typically provided by the apatite data, because temperatures sufficient to anneal zircon (ca. 200°C; 392°F) are often not encountered. Even when zircon is not annealed, however, it still can provide useful information. In addition to setting limits on maximum paleotemperatures, the ages of individual unannealed zircon grains reflect the age(s) of the source rocks of the sediments and therefore can give information valuable in provenance studies in the basin.

The fission-track method has been used to study the thermal history of sedimentary basins in California and Wyoming (Briggs et al., 1979, 1981a, 1981b; Naeser, 1984; Naeser, Naeser, and McCulloh, in prep.) and Australia (Gleadow and Duddy, 1981; Duddy and Gleadow, 1982, 1984; Gleadow et al., 1983). In this paper the method is applied to the thermal history of the Pinedale anticline area

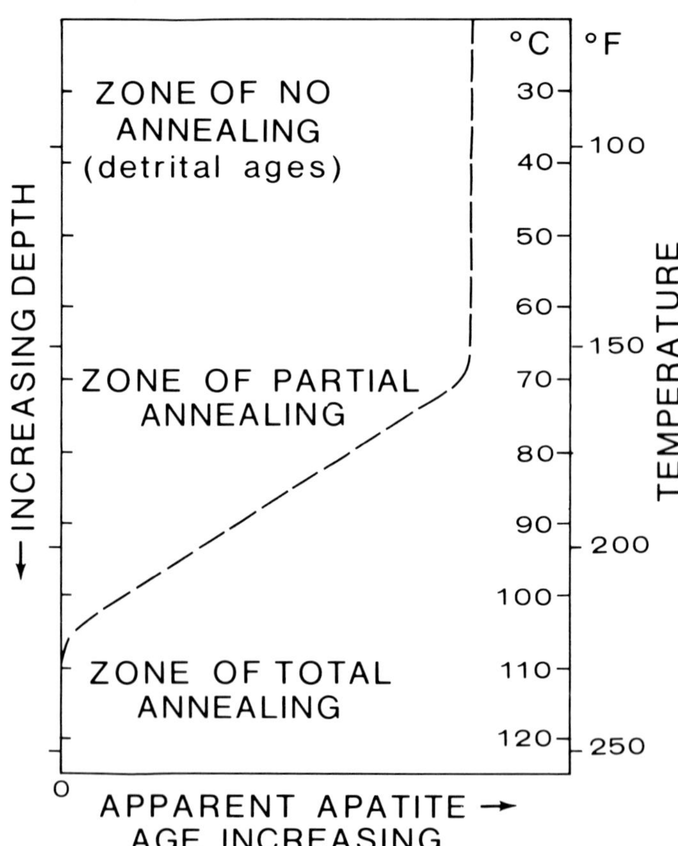

FIG. 2.—Expected distribution of apparent apatite ages in sediments at the time of maximum burial heating (modified from Naeser, 1979b). The temperature at which the apatite age decreases to zero depends on the duration of time the sediments have been heated near their present temperature (see Fig. 1). In this example, total annealing occurs at just over 105°C (221°F), indicating heating of 10^8 yr duration. See text for further explanation.

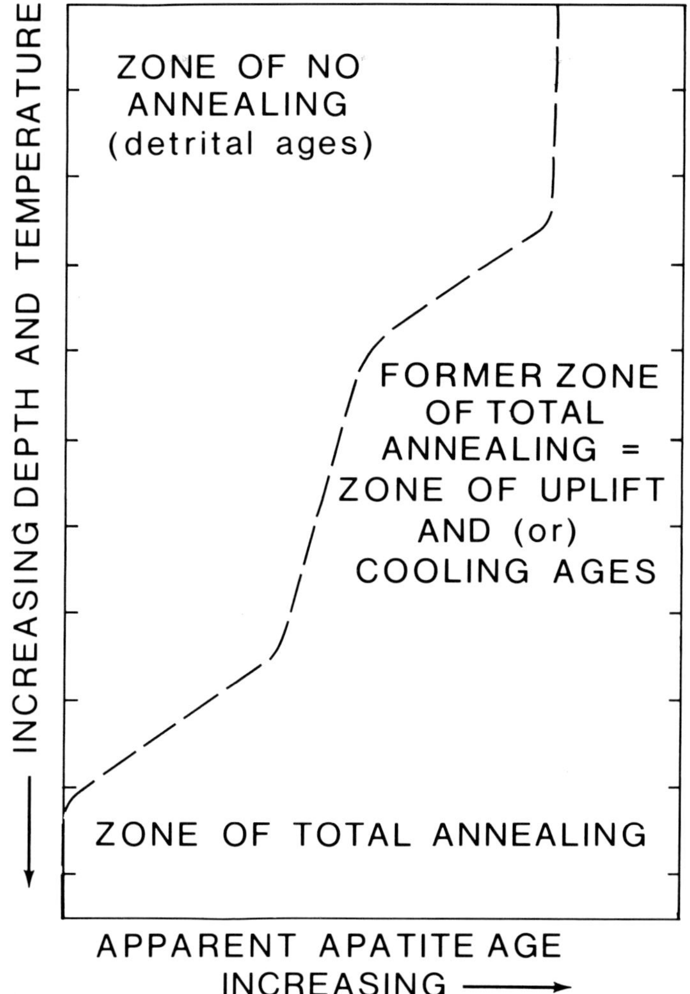

FIG. 3.—Expected distribution of apparent apatite ages with increasing depth (and temperature) in sediments after uplift and (or) cooling (modified from Naeser, 1979b).

of the northern Green River basin, Wyoming, using detrital apatite and zircon separated from Upper Cretaceous and lower Tertiary sedimentary rocks in the El Paso Natural Gas Wagon Wheel no. 1 well. Figure 4 shows the location of this well. The geologic setting of the northern Green River basin and the stratigraphy and structure of the Wagon Wheel no. 1 area are discussed in Law (1984b).

METHODS

Fission-track dating.—

Apatite and (or) zircon have been separated from 10 selected Upper Cretaceous and lower Tertiary sandstone core samples recovered from Wagon Wheel no. 1 over a depth interval of approximately 1,536 to 4,557 m (5,038 to 14,950 ft; Table 1).

Apatite and zircon were separated from the samples using the heavy liquid and magnetic separation techniques shown in Figure 5. Grains coarser than about 75 μm (200 mesh; fine sand size) can be dated by the fission-track method. Samples from Wagon Wheel no. 1 were sieved to 75 to 180 μm or 75 to 850 μm (+200-80 or +200-20 mesh) size fractions before treating with heavy liquids. This entire fraction was used when possible, but in several samples only the 75–125 μm (+200-115 mesh) size fraction yielded a sufficiently pure separation (i.e., greater than 50% apatite or zircon) for dating. In all samples, the methylene iodide light fraction required further treatment with the bromoform-methylene iodide mix (sp. gr. = 3.037) to improve the segregation of apatite from contaminating grains. Most samples required a further step of hand picking to isolate the apatite. The amount of core available for processing varied from 64 to 195 g.

Both apatite and zircon were dated by the external detector method (Naeser, 1976, 1979a), using a low uranium muscovite as the detector. The apatite and zircon fission-track ages quoted for each of the Wagon Wheel no. 1 samples were calculated by the conventional method of using the sums of the spontaneous and induced tracks counted in the individual grains in the sample (Fleischer et al., 1965; Naeser, 1967). The uncertainty in the ages was calculated by combining the Poisson errors on the spontaneous and induced counts and on the counts in the detector covering the dosimeter (Lindsey et al., 1975). The methods used to calculate age and uncertainty are standard methods commonly used in fission-track dating, but it should be recognized that they may not be strictly applicable to sedimentary rock samples in which the individual detrital grains were most likely derived from more than one age population. The "age" calculated for a sediment sample is in fact a composite of the individual grain ages and as such has

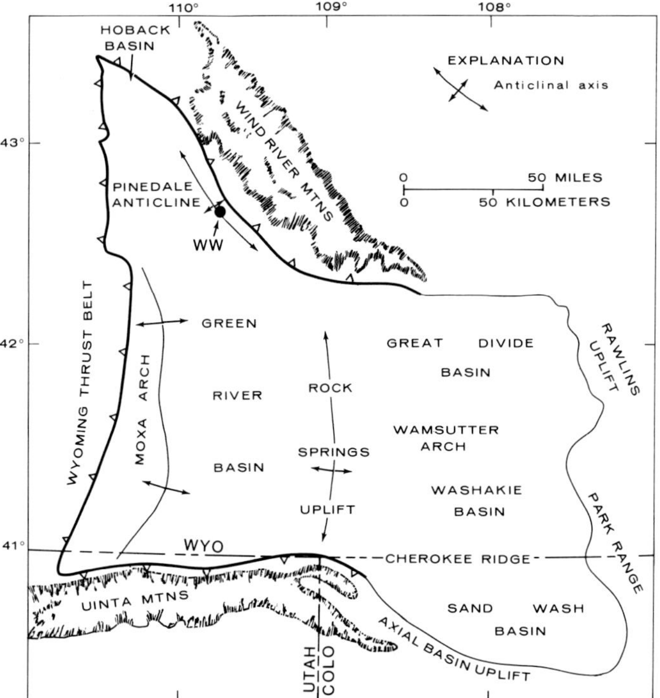

FIG. 4.—Map of the Green River basin, Wyoming, showing the major structural features of the basin and the location of the Wagon Wheel no. 1 well (WW) on the Pinedale anticline (modified from Law, 1984a).

TABLE 1.—DEPTH, STRATIGRAPHIC POSITION, AND PRESENT TEMPERATURE OF SAMPLES FROM WAGON WHEEL NO. 1, GREEN RIVER BASIN, WYOMING

	Depth			Present temperature (°C)	
Sample	feet	meters (approx.)	Stratigraphic unit[1]	measured	corrected[2]
WWH 1	5038	1536	Fort Union Fm.	44	50
WWH 3	7061	2152	unnamed Tertiary rocks	59	66
WWH 5	7340	2237	unnamed Tertiary rocks	61	68
WWH 9	8087	2465	Lance Fm.	67	74
WWH 13	10165	3098	Lance Fm.	81	90
WWH 16	11027	3361	Lance Fm.	84	96
WWH 24	11057	3370	Lance Fm.	87	96
WWH 17	13100	3993	Ericson Ss., upper part	101	109
WWH 26	13197	4022	Ericson Ss., lower part	102	110
WWH 21	14950	4557	Rock Springs Fm.	114	121

[1]From Law (1984b) and B. E. Law (pers. commun., 1984); "Fort Union Fm." = the interval described as "Fort Union and younger formations" in Law (1984b); "Lance Fm." = "Lance Formation and older rocks;" "Rock Springs Fm." = "Rock Springs Formation and older rocks."
[2]See text for explanation.

no real meaning in a stratigraphic sense. The range in the age of individual grains counted in each sample is given in Table 2.

Present-day temperatures, Wagon Wheel no. 1 well.—

Two sets of temperatures are given in Table 1:
(1) "measured" temperatures interpolated from log data, and

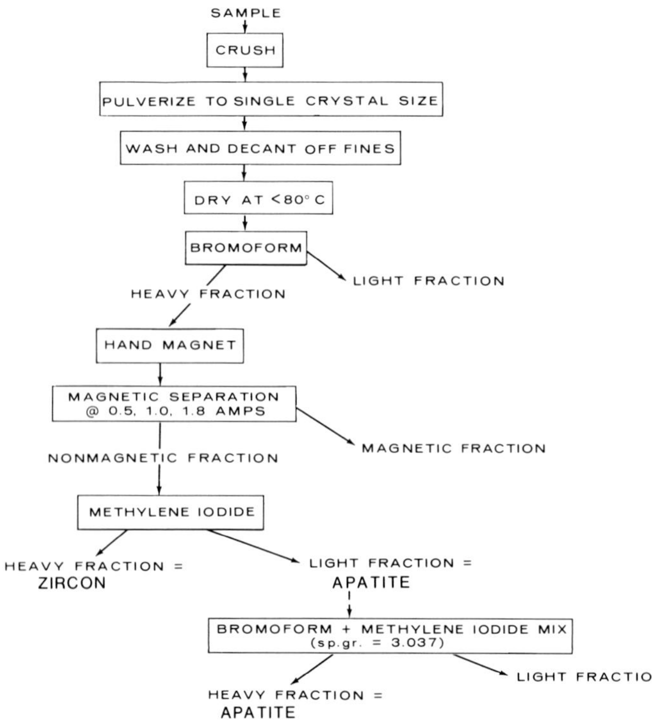

FIG. 5.—Steps involved in separating apatite and zircon from bulk rock samples for fission-track dating. The methylene iodide light fraction usually yields an adequate apatite separate, but in some samples further treatment with a bromoform-methylene iodide mix is necessary to remove contaminating grains.

(2) corrected temperatures that have been adjusted using the procedure outlined in McCulloh and Beyer (1979) (Fig. 6).

In Wagon Wheel no. 1, the McCulloh and Beyer (1979) correction yields temperatures that match very closely those that would be obtained using the correction procedures of the American Association of Petroleum Geologists Geothermal Survey of North America (Kehle et al., 1970; Kehle, 1972). The uncertainty associated with any individual corrected temperature is probably ca. ±3°C (5.4°F) from the surface to 10,000 ft (3048 m) and ca. ±5°C (9°F) from 10,000 ft to TD (19,000 ft; 5,791 m) (T. H. McCulloh, pers. commun., 1982). The corrected temperatures vary from 6°C to 12°C (ca. 11°F to 22°F) higher than the "measured" temperatures.

RESULTS

Zircon ages.—

Zircon ages have been determined for six of the Wagon Wheel no. 1 samples (Table 2; Figs. 7 and 8). Sample WWH 16 yielded insufficient zircon for an age determination, and WWH 1 cannot be dated because the zircon is metamict.

The time of deposition of the Cretaceous sedimentary units sampled in Wagon Wheel no. 1 is known with some confidence because the units can be correlated to ammonite zones whose ages have been determined by K-Ar dating of associated bentonite layers (Obradovich and Cobban, 1975; Weimer, 1961; Law, 1984b). The Tertiary samples from which zircons were dated (WWH 3 and WWH 5) are from a unit underlying the Paleocene Fort Union Formation and are thus older than the Eocene-Paleocene boundary (57.8 m.y.) and younger than the Tertiary-Cretaceous boundary (66.4 m.y.) (Geological Society of America, 1983). Comparison of these ages to the zircon fission-track ages (Table 2; Figs. 7 and 8) shows that the calculated fission-track ages of all of the Wagon Wheel no. 1 samples, as well as all but a few of the individual zircon grain ages, are equal to or older than the depositional age of the unit where they

TABLE 2.—APATITE AND ZIRCON FISSION-TRACK AGES FROM WAGON WHEEL NO. 1, GREEN RIVER BASIN, WYOMING

Sample[1]	Number of grains counted	ρ_s $\times 10^6$ t/cm^2	ρ_i[2] $\times 10^6$ t/cm^2	ϕ[3] $\times 10^{15}$ n/cm^2	Range in individual grain ages[4] $\times 10^6$ yr	Apparent age[4] $\times 10^6$ yr	$\pm 2s$[5] $\times 10^6$ yr
Apatite							
WWH 5	8	0.299	5.73	6.72	9–67	20.9	4.6
		(94)[6]	(902)[6]	(2354)[6]			
WWH 9	10	0.213	2.31	6.61	6–81	36.3	8.4
		(91)	(494)	(2354)			
WWH 13	9	0.093	2.53	6.51	0–35	14.3	5.6
		(28)	(380)	(2354)			
WWH 16	6	0.047	2.23	6.40	0–19	8.0	7.4
		(5)	(119)	(2354)			
WWH 24	6	0.036	2.05	3.67	0–17	3.8	2.8
		(8)	(230)	(2726)			
WWH 26	11	0.023	2.03	3.61	0–36	2.4	1.8
		(8)	(354)	(2726)			
WWH 21	9	0.025	3.50	6.30	0–10	2.6	2.4
		(5)	(357)	(2354)			
Zircon							
WWH 3	5	22.33	7.43	1.03	106–530	181.9	48.9
		(394)	(65)	(4424)			
WWH 5	9	12.48	6.53	0.889	57–581	100.9	11.4
		(1578)	(413)	(6340)			
WWH 9	9	10.28	6.60	1.01	63–288	94.4	12.0
		(107)	(346)	(4424)			
WWH 13	9	11.04	6.92	1.01	46–322	95.8	10.6
		(1472)	(461)	(4424)			
WWH 17	9	9.69	7.91	1.01	44–649	73.6	9.2
		(931)	(380)	(4424)			
WWH 21	4	13.55	10.12	1.01	47–275	100.7	19.5
		(477)	(142)	(4424)			

ρ_s = spontaneous track density; ρ_i = induced track density.
[1]Sample locality data given in Table 1.
[2]Reported induced track density = 2 × measured density.
[3]Thermal neutron fluence (ϕ) determined from calibrated muscovite detector covering glass dosimeters (for zircon: NBS glass standard SRM 962: 37.38 ± 0.08 ppm U; for apatite: U. S. Geological Survey glass standard calibrated against NBS glass standard SRM 963: 0.823 ± 0.002 ppm U) placed at top and bottom of each irradiation tube. Fluences were calibrated against the Cu value determined at NBS (Carpenter and Reimer, 1974). The fluence for each sample was calculated by interpolation between values determined for the standards.
[4]Calculated from the fission-track age equation (Fleischer et al., 1965; Naeser, 1967), using the sums of the spontaneous and the induced track counts obtained for all grains in the sample, and the following values: $\lambda_D = 1.551 \times 10^{-10}$ yr^{-1}; $\lambda_F = 7.03 \times 10^{-17}$ yr^{-1} (Roberts et al., 1968); $\sigma_F = 580 \times 10^{-24}$ cm^2; $I = 7.252 \times 10^{-3}$
[5]Standard error(s) calculated by combining the Poisson errors on the spontaneous and induced counts and the counts in the dosimeter (Lindsey et al., 1975).
[6]Number in parentheses = number of tracks counted.

were sampled. This suggests that there has been no significant annealing of the zircon in any of the dated samples since deposition.

Lack of significant annealing in the zircons is also suggested by the fact that through most of the Cretaceous and lowermost Tertiary section the zircon ages are indistinguishable from one another (at ± one sigma) (Table 2; Fig. 7). There is some variation in the range of the individual zircon grain ages in these samples (Fig. 7), but there is no obvious correlation between the calculated sample ages or the range in individual grain ages and the present temperature (or depth) of the samples. An exception is sample WWH 17, which did yield a significantly younger age than many of the samples and apparently contains a lower percentage of old (greater than ca. 110 m.y.) zircons (Fig. 7). This difference may be related at least in part to the provenance of this stratigraphic horizon, however, rather than to postdepositional annealing effects. Sample WWH 17 was collected from the "upper part of the Ericson Sandstone" (terminology of Law, 1984b), which has a different composition, and has been interpreted as having a different provenance, than the bulk of the Cretaceous sandstone in the Western Interior (Law, 1984b; Rice and Gautier, 1983, p. 4–24; D. L. Gautier, pers. commun., 1984). In Wagon Wheel no. 1, the "upper part of the Ericson Sandstone" was also distinguished from the remainder of the section by the fact that it yielded little or no apatite (see below).

In the upper part of the Tertiary section sampled in Wagon Wheel no. 1 (WWH 1 and WWH 3), the detrital zircon suites appear to be older than those observed in the underlying rocks. Zircons in WWH 1 are metamict, which implies that they are of considerable age, and in WWH 3 no zircon grains younger than 100 m.y. old were observed.

Apatite ages.—

Apatite has been dated from seven of the Wagon Wheel no. 1 samples (Table 2). Sample WWH 17, from the "upper part of the Ericson Sandstone," yielded insufficient apatite for an age determination (see discussion above).

All of the calculated apatite ages for the Wagon Wheel no. 1 samples, as well as almost all of the individual apatite grain ages, are younger than the stratigraphic age of the sedimentary units from which they were collected, indicating that significant annealing of the apatite has occurred during burial of the sediments. If the "zone of no anneal-

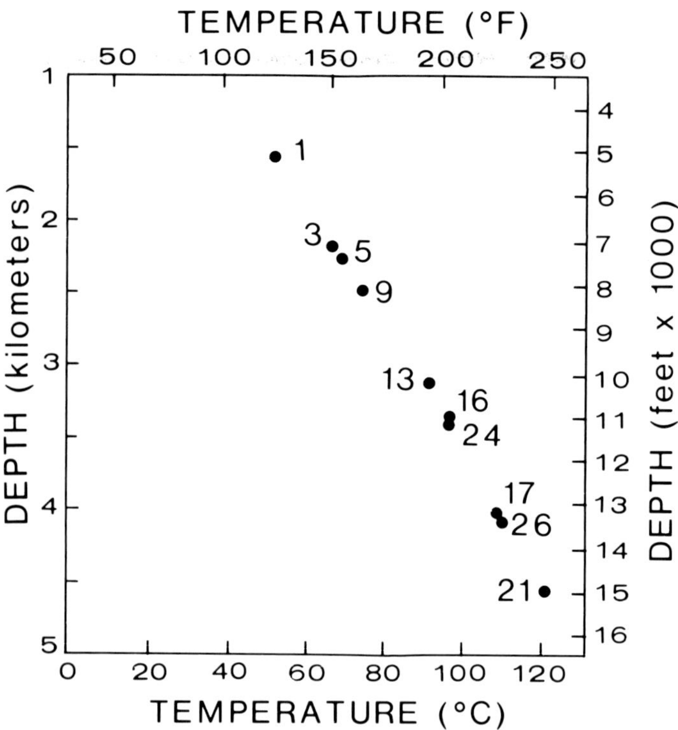

FIG. 6.—Depth plotted against present temperature (corrected) of the Wagon Wheel no. 1 samples. Sample numbers correspond to numbers in Table 1.

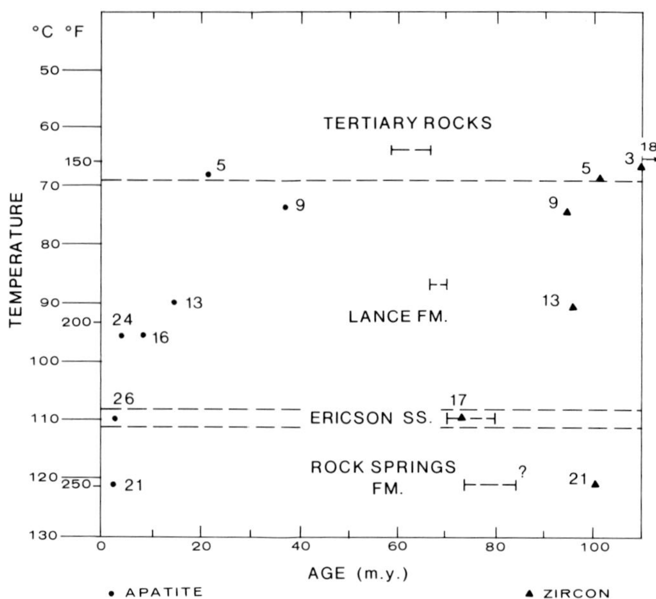

FIG. 8.—Fission-track ages of detrital apatite and zircon separated from sandstones in Wagon Wheel no. 1 plotted against present temperature (corrected). Sample numbers correspond to numbers in Table 1. Bars in each stratigraphic unit indicate the approximate time of deposition of the unit, as determined from K-Ar ages of Cretaceous ammonite zones (Obradovich and Cobban, 1975; recalculated for new K-Ar constants), the correlations shown in Weimer (1961) and Law (1984b), and the Geological Society of America (1983) time scale.

ing" (cf. Fig. 2) is preserved in Wagon Wheel no. 1, it must be above the shallowest sample analyzed (WWH 5; 2237 m, 7,340 ft depth) because even that sample contains partially annealed apatite.

The apatite fission-track ages calculated for the Wagon Wheel no. 1 samples fall into two distinct trends (Fig. 8). In the upper part of the sampled interval, the apatite ages generally decrease with increasing temperature and approach zero at just over 96°C (205°F). At this point, the trend in ages changes abruptly and the ages remain nearly constant, between about 2 to 4 m.y., over a temperature interval of at least 20°C (36°F), down through the lowest sample dated (WWH 21).

DISCUSSION

Thermal History in the Wagon Wheel No. 1 Area

Because of the time-temperature dependency in annealing (Fig. 1), the degree of annealing observed in the apatite in the upper part of Wagon Wheel no. 1 (down to the al-

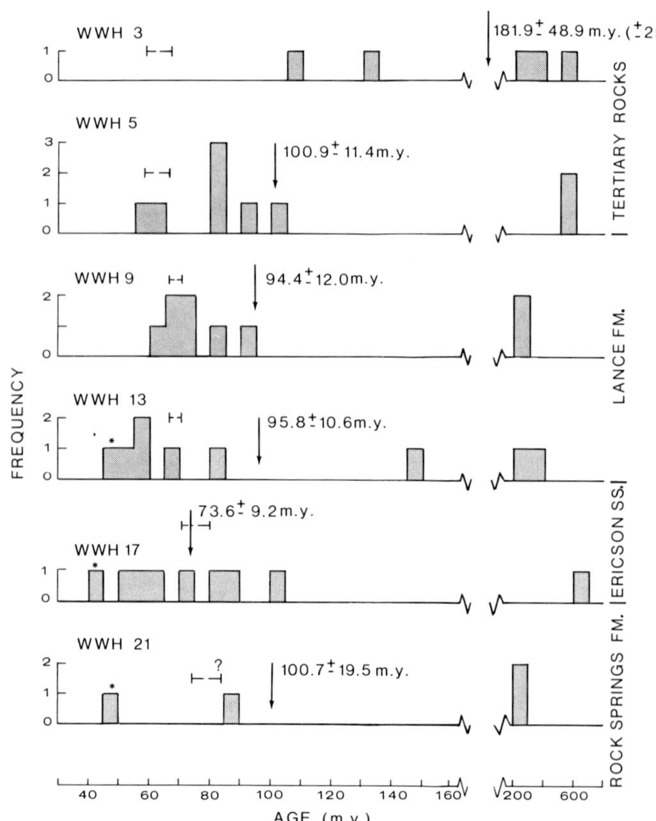

FIG. 7.—Histogram of the fission-track ages of individual zircon grains in each of the Wagon Wheel no. 1 samples. The "age" calculated for each sample (Table 2) is indicated by an arrow. Bars indicate the approximate time of deposition of the unit where the sample was collected (see text and Fig. 8 caption). Grain ages marked with an asterisk are younger than the stratigraphic age and do not overlap it at ± two sigma; all other ages are equal to or older than the stratigraphic age.

most totally annealed apatite at ca. 96°C; 205°F) could have been produced either by the rocks having been:

(1) near their present (or lower) temperature for a long time (at least 10^8 yr; see Fig. 1), or

(2) at a higher temperature for a shorter time, before cooling to their present temperature.

The rocks in this part of the drill hole (Lance Formation) are latest Cretaceous and younger in age (i.e., less than ca. 70 m.y.; Obradovich and Cobban, 1975; Weimer, 1961; Law, 1984b), and it is therefore unlikely that they have been near their present temperature for a period of time even approaching 10^8 yr. We are thus left with the second alternative.

On the basis of the annealing data from the upper part of the drill hole alone, it would be difficult to speculate much further on the thermal history in this area. The trend of the apatite ages below 96°C (205°F) (which is thought to be analagous to the "zone of uplift and (or) cooling ages" in Fig. 3) provides evidence, however, for the nature of the cooling. The apatite ages suggest that at least the latest phase of cooling in the Pinedale anticline was initiated only 2 to 4 m.y. ago and involved a relatively rapid decrease in temperature of 20°C (36°F) or more.

Pollastro and Barker (1984a, 1984b) have also studied the temperature history in Wagon Wheel no. 1, using several other paleotemperature indicators. Their depth-temperature curves determined from mean random vitrinite reflectance measured in thin coaly seams in the rocks and from illite-smectite transformation in the sandstone and shale layers are in good agreement, and together indicate that these rocks were subjected to temperatures 30°C to 50°C (54°F to 90°F) higher sometime in the past. For example, their data indicate a maximum paleotemperature of ca. 190°C (374°F) at 5486 m (18,000 ft). By comparison, the present (corrected) temperature at this depth is 141°C (286°F) (using the McCulloh and Beyer, 1979, correction method).

CONCLUSIONS

Vitrinite reflectance and illite-smectite transformation measurements in Wagon Wheel no. 1 suggest that sediments in the Pinedale anticline area of the northern Green River basin are presently ca. 30°C–50°C (54°F to 90°F) cooler than their maximum paleotemperature (Pollastro and Barker, 1984a, 1984b). The fission-track apatite ages from this drill hole support higher paleotemperatures in the past and suggest that at least the latest phase of cooling was initiated 2 to 4 m.y. ago (in the Pliocene) and has produced a relatively rapid temperature decrease of 20°C (36°F) or more.

Zircon data presently available from Wagon Wheel no. 1 suggest that the dominant source of sediments in this area remained unchanged through most of the Late Cretaceous and earliest Tertiary, but input from different source(s) is indicated for the "upper part of the Ericson Sandstone" (Cretaceous) and the upper part of the sampled Tertiary section.

ACKNOWLEDGMENTS

I thank R. A. Ullrich of El Paso Exploration Company and B. E. Law for providing samples from Wagon Wheel no. 1 well, L. R. Mahrt for mineral separations, and T. H. McCulloh for the corrected present-day temperatures for the samples. This paper was prepared in cooperation with the U.S. Department of Energy, Morgantown Energy Technology Center.

REFERENCES

BRIGGS, N. D., NAESER, C. W., AND MCCULLOH, T. H., 1979, Thermal history of sedimentary basins by fission-track dating: Geological Society of America Abstracts with Programs, v. 11, p. 394.

———, ———, AND ———, 1981a, Thermal history of sedimentary basins by fission-track dating: Nuclear Tracks, v. 5, p. 235–237.

———, ———, AND ———, 1981b, Thermal history by fission-track dating, Tejon Oil Field area, California: American Association of Petroleum Geologists Bulletin, v. 65, p. 906.

CARPENTER, B. S., AND REIMER, G. M., 1974, Standard reference materials: Calibrated glass standards for fission track use: National Bureau of Standards Special Publication 260–49, 16 p.

DUDDY, I. R., AND GLEADOW, A. J. W., 1982, Thermal history of the Otway Basin, southeastern Australia, from geologic annealing of fission tracks in detrital volcanic apatites (abs.): Fission-Track Dating Workshop, Fifth International Conference on Geochronology, Cosmochronology, and Isotope Geology, Nikko National Park, Japan, p. 13–16.

———, AND ———, 1984, The application of fission track thermochronology to sedimentary basins: Two case studies (abs.): Fourth International Fission Track Dating Workshop, Troy, New York, p. 11.

FLEISCHER, R. L., PRICE, P. B., AND WALKER, R. M., 1965, Tracks of charged particles in solids: Science, v. 149, p. 383–393.

GEOLOGICAL SOCIETY OF AMERICA, 1983, Decade of North America Geology, 1983 Geologic Time Scale, compiled by A. R. Palmer, 2 p.

GLEADOW, A. J. W., AND DUDDY, I. R., 1981, A natural long-term track annealing experiment for apatite: Nuclear Tracks, v. 5, p. 169–174.

———, ———, AND LOVERING, J. F., 1983, Fission track analysis: A new tool for the evaluation of thermal histories and hydrocarbon potential: Australian Petroleum Exploration Association Journal, v. 23, p. 93–102.

HARRISON, T. M., ARMSTRONG, R. L., NAESER, C. W., AND HARAKAL, J. E., 1979, Geochronology and thermal history of the Coast Plutonic Complex, near Price Rupert, British Columbia: Canadian Journal of Earth Sciences, v. 16, p. 400–410.

HURFORD, A. J., 1984, On the closure temperature for fission tracks in zircon (abs.): Fourth International Fission Track Dating Workshop, Troy, New York, p. 22.

KEHLE, R. O., 1972, Geothermal survey of North America 1971 annual progress report: American Association of Petroleum Geologists Research Committee, unpublished duplicated report, 31 p.

———, SCHOEPPEL, R. J., AND DEFORD, R. K., 1970, The AAPG geothermal survey of North America: Geothermics, Special Issue, p. 358–367.

LAW, B. E., 1984a, Introduction, *in* Law, B. E., ed., Geological Characteristics of Low-Permeability Upper Cretaceous and Lower Tertiary Rocks in the Pinedale Anticline Area, Sublette County, Wyoming: U.S. Geological Survey Open-File Report 84-753, p. 1–5.

———, 1984b, Structure and stratigraphy of the Pinedale anticline, Wyoming, *in* Law, B. E., ed., Geological Characteristics of Low-Permeability Upper Cretaceous and Lower Tertiary Rocks in the Pinedale Anticline Area, Sublette County, Wyoming: U.S. Geological Survey Open-File Report 84-753, p. 6–15.

LINDSEY, D. A., NAESER, C. W., AND SHAWE, D. R., 1975, Age of volcanism, intrusion, and mineralization in the Thomas Range, Keg Mountains, and Desert Mountain, western Utah: U.S. Geological Survey Journal of Research, v. 3, p. 597–604.

MÄRK, E., PAHL, M., PURTSCHELLER, F., AND MÄRK, T. D., 1973, Thermische Ausheilung von Uran-Spaltspuren in Apatiten, Alterskorrekturen und Beitrage zur Geochronologie: Tschermaks Mineralogische und Petrographische Mitteilungen, v. 20, p. 131–154.

MCCULLOH, T. H., AND BEYER, L. A., 1979, Geothermal gradients, *in* Cook, H. E., ed., Geologic Studies of the Point Conception Deep Stratigraphic Test Well OCS-CAL 78-164 No. 1, Outer Continental

Shelf, Southern California, United States: U.S. Geological Survey Open-File Report 79-1218, p. 43–48.

NAESER, C. W., 1967, The use of apatite and sphene for fission track age determinations: Geological Society of America Bulletin, v. 78, p. 1523–1526.

———, 1976, Fission track dating: U.S. Geological Survey Open-File Report 76-190, 65 p.

———, 1979a, Fission-track dating and geologic annealing of fission tracks, *in* Jäger, E., and Hunziker, J. C., eds., Lectures in Isotope Geology: New York, Springer-Verlag, p. 154–169.

———, 1979b, Thermal history of sedimentary basins: fission-track dating of subsurface rocks, *in* Scholle, P. A., and Schluger, P. R., eds., Aspects of Diagenesis: Society of Economic Paleontologists and Mineralogists Special Publication 26, p. 109–112.

———, 1981, The fading of fission tracks in the geologic environment—data from deep drill holes: Nuclear Tracks, v. 5, p. 248–250.

———, AND FAUL, HENRY, 1969, Fission track annealing in apatite and sphene: Journal of Geophysical Research, v. 74, p. 705–710.

NAESER, N. D., 1984, Thermal history determined by fission-track dating for three sedimentary basins in California and Wyoming: Geological Society of America Abstracts with Programs, v. 16, p. 607.

OBRADOVICH, J. D., AND COBBAN, W. A., 1975, A time-scale for the Late Cretaceous of the Western Interior of North America: Geological Association of Canada Special Paper 13, p. 31–54.

POLLASTRO, R. M., AND BARKER, C. E., 1984a, Comparative measures of paleotemperature: An example from clay-mineral, vitrinite reflectance, and fluid inclusion studies, Pinedale anticline, northern Green River basin, Wyoming (abs.): Society of Economic Paleontologists and Mineralogists Midyear Meeting, San Jose, California, p. 65–66.

———, 1984b, Geothermometry from clay minerals, vitrinite reflectance, and fluid inclusions—Applications to the thermal and burial history of rocks cored from the Wagon Wheel no. 1 well, Green River basin, Wyoming, *in* Law, B. E., ed., Geological Characteristics of Low-Permeability Upper Cretaceous and Lower Tertiary Rocks in the Pinedale Anticline Area, Sublette County, Wyoming: U.S. Geological Survey Open-File Report 84-753, p. 78–94.

REIMER, G. M., 1972, Fission track geochronology: Method for tectonic interpretation of apatite studies with examples from the central and southern Alps (Ph.D. Dissertation): Philadelphia, University of Pennsylvania, 85 p.

RICE, D. D., AND GAUTIER, D. L., 1983, Patterns of sedimentation, diagenesis, and hydrocarbon accumulation in Cretaceous rocks of the Rocky Mountains: Society of Economic Paleontologists and Mineralogists Short Course No. 11, 339 p.

ROBERTS, J. A., GOLD, R., AND ARMANI, R. J., 1968, Spontaneous-fission decay constant of ^{238}U: Physical Review, v. 174, p. 1482–1484.

SANFORD, S. J., 1981, Dating thermal events by fission track annealing, Cerro Prieto geothermal field, Baja California, Mexico (M.S. Thesis): Riverside, University of California, 105 p.

WEIMER, R. J., 1961, Uppermost Cretaceous rocks in central and southern Wyoming, and northwest Colorado: Wyoming Geological Association, Guidebook, Sixteenth Annual Field Conference, p. 17–28.

ZIMMERMANN, R. A., AND GAINES, A. M., 1978, A new approach to the study of fission-track fading, *in* Zartman, R. E., ed., Short Papers of the Fourth International Conference on Geochronology, Cosmochronology, and Isotope Geology: U.S. Geological Survey Open-File Report 78-701, p. 467–468.

APPLICATION OF CLAY-MINERAL, VITRINITE REFLECTANCE, AND FLUID INCLUSION STUDIES TO THE THERMAL AND BURIAL HISTORY OF THE PINEDALE ANTICLINE, GREEN RIVER BASIN, WYOMING

RICHARD M. POLLASTRO AND CHARLES E. BARKER

U.S. Geological Survey, Box 25046, MS 960 Denver Federal Center, Denver, Colorado 80225

ABSTRACT: Clay-mineralogy, mean random vitrinite reflectance (R_m), and fluid inclusion homogenization temperature (T_h) from host-rock and vein samples in cored intervals from 5,000–18,000 ft (1500–5500 m) indicate that paleotemperature was higher than the temperature currently measured in siliciclastic rocks of the Pinedale anticline, Green River basin, Wyoming. The cored intervals are from lower Tertiary and Upper Cretaceous rocks in the El Paso Natural Gas Wagon Wheel no. 1 well.

Compositional analyses of mixed-layer illite/smectite (I/S) clay from sandstone and shale in the Wagon Wheel core show a progressive increase in ordering and the number of illite layers with depth (and temperature). Temperatures determined from changes in the composition and ordering of I/S, and R_m data, imply that the rocks of Wagon Wheel reached about 200°C at 18,000 ft (5500 m). Uncorrected log temperature at this depth is about 135°–150°C. The thermal gradient calculated at maximum burial temperature (25°C/km), however, is similar to the thermal gradient established by present-day borehole temperature (24°C/km). These data suggest that maximum temperature was 30°–50°C higher than present-day uncorrected borehole temperature. The apparent temperature decrease can be explained by erosion of about 5,600 ft (1700 m) of section, as calculated from a surface R_m intercept of 0.33%, with a geothermal gradient that has remained constant since maximum burial. Major uplift of the Pinedale anticline occurred after apparent maximum burial and temperature were established in the Neogene.

T_h of 130°–150°C from aqueous fluid inclusions in quartz within calcite and quartz veins at 17,000 ft (5200 m), however, conform to the present temperature regime. Fractures probably formed during uplift of the Pinedale anticline and were later mineralized in a temperature regime much like that of the present.

Primary, hydrocarbon-bearing fluid inclusions in veins from 8,000–17,000 ft (2,400–5200 m) are evidence for petroleum migration occurring during the filling of these late fractures. Secondary oil fluid inclusions, trapped in healed microfractures that crosscut authigenic quartz in these veins at a depth of 17,000 ft (5200 m), indicate that petroleum migration continued sometime after these deep fractures were mineralized.

INTRODUCTION

A number of inorganic and organic techniques have been used to determine the burial and paleotemperature history of sedimentary basins (see Scholle and Schluger, 1979). When combined, data obtained from the same samples using several methods of geothermometry can provide more detailed information on the intensity and timing of thermal events than any single geothermometer.

Numerous studies of clay-rich rocks have demonstrated specific depth dependent mineralogical changes (see reviews by Weaver, 1979; and Hower, 1981). In particular, clay-mineral assemblages in sedimentary rocks are commonly used to determine the extent of burial diagenesis. The most significant reaction within the clay-mineral assemblages is the conversion of smectite to illite through a mixed-layer illite/smectite (I/S) series. As the percent of illite layers in I/S increases with burial and temperature, the alternation of illite and smectite layers in the I/S phase becomes progressively more predictable (less random, or more ordered). The extent of this reaction and the transition from a "random" interstratified I/S to an "ordered" interstratified I/S can be recognized in diffraction profiles (Reynolds and Hower, 1970). This specific reaction series is not reversible under burial conditions. Hoffman and Hower (1979), and Weaver (1979) have documented the temperatures at which such clay minerals are transformed. After a comprehensive analysis of the clay mineralogy in a given area, these clay-mineral phases can be used as maximum recording geothermometers (Pollastro and Bader, 1983).

Vitrinite reflectance (R_m) is the best optical technique for assessing the level of thermal maturity of sedimentary organic matter (OM) (Stach and others, 1982; Tissot and Welte, 1978). A sequence of increasing R_m values with depth in a well results from greater thermal maturity in response to the geothermal gradient (Dow, 1977). R_m is not reversible with decrease in temperature (Tissot and Welte, 1978). R_m is mainly set by peak burial temperature, and consequently, reaction duration can have little influence on the maturation of OM (Gretener and Curtis, 1982; Barker, 1983; Price, 1983). Barker and Pawlewicz (1986) have calibrated an empirical geothermometer based on the observation that R_m has a strong correlation ($r = 0.8$, $n \sim 650$) with temperature in more than 35 geologic systems that have known maximum temperatures. Reaction duration in these systems varies from 0.01 m.y. to greater than 30 m.y. They also conclude that the strong temperature dependence of R_m indicates that reaction duration is a neglible factor in the thermal maturation of OM. In this paper R_m is used to estimate the maximum temperature reached during the burial history of a rock.

Both aqueous and oil fluid inclusions are common in sedimentary rocks. Measurement of the phase transition temperatures in a fluid inclusion gives information on temperature, pressure, and the composition of fluids during diagenesis. The homogenization temperature (T_h) is a phase transition that is a quantitative property of the closed chemical system in the fluid inclusion. T_h can be directly interpreted as the minimum temperature of inclusion entrapment (see reviews by Roedder, 1984; Hollister and Crawford, 1981). Primary fluid inclusions in successive stages of cementation or crosscutting veins record a portion of that rock's thermal and fluid history; however, a sequence of diagenetic mineralization must be fixed in geologic time by burial reconstruction to interpret the significance of T_h in the thermal history. For example, Currie and Nwachukwu (1974) studied fluid inclusions in veins within the Cardium Sandstone, Central Alberta, Canada. They found that T_h varied in these inclusions from 45°–120°C. This variation in the minimum temperature of entrapment was attributed to successive mineral precipitation in fractures during cooling from regional uplift and erosional unloading.

The present study uses geothermometry from changes in clay-mineral phases and composition, vitrinite reflectance, and fluid inclusions to reconstruct the thermal history in gas-bearing rocks of the Pinedale anticline. These paleotemperature data and the present-day temperature records for the well, in conjunction with stratigraphy, reconstruct a thermal and burial history for the Pinedale anticline.

WAGON WHEEL NO. 1 WELL

The El Paso Natural Gas Wagon Wheel no. 1 well (Sec. 5, T. 30 N, 108W), located on the crest of the Pinedale anticline, northern Green River basin, Wyoming (Fig. 1), was a site originally planned for the nuclear stimulation of potential tight-gas sand units. In anticipation of this test, 12 sections of core, totalling about 900 ft (280 m), were taken from lower Tertiary and Upper Cretaceous rocks at depths ranging from 5,000–18,000 ft (1500–5500 m, Fig. 1). These rocks consist of interbedded, discontinuous, low-permeability sandstone and shale. With exception of the lower 2,800 ft (850 m), the rocks penetrated by Wagon Wheel were deposited in nonmarine environments. Details on the Wagon Wheel project and the geology of the Pinedale area can be found in Shaughnessy and Butcher (1974), and Law (1979, 1984a).

METHODS AND MODELS

Clay-mineral studies.—

Clay minerals from about 180 sandstone and shale samples were analyzed in this study. The clay minerals were initially characterized through X-ray powder diffraction (XRD) analyses of both the whole-rock and <2 μm clay-mineral fractions. Scanning electron microscopy and energy-dispersive elemental analysis were also used to help identify and characterize the clay minerals.

Composition and ordering of I/S clay were determined by the methods of Reynolds and Hower (1970) on oriented, glycol-saturated specimens of the <0.5 μm (equivalent spherical diameter) fraction. Specimens were prepared using the method described by Pollastro (1982). Ordering types were defined using the "Reichweite" (R) notation, as described by Jagodzinski (1949), Kakinoki and Komura (1952), and Reynolds (1980), where "R" signifies the most distant layer in an interstratified sequence that affects the probability of occurrence of the final layer. Diffractograms indicate that most naturally occurring I/S clay can be cate-

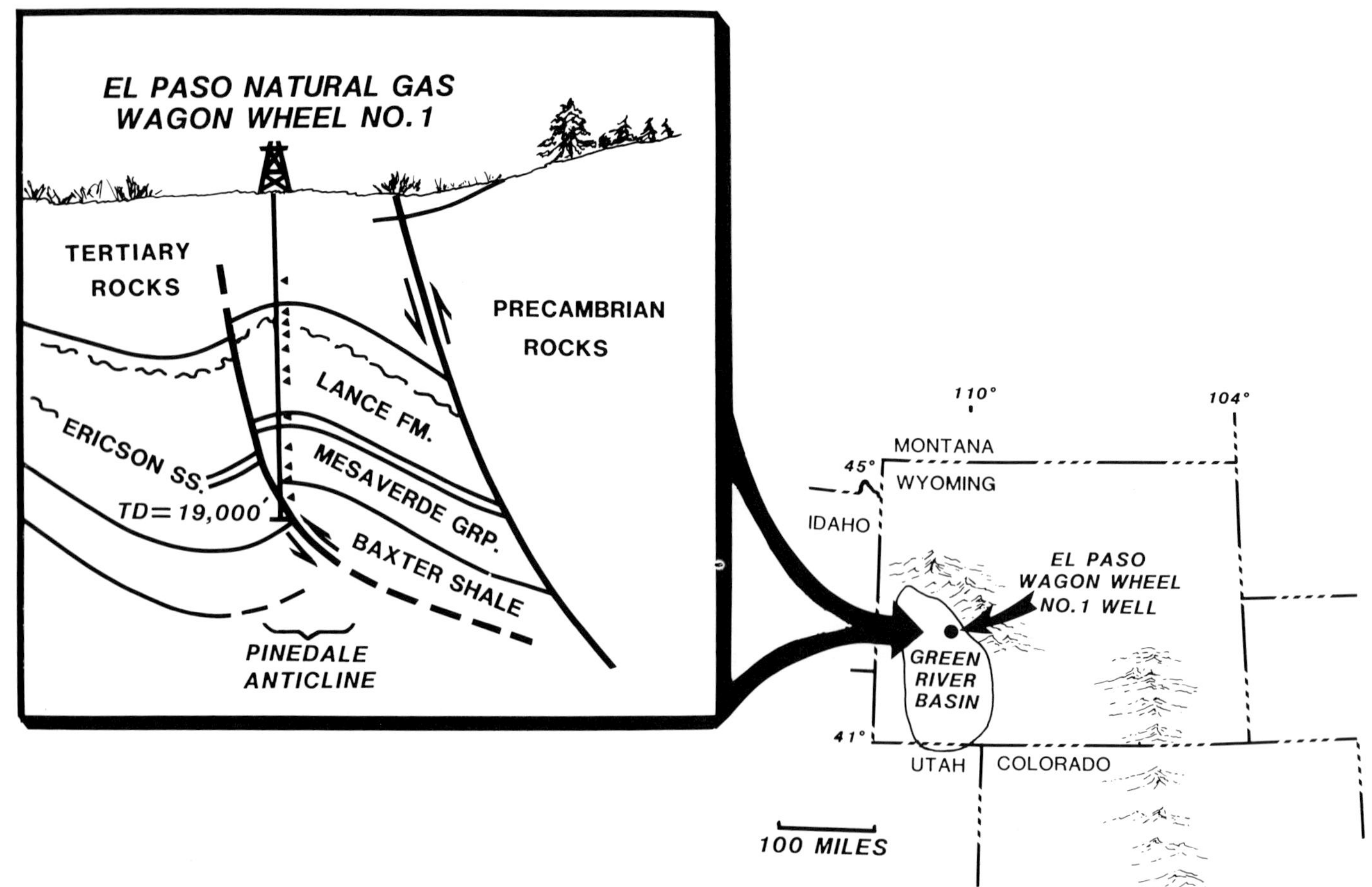

FIG. 1.—Location map and generalized stratigraphic cross section, Wagon Wheel no. 1 well, Wyoming. Solid triangles indicate approximate distribution of cored intervals; wavy lines indicate unconformable contacts. Stratigraphic section modified from Law (1984a). TD, total depth.

gorized as falling into one of three ordering types of interstratification (Reynolds and Hower, 1970). These three types are (1) random I/S ($R = 0$), (2) allevardite or short-range ordered I/S ($R = 1,2$), and (3) kalkberg or long-range ordered I/S ($R \geq 3$).

Paleotemperatures were assigned at specific depths in the Wagon Wheel well based on the first appearance or disappearance of different ordering types of I/S clay and corrensite using the temperatures over which these minerals are stable, as estimated by Hoffman and Hower (1979). The exact nature of the I/S reaction in the cores of the Wagon Wheel well has been discussed by Pollastro (1983; 1985).

Vitrinite reflectance.—

Vitrinite reflectance (R_m) was measured on 12 thin coal laminae in cores distributed over a depth interval from 5,000–18,000 ft (1500–5500 m). The coaly samples were ground to pass through a 0.5 mm mesh screen and prepared using the method of Baskin (1979). R_m is the mean calculated from the reflectance of 100 grains measured in random orientation. Calibration of the microscope photometer system is described by Barker and Halley (this volume). The coal laminae samples contained common, large monomaceralic and bimaceralic grains; structured vitrinite was rare. The vitrinite reflectance measurements from all samples have a tight unimodal distribution with a range no greater than 0.30% and a standard deviation of 0.10% or less. The moderate organic content of the whole-rock samples, averaging 1.2% total organic carbon (Dickinson, 1984), and gas-prone character of the kerogen (type III, Law, 1984b), indicate that suppression of R_m is not a significant factor (Price and Barker, 1985). Therefore, R_m was not corrected to compensate for polymodal reflectance distributions or suppression effects. The mean reflectance (as compared to median or modal reflectance) in these samples appears to be an adequate measure of thermal maturity.

Maximum burial temperature was estimated using the vitrinite reflectance model of Barker and Pawlewicz (1986). Vitrinite reflectance temperature estimates (T_{Rm}), calculated by their method, measure the peak burial temperature reached at the sampling point. The slope of the T_{Rm}-depth curve, when interpreted as the maximum geothermal gradient, requires that thermal continuity existed throughout the sedimentary section at the time peak temperature was reached. This condition is met in water-saturated rocks, where any temperature difference is equilibrated with time through a sedimentary column (Buntebarth, 1984).

Fluid inclusions.—

Veins were common throughout the cores of the Wagon Wheel well and have been described in detail by Dickinson (1984). Fluid inclusions were found in both calcite and calcite-quartz veins within the core samples. Doubly polished sections of the veins were prepared using the method of Barker and Reynolds (1984). Fluid inclusion homogenization temperatures (T_h) were determined in a U.S. Geological Survey gas-flow heating stage (Werre et al., 1979) calibrated using the procedure described by Barker and Halley (this volume). Fluid inclusions containing hydrocarbon gases were identified by cracking them open in kerosene with a crushing stage (Roedder, 1984). Homogenization to the vapor phase of some oil inclusions was difficult to observe in transmitted light because the thin fluid film of the inclusion was hidden by internal reflection in the inclusion wall. T_h in these vapor-rich inclusions was observed by the disappearance of fluorescent liquid into a less fluorescent vapor phase.

Hydrocarbon-bearing fluid inclusions were common in mineral veins from the Wagon Wheel core. Primary aqueous fluid inclusions contained considerable quantities of natural gas. These gas-rich inclusions, when in a calcite host, decrepitated or permanently deformed (stretched) with heating and no reliable T_h could be measured. Fluorescence petrography also identified trails of secondary oil inclusions in authigenic quartz. These characteristics are unfavorable for determining mineral formation temperature; however, these fluid inclusions were still an important record of fluid migration (including petroleum) through these rocks.

Aqueous inclusions in quartz were the only inclusions that would homogenize and provide acceptable T_h data. These inclusions are suspected to contain natural gas because of their coexistence with gas-bearing aqueous inclusions in calcite from the same vein. The T_h data from these inclusions were not pressure-corrected. The presence of natural gas can produce a large error in the pressure correction of T_h data to trapping temperature (Hanor, 1980). Hanor concluded, however, that T_h of natural-gas-bearing aqueous inclusions can be used as the minimum temperature of entrapment.

RESULTS AND INTERPRETATION

Clay mineralogy.—

I/S clay from both sandstone and shale show a progressive increase in the percent of illite layers with depth (Fig. 2). At shallow depths, I/S clay from sandstone contains much less illite than does I/S from shale. This difference in the amount of illite in I/S of sandstone versus shale can be explained by considering the origin of I/S in these two lithologies. I/S clay in these sandstones is largely authigenic cement that has formed in equilibrium with the physical and chemical conditions existing in the porous rock at the time of its precipitation. The range of starting composition of authigenic I/S is, therefore, relatively narrow. The composition of I/S can be modified later in the burial history if the environment is changed. In a vertical sequence of deeply-buried sandstones, the composition of authigenic I/S usually results from the maximum diagenetic conditions (primarily temperature) it has experienced. In shale, however, I/S clay is primarily detrital and has a large range in composition. It is from many sources and represents a wide range of conditions, including older, reworked rocks that have likely once undergone more extensive diagenetic reactions. Detrital I/S clay can also be a component of sandstone but its presence can be masked if authigenic I/S is more abundant. With increasing burial and temperature, the compositions of I/S from sandstone and shale begin to become more similar. Eventually, the depth-diage-

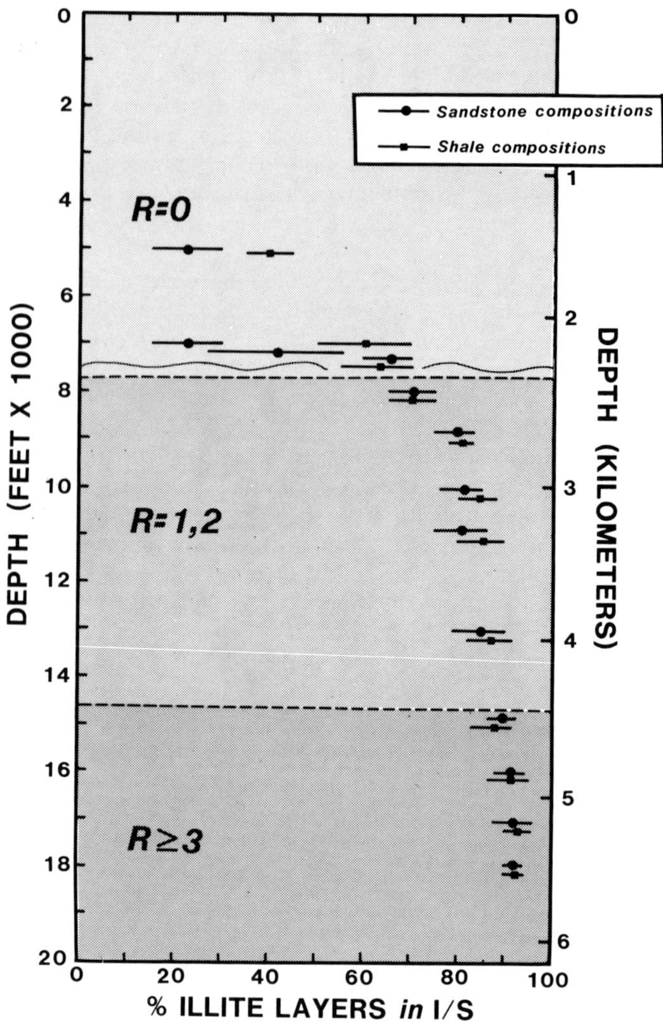

FIG. 2.—Composition of mixed-layer illite/smectite (I/S) clay from <0.5 μm fraction versus present burial depth in sandstone and shale, Wagon Wheel no. 1 well. Dashed lines show approximate depths where changes in ordering of I/S occur in the well. One hundred eighty samples are represented in the diagram. Bars indicate composition range of I/S in a group of samples analyzed; solid circles (sandstone) and squares (shales) on bars show average value for that group. Wavy line in diagram approximates Cretaceous-Tertiary unconformity in well. R, Reichweite notation for I/S ordering types (R = 0, random I/S; R = 1,2, short-range ordering; R ≥ 3, long-range ordering).

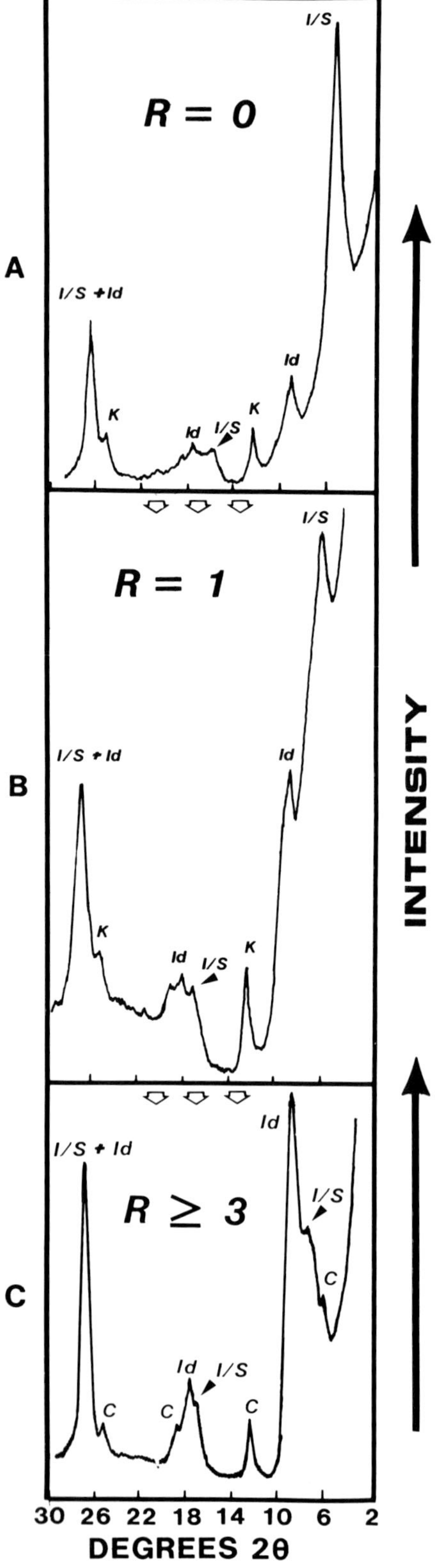

FIG. 3.—X-ray diffraction patterns showing changes in ordering of mixed-layer illite/smectite (I/S) clay. Patterns from oriented, glycol-saturated mounts of <0.5 μm clay. (A) Random I/S. (B) short-range ordered I/S. (C) long-range ordered I/S. Id, discrete illite; K, kaolinite; R, Reichweite notation for ordering types; CuK_α radiation.

netic reaction minimizes the difference in composition and produces similar I/S clays in both lithologies (Fig. 2).

As the amount of illite layers in I/S increase with increasing depth, I/S proceeds from a randomly interstratified phase (R = 0, Fig. 3A), to an ordered, allevardite-like mineral (R = 1). The transition is recognized on an XRD pattern after glycol-saturation by the disappearance of the 17-angstrom reflection, characteristic of glycol-smectite, and the appearance of a reflection at 13–14 angstroms, indicative of the short-range ordered I/S phase (allevardite-like, Fig. 3B). This change occurs in the Wagon Wheel core at about 7,800 ft (2400 m). The transformation from allevardite-like I/S to the kalkberg-like (R ≥ 3; > 85% illite layers in I/S, Fig. 3C) form occurs at about 15,000 ft (4600 m).

Corrensite, a regularly interstratified chlorite/smectite clay mineral, coexists with I/S clay in part of the Wagon Wheel core (Fig. 4). Corrensite is also useful as a geothermometer in sandstone. Corrensite first appears in the Wagon Wheel core at about 7,200 ft (2200 m) and persists in the cores to about 8,300 ft (2500 m).

Based on the first appearance or disappearance of ordering types of I/S clay or of corrensite, and the temperatures for which these minerals are stable (Hoffman and Hower, 1979), a maximum temperature curve was interpreted for the Wagon Wheel well (Fig. 4). The data calculated from the I/S reaction (Fig. 4), when compared to uncorrected borehole measurements, suggest that these rocks reached temperatures 30°–50°C higher in the past. The slope of the

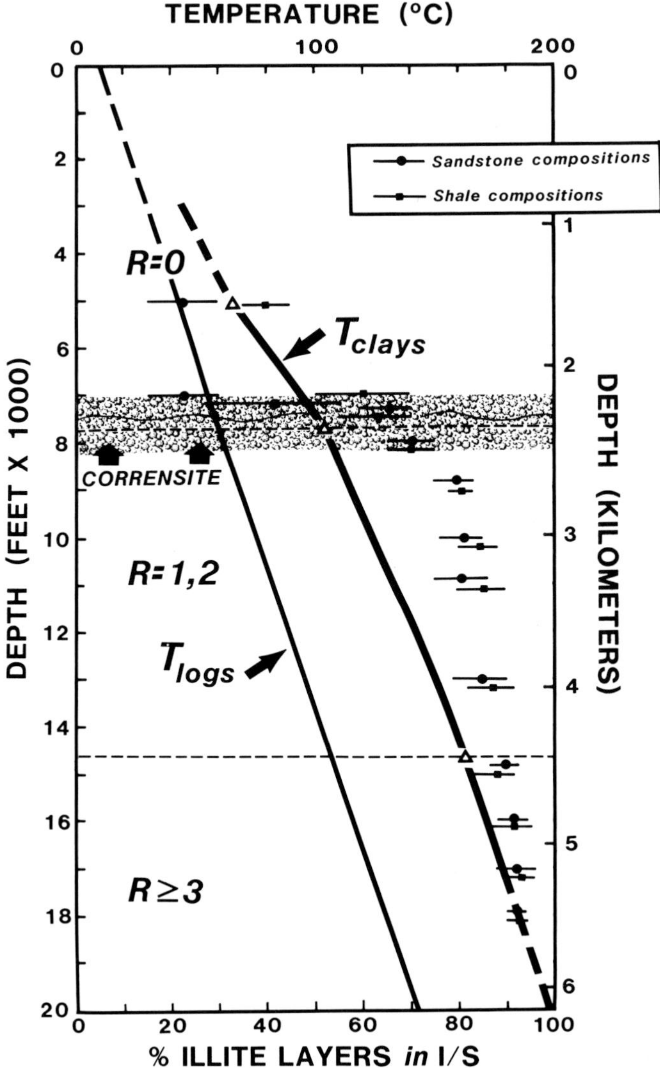

FIG. 4.—Paleotemperature gradient as determined by clay-mineral geothermometers and present-day thermal gradient from uncorrected borehole temperatures. Clay mineral temperatures were interpreted from changes in ordering and composition of mixed-layer illite/smectite (I/S) clay and the appearance of corrensite (Hoffman and Hower, 1979). R, Reichweite notation for ordering types; T, temperature.

paleotemperature curve established by the clay-mineral indicators is in good agreement with the present-day geothermal gradient at depths greater than 8,000 ft (2400 m). The disagreement in slope of the curve above this depth is because the I/S reaction likely begins at a shallower depth in the well; however, no core samples were available to test this hypothesis.

Vitrinite reflectance.—

Regression analysis of R_m measurements from the Wagon Wheel coal samples shows a strong linear relationship (r^2 = 0.86) with depth on a semilogarithmic plot (Fig. 5). This type of plot straightens the maturation curve, allowing an accurate R_m estimate at the existing ground surface, and any rock removed by erosion, by simple extrapolation of a least-square regression equation. Vitrinite reflectance near the surface in an uneroded terrain is about 0.20% R_m (Dow, 1977). The extrapolated intercept of the Wagon Wheel R_m-depth regression equation with the present surface is 0.33% R_m. Projection of this line to 0.20% R_m indicates 5,600 ft (1700 m) of rock has been eroded.

Calculation of the peak burial temperature (T_{Rm}) suggests the rocks of Wagon Wheel reached 200°C at 18,000 ft (5500 m). Comparison of T_{Rm}-depth curve (Fig. 6) with uncorrected borehole temperature (T_{log}) implies that temperature has decreased by some 50°C. The slope of a linear regression equation of T_{Rm}-depth (Fig. 6) is equivalent to a paleogeothermal gradient of about 25°C/km at maximum burial conditions. Comparison of the T_{Rm}-depth curve with the present-day geothermal gradient of 24°C/km established from the slope of uncorrected borehole measurements (Fig. 6) shows that the two curves are almost parallel. The similar slope of the two curves, and the erosion indicated by the surface R_m intercept, argue that burial temperature in the Pinedale area has been reduced by uplift and erosion, as suggested by Pollastro and Barker (1984), and Law et al. (in prep.), rather than by a major decrease in the geothermal gradient.

Fluid inclusions.—

The authigenic quartz that contained aqueous fluid inclusions, where T_h could be measured, coexists with calcite in fractures (Fig. 7) at about 17,000 ft (5200 m). These rare, 2–5 μm, gas-bearing(?) aqueous inclusions homogenized to the liquid phase at a T_h ranging from 130°–150°C. Undercooling of least 40°C was required to renucleate a vapor phase after homogenization, indicating that these are aqueous inclusions. We could not freeze these small inclusions to confirm their composition, but non-fluorescent inclusions in the coexisting calcite froze at about −40°C with contraction of the vapor bubble which is characteristic of a water-rich fluid (Roedder, 1984). The few homogenization temperature data from these inclusions in quartz is qualitatively supported by a similar range of decrepitation temperature (130°–170°C) from gas-rich aqueous inclusions in the adjacent calcite. Decrepitation temperature of low T_h inclusions (high fluid density) in soft minerals is usually near T_h (Roedder, 1984). Present uncorrected borehole temperature measured in Wagon Wheel at 17,000 ft (5200 m) is about

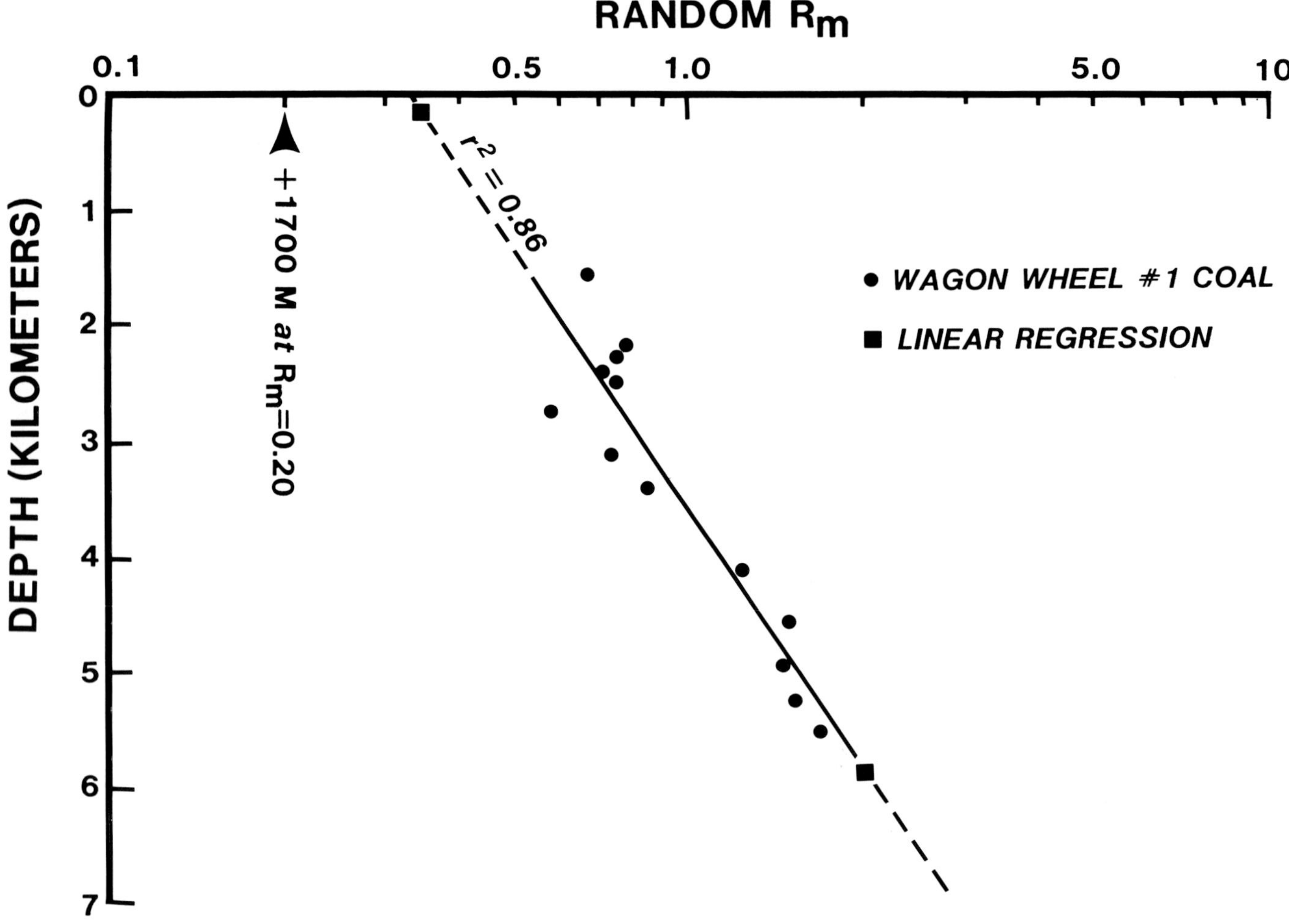

FIG. 5.—Semilogarithmic plot of mean vitrinite reflectance (R_m) versus depth from coaly laminae, Wagon Wheel no. 1 core. Intersection of linear regression line with surface occurs at R_m = 0.33%. Erosion estimate of 5600 ft (1700 m) was calculated by projection to a surface intersection of R_m = 0.20% (Dow, 1977).

125°C. The fluid inclusion data show that these calcite-quartz veins formed at temperatures much lower than the maximum temperature determined by clay-minerals and R_m.

Liquid- and vapor-rich oil inclusions, typically 20–40 μm long (Fig. 8A), occur in this same authigenic quartz. Both the liquid- and vapor-rich oil inclusions occur in the quartz as discontinuous trails, indicating that they formed after the mineral host (secondary inclusions). The liquid-rich types homogenize to the liquid phase over a T_h range of 60°–80°C. Vapor-rich oil inclusions crosscut the liquid-rich inclusion trails. Vapor-rich inclusions homogenize to the vapor phase at T_h ranging from 35°–80°C. After homogenization, cooling both types of oil inclusions renucleated a vapor phase within 10°C. The fluorescent character and rapid renucleation of the vapor phase with cooling from T_h are criteria for distinguishing oil inclusions from aqueous types. Both the liquid- and vapor-rich oil inclusions fluoresce bluish-white to white under ultraviolet excitation, suggesting that they contain similar oils with API gravity of at least 45° (Anderson, 1975). The secondary origin, similar upper limit of T_h, and fluorescence characteristics all suggest that these fluid inclusions trapped a gas-condensate that migrated into microfractures within the quartz. This is the youngest diagenetic event recorded in the fracture-filling cements deep within Wagon Wheel.

The most common type of inclusions found in the Wagon Wheel core are gas-bearing aqueous inclusions in calcite-filled fractures. They occur randomly over the interval from 8,000–17,000 ft (2400–5200 m) (Fig. 8B). These inclusions decrepitated or stretched near T_h and no quantitative temperature data could be obtained. A transition from only gas-rich aqueous inclusions to coexisting gas-rich aqueous inclusions and fluorescent oil inclusions was found at a depth of about 17,000 ft (5200 m). This change corresponds to the transition from continental to marine rocks in the Wagon Wheel well, possibly reflecting a change from gas-prone type III kerogen and coal to oil-prone type II kerogen.

THERMAL HISTORY AND BURIAL RECONSTRUCTION

Maximum burial temperature calculated from both clay-mineral phases and vitrinite reflectance indicate that the rocks of the Wagon Wheel core were 30°–50°C higher than those presently measured in the well (Fig. 9). The maximum tem-

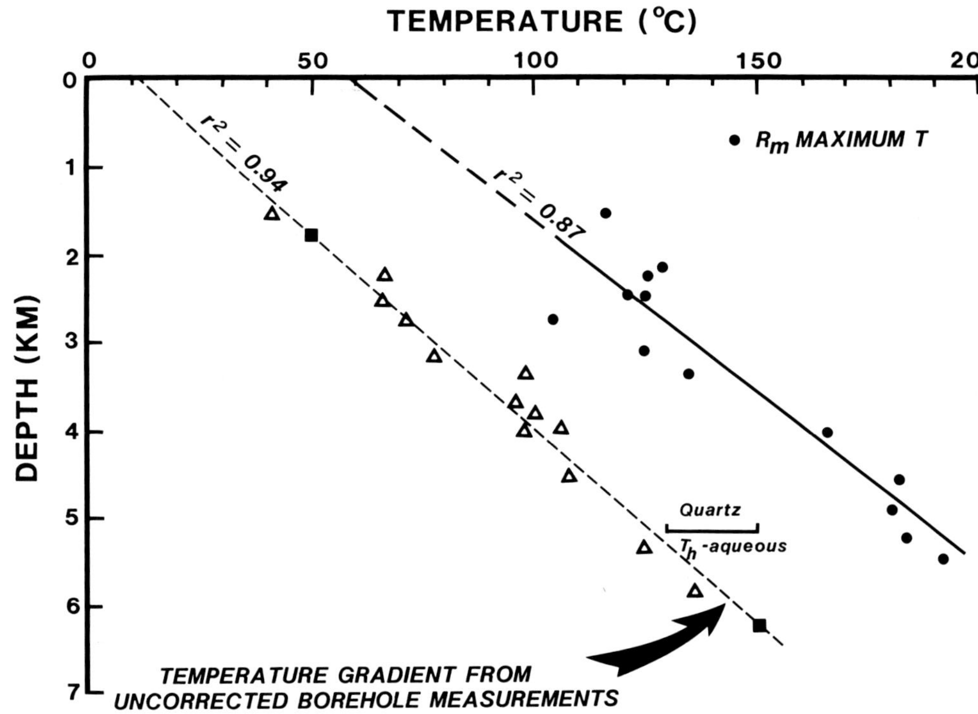

FIG. 6.—Maximum temperature (determined from R_m) versus depth and for present-day, uncorrected borehole temperatures versus depth, Wagon Wheel no. 1 well. Conversion of mean vitrinite reflectance (R_m) to maximum temperature is based on the vitrinite reflectance model of Barker and Pawlewicz (1986). The slopes of the linear regression curves indicate a geothermal gradient of about 25°C/km. Range of homogenization temperature measurements of aqueous fluid inclusions in quartz (T_h aqueous) at 17,000 ft (5200 m) agree with the present-day well log temperatures.

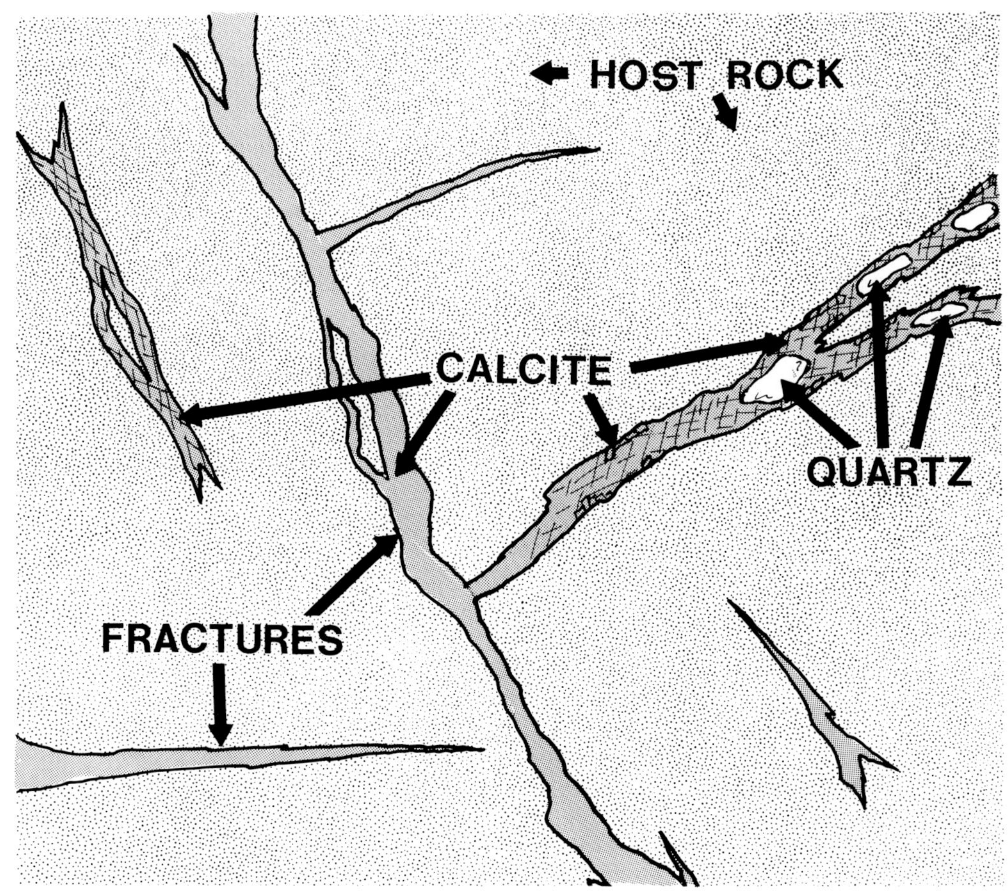

FIG. 7.—Textural relationships in calcite and quartz veins from core of sandstone at 17,000 ft (5200 m), Wagon Wheel no. 1 well.

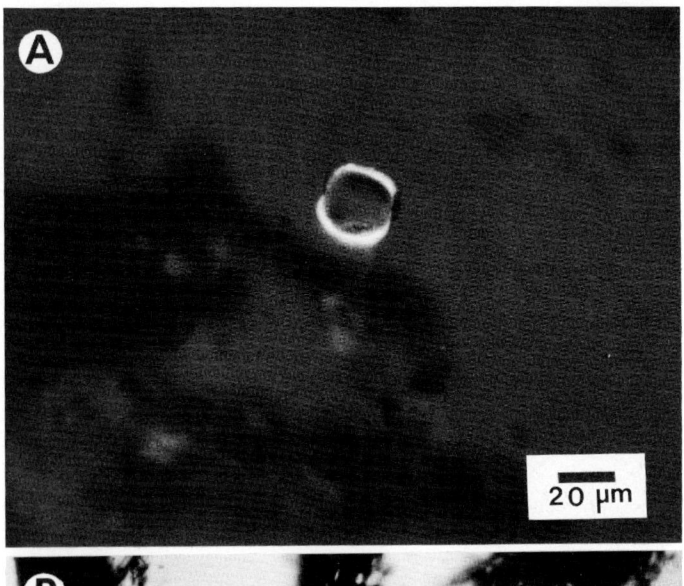

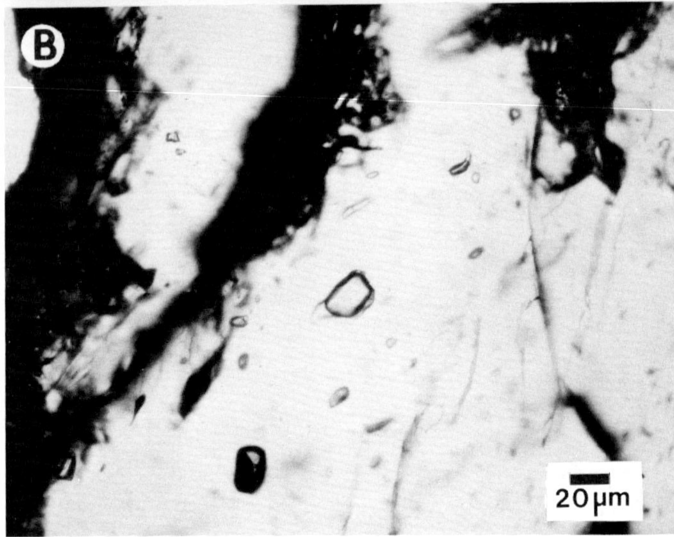

FIG. 8.—Polished thin section photomicrographs of fluid inclusions in calcite and quartz veins from sandstone core at 17,000 ft (5200 m), Wagon Wheel no. 1 well. (A) Fluorescent oil inclusion. (B) Gas-rich aqueous inclusions along a microfracture in quartz.

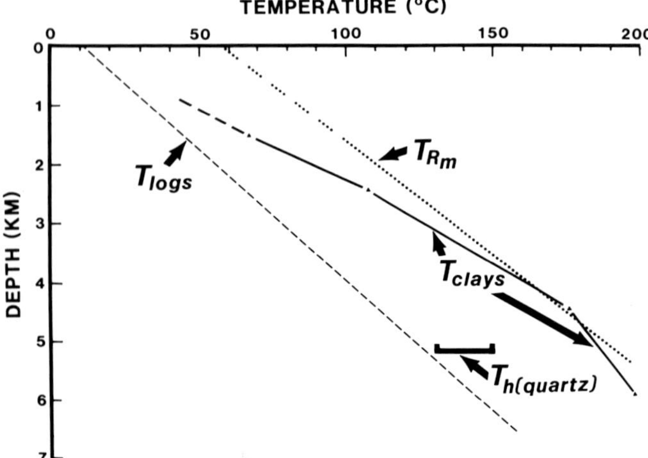

FIG. 9.—Composite diagram of temperature versus depth showing profiles from present-day borehole temperature (T_{log}) and from clay-mineral (T_{clay}) and vitrinite reflectance (T_{Rm}) geothermometry. Range of homogenization temperature ($T_{h-quarta}$) from aqueous fluid inclusions in vein quartz are shown.

perature in Wagon Wheel calculated from these methods was about 190°–200°C at 18,000 ft (5500 m) but has declined to a present, uncorrected bottom hole log temperature (T_{log}) of 135°C. Although there is no consensus on a method for correcting log temperature to actual undisturbed formation conditions, it is generally agreed that log temperatures are low. The American Association of Petroleum Geologists Geothermal Gradient Committee estimates that T_{log} should be increased by about 10°C at 18,000 ft (5500 m). Using this correction in Wagon Wheel gives an undisturbed formation temperature of about 145°C at 18,000 ft (5500 m). Thus, the T_h range of 130°–150°C from aqueous fluid inclusions in quartz at 17,000 ft (5200 m) is in agreement with present borehole temperatures (Fig. 9). Pollastro and Barker (1984) concluded that reduced temperatures at present were caused by uplift and erosional unloading during recurrent growth of the Pinedale anticline. The fractures apparently opened during this event and the fracture-filling minerals formed sometime after the present temperature regime was established. The thermal history determined by Pollastro and Barker (1984) is supported by new data from the Pinedale anticline. Lickus et al. (1984) showed that iso-reflectance lines conform to structure across the Pinedale anticline. Iso-reflectance lines established during peak burial-temperature should parallel major bedding planes in undeformed strata (Stach et al., 1982). If the sedimentary rock package is deformed later in the burial history, the iso-reflectance lines are deformed to conform with structure. The data of Lickus et al. (1984) are strong evidence that the Pinedale anticline formed after maximum temperature was established.

Burial reconstruction of the Baxter Shale/Mesaverde Group contact (Fig. 10) indicates that the rocks in the Wagon Wheel well were buried rapidly to a depth of about 7.3 km at 37 m.y. BP. Unconformable boundaries in this section can be neglected as they were insignificant in relation to time and the amount of sediment removed (Law, 1984a). After rapid burial, no post-Eocene (approximately 37 m.y. BP) net deposition of sediment occurred in the Pinedale area (Schuster and Steidmann, 1983). A recent study of the annealing of fission tracks in minerals from the Wagon Wheel core by Naeser (1984) confirmed the conclusion of Pollastro and Barker (1984) and Law et al. (in prep.) that temperature has declined in the Pinedale anticline. Naeser (1984) found that temperature decreased at least 20°C, commencing about 2–4 m.y. BP (Fig. 10).

A temperature decline of 30°–50°C in the Pinedale anticline is indicated by clay-mineral and vitrinite reflectance geothermometry. Furthermore, vitrinite reflectance data indicate that this decline resulted from uplift and erosion of 1700 m of section. A geothermal gradient of about 25°C/km has probably existed from the time of maximum burial. Erosion of 1700 m in a geothermal gradient of 25°C/km would decrease temperature by about 40°C, supporting the estimate of 30°–50°C from comparative geothermometry.

Primary hydrocarbon-bearing fluid inclusions occurring in these veins from 8,000–17,000 ft (2400–5200 m) show that oil and gas migration occurred throughout most of the Wagon Wheel well during mineralization of these fractures in the Neogene. Secondary oil fluid inclusions trapped in healed microfractures that crosscut the authigenic quartz in

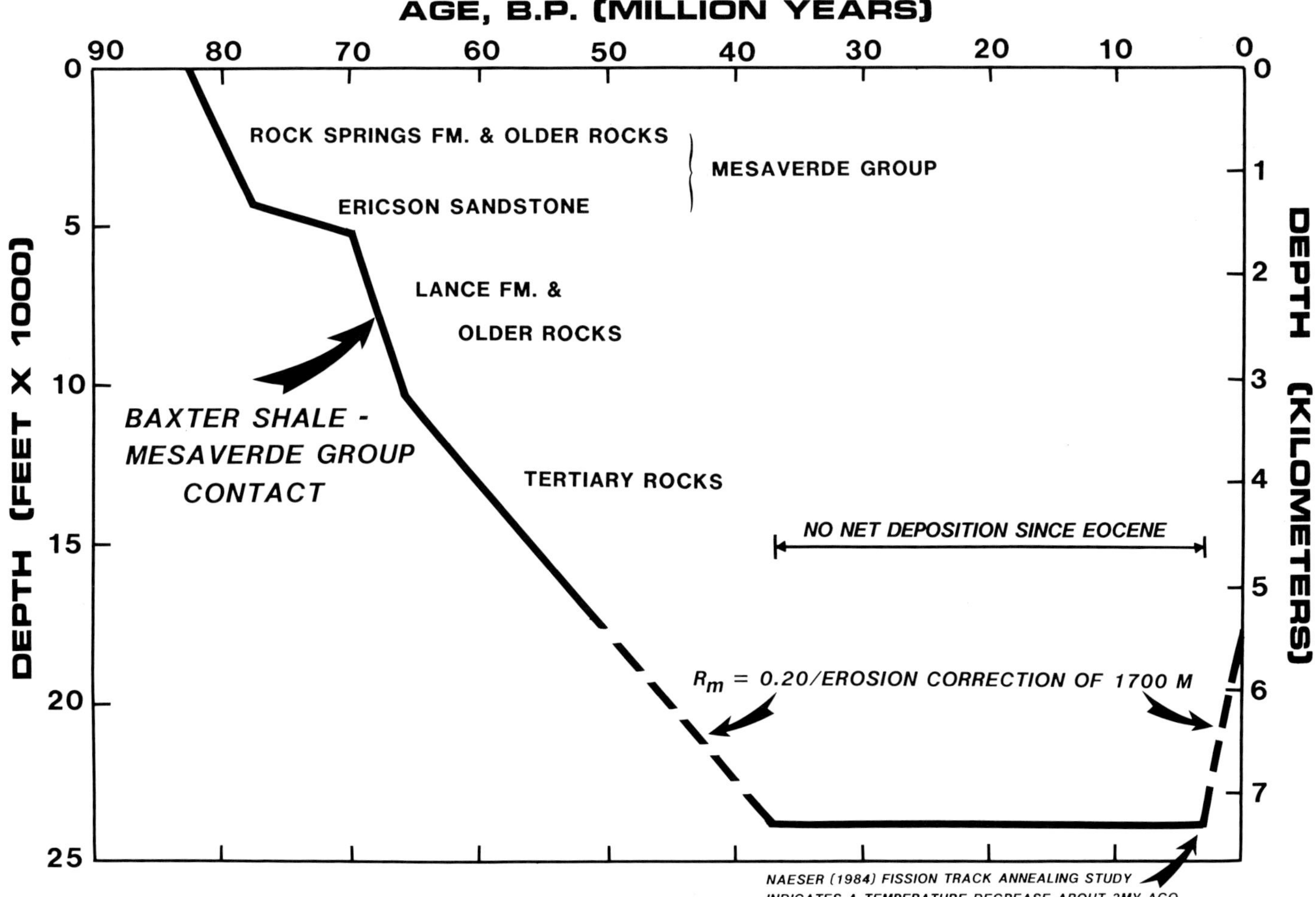

FIG. 10.—Burial reconstruction profile of the Baxter Shale-Mesaverde Group contact in the Pinedale anticline. The reconstruction is based on the stratigraphy and structure from Shaughnessy and Butcher (1974), Reynolds (1976), and Law (1979, 1984a). Absolute ages of the formation boundaries were taken from the potassium-argon ages of Reynolds (1976). Where no absolute ages were available for a particular rock unit, dates were assigned based on stratigraphy from the North American Geologic Time Scale (Geological Society of America, 1983). R_m, mean vitrinite reflectance.

veins at 17,000 ft (5200 m) are evidence that petroleum migration continued after these deep fractures were mineralized.

SUMMARY

Interpretation of inorganic and organic diagenesis in the lower Tertiary and Upper Cretaceous rocks of the Wagon Wheel well, Pinedale anticline, northern Green River basin, Wyoming, has established that:

1. I/S clay from sandstone and shale show a progressive increase in ordering and in the number of illite layers with depth and temperature.
2. Depth-temperature curves calculated from the appearance or disappearance of diagenetic clay-mineral phases and vitrinite reflectance both indicate that burial temperature was 30°–50°C higher in the past.
3. The slope of a linear regression equation of T_{Rm} versus depth indicates the paleogeothermal gradient at maximum burial conditions was about 25°C/km, which is in good agreement with the present geothermal gradient of 24°C/km. The similarity in slope between the two curves and the surface intercept of R_m = 0.33% argues that burial temperatures in the Pinedale anticline have been reduced by uplift and erosion rather than by a major decrease in the geothermal gradient.
4. Major uplift of the Pinedale anticline occurred after the apparent maximum temperature was established. This uplift and its associated erosional unloading apparently caused fractures in the rocks that were later mineralized in a temperature regime similar to the present.
5. Primary, hydrocarbon-bearing fluid inclusions in calcite veins show that petroleum migration occurred in the Neogene during mineral precipitation. Secondary oil fluid inclusions trapped in healed microfractures that crosscut authigenic quartz in veins at 17,000 ft (5200 m) are evidence that oil migration occurred after these deep fractures were mineralized.

ACKNOWLEDGMENTS

We thank B. E. Law and R. C. Burruss for their helpful discussions and critical reviews of the manuscript. B. L. Crysdale measured vitrinite reflectance on the coal samples colleted by W. W. Dickinson. This work was partly supported by the U.S. Department of Energy.

REFERENCES

ANDERSON, GENE, 1975, Coring and Core Analysis Handbook: Petroleum Publishing Company, Tulsa. 200 p.

BARKER, C. E., 1983, Influence of time on metamorphism of sedimentary organic matter in liquid-dominated geothermal systems, western North America: Geology, v. 11, p. 384–388.

BARKER, C. E., AND PAWLEWICZ, M. J., 1986, The correlation of vitrinite reflectance with maximum temperature in humic kerogen, *in* Buntebarth, Gunter, and Stegena, L. (eds.), Paleogeothermics: Springer-Verlag, New York. 16 p.

———, AND REYNOLDS, T. J., 1984, Preparing doubly polished sections of temperature-sensitive sedimentary rocks: Journal of Sedimentary Petrology, v. 54, p. 635–636.

BASKIN, D. K., 1979, A method of preparing phytoclasts for vitrinite reflectance analysis: Journal of Sedimentary Petrology, v. 49, p. 633–635.

BUNTEBARTH, GUNTER, 1984, Geothermics: Springer-Verlag, New York. 144 p.

CURRIE, J. B., AND NWACHUKWU, S. O., 1974, Evidence of incipient fracture porosity in reservoir rocks at depth: Bulletin of Canadian Petroleum Geologists, v. 22, p. 42–58.

DICKINSON, W. W., 1984, Isotope geochemistry of carbonate minerals in nonmarine rocks: Northern Green River Basin, Wyoming: Unpublished Ph.D. Thesis, University of Colorado, Boulder. 150 p.

DOW, W. G., 1977, Kerogen studies and geological interpretations: Journal of Geochemical Exploration, v. 7, p. 79–99.

GEOLOGICAL SOCIETY OF AMERICA, 1983, Decade of North America Geology, 1983 Geologic Time Scale, compiled by A. R. Palmer, 2 p.

GRETENER, P. E. AND CURTIS, C. D., 1982, Role of temperature and time on organic metamorphism: American Association of Petroleum Geologists Bulletin, v. 66, p. 1124–1129.

HANOR, J. S., 1980, Dissolved methane in sedimentary brines: Potential effect on the PVT properties of fluid inclusions: Economic Geology, v. 75, p. 603–617.

HOFFMAN, J., AND HOWER, J., 1979, Clay mineral assemblages as low grade metamorphic indicators: Application to the thrust faulted disturbed belt of Montana, U.S.A., *in* Scholle, P. A., and Schluger, P. K. eds., Aspects of Diagenesis: Society of Economic Paleontologists and Mineralogists Special Publication 26, p. 55–79.

HOLLISTER, L. S., AND CRAWFORD, M. L., eds., 1981, Short Course in Fluid Inclusions: Applications to Petrology: Mineralogical Association of Canada, Short Course Handbook, v. 6, 304 p.

HOWER, JOHN, 1981, Shale diagenesis, *in* Longstaffe, F. J., ed., Clays and the Resource Geologist: Mineralogical Association of Canada, Short Course Handbook, v. 7, p. 60–80.

JAGODZINSKI, H., 1949, Eindimensionale fehlordnung in kristallen und ihr einfluss auf die Rontginínterferenzen. I. berechnung des fehlordnunbsgrades aus der Rontgenintensitaten, Acta Crystallographica, v. 2, p. 201–207.

KAKINOKI, J., AND KOMURA, Y., 1952, Intensity of X-ray diffraction by a one-dimensional disordered crystal. I. General derivation in cases of the "Reichweite" S = 0 and 1: Journal of the Physical Society of Japan, v. 7, p. 30–35.

LAW, B. E., 1979, Section B–B'—Subsurface and surface correlations of some Upper Cretaceous and lower Tertiary rocks, northern Green River basin, Wyoming: U.S. Geological Survey Open-File Report 79-1689, 2 sheets.

———, 1984a, Structure and stratigraphy of the Pinedale anticline, Wyoming, *in* Law, B. E., ed., Geological Characteristics of Low-Permeability Upper Cretaceous and Lower Tertiary Rocks of the Pinedale Anticline Area, Sublette County, Wyoming: U.S. Geological Survey Open-File Report 84-753, p. 6–15.

———, 1984b, Source-rock evaluation of Upper Cretaceous and lower Tertiary rocks in the Pinedale anticline area, Sublette County, Wyoming, *in* Law, B. E., ed., Geologic Characteristics of Low-Permeability Upper Cretaceous and Lower Tertiary Rocks of the Pinedale Anticline Area, Sublette County, Wyoming: U.S. Geological Survey Open-File Report 84-753, p. 36–50.

LICKUS, M. R., PAWLEWICZ, M. J., LAW, B. E., AND DICKINSON, W. W., 1984, Thermal maturity map, northern Green River Basin, Wyoming, *in* Law, B. E., ed., Geologic Characteristics of Low-Permeability Upper Cretaceous and Lower Tertiary Rocks of the Pinedale Anticline Area, Sublette County, Wyoming: U.S. Geological Survey Open-File Report 84-753, p. 60–65.

NAESER, N. D., 1984, Fission-track ages from the Wagon Wheel no. 1 well, northern Green River Basin, Wyoming; evidence for recent cooling, *in* Law, B. E., eds., Geological Characteristics of Low-Permeability Upper Cretaceous and Lower Tertiary Rocks in the Pinedale Area, Sublette County, Wyoming: U.S. Geological Survey Open-File Report 84-753, p. 66–77.

POLLASTRO, R. M., 1982, A recommended procedure for the preparation of oriented clay-mineral specimens for X-ray diffraction analysis: Modifications to Drever's filter-membrane peel technique: U.S. Geological Survey Open-File Report 82–71, 10 p.

———, 1983, The formation of illite at the expense of illite/smectite: mineralogical and morphological support for a hypothesis (abs.): Clay Minerals Society, Thirty-second Annual Clay Minerals Conference, Buffalo, New York, p. 82.

———, 1985, Mineralogical and morphological evidence for the formation of illite at the expense of illite-smectite, Clays and Clay Minerals v. 33, p. 265–274.

———, and Bader, J. W., 1983, Clay-mineral relationships in some low-permeability hydrocarbon reservoirs and their use as predictive resource tools (abs.): American Association of Petroleum Geologists Bulletin, v. 67, p. 536.

———, and Barker, C. E., 1984, Comparative measures of paleotemperature—an example from clay-mineral, vitrinite reflectance, and fluid inclusion studies, Pinedale anticline, Green River basin, Wyoming (abs.): Society of Economic Paleontologists and Mineralogists First Annual Meeting, San Jose, California, p. 65–66.

Price, L. C., 1983, Geologic time as a parameter in organic metamorphism and vitrinite reflectance as an absolute paleogeothermometer: Journal of Petroleum Geology, v. 6, p. 5–38.

———, and Barker, C. E., 1985, Suppression of vitrinite reflectance in amorphous- and exinite-rich kerogen—a major unrecognized problem: Journal of Petroleum Geology, v. 8, p. 59–84.

Reynolds, M. W., 1976, Influence of recurrent Laramide structural growth on sedimentation and petroleum accumulation, Lost Soldier area, Wyoming: American Association of Petroleum Geologists Bulletin, v. 60, p. 12–33.

Reynolds, R. C., Jr., 1980, Interstratified clay minerals, *in* Brindley, G. W., and Brown, G. eds., Crystal Structures of Clay Minerals and Their X-ray Identification: Mineralogical Society, London, p. 249–303.

———, and Hower, John, 1970, The nature of interlayering in mixed-layer illite-montmorillonite: Clays and Clay Minerals, v. 18, p. 25–36.

Roedder, Edwin, 1984, Fluid Inclusions: Mineralogical Society of America, Reviews in Mineralogy, v. 12, 644 p.

Scholle, P. A., and Schluger, R. R., eds. 1979, Aspects of Diagenesis: Society of Economic Paleontologists and Mineralogists Special Publication 26, 443 p.

Schuster, M. W., and Steidman, J. R., 1983, Origin and development of northern Green River Basin—A stratigraphic and flexural study (abs.): American Association of Petroleum Geologists Bulletin, v. 67, p. 1356.

Shaughnessy, Jack, and Butcher, R. H., 1974, Geology of Wagon Wheel nuclear stimulation project, Pinedale field, Wyoming: American Association of Petroleum Geologists Bulletin, v. 58, p. 2250–2259.

Stach, Ernst, Mackowsky, M.-Th, Teichmuller, Marlies, Taylor, G. H., Chandra, D., and Teichmuller, Rolf, 1982, Stach's Textbook of Coal Petrology: Gebruder Borntraeger, Berlin, 535 p.

Tissot, B. P., and Welte, D. H., 1978, Petroleum Formation and Occurrence: Springer-Verlag, New York, 538 p.

Weaver, C. E., 1979, Geothermal alteration of clay minerals and shales: diagenesis: Office of Nuclear Waste Isolation Technical Report 21, 176 p.

Werre, R. C., Jr. Bodner, R. J., Bethke, P. M., and Barton, P. B., Jr. 1979, A novel gas-flow fluid inclusion heating/freezing stage: Geological Society of American Bulletin, Abstracts with Programs, v. 11, p. 539.

ORGANIC-INORGANIC INTERACTIONS AS A MECHANISM FOR POROSITY ENHANCEMENT IN THE UPPER CRETACEOUS ERICSON SANDSTONE, GREEN RIVER BASIN, WYOMING

JANELL D. EDMAN
Brown and Ruth Laboratories, Inc., 8985 E. Nichols Ave., Suite 300, Englewood, Colorado 80112

AND

RONALD C. SURDAM
Department of Geology and Geophysics, University of Wyoming, Laramie, Wyoming 82071

ABSTRACT: Geochemical and petrographic studies of core and cuttings samples from the Upper Cretaceous Ericson Sandstone in the Forest Oil, Jonah Gulch well, 9-26n-107w, reveal the presence of two distinct enhanced porosity zones that can be differentiated by their position relative to a paleo-oil/water contact. The evolution of enhanced porosity in these two zones can best be explained by experimentally predicted organic-inorganic chemical interactions. Both intragranular carbonate dissolution within chert clasts in the upper diagenetic zone and mixed-layer clay dissolution in the lower zone are apparently the result of organic-inorganic interactions. These interactions are the natural consequence of progressive burial of a sedimentary prism containing sand and shale.

The sequences of inorganic diagenesis for the Ericson Sandstone reservoir and organic maturation in the adjacent shales were determined independently. Then, spatial, temporal, and textural relationships in conjunction with reaction temperatures were used to integrate the two sequences. The synthesis of the organic and inorganic systems shows that the two major porosity enhancement events can be directly tied to organic solvents produced by progressive maturation reactions.

Dissolution of mixed-layer clay in the lower zone was associated with a pulse of organic acids generated prior to the formation and migration of most liquid hydrocarbons. These organic solvents—primarily carboxylic acids and phenols—complexed aluminum in the mixed-layer clay. This complexing action caused the destabilization and dissolution of the clay resulting in porosity enhancement. In the upper diagenetic zone, porosity enhancement occurred after hydrocarbon migration. The carbonic acid that dissolved carbonate within the chert grains was produced by thermal destruction of reservoired organic fluids in the upper Ericson. Thus, both porosity enhancement events can be related to organic solvents generated during progressive maturation.

The interpreted organic-inorganic interactions and associated porosity enhancement events can be diagrammed using a clastic reaction pathway flow chart. The flow chart uses the four component system of CO_2, organic acids, carbonates and aluminosilicates to illustrate the evolution of enhanced porosity through time. Use of such flow charts represents the initial step in predicting regions of maximum enhanced and preserved porosity in the subsurface.

INTRODUCTION

Two papers in this volume (Crossey et al.; Kharaka et al.) address development of the theoretical framework necessary to explain organic-inorganic interactions between shales and associated sandstones. These authors view organic-inorganic interactions as the natural consequence of progressive burial of a sedimentary prism containing both sand and shale. In particular, those organic-inorganic interactions that have the potential for transfer and redistribution of material in the subsurface have been emphasized.

The emphasis on these mass transfer reactions is directly related to the economic implications of porosity prediction in the subsurface. As discovery of hydrocarbon accumulations becomes increasingly difficult, the ability to predict zones of maximum enhanced and preserved porosity in the subsurface gains importance. The purpose of this paper is to establish the critical link between the theoretical framework developed by Surdam and Crossey (1985) and the ultimate goal of porosity prediction.

The approach used was first to determine the sequences of source rock maturation and Ericson Sandstone diagenesis independently in the Forest Oil, Jonah Gulch well, 9-26n-107w (Figs. 1, 2). Next, reaction temperatures and spatial, temporal, and textural relationships were used to integrate the two sequences. The observed sequence of enhanced porosity reactions was then explained using the previously developed framework of organic-inorganic interactions.

This research represents an initial attempt to demonstrate how an observed sequence of porosity enhancement events can be explained by experimentally documented organic-inorganic interactions. As such, it is one of the first in the series of case histories needed to achieve greater understanding of porosity enhancement/preservation reactions that occur in the subsurface. Data and knowledge accumulated in similar case histories may eventually enable geoscientists to predict zones of maximum porosity preservation and enhancement in the subsurface.

METHODS OF STUDY

The sequences of source rock maturation and inorganic diagenesis were derived independently and were then related using reaction temperatures and spatial, temporal and textural relationships. The sequences were determined from core and cuttings samples from the Forest Oil, Jonah Gulch well, 9-26n-107w, Sweetwater County, Wyoming (Fig. 1). The lithologies encountered in this well consist of a highly variable sequence of sand, shale, conglomerate, siltstone, and coal (Fig. 3). Cuttings samples were collected at 10 ft (3 m) intervals throughout the 10,570 ft (3221 m) depth of the well. The 54 ft (16 m) core was taken from the corrected depth of 10,132–10,186 ft (3088–3105 m). Core depths were corrected by using core porosity measurements in conjunction with porosity logs to adjust the field recorded core depths to the proper corresponding well log depths. Using this technique, 12 ft (3.7 m) were added to the field recorded core depth to obtain the corrected depth.

Slabbed cores (Fig. 4) were used to determine lithology, sedimentary structures and the inorganic paragenetic se-

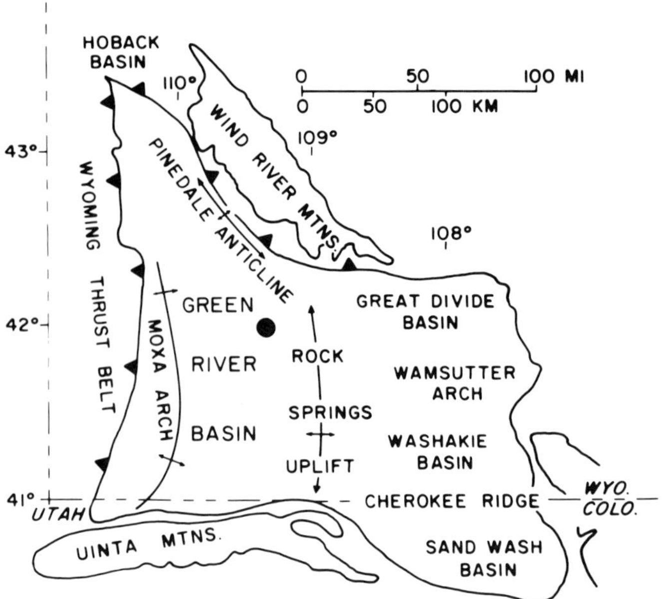

FIG. 1.—Location of Green River basin and associated structural features. Circle marks the location of the Jonah Gulch well. Modified from Law et al. (pers. commun.).

quence. Core samples were used for petrography, scanning electron microscopy, and both bulk and fine fraction X-ray diffraction analyses. Thin sections were impregnated with blue epoxy to show porosity. Initially, thin sections were stained for feldspars using amaranth and sodium cobaltinitrite and with Alizarin Red to indicate calcite. The initial staining process as well as bulk X-ray diffraction showed no feldspars or calcite in the thin sections and subsequent thin sections were not stained. A scanning electron microscope was used to identify clays, determine the distribution of pores, and study the alteration of framework grains.

Bulk X-ray diffraction analyses were performed on representative core samples to aid in determining the mineralogy of the samples. The $<2.0\ \mu m$ fraction from both core and cuttings samples was also analyzed using X-ray diffraction. The technique described by Drever (1973) was used to prepare the clay mounts. Glycolated $<0.5\ \mu m$ clay mounts were used to study composition and ordering of the mixed-layer clays (Hower, 1981). Most of the mixed-layer clay work was performed on cuttings samples. Whenever possible, the cuttings were handpicked to minimize caving problems.

The maturation and migration history of the organic-rich portion of the sedimentary sequence was determined from

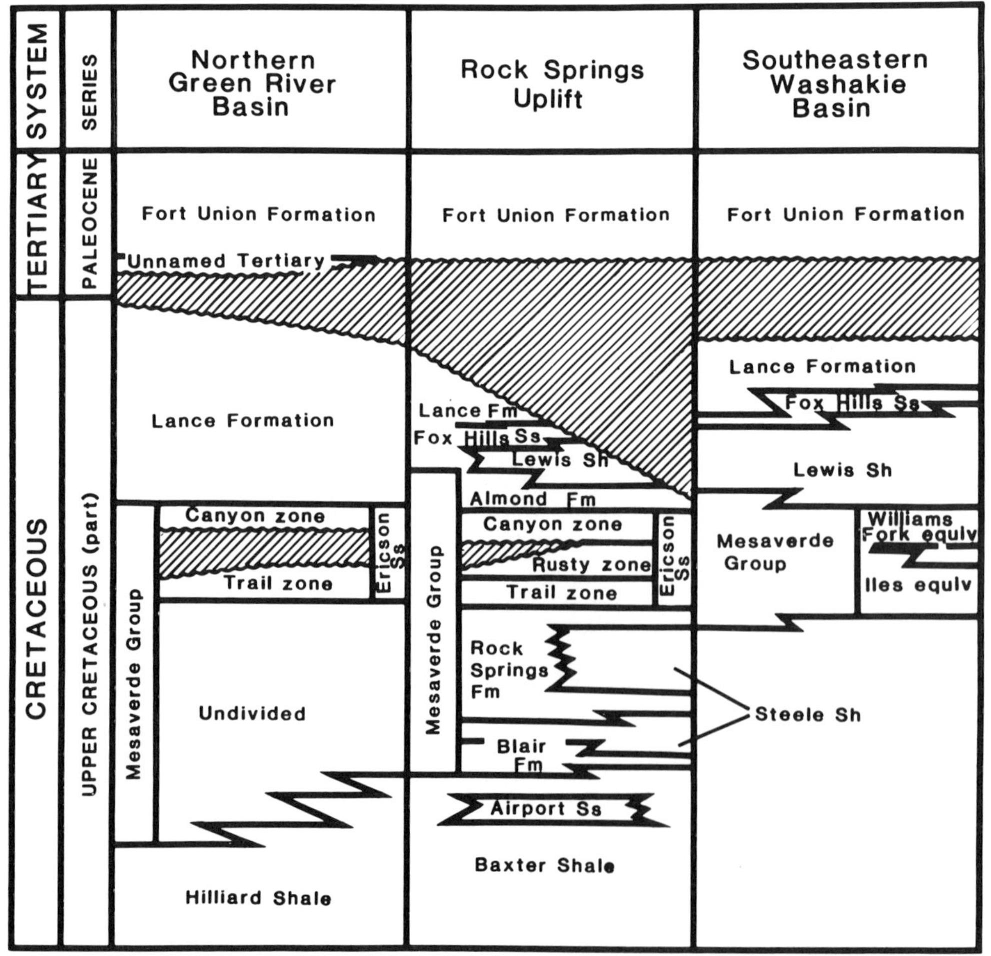

FIG. 2.—Stratigraphic column of part of the Upper Cretaceous and Tertiary rocks in the Green River basin. Modified from Law et al. (pers. commun.).

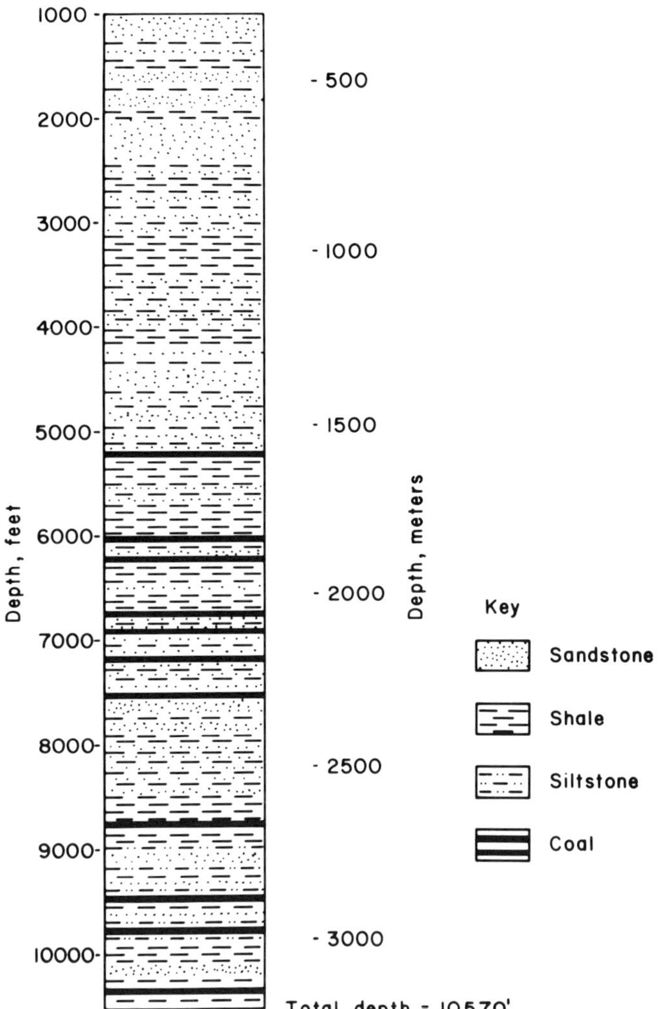

FIG. 3.—Simplified lithologic column for the Jonah Gulch well based on cuttings samples.

cuttings using total organic carbon, pyrolysis, vitrinite reflectance, and elemental analyses. Total organic carbon was measured by subtracting inorganic carbon from total carbon as described in Waples (1981). Pyrolysis was run using the methods given in Hagen and Surdam (1984). Hydrogen index, genetic potential, and production index were then calculated from the pyrolysis data (Tissot and Welte, 1984). The techniques for obtaining the vitrinite reflectance and elemental analysis data are also described in Waples (1981).

REGIONAL GEOLOGY

Structure and Stratigraphy

Core and cuttings samples from the Forest Oil, Jonah Gulch well in the Green River basin (Fig. 1) were used as the data base in this study. The 54 ft (16 m) core, which is the focal point of the inorganic diagenesis study, was taken in the Upper Cretaceous Ericson Sandstone which is part of the Mesaverde Group (Fig. 2). The Ericson Sandstone in this area is unconformably underlain by the Rock Springs Formation and is overlain by undivided Upper Cretaceous rocks that are temporal equivalents of the Almond Formation, Lewis Shale, Fox Hills Sandstone and Lance Formation (Law et al., pers. commun.)

The Green River basin (Fig. 1) is bounded on the west by the Wyoming thrust belt; the north by the Wind River thrust fault; the east by the Rock Springs Uplift; and the south by the Uinta Mountains thrust fault. The evolution of the Green River basin was intimately related to these major structural features. Jordan (1981) and Beaumont (1981) have suggested the Cretaceous foreland basin of North America formed as a flexural response to tectonic and sediment loading of the lithosphere. Shuster and Steidtmann (1983) showed that the lower Tertiary and perhaps some of the uppermost Cretaceous sediment accumulation of the northern Green River basin could be explained by infilling of the flexural depression caused by the Darby, Prospect, and Wind River

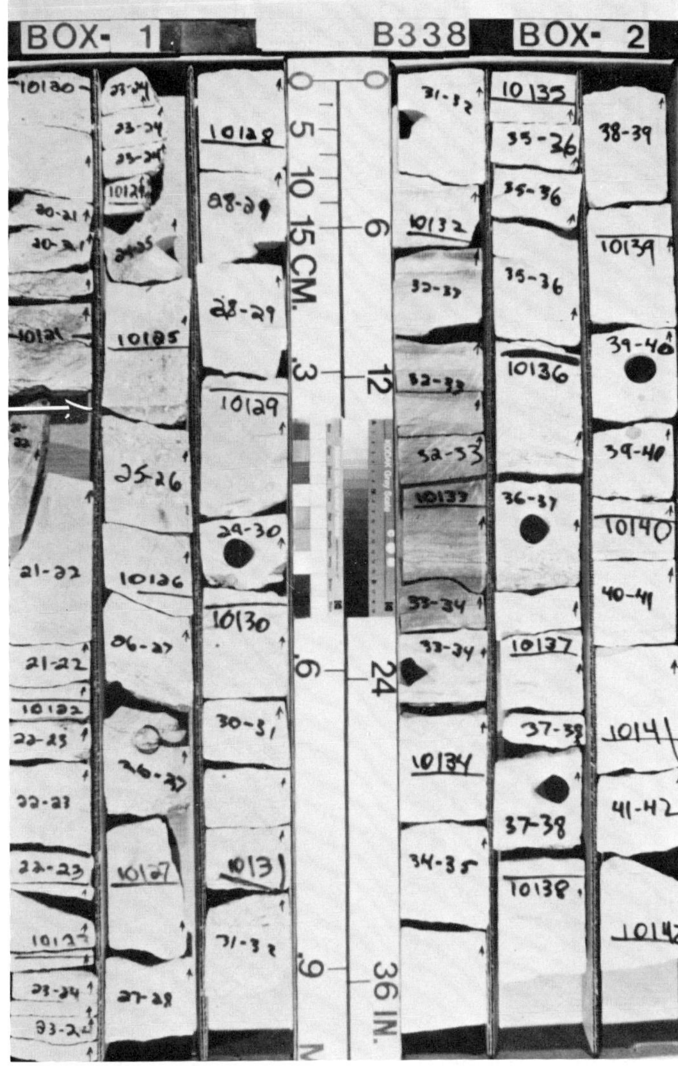

FIG. 4.—Photograph of the slabbed Ericson core. Arrow indicates the horizontal band of dead oil located between uncorrected depths of 10,125 and 10,126 ft (3085 m). This band is apparently a paleo-oil/water contact marking the base of an old hydrocarbon reservoir in the Ericson. The top of the dead oil layer also marks the boundary between the upper and lower diagenetic zones.

thrust loads. Thus, formation of the Green River basin can be interpreted as flexural depression of an elastically behaving lithosphere.

Many of the sediment packages deposited in the Green River basin are also related to thrusting events. Data summarized by Jordan (1981) indicate motion along the Absaroka thrust is recorded by synorogenic conglomerates in the Mesaverde Group. The Ericson Sandstone, which was deposited approximately 73 to 74 m. y. ago (B. E. Law, pers. commun., 1984), is included in this sequence of synorogenic sediments. The derivation of the Ericson Sandstone from thrust related orogenic highlands is substantiated by paleocurrent directions (Law et al., pers. commun.) and previous work by Hale (1955), Jacka (1965), and Roehler (1965). The Ericson also contains characteristic detrital clasts whose source can be correlated directly to lower Paleozoic rocks within the thrust belt containing these unique lithologies (Fig. 5). Abundant chert fragments within the Ericson, containing sponge spicules, apatite, and other materials characteristic of the Permian Phosphoria Formation, indicate a thrust belt provenance. Some of these Paleozoic clasts may have been recycled via older Cretaceous sediments that Dixon (1982) shows are exposed in the hanging wall of the Absaroka thrust.

The conglomerates and sandstones shed from the orogenic highlands were deposited in alluvial fan (Hale, 1955; Jacka, 1965), braided stream (Jacka, 1965; Roehler, 1965) and meandering stream (Schmitt, 1985) environments. This interpretation is consistent with the limited information contained in the slabbed Jonah Gulch core (Fig. 4). Apparently, the tectonic pulse responsible for Ericson deposition interrupted the stable pattern of shoreline deposition which dominated sedimentation during deposition of the Rock Springs Formation (Roehler, 1965).

The Ericson Sandstone differs from other Mesaverde units in several aspects that could significantly influence the diagenetic history. The Ericson is a nonmarine sandstone that has been subdivided into the Canyon zone, Rusty zone, and Trail zone (Fig. 2). The 54 ft (16 m) core in the Jonah Gulch well was cut from the Canyon zone, which is interpreted as a massive, fluvial sandstone (Smith, 1965, Roehler, 1965). These nonmarine Ericson sandstones are much

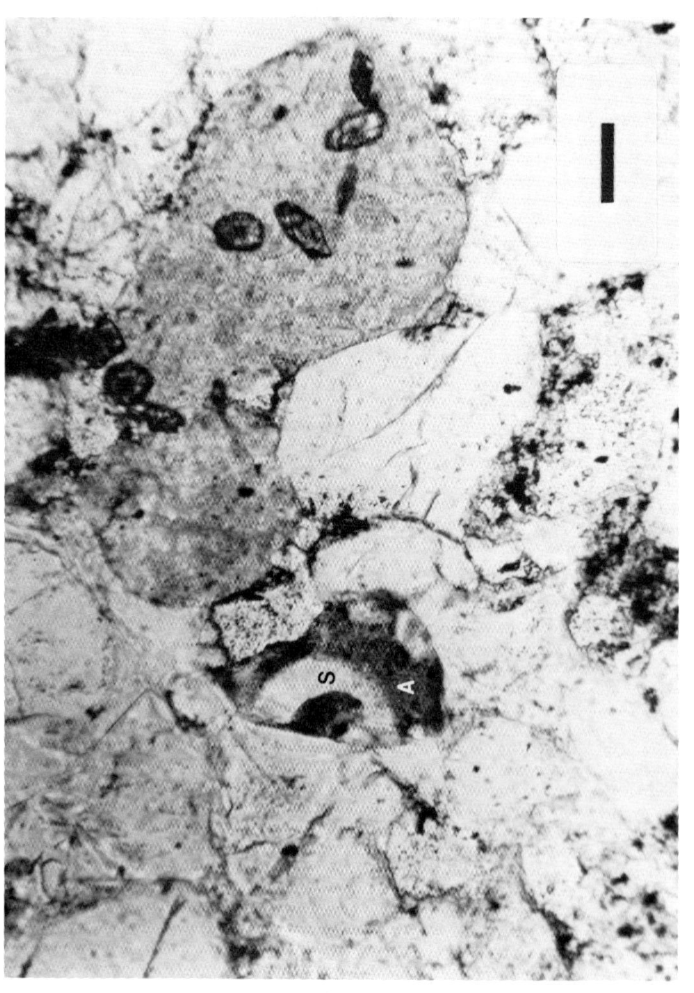

FIG. 5.—(a) Plane light photomicrograph of Ericson Sandstone that contains a detrital clast comprised of a sponge spicule fragment (S) and apatite (A) that are characteristic of the Phosphoria Formation. The presence of such unique Phosphoria lithologies within the Ericson indicates a Phosphoria source for many of the Ericson chert clasts. (b) Same as (a) but with crossed nichols. Apatite is uniaxial and is extinct. Scale bar = 0.1 mm.

more continuous than the intertonguing marginal marine sandstones and marine shales of the other Mesaverde units (B. E. Law, pers. commun., 1984). Consequently, the Ericson may have had better hydraulic transfer than the rest of the Mesaverde. The ability of the Ericson to transmit fluids is critical to hydrocarbon migration as well as the movement of diagenetic pore waters.

In addition, the Ericson is much cleaner compositionally than many of the other Mesaverde units. It is comprised almost exclusively of detrital quartz and chert clasts. Very few heavy minerals or highly reactive rock fragments were observed in thin section. The compositional maturity of the Ericson sediments also influenced the subsequent diagenetic evolution.

Hydrocarbon Shows

Hydrocarbon shows must be considered when evaluating organic-inorganic interactions in the Jonah Gulch well. In general, gas constitutes the majority of the hydrocarbon production from the Mesaverde Group in the Green River basin. A mud gas unit recorded mud gas shows from approximately 1100 ft (335 m) to total depth of the well at 10,570 ft (3222 m). The background gas recorded by this unit was very low (with the exception of some gas in coaly intervals) until about 9320 ft (2841 m) when minor gas shows were detected. Below 9500 ft (2896 m) a high level of methane and wet gas was encountered, and the scale on the mud gas unit had to be adjusted to keep the instrument recording in the track. High levels of wet gas were recorded from 9500 ft (2896 m) to total depth.

Two drill stem tests were run in the well. The first was from 9580 to 9618 ft (2920 to 2932 m) in undivided Upper Cretaceous rocks. This test recovered 2000 ft (610 m) of gas-cut water cushion and 100 ft (30 m) of gas-cut mud. The second drill stem test was in the Upper Cretaceous Lance Formation and Ericson Sandstone from 10,117 to 10,184 ft (3084 to 3104 m) and included most of the cored interval from 10,132 to 10,186 ft (3088 to 3105 m). This test recovered 2800 ft (853 m) of gas-cut water cushion.

In addition to gas shows in the mud and drill stem tests, the core also indicates the presence of hydrocarbons in the well. The core photograph (Fig. 4) shows a dark, horizontal band cutting across the core between the uncorrected depths of 10,125 and 10,126 ft (3086 m). This band is interpreted as a paleo-oil/water contact formed within a hydrocarbon reservoir. Dead oil staining is observed in thin sections from core samples cut above this paleo-oil/water contact but no staining is seen below the contact. The hydrocarbons that were trapped in this thin reservoir are interpreted to have been stratigraphically trapped by the immediately overlying shales (Fig. 6). Unpublished regional seismic data show the Mesaverde interval in the Jonah Gulch well is in a regional structural low between the Rock Springs uplift and Moxa arch. Consequently, structure probably played a minor role in hydrocarbon entrapment.

Overpressuring

Overpressuring may affect the hydraulic transfer characteristic (Dickinson, 1984; Bruce, 1984) and thermal conductivity of rocks (Lewis and Rose, 1970). Although precise quantitative expressions relating overpressuring to various rock properties have yet to be derived, recognition of overpressured zones may be helpful in explaining diagenetic and maturation patterns. The Jonah Gulch well lies within a region defined by Law (1984) and Spencer (1983) as containing overpressured Upper Cretaceous and Tertiary rocks. Law (1984) states, "Thermal generation of gas is the main cause of overpressuring and is directly related to organic richness, level of organic maturation, and temperature." He used drilling mud weights and/or drill stem tests to recognize overpressured zones. Law et al. (pers. commun.) also noted that coincident with the top of overpressuring, in many cases, is a significant increase in the volume of gas, as indicated by gas shows in the mud log data. Law et al. (1979, 1980) observed reversed spontaneous potential curves commonly associated with overpressuring throughout the basin. Applying these criteria to the Jonah Gulch well indicates the Ericson Sandstone interval near the Jonah Gulch well is probably overpressured. Although mud weight through the region of gas shows was not increased significantly and exhibits the same variation in weight used further uphole, approximately 9.4 to 9.6, other data do suggest overpressuring. Both a reversal in the spontaneous potential log at the top of the Ericson (Fig. 6) and an abrupt increase in gas shows below 9500 ft (2895 m) indicate the onset of overpressuring near the Ericson Sandstone interval.

INORGANIC DIAGENESIS

The paragenetic sequence for the Ericson Sandstone was determined from core samples. As discussed in the hydrocarbon shows section, there is a dark, horizontal band of dead oil interpreted as a paleo-oil/water contact cutting across the core between the uncorrected depths of 10,125 and 10,126 ft (3086 m) (Fig. 4). This paleo-oil/water contact is the boundary between two very distinct zones of diagenesis observed in the Ericson core. The upper zone lies above the paleo-oil/water contact, while the lower zone includes the band and the underlying rock. The upper zone originally contained some liquid hydrocarbons and may still contain gas, although not in commercial quantities. In contrast, the only hydrocarbons in the lower zone are present in the band of dead oil; no other staining was observed in this portion of the core. Throughout most of its history, the pores in the lower zone probably were filled with formation water.

Figure 7 illustrates the difference in diagenesis between the upper and lower zones. The different diagenetic pathways followed by these two zones make it necessary to develop and describe their diagenetic histories separately. The distinct paragenetic sequences for these two zones are given below, beginning with the upper zone.

Upper Zone

Figure 8 is a schematic diagram illustrating the paragenetic sequence for the upper zone. Details of the sequence are included in the following discussion:

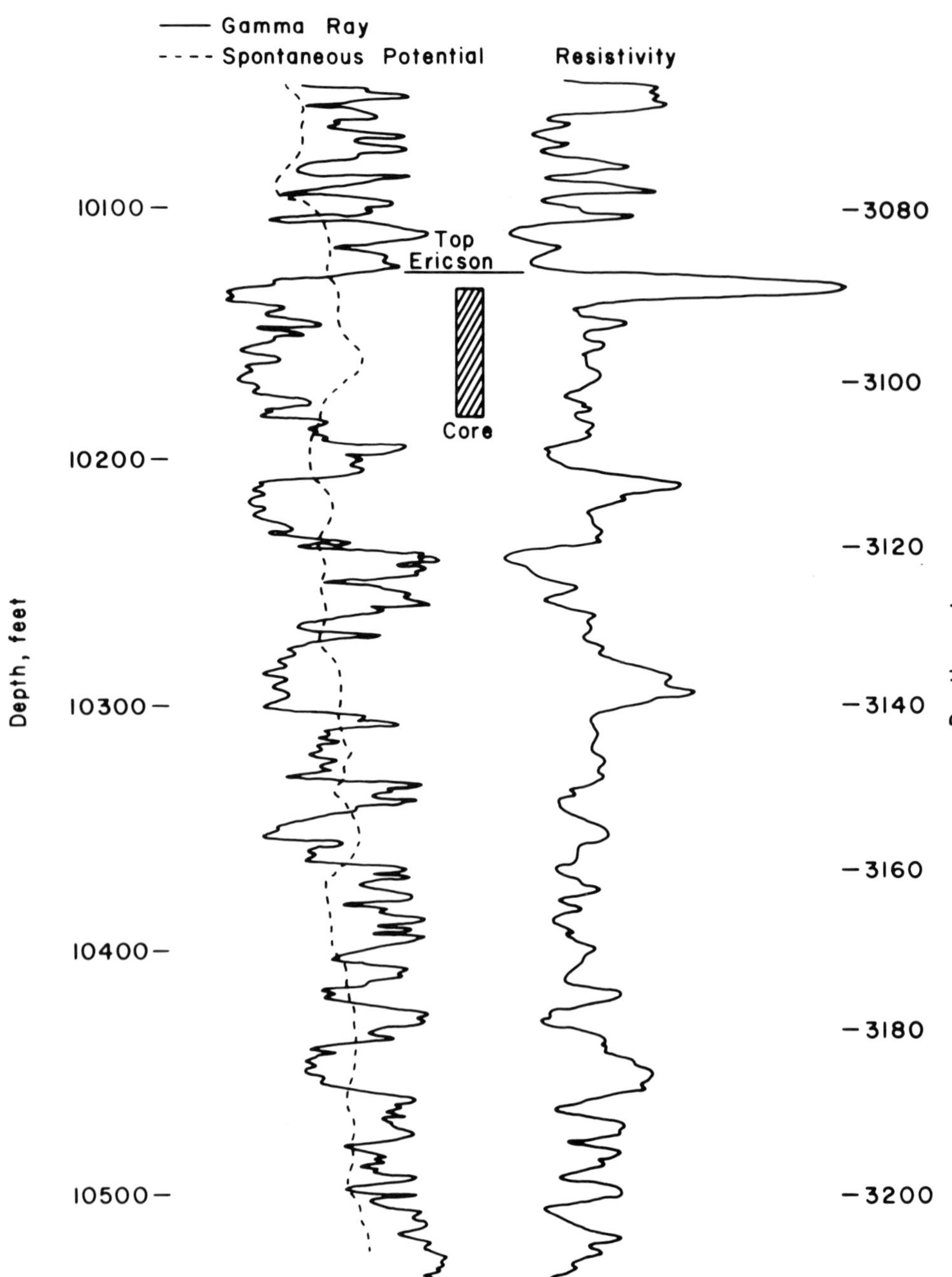

FIG. 6.—Portion of a well log through the Ericson interval from the Forest Oil No. 1 Jonah Gulch well, 9-26N-107W. Gamma ray (solid line) and spontaneous potential (dashed line) are on the left, resistivity is on the right. The layer of shale/siltstone immediately overlying the Ericson may have acted as a permeability barrier to prevent upward migration of hydrocarbons out of the Ericson. The reversal of the spontaneous potential curve at the top of the Ericson has been interpreted by Law et al., (1979, 1980) to indicate overpressuring.

1a) Quartz overgrowths.—

1b) Pressure solution.—

The presence of abundant syntaxial quartz overgrowths (Figs. 7, 9) is the distinguishing characteristic of the upper zone. The euhedral crystal faces and mosaic of interlocking overgrowths are used to infer relatively early precipitation of these overgrowths while pore spaces were still open. There were several periods when quartz overgrowths were precipitated. Figure 10 shows the sequence of quartz overgrowth precipitation through time. Most of the quartz overgrowth layers in this figure were clearly formed prior to the precipitation of kaolinite, although minor overgrowth layers did form after kaolinite emplacement. The early precipitation of quartz overgrowths exerted the dominant control on the overall pore geometry present during hydrocarbon migration (Fig. 11).

Pressure solution is observed at both quartz-to-quartz (Fig. 12) and quartz-to-chert (Fig. 13) grain contacts. Figure 12 documents the onset of pressure solution at quartz-to-quartz grain contacts after the precipitation of quartz overgrowths. In Figure 13, the chert grain that is now partially dissolved appears to have been involved in pressure solution with both the detrital quartz grain and the overgrowth on that detrital grain. The chert grain must have been intact at the time of pressure solution as well as when the kaolinite that is now stained with dead oil was precipitated in the adjacent pore.

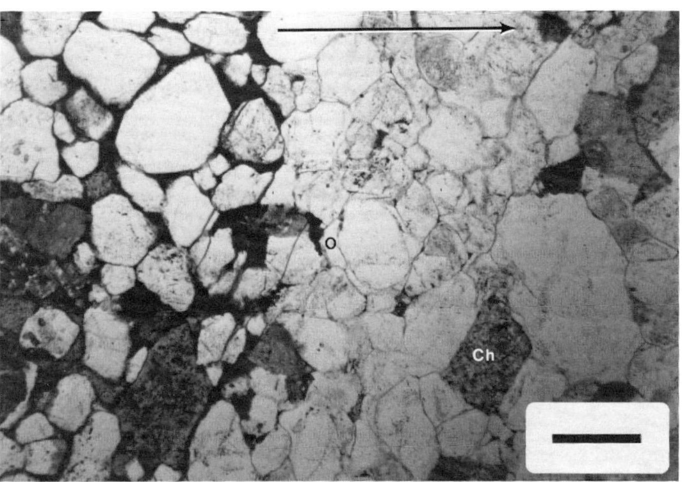

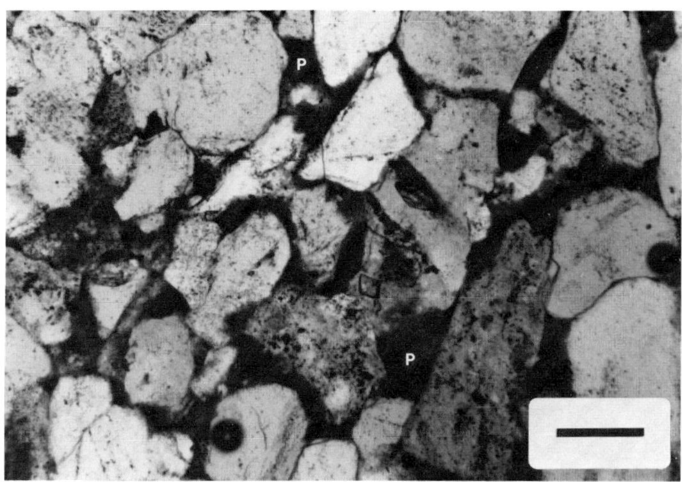

FIG. 7.—These two photomicrographs summarize the major differences in diagenesis between the upper and lower zones. (a) is a vertical slice (arrow indicates up) across the band of dead oil shown in Figure 4. The upper zone, above the dead oil, is characterized by abundant syntaxial quartz overgrowths (O) and enhanced porosity created by chert dissolution (Ch). In contrast, the lower zone in (b) contains abundant enhanced porosity created by the dissolution of mixed-layer clay cement (P). Quartz grains are embayed where mixed-layer clay that replaced the quartz was later dissolved to create enhanced porosity. Only minor quartz overgrowths are present in the lower zone, and the chert grains are relatively unaltered. a) Scale bar = 0.5 mm; b) Scale bar = 0.2 mm.

Upper Zone

Diagenesis

Early → Late

Quartz Overgrowths

Pressure Solution

Minor chert dissolution

Kaolinite

Hydrocarbon migration

Significant chert dissolution

FIG. 8.—Paragenetic sequence for the upper zone.

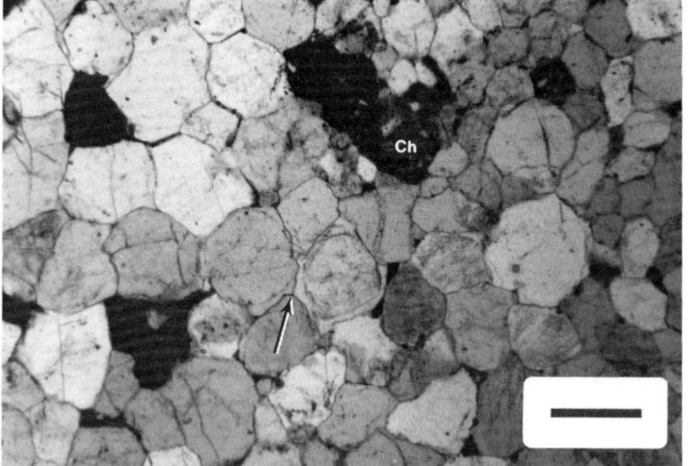

FIG. 9.—The upper zone contains abundant syntaxial quartz overgrowths (arrows). Although the detrital quartz grains and overgrowths are intact, chert clasts (Ch) are now partially dissolved and filled with blue epoxy. Scale bar = 0.5 mm.

Furthermore, the contact between the dissolved chert grain, quartz overgrowth, and dead oil indicates the pore now filled with kaolinite and dead oil attained its configuration after the onset of pressure solution. Kaolinite, followed by oil, then filled the pore after pressure solution. This textural evidence is used to infer the onset of pressure solution after precipitation of quartz overgrowths but prior to the precipitation of authigenic kaolinite.

2a) Minor chert dissolution.—

2b) Precipitation of kaolinite in remnant primary pores.—

Following emplacement of most of the syntaxial quartz overgrowths and the occurrence of pressure solution, authigenic kaolinite precipitated in some of the remnant primary pore space (Fig. 10). Kaolinite also precipitated in small pore spaces created by very minor dissolution of some of the chert grains.

FIG. 10.—Scanning electron photomicrograph illustrating the relationship between quartz overgrowths and kaolinite in the upper zone. Quartz overgrowths (O) precipitated onto the detrital quartz grain (Q). This early precipitation of overgrowths exerted primary control on the pore space configuration present during hydrocarbon migration. After precipitation of most of the overgrowths, kaolinite (K) precipitated in some of the remnant pore space. Minor quartz precipitation continued after kaolinite formed. Scale bar = 10 μm.

FIG. 11.—Scanning electron photomicrograph showing detrital quartz grains, overgrowths, and pores. The porosity configuration in this sample is probably similar to the pore distribution that was present during migration. Scale bar = 100 μm.

3) Hydrocarbon migration.—

Two diagenetic events can be used to bracket the timing of hydrocarbon migration within the upper zone. First, the kaolinite that precipitated in remnant primary pore space is frequently stained with dead oil (Fig. 13). The oil apparently filled microporosity between the kaolinite booklets. This indicates kaolinite precipitated prior to hydrocarbon migration. Second, dead oil rims remnants of partially dissolved chert grains (Figs. 13, 14) but is generally excluded from enhanced porosity created by chert dissolution. Therefore, the chert was apparently intact during migration, and dissolution occurred after hydrocarbon migration. Thus, hydrocarbon migration took place after kaolinite precipitation but before the dissolution of chert grains.

4) Significant chert dissolution.—

As mentioned above, significant dissolution of chert grains occurred after hydrocarbon migration (Figs. 13, 14). Photomicrographs illustrating partially dissolved chert grains are given in Figure 15. Chert dissolution comprises the major porosity enhancement event in the upper zone. This enhanced porosity is responsible for most of the porosity currently present in this zone.

Lower Zone

Figure 16 shows the schematic paragenetic sequence for the lower zone. Detailed description of this sequence is given below:

1) Minor quartz overgrowths.—

As in the upper zone, the precipitation of syntaxial quartz overgrowths was the earliest significant diagenetic event. Before these overgrowths could become spatially extensive, however, the fluid chemistry changed, and clay cement began to precipitate instead of quartz.

2) Clay cement.—

The precipitation of mixed-layer clay cement apparently arrested further development of quartz overgrowths (Fig. 17). In places, the clay cement actually replaced the earlier overgrowths (Fig. 18). Subsequent dissolution of the replacement clay cement resulted in embayed quartz grains (Fig. 7). The clay cement is interpreted to be an early, authigenic pore-filling material. Figure 19 shows that some of the clastic grains almost appear to be "floating" in the clay cement. Such textural evidence indicates relative early emplacement of the clay cement before much compaction had occurred.

X-ray diffraction analysis shows this clay is an authigenic illite/smectite mixed-layer clay. This cement is dull blue-green in uncrossed nichols and can be difficult to distinguish from the blue epoxy impregnating the pore space. In crossed nichols it has very low birefringence and a faint green tint. Expansibility and ordering of this clay are discussed later in a section on clay diagenesis.

3) Pressure solution.—

Pressure solution occurred after precipitation of clay cement. The presence of clay cement seems to have inhibited pressure solution because it only occurs at those grain-to-grain contacts where no clay was emplaced. The ability of the clay cement to inhibit pressure solution indicates the cement was in place prior to pressure solution.

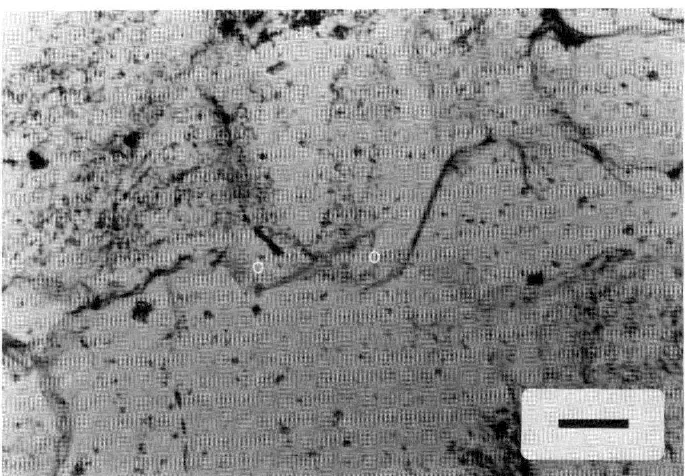

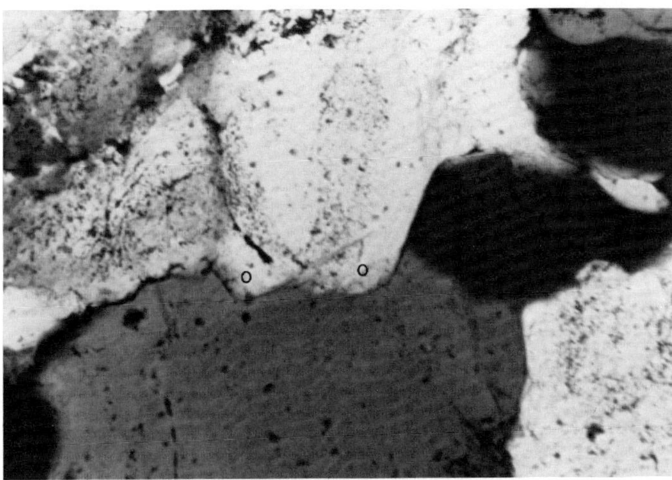

FIG. 12.—Photomicrographs showing quartz overgrowths (O) involved in pressure solution. The overgrowths are now penetrating into a detrital grain, demonstrating that the overgrowths were precipitated prior to pressure solution. (a) plane polarized light; (b) crossed nichols. Scale bar = 0.5 mm.

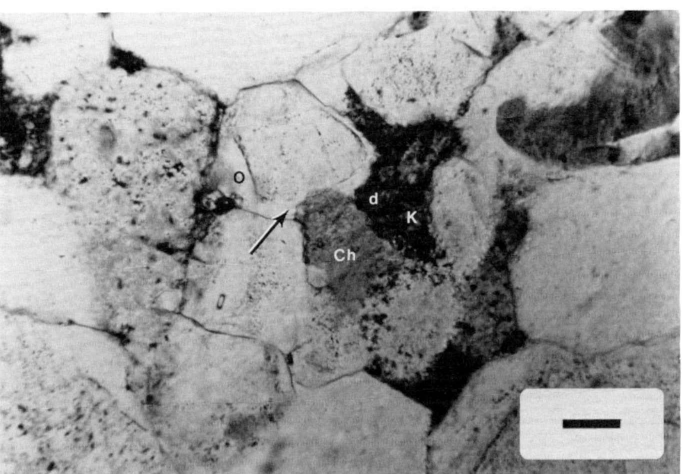

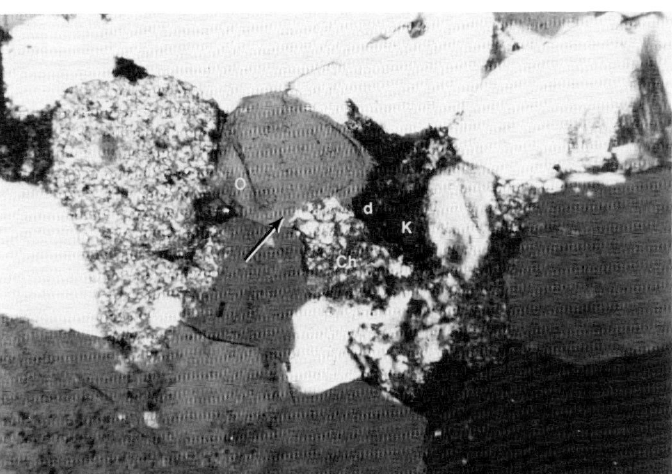

FIG. 13.—Photomicrographs summarizing the paragenetic sequence in the upper zone. First, syntaxial quartz overgrowths (O) precipitated on detrital grains. Pressure solution (arrow) occurred after the initial precipitation of quartz overgrowths. Following the overgrowths and pressure solution, authigenic kaolinite (K), filled some of the remnant primary pore space. Later migrating hydrocarbons (d) filled microporosity between the kaolinite booklets. Dissolution of the chert grains to create enhanced porosity took place after hydrocarbon migration, as shown by the exclusion of hydrocarbon from the chert grain (Ch) that is now dissolved. (a) plane polarized light; (b) crossed nichols. Scale bar = 0.1 mm.

4) Clay dissolution/minor chert dissolution.—

The significant porosity enhancement event in the lower zone was the dissolution of the pore-filling, mixed-layer clay cement (Fig. 20). This dissolution occurred prior to hydrocarbon migration. In many places the cement was dissolved almost completely, leaving only minor traces of the cement around the pore edges (Fig. 20). Minor chert dissolution may also have taken place at approximately the same time.

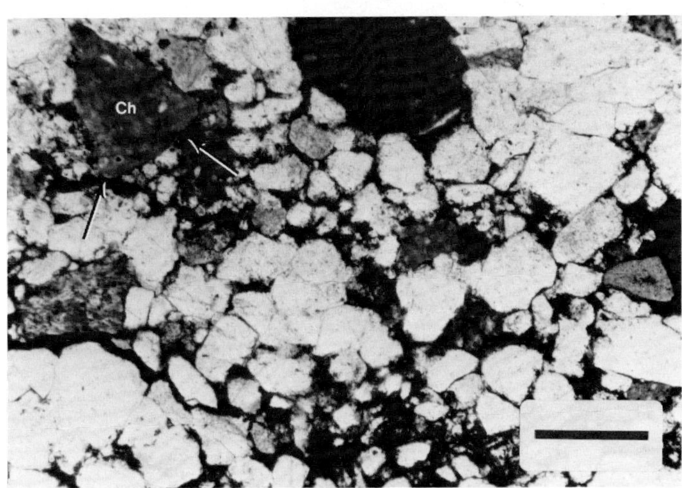

FIG. 14.—Photomicrograph of partially dissolved chert grain (Ch) rimmed by dead oil (arrows). Exclusion of the oil from enhanced porosity in the chert demonstrates the chert grain was intact at the time of migration. Therefore, chert dissolution occurred after hydrocarbons were emplaced. Scale bar = 0.5 mm.

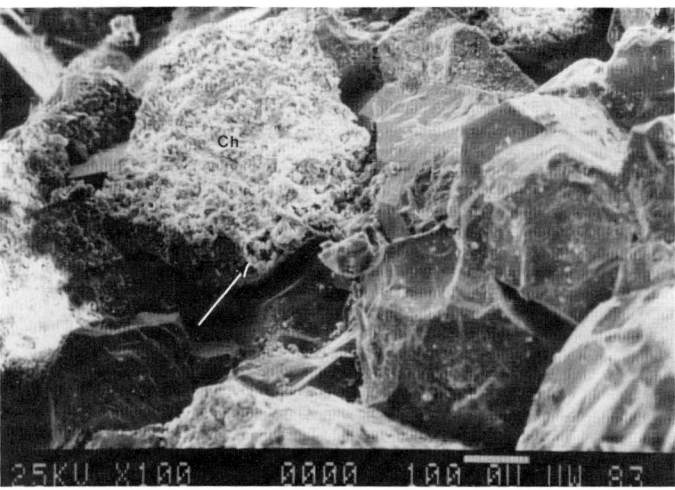

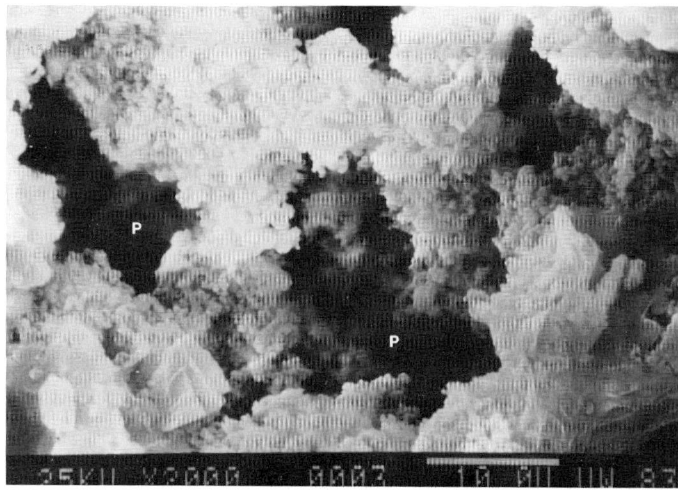

FIG. 15.—Scanning electron photomicrographs showing dissolution porosity (P) in detrital chert clasts (Ch). (b) is a higher magnification photo of the area indicated in (a) by the arrow. Note the pores are open and not blocked by authigenic clays. a) Scale bar = 100 μm; b) Scale bar = 10 μm.

Lower Zone

Diagenesis

Early → Late

Minor quartz overgrowths

Mixed layer clay

Pressure Solution

Clay dissolution/minor chert dissolution

Kaolinite

Hydrocarbon migration

Illite

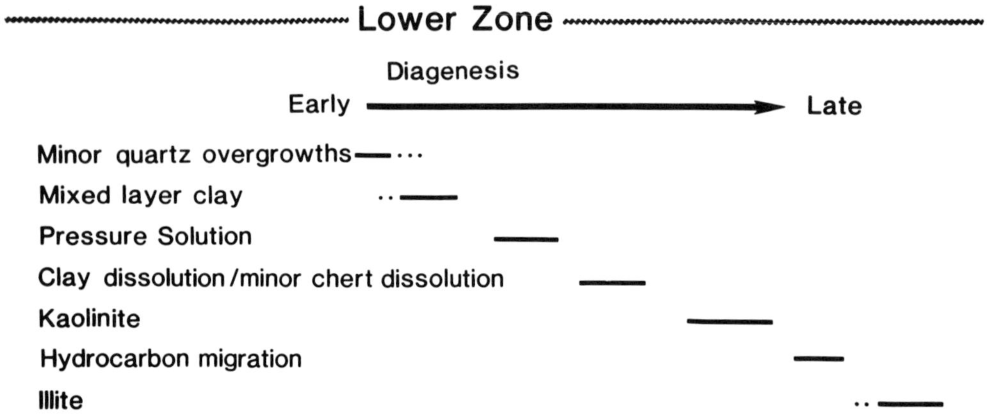

FIG. 16.—Paragenetic sequence for the lower zone.

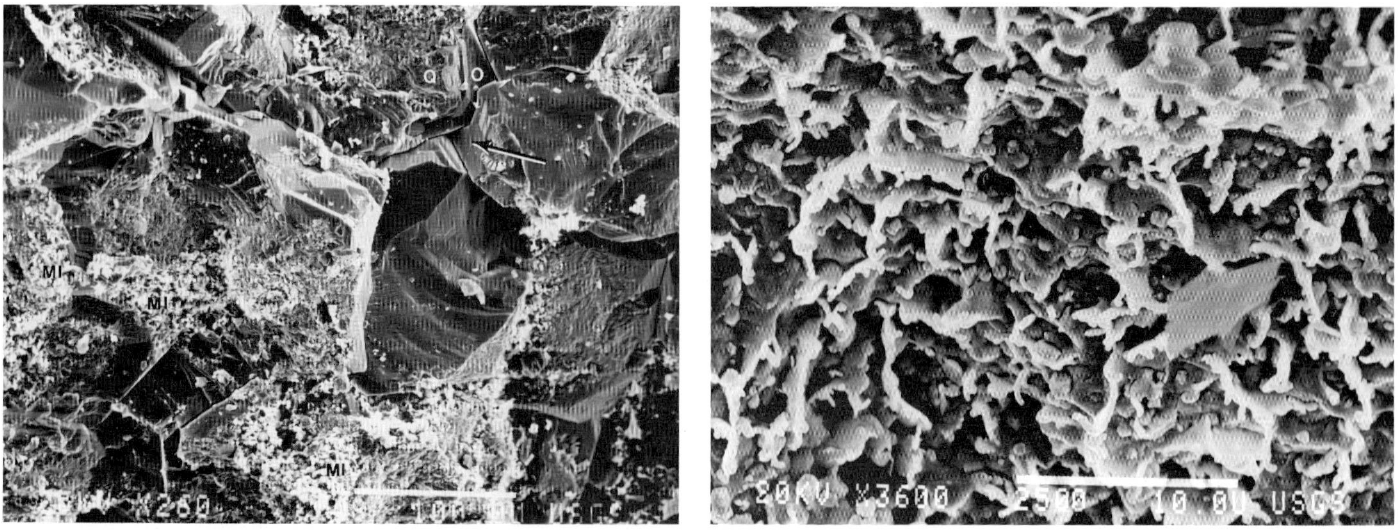

FIG. 17.—The scanning electron photomicrograph in (a) shows mixed-layer clay cement (Ml), quartz overgrowths (O), and detrital quartz (Q) in the lower zone. Precipitation of clay cement seems to have inhibited development of quartz overgrowths. Overgrowths are developed best in those areas where clay cement is absent (arrow). (b) is a higher magnification photograph of this same cement. a) Scale bar = 100 μm; b) Scale bar = 10 μm.

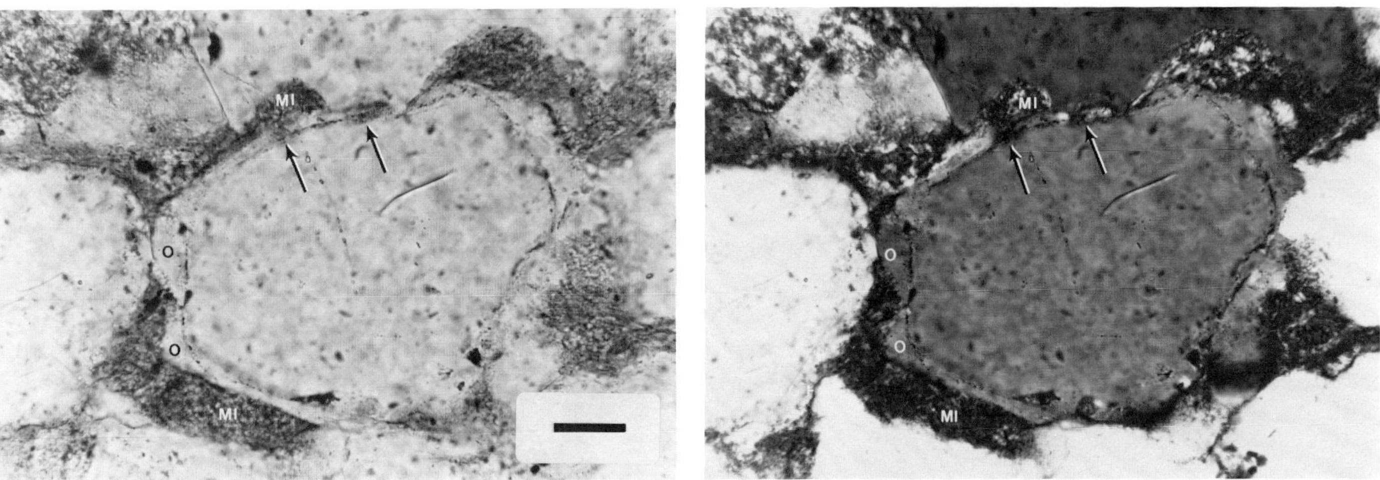

FIG. 18.—Photomicrograph of quartz overgrowth (O) and mixed-layer clay cement (Ml) in the lower zone. The quartz overgrowth appears to have been partially replaced by the clay (arrows) creating embayments and irregularities in the overgrowth. (a) plane polarized light; (b) crossed nichols. Scale bar = 0.05 mm.

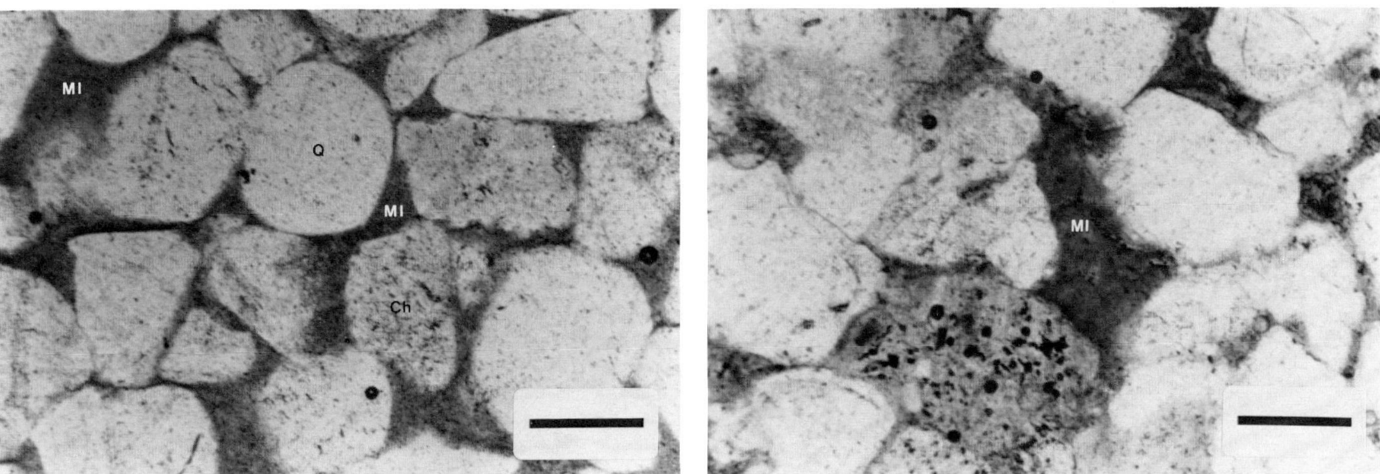

FIG. 19.—Photomicrographs of mixed-layer clay cement. In (a), mixed-layer clay cement (Ml) is filling pores between detrital quartz (Q) and chert (Ch) clasts. Grains that appear to be floating in the clay cement and point contacts between some grains are used to infer relatively early precipitation of the clay prior to significant compaction. Photomicrograph in (b) shows a large patch of remnant mixed-layer clay cement (Ml). The clay cement is dull blue-green in plane polarized light and exhibits low birefringence under crossed nichols. Scale bar = 0.5 mm.

5) Precipitation of kaolinite.—

Authigenic kaolinite observed in the lower zone apparently precipitated in the centers of the pores created by dissolution of mixed-layer clay cement (Figs. 21, 22). The clay cement still rims the center of many of the pores that now contain kaolinite (Fig. 21).

6) Hydrocarbon migration.—

The dead oil at the top of the lower zone apparently was emplaced in pore spaces that originated by dissolution of mixed-layer clay cement (Fig. 23). The only other explanation for the observed geometry in Figure 23 would be to infer another earlier phase of hydrocarbon migration into some type of primary intergranular pore space. This latter explanation conflicts, however, with the observed geometry of the pyrobitumen band.

7) Precipitation of late illite.—

Scanning electron photomicrographs reveal the presence of minor amounts of late illite in the lower zone (Fig. 24). This photomicrograph shows illite flames that have grown across euhedral kaolinite booklets. Such textural evidence indicates illite precipitated after kaolinite. There is no dead oil staining in the lower zone where late illite is present, so the relative timing of illite and hydrocarbon migration is ambiguous. Placement of illite after hydrocarbon migration in the paragenetic sequence is inferred rather than observed.

Two additional topics also need to be considered within

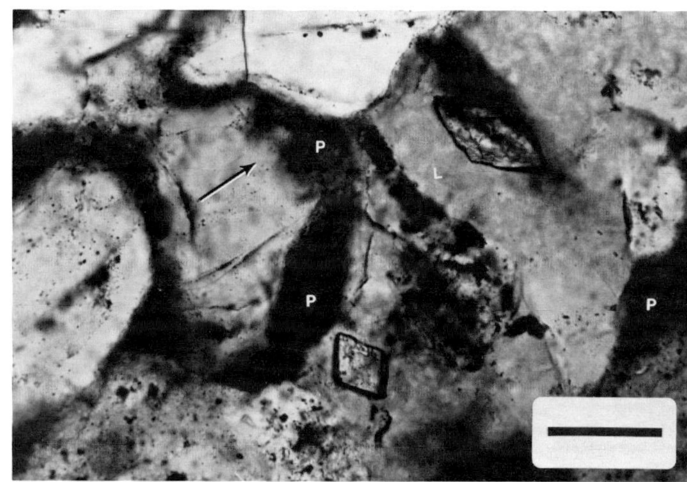

FIG. 20.—Two photomicrographs illustrating enhanced porosity (P) in the lower zone that was created by dissolution of mixed-layer clay cement. In (a) remnants of the clay cement (arrows) still rim the periphery of some of the enhanced pores. Embayed quartz grain (arrow) in (b) was created when the mixed-layer clay cement that replaced part of the quartz grain was dissolved. Scale bar = 0.1 mm.

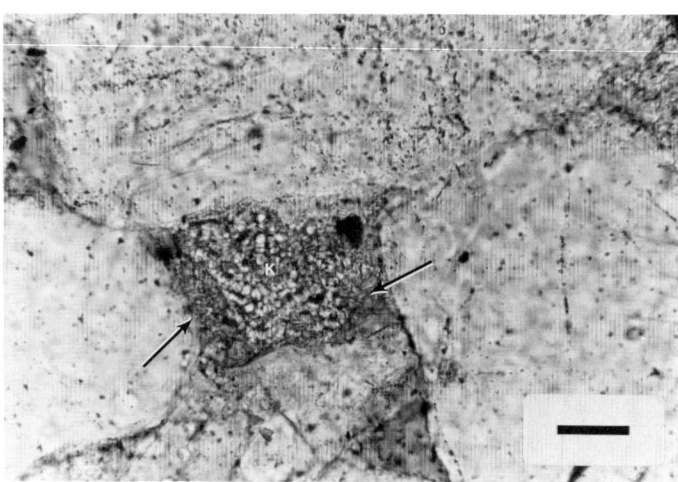

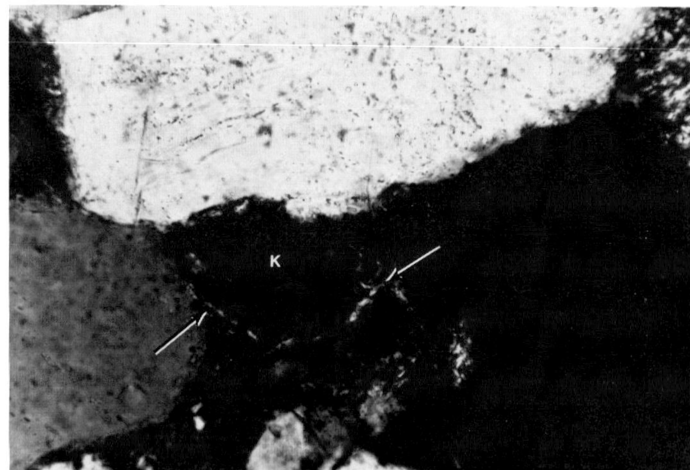

FIG. 21.—Photomicrograph of kaolinite (K) filling a pore in the lower zone. Kaolinite in the lower zone typically precipitated in the centers of pores interpreted to have been created by dissolution of mixed-layer clay cement (see Fig. 20). The remnant mixed-layer clay frequently rims the pore now filled with kaolinite (arrows). This is particularly noticeable in the crossed nichols photomicrograph (b), where the mixed-layer clay is more birefringent than the kaolinite and appears as white strands near the edge of the pore. Scale bar = 0.05 mm.

FIG. 22.—Scanning electron photomicrograph showing relationship between mixed-layer clay cement (arrow) and kaolinite (K) in the lower zone. The mixed-layer clay is a remnant left after dissolution of most of the adjacent clay cement occurred. Kaolinite (K) precipitated in the pore created by dissolution of the mixed-layer clay and now surrounds the remnant of the earlier clay. Scale bar = 10 μm.

the context of inorganic diagenesis. These are fracturing and clay diagenesis.

Fracturing

Various fluids have moved through the Ericson reservoir. Dead oil staining indicates hydrocarbon migration. Formation fluid had to move in order for material generated by dissolution of the clay cement and chert to be flushed out of the system allowing enhancement of the previously existing porosity. Although the present porosity in these rocks

FIG. 23.—Photomicrograph illustrating the distribution of dead oil (black) in the paleo-oil/water contact. Note the similar geometry between the dead oil distribution shown here and the geometry of the pore-filling mixed-layer clay cement in Figure 19. The dead oil is interpreted to have filled pore space created by dissolution of the mixed-layer clay cement. Scale bar = 0.5 mm.

FIG. 24.—Scanning electron photomicrograph showing late illite (I) that has grown across the top (arrow) of earlier kaolinite (K). Scale bar = 10 μm.

is approximately 7–12% in many places, however, the permeability is often less than 0.5 md. Even though the Ericson may have had better hydraulic transfer than other Mesaverde units, fluid flow through these low permeability sandstones may have been a problem. Figure 25 shows a photomicrograph of hydrocarbons that migrated along fratures in the cored interval. The volumes of fluid necessary to create the observed enhanced porosity may have migrated along fractures similar to these.

Clay Diagenesis

Several lines of evidence suggest clay diagenesis may play a key role in organic-inorganic interations and consequently in the enhancement and preservation of porosity. Johns and Shimoyama (1972) demonstrated that the presence of clay minerals (especially mixed-layer clays) can be a catalyst to liquid hydrocarbon cracking reactions. Almon (1974) showed the chemical mechanism involved in the catalytic process might involve electron transfer from kerogen to Fe^{+3} within the clay lattice. This electron transfer may in turn be related to the oxidation of kerogen to cleave organic solvents such as phenols and carboxylic acids from complex kerogen molecules (Surdam and Crossey, 1985). In addition, the changes in composition and ordering in mixed-layer clays with progressive burial result in the generation of large quantities of water. This water may be necessary to transport the water soluble organic acids and their dissolved or chelated load. For these reasons, clay diagenesis was examined in the Jonah Gulch well.

Results of the $<0.5\mu m$ fraction clay analyses show that an abrupt change in illite/smectite composition as well as ordering occurs at approximately 8000 ft (2438 m) (Fig. 26). This depth is slightly shallower than the Tertiary-Cretaceous unconformity picked by B. E. Law (pers. commun., 1985) in this well. Pollastro and Barker (1984) also observed the conversion from random interstratification of the illite/smectite clay to short-range ordering of illite/smectite over a short depth interval immediately above the unconformity to the north of Jonah Gulch in the Wagon Wheel well.

Summary of Inorganic Diagenesis

Two distinct zones of diagenesis are observed in the Ericson Sandstone core in the Jonah Gulch well. These two zones are separated by a paleo-oil/water contact that is present immediately below the upper zone. The upper and lower zones exhibit different petrographic characteristics, and the porosity enhancement events in the two zones are distinct.

In the upper zone, characterized by abundant quartz overgrowths, porosity enhancement by dissolution of chert clasts occurred after hydrocarbon migration. The upper zone

FIG. 25.—Photomicrograph of fracture filled with hydrocarbon. Fracture split the grain indicated by arrows. Fracturing may have played an important role in the migration of both hydrocarbons and the organic acid solvents. Scale bar = 0.5 mm.

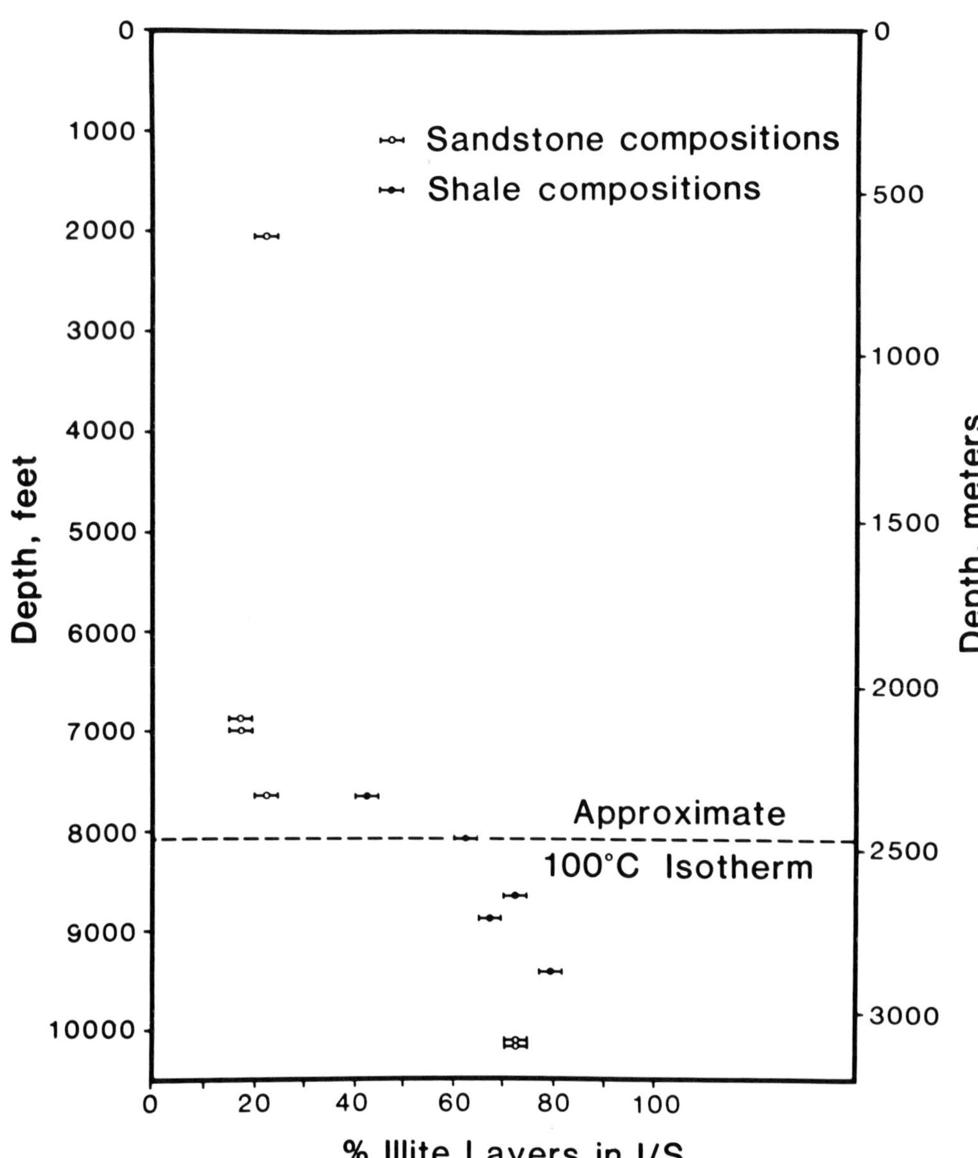

FIG. 26.—Composition of mixed-layer illite-smectite clay from < 0.5 μm fraction versus present burial depth in sandstones and shales from the Jonah Gulch well. The conversion from random interstratification of the illite-smectite clay to short-range ordering occurs at approximately 8000 ft (2438 m). The approximate location of the 100° C (212° F) isotherm is also shown.

may presently be a small gas reservoir. In contrast, the chert in the lower zone is relatively untouched and quartz overgrowths are minor. Porosity enhancement in the lower zone occurred prior to hydrocarbon migration and involved an aluminum mobility problem related to the dissolution of mixed-layer clay cement. With the exception of the band of dead oil in the top of the lower zone, there are no hydrocarbons in this interval.

ORGANIC GEOCHEMISTRY

According to Law (1984) the most probable sources for hydrocarbons in Upper Cretaceous and Tertiary reservoirs in the Green River basin are the interbedded coals and other carbonaceous lithologies. Based on this assumption, cuttings samples from the Jonah Gulch well were used to evaluate the quality, quantity, and level of thermal maturity of organic material in the Tertiary and Upper Cretaceous stratigraphic section. A suite of organic analyses including TOC, pyrolysis, vitrinite reflectance, and elemental analysis was performed to provide the necessary information. This suite of analyses also permits cross-checking of quality and thermal maturity indicators to insure reliability and internal consistency of the data. Data from these analyses are given in Tables I–III. The S_1 and S_2 values used in the hydrogen index, genetic potential, and production index calculations (Table I) were determined from pyrolysis data. S_1 represents the fraction of original kerogen that has been converted to hydrocarbon, whereas S_2 represents the residual potential of the kerogen not yet converted to hydrocarbon (Tissot and Welte, 1984). Results of the individual analyses are discussed below.

Total Organic Carbon

Total organic carbon (TOC) provides a means for determining the quantity of organic material in a source rock. Mature source rocks have already lost some organic carbon

TABLE I.—Total Organic Carbon and Pyrolysis Data.

Sample Depth (feet)	Age	Lithology	Total Organic Carbon (wt %)	Hydrogen Index (S2/Corg)	Genetic Potential (S1 + S2)	Production Index ((S1)/(S1 + S2))
1950	Tertiary	Shale	2.21	*	*	*
3320	Tertiary	Shale	0.38	*	*	*
4300	Tertiary	Shale	0.28	*	*	*
5480	Tertiary	Shale	0.36	*	*	*
6610	Tertiary	Shale	9.90	330	34.30	0.05
6940	Tertiary	Coal	14.10	299	44.94	0.06
7120	Tertiary	Coal	42.40	462	201.00	0.02
7420	Tertiary	Coal	58.20	397	250.74	0.08
7430	Tertiary	Coal	59.00	366	220.00	0.02
7860	Tertiary	Shale	2.92	175	5.40	0.06
8660	Cretaceous	Shale	1.64	131	2.43	0.12
9770	Cretaceous	Coal	36.04	316	141.31	0.12
10310	Cretaceous	Coal	1.99	156	3.60	0.13
10410	Cretaceous	Coal	4.15	209	10.08	0.14

*Contaminated by migrated hydrocarabons

TABLE II.—Vitrinite Reflectance Data.

Depth (feet)	Age	Lithology	% Ro
1980	Tertiary	Shale	No vitrinite
3340	Tertiary	Shale	0.53
5440	Tertiary	Shale	0.67
6950	Tertiary	Coal	0.79
7430	Tertiary	Coal	0.72
8670	Cretaceous	Shale	0.73
9780	Cretaceous	Coal	0.91
10420	Cretaceous	Coal	0.94

TABLE III.—Elemental Analysis Data.

	Sample Interval (feet)	Lithology	Age	*Approximate TOC wt %	Atomic Ratio H/C	Atomic Ratio O/C
1)	7130–40	Tertiary	Coal	42.40	1.069	0.188
2)	7850–60 7860–70	Tertiary	Shale	2.92	0.794	**
3)	8640–50 8670–80	Cretaceous	Shale	1.64	0.783	0.143
4)	9780–90	Cretaceous	Coal	39.40	0.796	0.140
5)	10300–10 10320–30	Cretaceous	Coal	1.99	0.774	0.071

*Due to limited sample, the number given here is the TOC value for the interval closest to the depth listed in this table for the H/C, O/C data. In general, the cuttings samples used for the TOC determinations are at most 10 ft (3.05 m) from the samples used for H/C, O/C; see Table I for details.
**Insufficient material for analysis.

through the migration process, and consequently TOC values in these rocks are minimum values. Hunt (1979) concludes that source rocks should contain at least 0.5 weight % TOC before they are capable of generating commercial hydrocarbon deposits. Fine-grained rocks with TOC values in the 1 to 2 weight % category are usually considered good source rocks. In the Jonah Gulch well the samples from the upper 5000 ft (1524 m) have relatively low TOC values (Table I). In the lower half of the well, however, from approximately 6,610 to 10,410 ft (2014 to 3172 m) in the Fort Union and Upper Cretaceous section, there are a significant number of organic-rich intervals. Samples from these organic-rich intervals range from 2 to 59 weight % TOC (Table I). According to the mud log, most of this organic-rich material comes from coaly intervals. This 5000-ft-thick zone contains some very organic-rich samples which are ample source rocks for the section.

The decision to include the coaly intervals as possible source rocks for the wet gas and dead oil in the Ericson is based on the work of Tissot (1984), Snowdon and Powell (1982), Durand and Oudin (1979), and Vandenbroucke et al. (1983). Tissot (1984) states, "Coal is widely considered as a potential source for gas, but some coals may generate sizeable amounts of oil as well." Both Tissot (1984) and Snowdon and Powell (1982) suggest that the different generation potential of coals may be due to variations in concentration of hydrogen-rich organic matter. Those coals containing greater proportions of resinite and other macerals of the liptinite/exinite group will have greater potential for generating liquid hydrocarbons than coals containing predominantly vitrinite. Hydrogen indices determined from pyrolysis analyses (Table I) show some of the coals in the Jonah Gulch well have unusually high concentrations of hydrogen-rich organic matter. Consequently, as will be discussed in more detail in the next section, some of these coals may have been the source of the wet gas and oil observed at Jonah Gulch.

Hydrogen Index

Hydrogen index (HI) is defined as the ratio of S_1 to organic carbon (Tissot and Welte, 1984). HI can be used to determine whether kerogen in the sample is oil, oil/gas, or gas prone. Type I kerogen, which is primarily a source of liquid hydrocarbons (Tissot and Welte, 1978), generally has an HI greater than 500 (Hunt, 1979). Type II kerogen, which generates both liquid hydrocarbons and gas, commonly has an HI between 300 and 450. Type III kerogen is primarily a gas source and typically has an HI of 150 or less. By determining the HI of potential source rocks, it is possible to predict roughly the type of hydrocarbons that will be generated during maturation. In this study, HI was only used to evaluate the type of hydrocarbon that would be evolved during maturation; no attempt was made to type the kerogen genetically.

The analyses in Table I show that samples with both relatively low and high HI values are present in the Jonah Gulch well. Low HI values generally indicate a low level of organic matter, and consequently those samples probably were the source of only minor amounts of hydrocarbons.

The most surprising result of the source rock analyses in the Jonah Gulch well is that many of the samples, although appearing coaly, have relatively high HI values (299 to 462). Some of these samples are conventional coals, whereas others, i.e., at 7120 ft (2169 m) where TOC = 42 weight % and HI = 462, are clearly coals containing higher proportions of the hydrogen-rich macerals of the liptinite/exinite group. This interpretation is substantiated by the van Krevelen diagram given in Figure 27, where coals from this interval plot between Type II and Type III kerogen. The samples with HI values greater than 300 can generate liquid hydrocarbons as well as gas during maturation. Wet gas in mud log shows and the band of dead oil at the base of the upper zone substantiate the capability of these source rocks to generate C_{2+} hydrocarbons.

Genetic Potential

Genetic potential (GP) is the sum of the S_1 and S_2 values obtained from pyrolysis. GP provides a semi-quantitative evaluation of the abundance and type of organic material in a source rock (Tissot and Welte, 1984) and is expressed in kilograms of hydrocarbon per ton of rock. The presence

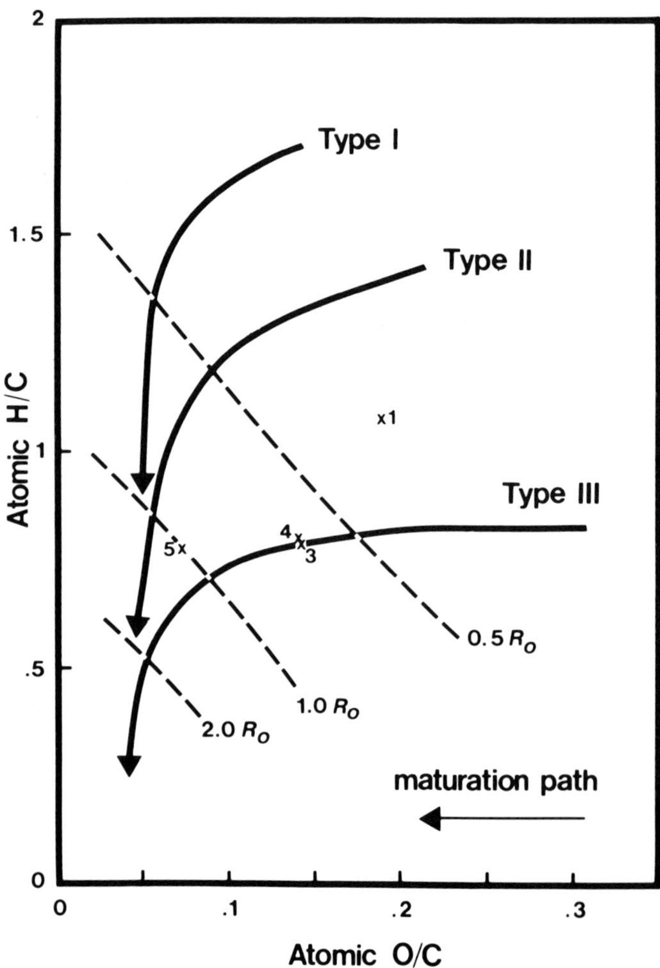

FIG. 27.—Van Krevelen diagram for the data listed in Table III. Isomaturity lines are in % R_o.

TABLE IV.—Criteria for Evaluating Genetic Potential Based on Tissot and Welte (1984).

Genetic Potential Value	Classification
<2 kg/ton	No oil, some gas potential
2 to 6 kg/ton	Moderate oil and gas potential
7 to 99 kg/ton	Good oil and gas potential
100 to 200 kg/ton	Excellent source rock

of significant quantities of migrated hydrocarbon can bias the genetic potential determinations.

The GP classification scheme presented in Table IV provides a means for evaluating the GP data from the Jonah Gulch well. According to this scheme, the section in the lower half of the well contains moderate to excellent source rocks. The samples from the 7120–7430 ft (2169–2264 m) interval are outstanding source rocks. To summarize, the total organic carbon, hydrogen index, and genetic potential data all indicate there is sufficient high quality organic material in the Jonah Gulch well to be a source of both wet gas and liquid hydrocarbons as well as dry gas.

Production Index

Pyrolysis data can be used to determine the position of a sample within the liquid hydrocarbon window. In particular, the production index (PI) or transformation ratio $[S_1/(S_1 + S_2)]$ can be used to make this determination. Typically, source rocks will yield a PI value of 0.1 at the entrance to the liquid window and a value of 0.4 at the exit to the window (Tissot and Welte, 1984; Hunt, 1979). Significant migration of hydrocarbons, either into or out of the interval of interest, can affect the accuracy of the PI numbers.

The PI data (Table I) are probably the most definitive of all the organic geochemical indicators in delineating the top of the liquid window. Despite considerable variability in the other geochemical data—TOC, HI, GP—within the same interval, a clear, consistent change in the PI data occurs between 7860 ft (2395 m) and 8660 ft (2639 m). The PI values are all less than 0.1 above 8660 ft (2639 m) and greater than 0.1 below this depth. This is true for both lean and rich source rocks as well as for hydrogen-poor and hydrogen-rich organic matter. Based on these data, the entrance to the liquid window is placed close to the unconformity separating the Tertiary and Cretaceous sections. In addition, the PI values close to 0.1 support the assumption that the samples have not been contaminated by migrating hydrocarbons (Tissot and Welte, 1984).

Vitrinite Reflectance

Seven vitrinite reflectance measurements (R_o) were made on coaly intervals in the Jonah Gulch well (Table II). A plot of depth vs. log % R_o is given in Figure 28. The data appear to plot as two populations—one above and one below the unconformity. A similar trend can be interpreted in vitrinite data gathered by Pollastro and Barker (1984) in the Wagon Wheel No. 1 well to the north. The different populations may be explained by either a change in kerogen composition across the unconformity or by bitumen impregnation of kerogen below the unconformity. Bitumen

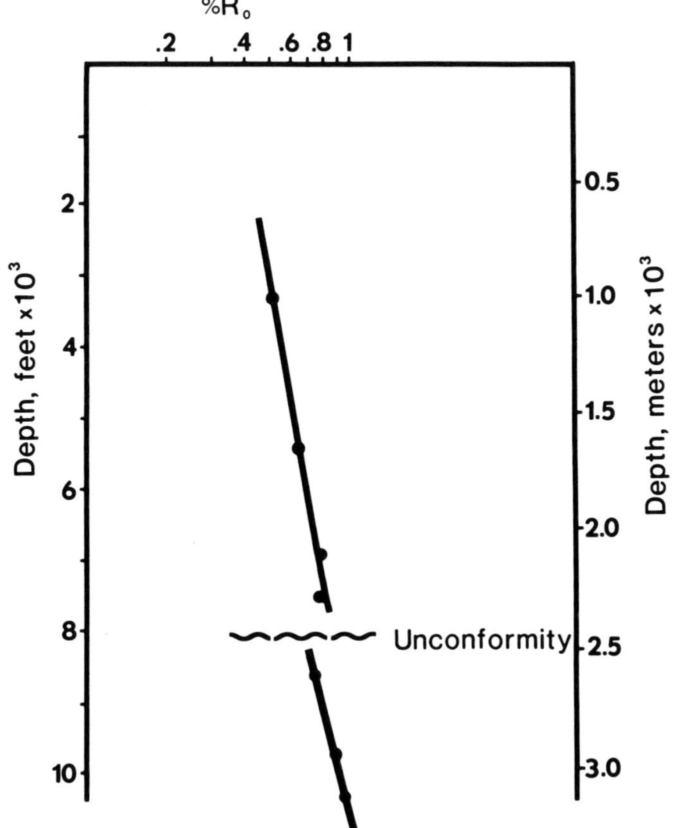

FIG. 28.—Plot of depth vs. log % R_o. The break in slope at the Tertiary-Cretaceous unconformity is interpreted on the basis of additional data from Pollastro and Barker (1984) and on reasoning explained in the text. The slope of the Cretaceous data gives a value of R_o = 0.93% at the top of the Ericson Sandstone.

contamination is not very common (H. Dembicki, pers. commun., 1984), and so a change in kerogen type seems more plausible.

Several considerations indicate the slope of the Tertiary data is probably too steep. A single, best-fit straight line through all the data points would also be too steep and therefore subject to these same limitations. First, using R_o values in the range of 0.65% (Waples, 1980) to 0.60% (Hunt, 1979) to indicate the entrance to the liquid window, source rocks between 4500 and 5300 ft (1371–1615 m) should be within the liquid window according to the steeper line. This conflicts with the maturity levels predicted by production index and elemental analysis data. Second, if source beds at this shallow level are in the liquid window, then either unreasonably high erosion and/or geothermal gradient is required to elevate the organic matter to the necessary level of maturity. For these reasons, the line for the Cretaceous data is believed to give a better approximation of the true maturity level—especially for the Ericson. The Cretaceous line is certainly in better agreement with the other maturity indicators. Using the Cretaceous data to determine the maturity of source rocks near the top of the Ericson gives an observed value of R_o = 0.93% at this stratigraphic level. These same data indicate the present top of the liquid window lies between 8100 and 8700 ft (2468–2651 m).

Elemental Analysis

Elemental analysis was performed on five selected source rock intervals to validate and check the results of the other geochemical data. Both shales and coals were included in the samples selected for analysis. The elemental data were then used to calculate atomic H/C and O/C ratios (Table III). Data for these samples are plotted in the van Krevelen diagram shown in Figure 27.

The van Krevelen diagram confirms the previous interpretations of the HI, PI, and vitrinite reflectance data. The lines of equal vitrinite reflectance plotted on the diagram were derived from Patton (1984). Only point 1 from a depth of 7130–40 ft (2173 to 2176 m) lies above the zone of initial oil generation. The other points are all within the liquid window. Point 5, from a depth range of 10,300–10,330 ft (3138–3148 m) lies close to the R_o = 1.00% contour, in good agreement with an R_o value of 0.93% at the top of the Ericson (10,126 ft, 3085 m). Furthermore, both points 1 and 5, although they are from coaly intervals, plot in between the Type II and Type III kerogen contours. This supports the hypothesis that some of the coals in the Jonah Gulch well are hydrogen-rich and may have liquid hydrocarbon generation potential.

Time-Temperature History

A burial history curve for the Ericson Sandstone was constructed using regional work by Law (1984) and formation tops picked by Law in the Jonah Gulch well (B. E. Law, pers. commun., 1985). Data used to construct the burial history curve are given in Table V. Source rocks adjacent to or slightly below the Ericson reservoir should have burial history curves similar to the Ericson. This burial history curve and the present geothermal gradient of 24.2 °C/km (2.3°F/100 ft), (Law and Smith, 1983) were used in conjunction with techniques described by Waples (1980) to develop a time-temperature diagram for the probable source rocks. Accuracy of the time-temperature diagram is subject to uncertainty in evaluating both the paleogeothermal gradient and the amount of erosion that occurred at two unconformities in the section. Work by Naeser (1984) and Pollastro and Barker (1984) on the Pinedale anticline area to the north of the Jonah Gulch well indicates the geothermal gradient was not constant through time. Instead, a relatively rapid cooling phase was initiated in the area approximately 2–4 m.y. ago. Evidence for this cooling event can be inferred in the Jonah Gulch well. Clay analysis places the 100°C (212°F) isotherm determined from clay diagenesis reactions (Fig. 26) at approximately 8000 ft (2438 m). The present bottom hole temperature in the well at 10,570 ft (3221 m), however, is calculated to be 83°C (observed

TABLE V.—Burial History Data.

Observed vitrinite reflectance at top of Ericson = 0.93%
Surface temperature = 5°C
Ericson age = 73 m.y.
Depth to top of Ericson = 10,126 ft
Lance thickness = 1588 ft
Fort Union thickness = 815 ft
Wasatch and Green River thickness = 7723 ft

BHT = 186°F). The apparent contradiction between these two data points (temperature, depth) can be explained by a relatively recent cooling event.

To evaluate the effect of uncertainties related to the geothermal gradient and erosional unconformities, a series of time-temperature curves were calculated. For each of these eight curves the thicknesses of the present-day formations were held constant but the gradient and amount of erosion at the two major unconformities were systematically varied (Table VI). Selection of 32.8°C/km (2.8°F/100 ft) as the upper limit for the geothermal gradient was based on the successful application of this gradient to time-temperature modeling in the thrust belt to the west by Edman and Surdam (1984). A value of R_o = 0.93% at the top of the Ericson, interpolated from the Cretaceous vitrinite data in Figure 28, was used to compare the calculated vitrinite values in Table VI to the observed value.

This series of time-temperature models demonstrates three critical factors. First, the amount of erosion at the Tertiary-Cretaceous boundary is not a significant parameter influencing the final calculated vitrinite value. Second, even with a higher gradient of 32.8°C/km (2.8°F/100 ft), more than 200 ft (61 m) of Tertiary erosion is necessary to match successfully the observed vitrinite value. Finally, the models clearly demonstrate the geothermal gradient must have been higher in the past than it is now. This is in agreement with the work of Naeser (1984), Pollastro and Barker (1984), and the observed diagenetic reactions in the well.

TABLE VI.—Preliminary Time Temperature Models for Jonah Gulch Well.

Model	Gradient	Tertiary-Cretaceous boundary erosion (feet)	Intra-Tertiary erosion (feet)	TTI	% Ro
1	24.2°C/km	1000	200	8.78	0.58
2	24.2°C/km	5500	200	8.86	0.58
3	32.8°C/km	1000	200	44.28	0.87
4	32.8°C/km	5500	200	44.43	0.87
5	24.2°C/km	1000	2000	14.58	0.65
6	24.2°C/km	5500	2000	14.66	0.65
7	32.8°C/km	1000	2000	109.38	1.15
8	32.8°C/km	1000	1500	85.88	1.05

Using the information obtained from the eight preliminary time-temperature models, a best fit time-temperature curve was constructed (Fig. 29). This curve incorporates the same basic data (Table V) as the previous models but uses 3250 ft (990 m) of erosion at the Tertiary-Cretaceous boundary, 2000 ft (609 m) of Tertiary erosion (Love, 1960),

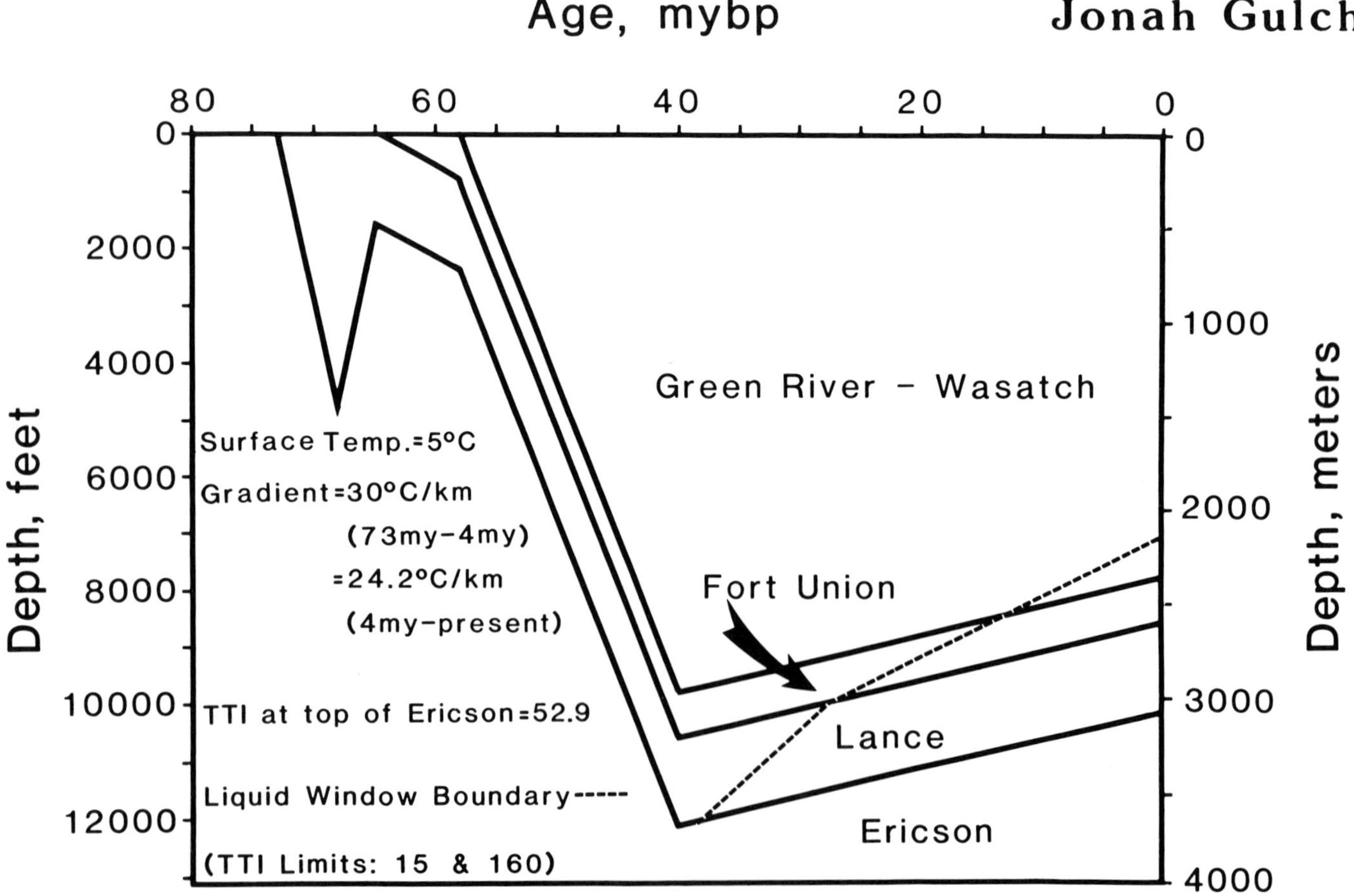

FIG. 29.—Time-temperature diagram for the Ericson Sandstone. Top of the liquid window (TTI = 15) is indicated by the dashed line. Source rocks adjacent to the Ericson should have approximately the same time-temperature history that is shown here. According to this diagram, these soure rocks entered the liquid window approximately 38 m.y.BP. At the present time the liquid window is at about 7000 ft (2133 m). The present time-temperature index for source rocks near the top of the Ericson is 52.9. Using tables from Waples (1980) to convert time-temperature index to vitrinite reflectance yields a corresponding R_o value of 0.92%. This compares well with the observed value of R_o = 0.93%.

and a gradient of 30°C/km (2.6°F/100 ft) from 73 to 74 m.y. BP and then a lower, present-day gradient of 24.2°km (2.3°F/100 ft) from 4 m.y. BP to the present. This best fit model is not a unique path but does provide valuable constraints on estimating the maturation history of source beds adjacent to the Ericson. The best fit model predicts R_o = 0.92% at the top of the Ericson, which is in good agreement with the observed value of R_o = 0.93%. Using a TTI value of 15 to represent the entrance to the liquid window (Waples, 1980), the source adjacent to the Ericson Sandstone entered the liquid window approximately 38 m.y. BP. The present top of the liquid window in the Jonah Gulch well is calculated to occur at 7300 ft (2224 m). This prediction concurs with observed PI, R_o, and elemental analysis data that show the top of the present liquid window lies between 7860 and 8640 ft (2395–2633 m).

Organic Acids

In addition to the generation of hydrocarbons, production of organic acids also needs to be considered in the synthesis of organic-inorganic interactions. These organic acids produced prior to the generation of liquid hydrocarbons (Fig. 30), have been shown by Surdam et al. (1984) to be powerful solvents, capable of complexing and mobilizing aluminum. Carothers and Kharaka (1978) have found these organic acids in oil field brines, and it is possible that these organic solvents play an important role in porosity enhancement and preservation. These water soluble organic acids are not only effective complexing agents for metals, but they can also buffer the alkalinity of formation waters over the 80° to 120°C temperature range (Surdam et al., 1984). Therefore, the organic acids not only have a pronounced effect on the stability of aluminosilicate framework grains, but they also affect carbonate grains and cements.

According to Surdam and Crossey (1985), peak production of these organic acids occurs between 80°C (176°F) and the onset of liquid hydrocarbon generation. This is in agreement with the observations of Franks and Forester (1984) in the Gulf Coast. Based on the organic geochemical data described previously, it is apparent that the Upper Cretaceous rocks in the Jonah Gulch area have gone through

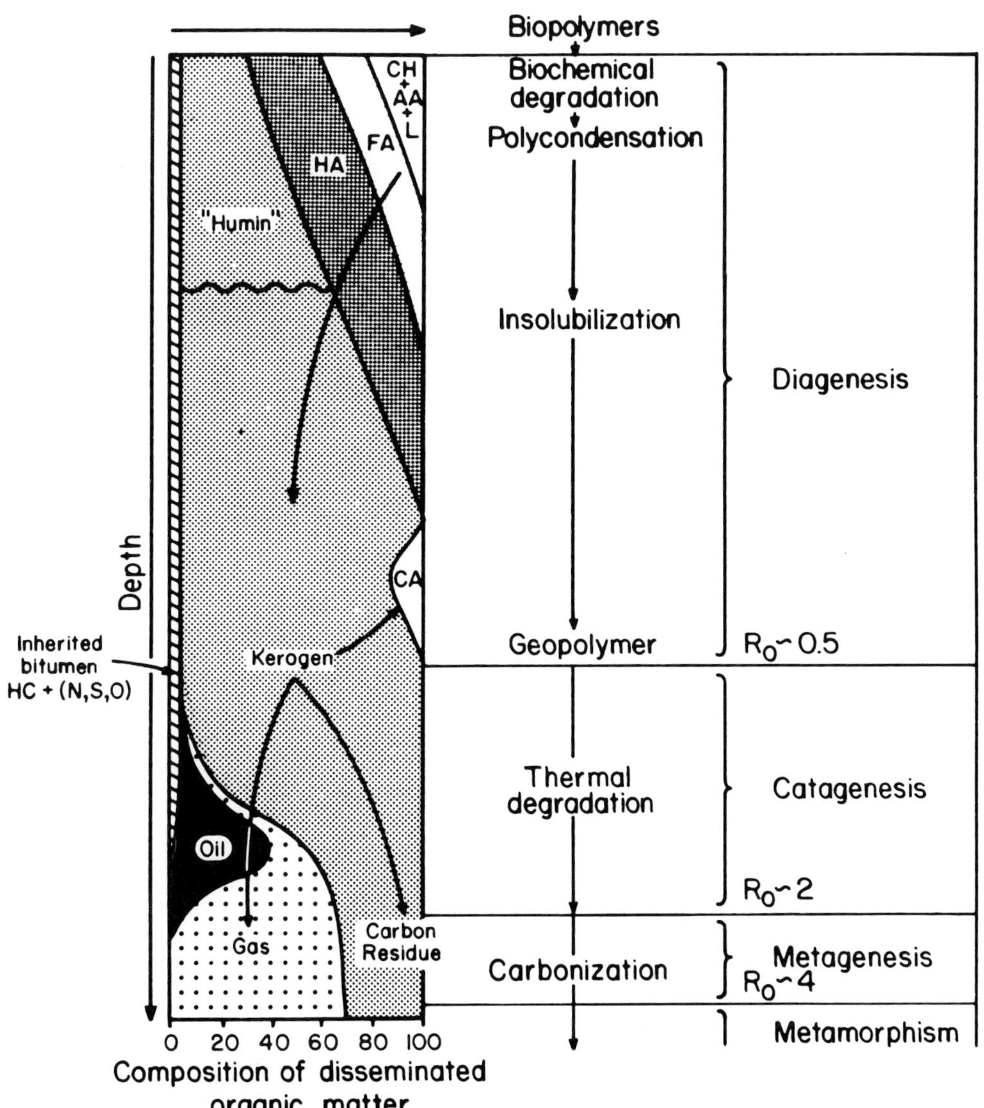

FIG. 30.—Diagram from Tissot and Welte (1984) showing the general scheme of evolution of organic matter. Note that the various organic acid solvents such as CA, carboxylic acid, are produced just prior to liquid hydrocarbon generation. Used with permission, copyright 1984, Springer-Verlag, Heidelberg.

the 80°C (176°F) to liquid hydrocarbon generation interval. Therefore, the kerogen-rich rocks within the Upper Cretaceous section have probably already generated significant amounts of these organic acids. The interaction of organic acids with the inorganic diagenesis in the adjacent sandstones is described in the next section.

Summary of Organic Geochemistry

The organic geochemical data and time-temperature analysis indicate the source rocks laterally adjacent to the Ericson Sandstone had sufficient high quality organic matter at the proper maturity level to generate the wet gas and dead oil present in the Ericson reservoir. The time-temperature analysis, vitrinite, and elemental analysis data all place the present top of the liquid window between 7300 and 8700 ft (2224–2651 m). Thus, the source rocks have already generated both the organic solvent precursors as well as the hydrocarbons now present in the Ericson reservoir. Using this information, the organic maturation sequence can be integrated with the inorganic diagenesis.

DISCUSSION

Synthesis of Organic and Inorganic Diagenesis

The series of diagrams in Figure 31 provide a schematic interpretation of how the integrated organic-inorganic diagenesis can be used to explain the observed sequence of porosity enhancement events. The first diagram shows the hypothesized configuration of the organic-inorganic system at a time T_1 prior to hydrocarbon migration. At this time in the upper zone, the abundant syntaxial quartz overgrowths had been emplaced and some pressure solution had occurred. The chert grains, with their carbonate constituents illustrated schematically by rhomb-shaped crystals, were still intact. Porosity is shown in white. In the lower zone, minor quartz overgrowths had formed on the clastic quartz grains prior to the precipitation of the mixed-layer clay cement. This cement then filled most, but not all, of the remaining primary pore space. Following precipitation of the clay cement, some pressure solution occurred. As in the upper zone, the chert grains were unaffected at this time.

The diagrammatic portrayal of a complex kerogen molecule in the lower portion of the figure is based on the experimental work of Surdam and Crossey (in prep.). In their working model, kerogen consists of an inner aromatic core surrounded by an aliphatic portion. Mono- and difunctional carboxylic acids as well as phenols are bonded peripherally to the core of the molecule. As temperature is increased with greater burial, the bonds between these peripheral functional groups are broken. Surdam and Crossey (1985) have used the coincidence in time, temperature, and space of mixed-layer smectite/illite ordering reactions in shales (100–110°C, 212–230°F) and the peak concentrations of organic acids (80–120°C, 176–240°F) in oil field brines to propose a possible mechanism for the generation of difunctional organic solvents. They suggest the reduction of mineral oxidants and simultaneous oxidation of kerogen may

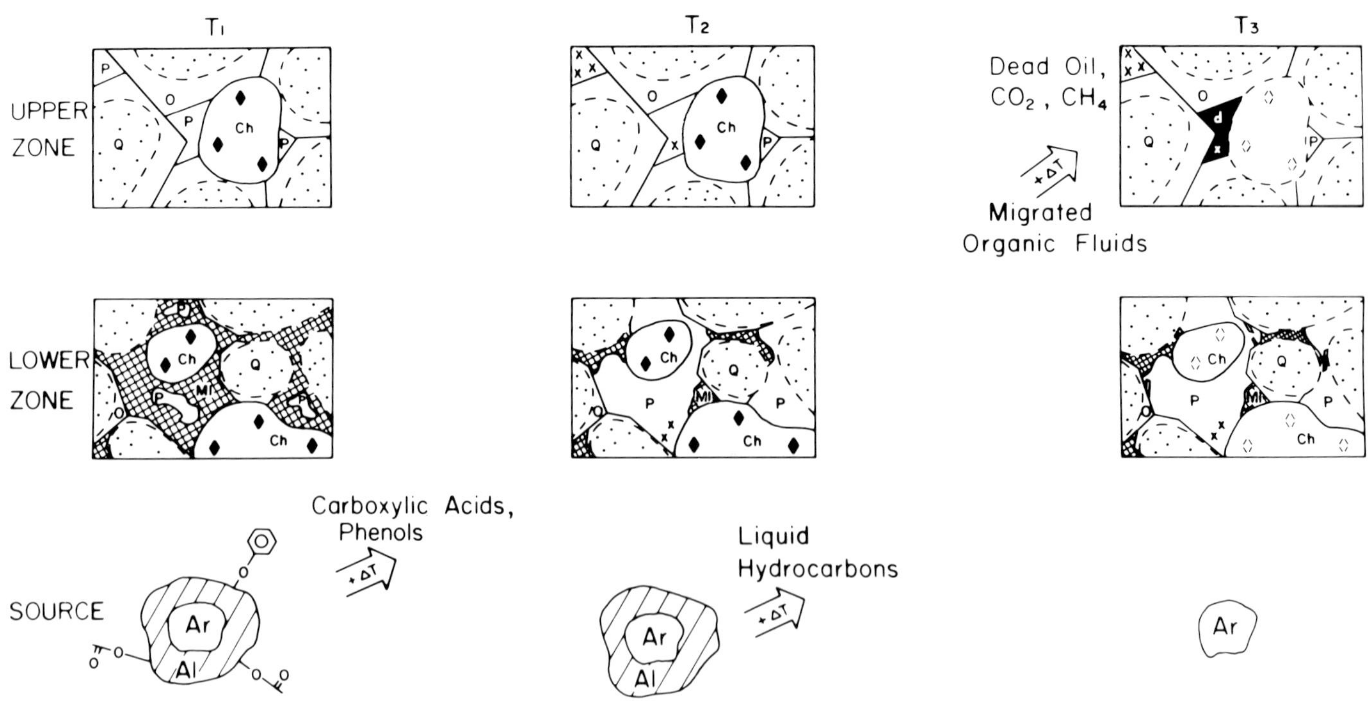

FIG. 31.—These three time slices provide a schematic view of the use of organic-inorganic interactions to explain the observed sequence of enhanced porosity generation in the Ericson Sandstone. Detailed explanation of this diagram is given in the text.

be an effective process for increasing the release of peripheral organic acid groups during thermal degradation. These water soluble acids could be transported from the source rocks to adjacent reservoir sandstones by water released during clay diagenesis. This process—the generation and migration of organic acid solvents from source rocks to adjacent sandstones—is illustrated in the lower third of Figure 31.

At time T_2 the organic solvents previously generated have already migrated into the Ericson Sandstone. In the lower zone, these difunctional organic acids complexed aluminum (Surdam et al., 1984) contained in the mixed-layer clay cement and transported this chelated load higher into the sandstone unit. Complexation of aluminum from the mixed-layer clay cement resulted in the creation of enhanced porosity in the lower zone. Although clay cement was the primary mineral phase being attacked by the organic solvents at this time, minor dissolution of chert grains in both the upper and lower zones apparently also took place during migration of the organic solvents. As the fluid-rock chemistry changed and the aluminum-organic complexes became unstable, kaolinite precipitated into available pore space in both the upper and lower zones. Kaolinite abundance increases stratigraphically higher in the sandstone and is most common in the upper zone. Thus, the mass transfer of complexed aluminum from the lower zone to the upper zone seems to have been fairly efficient.

At the beginning of time T_2 the kerogen molecule was an aromatic core surrounded by the aliphatic region; the peripheral organic acid groups had already been cleaved off and transported. As the temperature increased further, bonds between aliphatic hydrocarbons and the aromatic core were broken, and hydrocarbons migrated into the Ericson Sandstone reservoir.

The last time slice shown is at a time T_3, after hydrocarbon migration. All that remains of the complex kerogen molecule is the aromatic core. The only significant change in the lower zone since time T_2 has been the precipitation of minor amounts of late illite. In contrast, dramatic changes have taken place in the upper zone. Oil that migrated into the upper zone filled some of the remnant primary pore space. In places, the oil also migrated into microporosity between authigenic kaolinite booklets. As burial of the reservoir continued, the temperature increased and the reservoired organic fluids altered in place. Thermal decarboxylation of organic acids produced methane and CO_2. Thermal alteration of hydrocarbons also generated methane and left remnant dead oil in a paleo-oil/water contact. During this same post-hydrocarbon migration phase of the reservoir, significant dissolution of chert grains in the upper zone occurred. This late dissolution of chert constitutes the major porosity enhancement event in the upper zone. Chert dissolution poses a problem, however: what mechanism and formation fluids were responsible for the apparent mobilization of silica and the resulting enhanced porosity?

To resolve this problem, laboratory dissolution experiments using organic solvents were performed on chert samples. A detailed description of the experiments summarized here is given in Edman et al. and Surdam (in prep.). The chert samples used in the dissolution experiments were from Phosphoria cores taken in Mobil's Tip Top field located in the Overthrust belt to the west of the Jonah Gulch well. These Phosphoria samples were used based on petrographic work, discussed earlier, which shows the chert clasts in the Ericson Sandstone were originally derived from the Phosphoria Formation.

Preliminary X-ray diffraction and petrographic analyses of the Phosphoria chert samples revealed the presence of abundant carbonate—both calcite and dolomite—in the Phosphoria cherts (Fig. 32). Inorganic chemical analyses confirmed the presence of approximately 25–30% carbonate in the Phosphoria cherts. To evaluate the role of the carbonate in chert dissolution, both acidized and untreated chert samples were used in the dissolution experiments. Experimental results showed it was the carbonate within the chert that was dissolved by the organic solvents, not the silica.

These experimental results allow an easier explanation of the observed chert dissolution in the upper zone. Apparently the abundant CO_2 generated by thermal decarboxylation of the upper zone organic fluids elevated CO_2 concentration until $CO_2 >$ carbonate. The weak acids present in the CO_2-dominated system attacked carbonate within the chert grains to create late enhanced porosity in the upper zone. In the lower zone, where there was not an abundant source of reservoired organic fluid to form CO_2, only very minor late chert dissolution is observed.

The synthesis of organic-inorganic interaction summarized in the diagrams of Figure 31, provides a simple, unified mechanism for explaining the observed sequence of porosity enhancement events in the Ericson Sandstone. Clearly, it is not the only possible explanation, but the advantages in the integrated and cohesive approach presented here make this mechanism very attractive.

FIG. 32.—Photomicrograph of the Phosphoria Formation in the Overthrust belt. Note sponge spicules characteristic of the Phosphoria (arrow). Of greater importance is the abundant carbonate, both calcite (C) and dolomite (D). Chemical analyses show that approximately 25–30% of the Phosphoria samples used in the organic solvent dissolution experiments is carbonate. It is this abundant carbonate within the detrital chert clasts that is being dissolved to create late enhanced porosity in the upper zone. Scale bar = 0.5 mm.

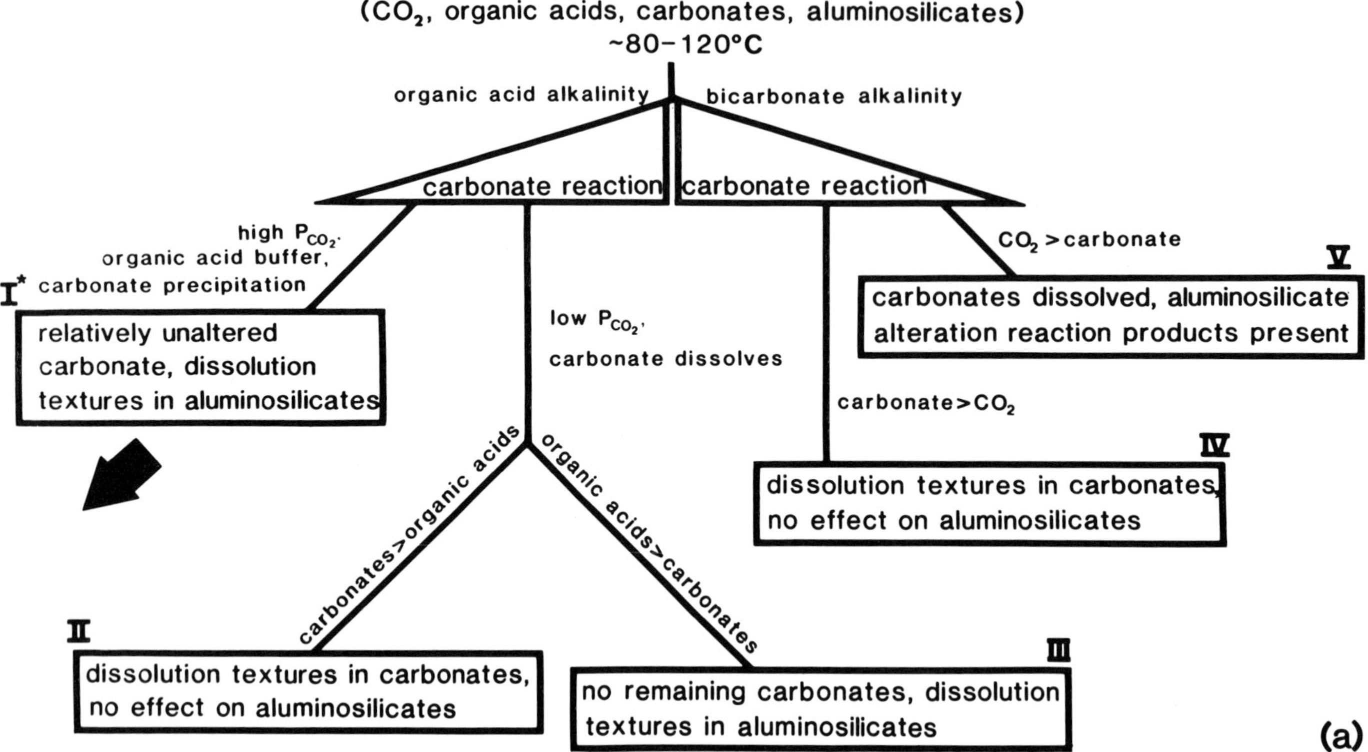

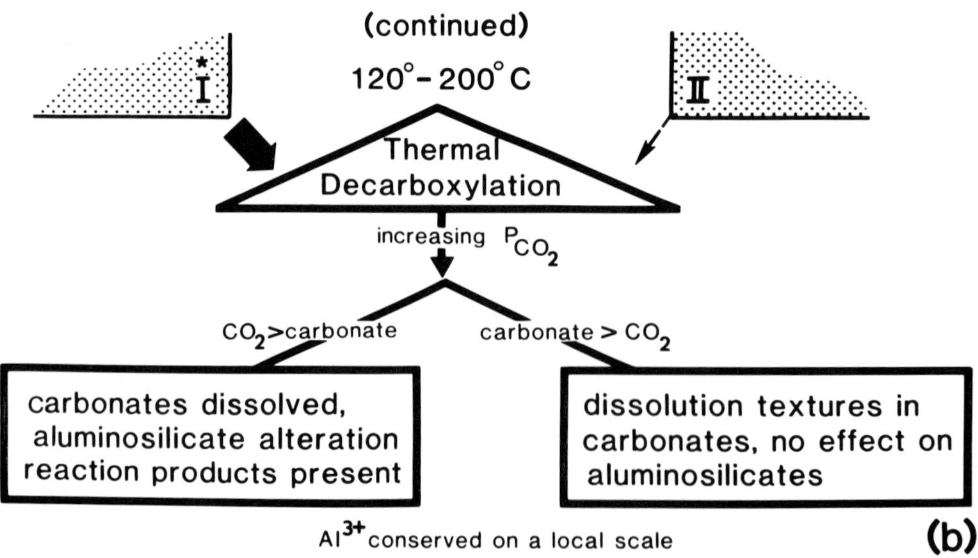

FIG. 33.—Generalized clastic reaction pathway flow charts for the system CO_2, organic acids, carbonates, and aluminosilicates modified from Surdam and Crossey 1985. Interpreted pathways for the Ericson Sandstone are indicated. Detailed discussion of these reaction pathways is included in the text.

Clastic Reaction Pathways

Figure 33 is a flow chart of possible clastic reaction pathways involving CO_2, organic acids, carbonates, and aluminosilicates. These four components adequately describe the Ericson Sandstone system. The generalized flow charts, which have been modified from Surdam and Crossey (1985), have the interpreted pathways for the Ericson Sandstone highlighted. The ability to diagram a reaction path for a given clastic unit represents the first step toward predicting zones of maximum enhanced and preserved porosity in the subsurface.

Depending upon the fluid chemistry and mineral constituents, any of the five reaction pathways shown in Figure 33a are possible. The paths on the right side of the chart are appropriate for rock-fluid systems where bicarbonate controls the alkalinity of the system. The paths on the left are for systems where organic acids control the alkalinity. Textural evidence described in the inorganic diagenesis section suggests it is the left-hand pathways that apply to the Ericson Sandstone.

Initially, organic acids controlled the alkalinity in the Ericson and reaction path I was followed (Fig. 33a). In this system, even though the P_{CO_2} was high, the fluid was buffered by organic acids and carbonate was stable (Surdam and Crossey, 1985. Aluminosilicates, however, were not stable in this system because of the aluminum complexing ability of the organic acids. Thus, at this point in the early diagenesis of the Ericson, the mixed-layer clay cement in the lower zone dissolved while the carbonates in the chert grains of both zones remained relatively unaltered.

Progressing along the flow chart to later diagenesis (Fig. 33b), the onset of thermal decarboxylation was reached via pathway I on the left side of the previous early diagenesis diagram. At this time, thermal decarboxylation of the reservoired organic fluids in the upper zone significantly increased the P_{CO_2} in this zone. This caused the level of CO_2 to be greater than the amount of carbonate available. In the lower zone, however, which was not a hydrocarbon reservoir, organic fluids were unavailable for thermal decarboxylation reactions, and carbonate remained greater than CO_2. Because of the difference in availability of organic fluids at this stage in the flow chart, the pathway for the upper zone branches to the left while the lower zone follows the right-hand branch. Only minor dissolution of the carbonates in the chert grains occurred at this time in the lower zone, and the aluminosilicates were untouched. In contrast, where $P_{CO_2} >$ carbonate in the upper zone, carbonate in the chert was completely dissolved and flushed from the system. This pervasive carbonate dissolution in the upper zone cherts created abundant late enhanced porosity. The dissolution of carbonate, however, did not improve the permeability, and the Ericson remains a tight gas sand.

The clastic reaction pathways described above are consistent with both the observed textural evidence and the hypothesized sequence of organic-inorganic interactions. By following the appropriate pathways, the diagenetic history of the upper and lower zones can be traced. In the lower zone, the pathway follows a route where early dissolution of mixed-layer clay cement occurred as the result of aluminum complexation by organic acids. After dissolution of the clay cement, the lower zone remained relatively unchanged with only minor carbonate dissolution taking place during late diagenesis. In the upper zone, porosity enhancement occurred only after hydrocarbon migration when thermal decarboxylation of reservoired organic fluids produced abundant CO_2. This resulted in two distinct diagenetic zones; a lower zone of enhanced porosity due to the early dissolution of clay cement and an upper zone of enhanced porosity created by late carbonate dissolution.

CONCLUSIONS

This study shows how experimentally verified organic-inorganic chemical reactions can be used to explain the observed sequence of porosity enhancement events in the Ericson Sandstone. In particular, there are two distinct porosity enhancement events in the Ericson that can be explained by organic-inorganic interactions. The first porosity enhancement occurred prior to hydrocarbon migration and involved the dissolution of mixed-layer clay cement. The dissolution mechanism is interpreted to be complexation of aluminum in the clay by carboxylic acids and phenols cleaved from kerogen molecules prior to hydrocarbon migration. The second porosity enhancement event occurred after organic fluids were stratigraphically trapped in the upper Ericson. Porosity enhancement in the upper zone involved late intragranular dissolution of carbonate from chert clasts by carbonic acid generated during thermal destruction of reservoired organic fluids.

The synthesis of organic-inorganic reactions to explain the evolution of enhanced porosity presents a powerful new tool for predicting zones of maximum enhanced and preserved porosity in the subsurface. The technique described here provides a unified theory for interrelating the concurrent development of the organic and inorganic geochemical systems. The chemical reactions involved in this integrated approach are the natural consequence of the progressive burial of a sedimentary sand/shale prism. As such, they should have universal applicability. The derivation of generalized clastic reaction flow path diagrams that are appropriate for any system containing the four components CO_2, organic acids, carbonates, and aluminosilicates, illustrates the wide applicability of the organic-inorganic interaction concept. These generalized flow charts were successfully used in this study to diagram the evolution of enhanced porosity in the Ericson. The next step is to extend this, or similar investigations, to a basin-wide scale. By developing such additional case studies, the ultimate goal of subsurface porosity prediction may become a reality.

ACKNOWLEDGMENTS

We thank John Robinson at Forest Oil for providing us with data from the Jonah Gulch well. J. D. Edman used facilities at the U.S. Geological Survey in Denver for many of the inorganic analyses. Janet Pitman and Dudley Rice were instrumental in arranging for access to these facilities. Rich Pollastro generously allowed the use of his clay mineralogy laboratory. R. C. Surdam gratefully acknowledges financial support from Mobil, Conoco, Texaco, Sohio, and Amoco for various aspects of this work. Janet Pitman, Rich Pollastro, and Ben Law all provided many helpful comments during the course of the research. Reviews by Ben Law and an anonymous reviewer improved the manuscript. We also thank Sharon Stonecipher, Harry Dembicki, Al Kaltenback and Judy Fredrickson for their suggestions and advice.

REFERENCES

ALMON, W. R., 1974, Petroleum-forming reactions: clay catalyzed fatty acid decarboxylation: Ph.D. Thesis, University of Missouri, 117 p.

BEAUMONT, C., 1981, Foreland basins: Geophysical Journal of the Royal Astronomical Society, V. 65, p. 291–329.

BRUCE, C. H., 1984, Smectite dehydration—its relaton to structural development and hydrocarbon accumulation in northern Gulf of Mexico basin: American Association of Petroleum Geologists Bulletin, v. 68 p. 673–683.

CAROTHERS, W. W., AND KHARAKA, Y. K., 1978, Aliphatic acid anions in oil-field waters—Implications for origin of natural gas: American Association of Petroleum Geologists Bulletin, v. 62, p. 2441–2453.

DICKINSON, W. W., 1984, Isotope geochemistry of carbonate minerals in nonmarine rocks: Northern Green River basin, Wyoming: Ph.D. Thesis, University of Colorado, Boulder, 150 p.

DIXON, J. S., 1982, Regional structural synthesis, Wyoming salient of western overthrust belt: American Association of Petroleum Geologists Bulletin, v. 66, p. 1500–1531.

DREVER, J. I., 1973, The preparation of oriented clay mineral specimens for X-ray diffraction analysis by a filter-membrane peel technique: American Mineralogist, v. 58, p. 553–554.

DURAND, B., AND OUDIN, J. L., 1979, Exemple de migration des hydrocarbures dan une se'rie deltaique: le delta de la Mahakam, Kalimantan, Indone'sie, *in* Exploration, Supply and Demand: 10th World Petroleum Congress, Proceedings, v. 2, p. 3–11.

EDMAN, J. D., AND SURDAM, R. C., 1984, Influence of overthrusting on maturation of hydrocarbons in Phosphoria Formation, Wyoming-Idaho-Utah Overthrust belt: American Association of Petroleum Geologists Bulletin, v. 68 p. 1803–1817.

FRANKS, S. G., AND FORESTER, R. W., 1984, Relationships among secondary porosity, pore-fluid chemistry and carbon dioxide, Texas Gulf Coast, *in* McDonald, D. A., and Surdam, R. C., eds., Clastic Diagenesis: American Association of Petroleum Geologists Memoir 37, p. 63–80.

HAGEN, E. S., AND SURDAM, R. C., 1984, Maturation history and thermal evolution of Cretaceous source rocks of the Bighorn basin, Wyoming and Montana, *in* Woodward, J., Meissner, F. F., and Clayton, J. L., eds., Hydrocarbon Source Rocks of the Greater Rocky Mountain Region: Rocky Mountain Association of Geologists, p. 321–338.

HALE, L. A., 1955, Stratigraphy and facies relationships of the Montanan Group in south-central Wyoming, northeastern Utah and northwestern Colorado, *in* Wyoming Geological Association Guidebook, 10th Annual Field Conference, Green River Basin, p. 89–94.

HOWER, J., 1981, X-ray diffraction identification of mixed-layer clay minerals, *in* Longstaffe, F. J., ed., Clays and the Resource Geologist, Short Course Handbook No. 7: Mineralogical Association of Canada, p. 39–59.

HUNT, J. M., 1979, Petroleum geochemistry and geology: San Francisco, W. H. Freeman, 617 p.

JACKA, A. D., 1965, Depositional dynamics of the Almond Formation, Rock Springs uplift, Wyoming, *in* Wyoming Geological Association Guidebook, 19th Field Conference, Rock Springs Uplift, p. 81–100.

JOHNS, W. D., AND SHIMOYAMA, A. 1972, Clay minerals and petroleum-forming reactions during burial and diagenesis: American Association of Petroleum Geologists Bulletin, v. 56, p. 2160–67.

JORDAN, T. E., 1981, Thrust loads and foreland basin evolution, Cretaceous, Western United States: American Association of Petroleum Geologists Bulletin, v. 65, p. 2500–2520.

LAW, B. E., 1984, Relationship of source-rock, thermal maturity, and overpressuring to gas generation and occurrence in low-permeability Upper Cretaceous and Lower Tertiary rocks, greater Green River basin, Wyoming, Colorado and Utah, *in* Woodward, J., Meissner, F. F., and Clayton, J. L., eds., Hydrocarbon Source Rocks of the Greater Rocky Mountain Region: Rocky Mountain Association of Geologists, p. 469–490.

———, AND SMITH, C. R., 1983, Subsurface temperature map showing depth to 180° Fahrenheit in the Greater Green River basin, Wyoming, Colorado, and Utah: United States Geological Survey Miscellaneous Field Studies Map MF-1504.

———, SPENCER, C. W., AND BOSTICK, N. H., 1979, Preliminary results of organic maturation, temperature, and pressure studies in the Pacific Creek area, Sublette County, Wyoming, *in* 5th Department of Energy Symposium on Enhanced Oil and Gas Recovery and Improved Drilling Methods, Proceedings, v. 33—Gas and Drilling: Tulsa, The Petroleum Publishing Co., p. 2/1–2/13.

———, 1980, Evaluation of organic matter, subsurface temperature, and pressure with regard to gas generation in low-permeability Upper Cretaceous and Lower Tertiary Sandstones in Pacific Creek area, Sublette and Sweetwater Counties, Wyoming: The Mountain Geologist, v. 17, p. 23–35.

LEWIS, C. R., AND ROSE, S. C., 1970, A theory relating high temperatures and overpressures: Journal of Petroleum Technology, v. 22, p. 11–16.

LOVE, J. D., 1960, Cenozoic sedimentation and crustal movement in Wyoming: American Journal of Science, v. 258-A p. 204–214.

Naeser, N. D., 1984, Fission-track ages from the Wagon Wheel No. 1 well, northern Green River basin, Wyoming: Evidence for recent cooling, *in* Law, B. E., ed., Geological Characteristics of Low-Permeability Upper Cretaceous and Lower Tertiary Rocks in the Pinedale Anticline Area, Sublette County, Wyoming: United States Geological Survey Open-File Report 84-753, p. 66–77.

PATTON, J. W., 1984, Geochemistry, site 535, from Deep Sea Drilling Project Cores, southeastern Gulf of Mexico, *in* Buffler, R. T., Schlager, W., et al., Initial Reports of the Deep Sea Drilling Project, 77: Washington, United States Government Printing Office p. 58–64.

POLLASTRO, R. M., AND BARKER, C. E., 1984, Geothermometry from clay minerals, vitrinite reflectance, and fluid inclusions—Applications to the thermal and burial history of rocks cored from the Wagon Wheel No. 1 well, Green River basin, Wyoming, *in* Law, B. E., ed., Geological Characteristics of Low-Permeability Upper Cretaceous and Lower Tertiary Rocks in the Pinedale Anticline Area, Sublette County, Wyoming: United States Geological Survey Open-File Report 84-753, p. 78–94.

ROEHLER, H. W., 1965, Summary of Pre-Laramide Late Cretaceous Sedimentation in the Rock Springs uplift area, *in* Wyoming Geological Association Guidebook, 19th Field Conference, Rock Springs Uplift, p. 11–13.

SCHMITT, J. G., 1985, Synorogenic sedimentation of Upper Cretaceous Frontier Formation conglomerates and associated strata, Wyoming-Idaho-Utah thrust belt: The Mountain Geologist, v. 22, p. 5–16.

SHUSTER, M. W., AND STEIDTMANN, J. R., 1983, Origin and development of northern Green River basin: A stratigraphic and flexural study: American Association of Petroleum Geologists Bulletin, v. 67, p. 1356.

SMITH, J. H., 1965, A summary of stratigraphy and paleontology, Upper Colorado and Montanan Groups, southcentral Wyoming, northern Utah, and northwestern Colorado, *in* Wyoming Geological Association Guidebook, 19th Field Conference, Rock Springs Uplift, p. 13–26.

SNOWDON, L. R., AND POWELL, T. G., 1982, Immature oil and condensate—Modification of hydrocarbon generation model for terrestrial organic matter: American Association of Petroleum Geologists Bulletin, v. 66, p. 775–788.

SPENCER, C. W., 1983, Geologic aspects of tight gas reservoirs in the Rocky Mountain region, *in* Society of Petroleum Engineers/Department of Energy Symposium on Low-Permeability Gas Reservoirs, Proceedings, 1983, Denver: Society of Petroleum Engineers, p. 399–404.

SURDAM, R. C., BOESE, S. W., AND CROSSEY, L. J., 1984, The chemistry of secondary porosity, *in* McDonald, D. A., and Surdam, R. C., eds., Clastic Diagenesis: American Association of Petroleum Geologists Memoir 37, p. 127–149.

SURDAM, R. C., AND CROSSEY, L. J., 1985, Organic-inorganic reactions during progressive burial: key to porosity and permeability enhancement and preservation: Philosophical Transactions, Royal Society of London, A, v. 315, p. 135–156.

TISSOT, B. P., 1984, Recent advances in petroleum geochemistry applied to hydrocarbon exploration: American Association of Petroleum Geologists Bulletin, v. 68, p. 545–563.

———, AND WELTE, D. H., 1985, Petroleum formation and occurrence: New York, Springer-Verlag, 699 p.

VANDENBROUCKE, M., DURAND, B., AND OUDIN, J. L., 1983, Detecting migration phenomena in a geological series by means of C_1–C_{35} hydrocarbon amounts and distribution, *in* Bjorøy, M., Albrecht, C., Corning, C., de Groot, K., Eglinton, G., and Speers, G., eds., Advances in Organic Geochemistry 1981: Chichester, John Wiley, p. 147–155.

WAPLES, D. W., 1980, Time and temperature in petroleum formation: Application of Lopatin's method to petroleum exploration: American Association of Petroleum Geologists Bulletin, v. 64, p. 916–926.

———, 1981, Organifc Geochemistry for Exploration Geologists: Minneapolis, Burgess, 151 p.

ROLE OF ORGANIC SPECIES DISSOLVED IN FORMATION WATERS FROM SEDIMENTARY BASINS IN MINERAL DIAGENESIS

YOUSIF K. KHARAKA, LEROY M. LAW, WILLIAM W. CAROTHERS AND DONALD F. GOERLITZ

U.S. Geological Survey, 345 Middlefield Road, Menlo Park, California 94025

ABSTRACT: Dissolved organic species are present in high concentrations in almost all subsurface waters, especially those associated with petroleum. The highest concentrations (as much as 10,000 mg/l) are present in formation waters from relatively young reservoir rocks with subsurface temperatures of 80 to 100° C. As a result of thermal decarboxylation, concentrations of dissolved organics decrease with increasing temperatures above 100° C. The concentrations of organic species in subsurface waters with temperatures lower than 80° C are generally less than about 100 mg/l because of bacterial degradation. Short-chain (C_2—C_5) aliphatic acid anions (principally acetate and propionate) form as much as 90% of these dissolved organic species. Detailed chemical analysis using GC and GC/MS techniques has shown the presence of longer-chain aliphatic acid anions (C_6—C_9), dicarboxylic acid anions (C_4—C_{10}), phenols and other organic species in formation waters from High Island field, offshore Texas.

Dissolved organic species may play an important role in mineral diagenesis. This role is related to the following chemical properties. (1) They can be the dominant source or sink for the hydrogen ion (H^+), and thus, directly or indirectly control the pH and buffer capacity of subsurface waters. (2) They behave as reducing agents controlling the oxidation state (Eh) of subsurface waters and the concentrations of multivalent elements (e.g., iron). (3) They can be decarboxylated, thermally or with the aid of bacteria, into carbon dioxide and hydrocarbon gases. (4) They form soluble complexes with metals and other inorganic species, but quantitatively these complexes are not as important as previously reported. Computations with a geochemical model show that when organic as well as inorganic species are considered, the diagenetic mineral sequences predicted by the model are in general agreement with those observed in the field.

INTRODUCTION

In the last 10 years organic species dissolved in subsurface waters and their inorganic interactions at low and intermediate temperatures have become one of the most intensively studied fields in geochemistry. This increased interest stems from the realization that organic species, which may be present at very high concentrations (as much as 10,000 mg/l) in subsurface waters (Carothers and Kharaka, 1978; Surdam et al., 1984), can play an important role in mineral diagenesis (Kharaka and Mariner, 1977; Curtis, 1978; Schmidt and McDonald, 1979; Lundegard et al., 1984) and may be precursors for an important portion of natural gas (Cordell, 1972; Kharaka et al., 1983a). Organic species have also been studied for possible use as proximity indicators in petroleum exploration (Lochte et al., 1949; Gullikson et al., 1961; Zinger and Kravchik, 1973; Kartsev, 1974; Collins, 1975; Carothers and Kharaka, 1978).

Dissolved organic species, including the short-chain aliphatic acid anions (C_2—C_{10}), the dominant species present in oil field waters, can play an important role in mineral diagenesis. This role is related to the following chemical properties. (1) They can be the dominant source or sink for the hydrogen ion (H^+), and thus, directly or indirectly control the pH and buffer capacity of subsurface waters. The pH of subsurface water is probably the most important chemical component that affects mineral diagenesis (Kharaka et al., 1980). (2) They form soluble complexes with metals and other inorganic species, thus enhancing their transport and modifying their diagenetic interaction. (3) They behave as reducing agents controlling the oxidation state (Eh) of subsurface waters and the concentration of multivalent elements (e.g., iron). (4) They can be decarboxylated, thermally or with the aid of bacteria, into carbon dioxide and hydrocarbon gases (Kharaka et al., 1983a). Solution of the produced carbon dioxide lowers the pH and decreases the activity of carbonate (CO_3^{--}) species in the formation waters and thus may cause dissolution of carbonate and other minerals (Curtis, 1978; Schmidt and McDonald, 1979; Kharaka et al., 1980). Precipitation of minerals and reduced porosity, on the other hand, can result from higher pH and activity of carbonate species caused by degassing of CO_2 from formation water (Blatt, 1979; Lundegard et al., 1984).

This paper summarizes the nature and distribution of organic species dissolved in formation waters from several sedimentary basins in North America. In particular, new data on the organic species dissolved in oil field waters from High Island field, offshore Texas, are presented. The formation waters in this field were obtained from the middle and lower Pleistocene reservoir rocks. The waters contain not only short-chain (C_2—C_5) aliphatic acid anions identified by previous workers, but also longer-chain monocarboxylic acid anions (C_6—C_9), dicarboxylic acid anions (C_4—C_{10}), phenols and other organic species. Computations with SOLMNEQ II (Kharaka et al., in prep.) show that when organic as well as inorganic species are considered, a better correlation is obtained between diagenetic mineral sequences predicted by the model and those observed in the field.

ORGANIC SPECIES IN SUBSURFACE WATERS

The median concentration of dissolved organic species in shallow groundwaters is about 1 mg C/l (Leenheer et al., 1974; Thurman, 1984). An important fraction (20 to 40% and as much as 90%) of this dissolved carbon consists of high molecular weight fulvic and humic acids (Thurman, 1984). The proportion of these humic substances in organic species dissolved in the waters of the unsaturated zone is even higher, amounting to 50 to 60%. The median concentrations of dissolved organic carbon, however, are also higher, ranging from about 20 mg/l in the A horizon to 2 mg/l in the C horizon (Thurman, 1984).

Concentrations of dissolved organic species in oil field waters (40° to 200° C) are much higher than in groundwaters and may reach values as high as 10,000 mg/l (as acetate). Acetate, propionate, butyrate and valerate were identified as the dissolved species that formed as much as 90% of the organic carbon in these waters (Willey et al., 1975; Carothers and Kharaka, 1978). Carothers and Kharaka (1978) and Kharaka et al. (1983a) showed that the concentrations of these aliphatic acid anions were controlled mainly by subsurface temperatures and ages of the reservoir rocks.

The distribution of aliphatic acid anions in oil field waters from Central Valley, California, northern Gulf of Mexico basin and North Slope, Alaska are shown in Figure 1 (Kharaka et al., 1983a). Three distinct temperature regimes are depicted in Figure 1. Zone 1 is characterized by concentrations of acid anions less than about 100 mg/l and temperatures less than about 80° C. The concentrations of acetate in this zone are below detection limit and propionate predominates. Bacterial degradation is believed to be responsible for the absence of acetate from Zone 1 (Carothers and Kharaka, 1978).

The highest concentrations of these acid anions are present in the youngest and shallowest reservoir rocks of Zone 2 (temperature higher than about 80° C) and decrease with increasing subsurface temperatures (Fig. 1) and age of the rocks. Acetate constitutes more than 90% of aliphatic acid anions with propionate forming about 5% of anions (Carothers and Kharaka, 1978). The boundary of Zone 3, where no acid anions are present, is placed at about 200° C and is obtained by extrapolation of the least-squares line of Figure 1 (Kharaka et al., 1983a).

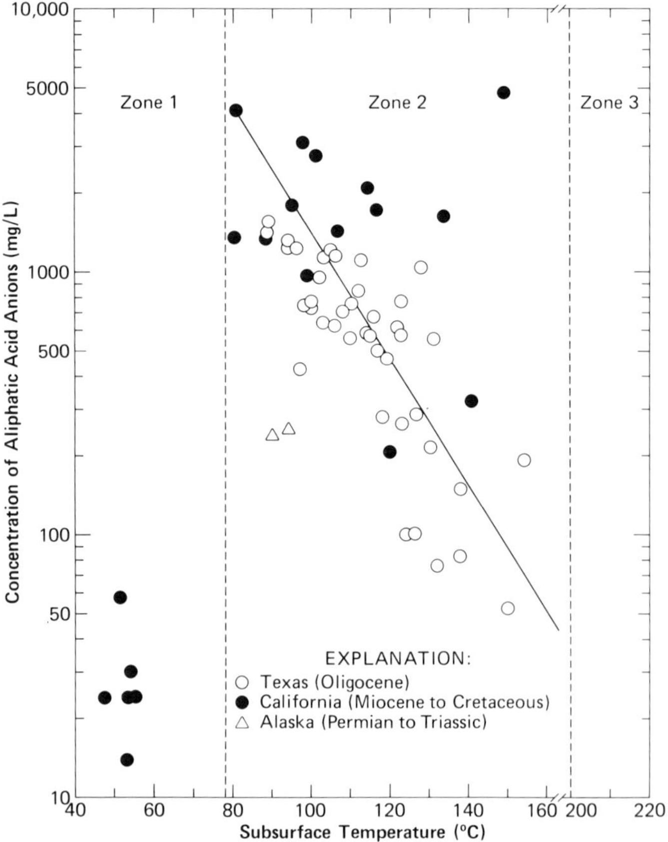

FIG. 1.—Concentrations of aliphatic acid anions (C_2—C_5) in formation waters from three sedimentary basins. Note that the highest concentrations are at about 80°C and decrease with increasing temperatures. Note also the sudden decrease between Zone 2 and Zone 1 (from Kharaka et al., 1983a).

The decrease in the concentrations of acid anions with increasing temperatures in Zone 2 (Fig. 1) and age of the reservoir rocks (Carothers and Kharaka, 1978) indicates thermal decarboxylation of these acid anions. Further evidence for thermal decarboxylation of these acid anions to CO_2 and hydrocarbon gases is obtained from the $\delta^{13}C$ values of dissolved bicarbonate, CO_2 and CH_4 in natural gas and diagenetic calcite and ankerite (Boles, 1978; Carothers and Kharaka, 1980; Milliken et al., 1981; Lundegard et al., 1984). The $\delta^{13}C$ values reported above showed that the carbon in CO_2 gas, dissolved bicarbonate and carbonate cements was largely from an organic source.

Thermal decarboxylation of acetic acid was investigated experimentally by Kharaka et al. (1983a). They showed that the concentrations of acetic acid, and not total acetate in solution, control the reaction rates (Fig. 2), which follow a first-order equation based on decreasing concentrations of acetic acid with time. The decarboxylation rates obtained gave half-lives that ranged from about 100 days (300° C) to 100 years (50° C). The decarboxylation rates obtained by Drummond and Palmer (pers. commun., 1985) using stainless steel vessels were similar to those of Kharaka et al. (1983a). Drummond and Palmer (pers. commun., 1985), however, obtained rates that were several orders of magnitude lower when using catalytically less active gold and titanium vessels. Nevertheless, all the experimental data indicate that the organic acid anions originally present in the oil field waters may have been much higher than those shown in Figure 1.

Other organic species identified in oil field waters include several amino acids (Degens et al., 1964; Rapp, 1976). The major amino acids found in both studies were serine, glycine, alanine and aspartic acid. The concentrations of these acids, however, were low, ranging to as much as only about 0.3 mg/l.

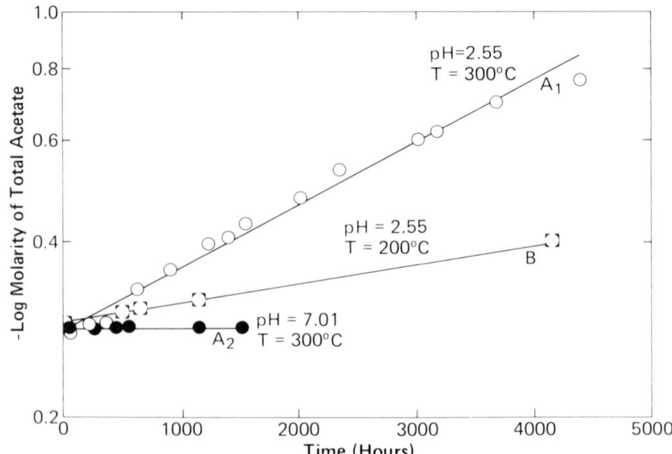

FIG. 2.—Experimental decarboxylation of acetic acid at 200 and 300°C. The absence of decarboxylation in the solution with pH = 7.01 shows that acetic acid and not total acetate controls the decarboxylation rates (from Kharaka et al., 1983a).

Gaseous and liquid hydrocarbons dissolved in formation waters have been studied extensively (Collins, 1975; Hunt, 1979; Price et al., 1983). They have been used as proximity indicators, in studies related to the origin and migration of petroleum, and as a possible new source of energy (Collins, 1975; Price, 1978; Price et al., 1983). These hydrocarbons are not covered in detail in this paper because their importance in mineral diagenesis is probably minor. It suffices to say that concentrations of methane and other hydrocarbon gases in subsurface waters generally increase with increasing pressure, decreasing water salinity, and at temperatures higher than about 75° C, with increasing temperature (Hass, 1978; Randolph, 1978). Also, the concentrations of dissolved liquid hydrocarbons generally increase with increasing subsurface temperatures and decrease with increasing water salinities (Price, 1978; Price et al., 1983).

ORGANIC SPECIES IN FORMATION WATERS FROM HIGH ISLAND FIELD

The distribution of organic species dissolved in the formation waters of the High Island field were studied because: (1) the reservoir rocks are geologically young (middle and lower Pleistocene sandstones); (2) the subsurface-temperature range (50 to 100° C) had not been examined in detail in previous studies (Fig. 1), and (3) the concentration of short-chain aliphatic acid anions and other organic species were expected to be very high because of (1) and (2). The field is located offshore about 200 km southeast of Houston, Texas. Petroleum is obtained from a number of production zones in a complex, highly faulted roll-over structure. A salt dome, located below the field at a depth of about 5 km below sea level, is probably responsible for the structure of this field.

The inorganic chemical compositions of waters from High Island field (Table 1) show a salinity that ranges from about 39,000 mg/l (essentially that of ocean water) to about 160,000 mg/l total dissolved solids. All the waters are Na—Cl type waters with Ca and Sr concentrations increasing with increasing salinities. Concentrations of Mg generally decrease with increasing water salinities, but they are generally higher, relative to water salinities, than those reported (Kharaka et al., 1980) in formation water from older reservoir rocks.

The concentrations of short chain aliphatic acid anions (C_2—C_5) in the formation waters of High Island field are given in Table 2. The analyses were obtained by gas chromatography (GC) using direct aqueous injection on a packed column. The concentrations of acetate range from 2.4 mg/l to 1,525 mg/l and constitute from 40% to 82% by weight of these anions. Propionate and iso-valerate are, in general, the second and third most important organic anions in these waters. The concentrations of these acid anions and the proportions of acetate generally increase with increasing subsurface temperatures (Fig. 3).

The samples in this study, with the exception of sample 83-TX-12, plot in Zone 1 (Fig. 1). There are, however, three major differences between the samples of this study and those from previous investigations (Fig. 1). First, the concentrations of C_2—C_5 acid anions in Zone 1 are generally higher, reaching values of as much as about 300 mg/l. Second, acetate, which is essentially absent from the waters of Zone 1 in previous studies, is the dominant spe-

TABLE 1.—CHEMICAL COMPOSITION (MG/L) OF FORMATION WATERS FROM HIGH ISLAND FIELD,[1] OFFSHORE TEXAS.

Sample No. Well Name	83-TX-1 A-11A	83-TX-2 A-11B	83-TX-3 A-10A	83-TX-5 A-9A	83-TX-6 C-8A	83-TX-7 C-14A	83-TX-8 B-14A	83-TX-9 B-2A	83-TX-10 A-14B	83-TX-11 B-12A	83-TX-12 A-45-1
TDS[2]	74000	80000	61000	159000	44000	39000	138000	92000	63000	88000	91000
Li	1.79	1.97	1.74	5.03	0.63	0.58	4.89	2.21	1.79	2.25	5.50
Na	26300	28400	19800	53700	15400	13300	47700	31800	22300	30400	30900
K	160	172	139	244	110	103	250	167	143	183	271
Rb	0.15	0.16	0.12	0.24	0.08	0.10	0.19	0.14	0.17	0.16	0.39
Cs	0.09	0.07	0.07	0.29	0.03	0.04	0.24	0.15	0.07	0.14	0.21
Mg	788	869	287	1270	499	349	1040	1010	363	873	482
Ca	1320	1370	1030	3810	679	850	2840	2250	1130	2010	2690
Sr	97	119	59	346	36	27	238	147	66	129	146
Ba	77	102	75	316	40	52	256	171	90	113	200
Mn	0.28	0.19	0.16	0.60	0.09	0.23	0.25	0.43	0.29	0.43	2.63
Fe	7.4	5.8	1.7	17.2	4.3	3.4	11.3	4.4	13.9	5.0	6.5
NH_3	140	152	97	206	73	83	182	170	112	182	121
B	23	25	31	33	16	16	40	18	37	24	42
Al-µg/L	—	—	—	<1	15	—	—	—	<1	<1	3
H_2S-Titration	0.25	0.68	0.85	0.08	0.00	0.51	0.68	0.51	0.85	0.93	0.76
H_2S-Electrode	0.04	0.02	0.06	<0.01	0.05	0.04	0.08	0.09	<0.01	0.08	<0.01
Organic (C)	54	60	162	78	73	82	—	36	209	47	933
Inorganic (C)	89	94	71	34	84	76	—	64	44	75	38
F	2.3	2.9	1.9	0.53	2.0	5.9	0.81	0.50	13.1	0.81	0.51
Cl	44400	48400	33400	99000	26200	23000	84600	56200	37700	53800	53300
Br	143	88.9	102	177	102	287	90.9	108	185	113	118
I	4.0	4.7	4.5	3.1	5.1	4.3	4.6	4.4	5.6	4.3	6.2
SO_4	12.4	9.9	16.0	7.4	9.9	9.9	0.5	0.4	14.8	15.3	11.5
SiO_2	40.1	38.8	45.4	30.0	40.8	60.0	31.6	32.9	162	41.2	65.8
Alk.(HCO_3^-)[3]	494	546	568	308	480	443	293	350	460	402	2140
pH	7.51	7.39	7.34	7.25	7.59	7.54	7.09	7.11	7.60	6.97	6.83

[1]Samples from High Island 573 field (Mobil Oil) except sample 83-TX-12 which is from High Island 571 field (Columbia Gas).
[2]TDS-mg/L total dissolved solids.
[3]Field titrated alkalinity as HCO_3^-. It comprises organic as well as inorganic species.

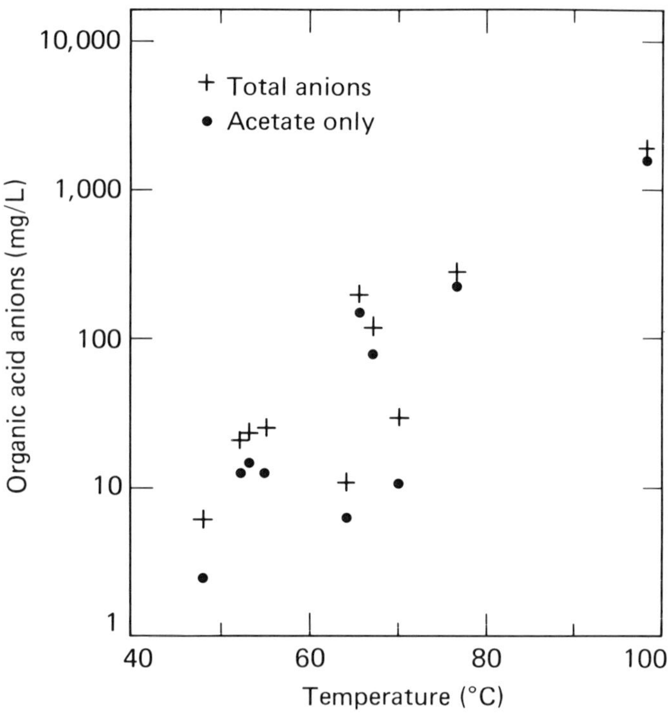

FIG. 3.—Concentrations of C_2—C_5 aliphatic acid anions (+) and acetate (.) in formation waters from High Island field, offshore Texas. Note that the proportions of acetate are lower at lower temperatures.

cies comprising as much as 80% of organic anions. And third, the concentrations and proportions of iso-valerate are much higher (Table 2).

The concentrations and proportions of organic acid anions in sample 83-TX-12 are essentially similar to those from Zone 2 (Fig. 1) at a subsurface temperature of 98° C. The concentrations of acid anions in sample 83-TX-12, however, are much lower than the expected values when both the age of the reservoir rocks (Pleistocene) and the subsurface temperature (98° C) are considered (see Carothers and Kharada, 1978, Fig. 5).

The rather unusual composition and distribution of organic acid anions in these waters (also their inorganic chemical compositions) are explained by the origin of waters and petroleum in this field. Carothers et al. (in prep.) show that the formation waters are obtained by proportional mixing (Fig. 4) of a Pleistocene connate water (represented by samples 83-TX-6 and 83-TX-7) and a highly modified brine (sample 83-TX-5) that has migrated upward from a subsea depth of about 5,000 m (130° C). They obtained essentially the same mixing proportions as those shown in Figure 4 when using δD and $\delta^{18}O$ values of the waters or the concentrations of Cl, Na, Ca or Li.

This mixing model can explain the distribution of acid anions in these waters. The concentrations of organic acid anions in sample 83-TX-12 in this model can result from mixing of waters originating in Miocene sediments at about 130° C and containing about 4,000 mg/l total acid anions with about an equal portion (Fig. 4) of Pleistocene water with minor amounts of acid anions. Bacterial action probably has not affected the concentration of acid anions in this sample as the subsurface temperature (98° C) is higher than the survival temperature for bacteria in sedimentary basins (Davis, 1967; Carothers and Kharaka, 1978). Bacterial degradation, on the other hand, is responsible for reducing the concentrations of acid anions in general and acetate in particular in the samples that are present at subsurface temperatures lower than about 80° C. Bacterial action probably will eventually consume all the acetate in these latter samples.

Newly identified dissolved organic species.—

The concentration of dissolved organic carbon in the waters from High Island field was also measured and is compared with dissolved organic carbon from C_2—C_5 acid anions (Fig. 5). The proportions of organic carbon that can be attributed to these acid anions generally increase as the concentrations of organic carbon increase; however, 20% to 90% of dissolved organic carbon in these waters cannot be attributed to previously reported organic acid anions (Carothers and Kharaka, 1978).

The uncharacterized portion of dissolved organic carbon in samples of this study was investigated further. The organic species were esterified with a mixture of methanol and sulfuric acid and extracted into an equal volume of dichloromethane (Holdeman et al., 1977). The extract was dried with sodium sulfate and analyzed using a Finnigan gas chromatograph/mass spectrometer (GC/MS) fitted with an open tubular capillary column. A recover factor of 60%, obtained with butanedioic (succinic) acid, was applied to all the dicarboxylic acids obtained by this method.

The concentrations of dicarboxylic acid anions obtained in these waters are shown in Table 3. Methylbutanedioic

TABLE 2.—CONCENTRATION OF SHORT-CHAIN ALIPHATIC ACID ANIONS IN FORMATION WATERS FROM HIGH ISLAND FIELD.

Sample Number	Subsurface Temperature (°C)	Monocarboxylic acid anion (mg/L)						
		Acetate	Propionate	i-Butyrate	n-Butyrate	3-me-Butyrate	i-Valerate	n-Valerate
83-TX-1	70	10.3	1.6	0.6	1.1	0.0	17.0	0.0
83-TX-2	55	12.4	1.9	0.0	0.0	0.0	10.8	0.0
83-TX-3	66	147.9	27.5	3.6	3.6	1.2	10.1	0.0
83-TX-5	67	77.2	26.1	1.2	0.0	0.0	10.0	0.0
83-TX-6	53	14.5	1.6	0.0	0.0	0.0	7.4	0.0
83-TX-7	52	12.7	2.6	0.6	0.9	0.0	4.2	0.0
83-TX-9	49	2.4	0.0	0.0	0.0	0.0	3.7	0.0
83-TX-10	77	221.9	34.6	4.3	4.9	2.1	5.9	5.2
83-TX-11	67	6.1	1.7	0.0	0.0	0.0	2.9	0.0
83-TX-12	98	1525.00	226.2	25.5	34.4	7.8	19.1	12.2

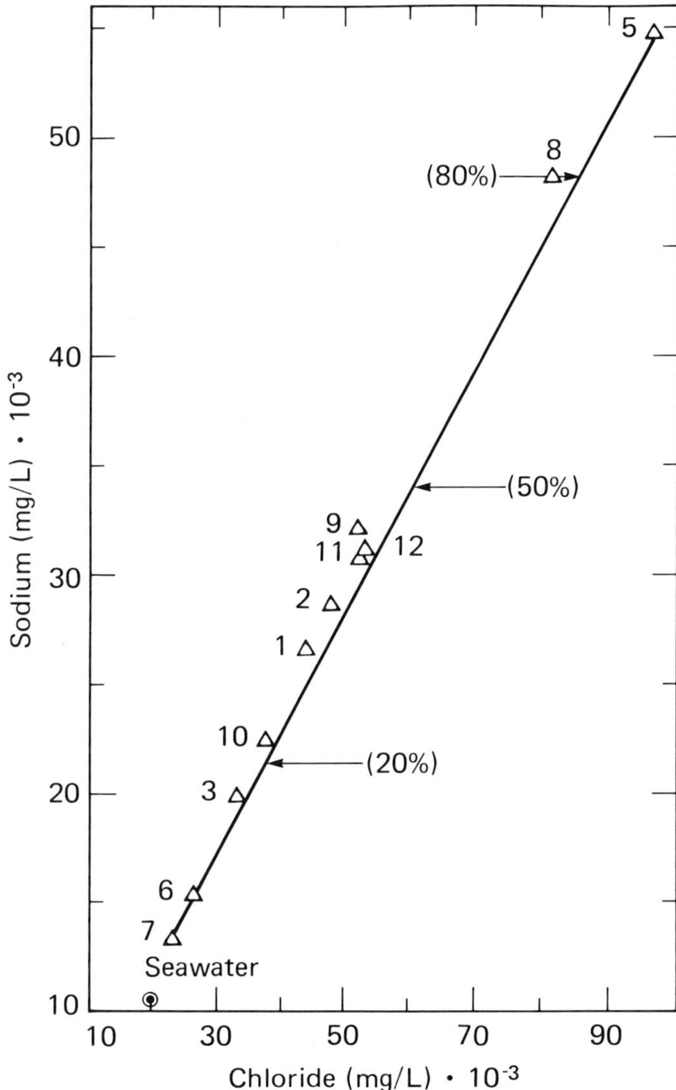

FIG. 4.—Concentrations of Na and Cl in formation waters from High Island field, offshore Texas showing proportional mixing between a highly modified water represented by sample 83-TX-5 and an only slightly modified Pleistocene connate water represented by sample 83-TX-7.

and 2-methylpentanedioic acids were also identified but not quantified in samples 83-TX-1 and 83-TX-12, the only samples tested. Table 3 shows that the quantified acid anions were present in measurable amounts in only three of the samples. These samples also have the highest concentrations of C_2—C_5 monocarboxylic acid anions (Table 2) and were obtained from greater depths than the other samples. The highest concentrations and proportions of dicarboxylic acid anions are present in sample 83-TX-12, where they comprised about 6% by weight of total acid anions. In this sample, butanedioic (succinic) and pentanedioic (glutaric) acids are the dominant acids (Table 3). It is noteworthy to mention that ethanedioic (oxalic) and propanedioic (malonic) acids are not present in measurable quantities in any of the samples. Their absence is probably related to the fact that the concentrations of Ca in these waters are

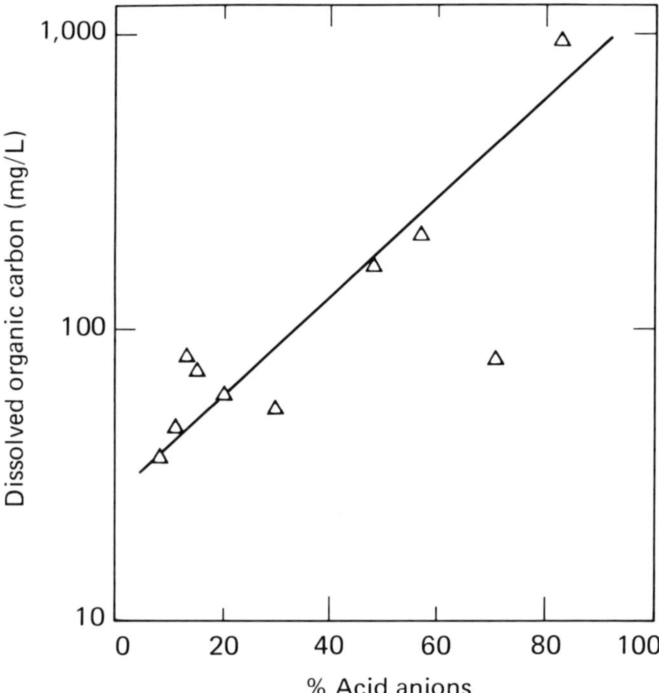

FIG. 5.—Concentrations of dissolved organic carbon in formation waters from High Island field, offshore Texas, and the fraction of organic carbon that can be attributed to C_2—C_5 monocarboxylic acid anions.

high, and solubilities of calcium oxalate and malonate are very low.

The organic species in acidified (pH ≃ 2) aliquots of samples 83-TX-1 and 83-TX-12 were extracted into an equal volume of dichloromethane and analyzed by GC/MS fitted with a capillary column. A number of dissolved organic species were identified but not quantified. These include C_6-C_9 monocarboxylic acid anions, phenol, 2-, 3-, and 4-methylphenol, 2-ethylphenol, 3, 4- and 3, 5-dimethylphenol, cyclohexanone, and 1, 2- and 1, 4-dimethylbenzene.

The characterized portion of dissolved organic carbon in sample number 83-TX-12 is about 90%. An important part of the unidentified fraction in this sample is probably due to the dissolved hydrocarbons (Price, 1978) and emulsified oil that are difficult to separate from the formation water. Because the concentrations of dissolved organic carbon in the remaining samples are substantially lower than that in sample 83-TX-12 (Table 2), it is probable that most of the

TABLE 3.—CONCENTRATION OF DICARBOXYLIC ACIDS IN FORMATION WATER FROM HIGH ISLAND FIELD, OFFSHORE TEXAS.

Dicarboxylic acid	Carbon Number	83-TX-3	Sample Number 83-TX-10 Concentration (mg/L)	83-TX-12
Butanedioic Acid	4	1.8	0.2	63
Pentanedioic Acid	5	1.7	1.2	36
Hexanedioic Acid	6	0.5	0.5	—
Heptanedioic Acid	7	—	0.6	0.2
Octanedioic Acid	8	—	0.7	5.0
Nonanedioic Acid	9	—	1.4	6.0
Decanedioic Acid	10	—	—	1.3

unidentified fraction of organic carbon (Fig. 5) is also due to dissolved hydrocarbons and oil globules.

ORIGIN OF ORGANIC SPECIES IN SUBSURFACE WATERS

Aliphatic acids, the most important group of organic compounds identified in subsurface waters, are present in such abundance in organisms and sedimentary rocks that they have been proposed as an important source for petroleum (Eisma and Jurg, 1967; Kvenvolden and Weiser, 1967; Cordell, 1972; Hunt, 1979). The fact that the concentrations of dissolved organic carbon are relatively minor in shallow groundwater and formation water at temperatures lower than about 80°C indicates that any dissolved organic species (especially acetate) produced from organisms or organic matter in these rocks are rather rapidly degraded by bacterial action. This bacterial degradation of dissolved organics to carbon dioxide and hydrocarbon gases probably ceases at temperatures of about 80°C (Davis, 1967; Carothers and Kharaka, 1978).

The highest concentrations of acid anions are present in formation waters with subsurface temperatures higher than about 80°C (Fig. 1). The most likely source of these acid anions is from thermocatalytic degradation of kerogen that has an abundance of aliphatic acids incorporated in it (Eisma and Jurg, 1967; Kharaka et al., 1983a). The amount of oxygen present in kerogen (especially type III) is high (atomic ratio of O/C as much as 30%) and adequate to account for the highest organic acid anions reported in this study. Also, high concentrations of acid anions have been reported from many gas fields where liquid hydrocarbon, the other likely source, probably never existed (e.g., Sacramento Valley, California). Kharaka et al. (1983a) presented evidence indicating that an important portion of natural gas in sedimentary basins is from decarboxylation of these acid anions.

The source of an unknown portion of these acid anions could be the liquid petroleum. The oxygen (also nitrogen and sulfur) compounds are present in small amounts throughout petroleum with the highest proportions present in the lubricating oil and the residuum fractions (Hunt, 1979). Oxygen is present mainly in the resins, asphaltanes and high molecular weight waxes with oxygen comprising more than 5% of the resins (Hunt, 1979). The proportion of these heavy heteroatomic compounds is generally less than 10% of the total oil, but increases with decreasing API gravity of the oil reaching values of over 50% in tar sands (Tissot and Welte, 1978). Organic acid anions could be produced, as in the case of kerogen, from thermocatalytic or bacterial degradation of the resins and asphatenes.

MINERAL DIAGENESIS

In this paper we consider only the chemical aspects of mineral diagenesis. The main processes involved are dissolution of detrital and authigenic minerals, precipitation of authigenic minerals and mineral transformations including recrystallization, hydration/dehydration and replacement. Understanding these processes is important for the petroleum industry because they lead to either increased or decreased porosity and permeability of reservoir rocks.

Method of computation.—

The chemical composition of pore water, the mineral composition of the reservoir rocks and the subsurface temperature and pressure determine the possible water/mineral interactions. The thermodynamic properties of pore water determine whether a mineral present in a given reservoir will dissolve or whether it and or/other minerals will precipitate. The number of aqueous species in subsurface waters and possible minerals in reservoir rocks is large and requires computer models for a complete description.

The geochemical model SOLMNEQ II (Kharaka et al., in prep.) is used by us to study mineral diagenesis. SOLMNEQ II, written in Fortran F/77, is an updated version of SOLMNEQ (Kharaka and Barnes, 1973). The new version has a revised thermodynamic data base, more inorganic and organic aqueous species (250) and minerals (300), computes pH at subsurface conditions (see below) and computes the solubilities of minerals at subsurface pressure. Higher hydrostatic pressure almost always increases the solubility of minerals. The increase becomes especially significant (>0.2 kcal) in geopressured zones of sedimentary basins and at temperatures that are higher than about 100°C (Fig. 6).

The Gibbs free energy difference (ΔG_{diff}) between the actual and equilibrium states of a mineral in a given subsurface water (sometimes denoted by the chemical affinity A) is used in these programs as the test for possible dissolution or precipitation of that mineral. ΔG_{diff} is given by

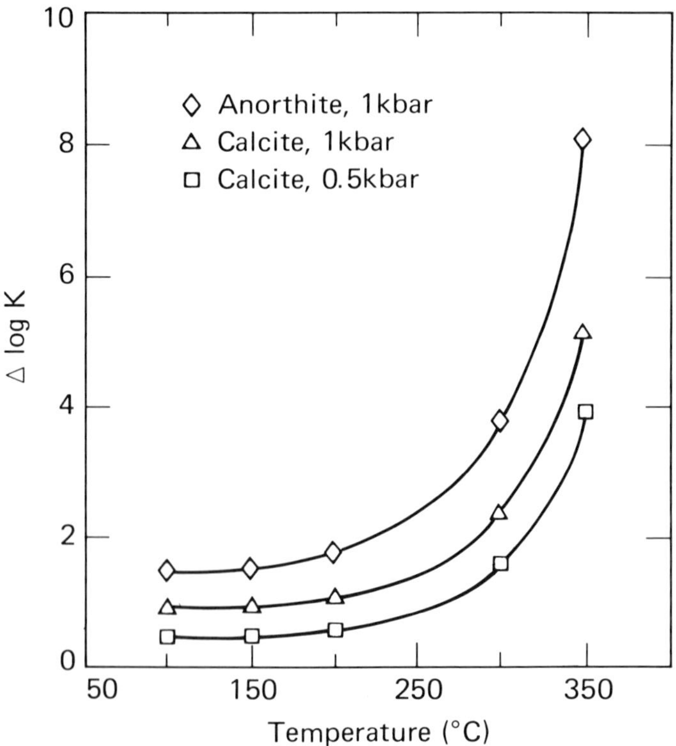

FIG. 6.—Effects of pressure on the solubilities of calcite and anorthite. Note that the difference between the solubilities of minerals at higher pressure and at vapor pressure of water (Δ log K) becomes very large at temperatures higher than about 200°C.

$$\Delta G_{diff} = -RT\ln K - (-RT\ln Q) = RT\ln(Q/K) \quad (1)$$

where R is the gas constant, T is temperature in degrees Kelvin, Q is the reaction quotient, and K is the equilibrium constant at subsurface temperature and pressure.

The formal reactions used in SOLMNEQ II are for complete hydrolysis of minerals; no incongruent reactions which require additional assumptions (e.g., conserving Al) are used. For anorthite the reaction is

$$CaAl_2Si_2O_8 + 8H^+ \rightleftarrows Ca^{++} + 2Al^{+3} + 2H_4SiO_4. \quad (2)$$

The reaction quotient (Q) for reaction (2) is given by

$$Q_{anorthite} = \frac{a_{Ca^{++}} \cdot a_{Al}^2 + 3 \cdot a_{H_4SiO_4^0}^2}{a_{H^+}^8 \cdot a_{anorthite}} \quad (3)$$

where (a) is the activity of the specified species in the given subsurface water at the specified temperature and pressure; $a_{anorthite}$ is generally assumed to be equal to 1.0.

At equilibrium $Q = K$ and $\Delta G_{diff} = 0$. In this case the subsurface water is in equilibrium with anorthite and no dissolution or precipitation should take place. Where $\Delta G_{diff} < 0$, anorthite cannot precipitate from that subsurface water because of undersaturation, but anorthite probably will dissolve if the mineral is present in the reservoir rocks. Where $\Delta G_{diff} > 0$, anorthite will not dissolve because of supersaturation but most likely will precipitate and form authigenic anorthite.

Distribution of aqueous species.—

SOLMNEQ II computes the equilibrium distribution of 300 inorganic and organic aqueous species of major and minor elements generally present in natural waters from the reported chemical analyses, temperature (0 to 350°C), pH, and Eh (optional). The distribution of species is carried out to compute the activities of aqueous species used in mass-action equations such as equation (3). The distribution of aqueous species is computed using mass-action and oxidation-reduction reactions, and mass balance relationships.

The mass balance relationships are of the type

$$m_{i,t} = \sum_i n_i m_i \quad (4)$$

where $m_{i,t}$ is the total reported or computed molality of the species (i), n_i is the number of molecules of (i) in the formula weight of species (i), and the summation Σ_i covers all species containing (i). For aqueous species of Ca, for example, the mass balance equation is

$$m_{Ca,t} = m_{Ca^{++}} + m_{CaHCO_3^+} + m_{CaCO_{3(aq)}} + m_{CaCH_3COO^+} + \ldots \quad (5)$$

The mass-action equations for the dissociation of $CaHCO_3^+$ and $CaCO_{3(aq)}$, for example, lead to

$$K_{CaHCO_3^+} = \frac{m_{Ca^{++}} \cdot \gamma_{Ca^{++}} \cdot m_{HCO_3^-} \cdot \gamma_{HCO_3^-}}{m_{CaHCO_3^+} \cdot \gamma_{CaHCO_3^+}} \quad (6)$$

$$K_{CaCO_{3(aq)}} = \frac{m_{Ca^{++}} \cdot \gamma_{Ca^{++}} \cdot m_{CO_3^{--}} \cdot \gamma_{CO_3^{--}}}{m_{CaCO_{3(aq)}} \cdot \gamma_{CaCO_{3(aq)}}} \quad (7)$$

where (γ_i) is the activity coefficient of (i).

The mass-action and mass-balance equations are combined and rearranged after the activity coefficients have been computed to give the value of $m_{Ca^{++}}$ from the following:

$$m_{Ca^{++}} = \frac{m_{Ca,t}}{1 + \gamma_{Ca^{++}}\left(\dfrac{m_{HCO_3^-} \cdot \gamma_{HCO_3^-}}{K_{CaHCO_3^+} \cdot \gamma_{CaHCO_3^+}} + \dfrac{m_{CO_3^{--}} \cdot \gamma_{CO_3^{--}}}{K_{CaCO_{3(aq)}} \cdot \gamma_{CaCO_{3(aq)}}} + \ldots\right)}. \quad (8)$$

The activity of Ca^{++} is computed from

$$a_{Ca^{++}} = m_{Ca^{++}} \cdot \gamma_{Ca^{++}}. \quad (9)$$

$m_{CaHCO_3^+}$ and $m_{CaCO_{3(aq)}}$ are then computed from equations (6) and (7), respectively, after substituting the computed $m_{Ca^{++}}$.

Oxidation-reduction reactions may be used in conjunction with mass-action reactions and mass-balance expressions to compute the distribution of multivalent elements where the Eh of the solution is known. The reaction for iron is

$$Fe^{++} \rightleftarrows Fe^{+3} + e^- \quad (10)$$

where e^- represents an electron. The concentrations of Fe^{++} and Fe^{+3} are related by

$$Eh(T) = E^0(T) + \frac{RT}{n\mathrm{F}} \ln \frac{m_{Fe^{+3}} \cdot \gamma_{Fe^{+3}}}{m_{Fe^{++}} \cdot \gamma_{Fe^{++}}} \quad (11)$$

where Eh is the oxidation potential (at temperature T) referred to the hydrogen half cell, E^0 is the potential in which all reactants and products are in their standard states, n is the number of electrons involved, and F is the Faraday constant.

Computations based on measured Eh have been kept to a minimum in SOLMNEQ II because of uncertainties in the measurement and interpretation of oxidation potentials (Barnes and Clarke, 1969). The program computes the distribution of species for elements with multiple redox states, however, by assuming equilibrium between the species HS^- and SO_4^{--}. Kharaka et al. (1980) have shown that oil field waters are strongly reducing (Eh < −200 mV) and that the sulfide-sulfate redox reaction

$$HS^- + 4H_2O \rightleftarrows SO_4^{--} + 9H^+ + 8e^- \quad (12)$$

is probably the best of the possible redox reactions for computing the Eh of these waters.

Pore water pH at subsurface conditions.—

The pH of subsurface waters measured at the wellhead generally is different (as much as 3 pH units) from its value in the reservoir rocks. The main change results from loss of volatile gases such as CO_2 and H_2S, but changes in temperature and pressure alone will result in pH changes due to variations in the stabilities of hydrogen-bearing aqueous species (Kharaka et al., 1980). An accurate value for the subsurface pH is critical because the activity of the proton (a_{H^+}) raised to a power of 2 or higher appears in many equations used to determine the distribution of species (e.g., equation 8) and to calculate saturation states (ΔG_{diff}) of minerals (e.g., equation 3).

SOLMNEQ II can compute the pH at subsurface conditions using the mass balance for hydrogen at surface and subsurface conditions. In this method, which is a modification of that used by Truesdell and Singer (1974) and Bischoff and Dickson (1975), the distribution of species is carried out at surface conditions using the measured pH. The total hydrogen (m_{H,t,T_1}) and hydroxyl (m_{OH,t,T_1}) are then computed at the surface conditions from the equations

$$m_{H,t,T_1} = m_{H^+} + m_{HCO_3^-} + 2m_{H_2CO_3^0} + m_{HS^-} + 2m_{H_2S^0} + m_{CH_3COOH^0} + \ldots \quad (13)$$

$$m_{OH,t,T_1} = m_{OH^-} + m_{Ca(OH)^+} + m_{Mg(OH)^+} + m_{NH_4OH^0} + 2m_{Fe(OH)_2^+} + \ldots \quad (14)$$

The distribution of species is then computed at the subsurface temperature assuming that the subsurface pH is equal to the measured surface pH. The total molalities of hydrogen (m_{H,t,T_2}) and hydroxyl (m_{OH,t,T_2}) at subsurface temperature (T_2) are then computed. The value of Y given by

$$Y = m_{H,t,T_2} - [m_{H,t,T_1} + (m_{OH,t,T_2} - m_{OH,t,T_1})] \quad (15)$$

is then computed. The term ($m_{OH,t,T_2} - m_{OH,t,T_1}$) is in equation (15) to include the changes in the hydrogen mass balance from hydrolysis of water in the reaction

$$H_2O\,(l) \rightleftarrows H^+ + OH^-. \quad (16)$$

Iterations are then carried out changing the pH by small increments until the value of Y is less than 0.5 percent of the value of m_{H,t,T_2}. The final computed pH is then used to compute the distribution of species and the states of reaction of the solution with respect to minerals.

The subsurface pH may be computed for samples where known quantities of CO_2 and H_2S are lost prior to surface pH measurement. CO_2 and H_2S are added to the chemical analysis as $H_2CO_3^0$ and H_2S^0 and included as $2m_{H_2CO_3}$ and $2m_{H_2S}$, respectively, in the computed value of m_{H,t,T_2}.

ORGANIC SPECIES AND MINERAL DIAGENESIS

Detailed studies of organic species in oil field waters were initiated (Willey et al., 1975), to a large degree, because computations using SOLMNEQ (Kharaka and Barnes, 1973) had shown that oil field waters at Kettleman North Dome, California, were supersaturated by as much as 3 kcal with respect to calcite (Willey et al., 1975; Merino, 1975). Willey et al. (1975) showed that the error resulted because the alkalinites of the waters were attributed only to inorganic species (mainly HCO_3^-, CO_3^{--} and HS^-). They further showed that organic acid anions (C_2—C_5) contributed as much as 99% of the measured alkalinities. Carothers and Kharaka (1978) and Lundegard et al. (1984) showed that a variable portion (as much as 100%) of field-determined alkalinities in oil field waters may be attributed to organic species (Fig. 7).

The proportion of alkalinity due to organic acid anions in the formation waters from High Island field (Table 1) ranges from about 10% (several samples) to about 90% (sample 83-TX-12). These proportions are obtained by comparing carbonate alkalinity (obtained from dissolved organic carbon) with total alkalinity. Approximately the same proportions (±5%) are obtained when comparing field-measured alkalinity with alkalinity due to all the dissolved organic species listed in Tables 2 and 3.

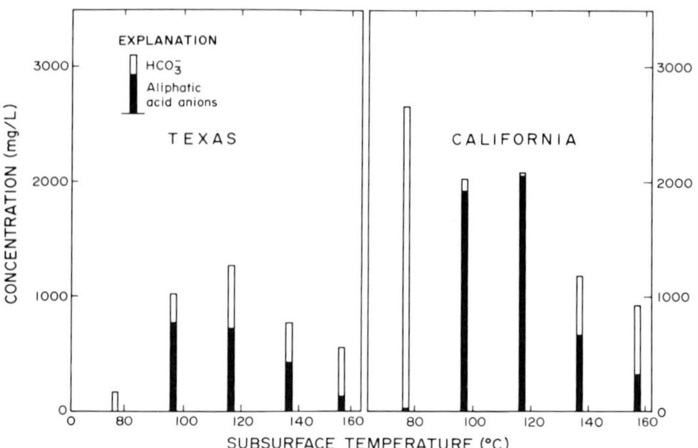

FIG. 7.—Average concentrations of C_2—C_5 aliphatic acid anions and total bicarbonate in oil field waters from onshore Texas and California. All samples with temperatures lower than 80°C have been averaged together; for samples with temperatures higher than 80°C, averaging is over 20°C intervals (from Carothers and Kharaka, 1978).

Effect on pH at reservoir conditions.—

The pH ($-\log a_{H^+}$) of any water and its buffer capacity (a measure of the ability of the solution to resist pH changes when titrated with acid or base) are controlled mainly by the concentrations of the dominant aqueous species resulting from the dissociation of weak acids (Morel, 1983). In oil field waters, the most important inorganic weak acid is $H_2CO_3^0$ (HCO_3^-, CO_3^{--}) with H_2S^0 (HS^-, S^{--}) and $H_4SiO_4^0$ ($H_3SiO_4^-$, $H_2SiO_4^{--}$) being of secondary importance. Weak organic acids, as we have stated previously, are the dominant species in oil field waters at temperatures higher than about 80°C. The dissolved species of these organic acids (especially CH_3COOH^0, CH_3COO^-) control the pH and buffer capacity of these waters.

As stated previously, the pH of pore water in reservoir rocks is the most important chemical parameter needed to study mineral diagenesis. Obtaining reliable pH values at the wellhead is straightforward and the values generally range from 5.5 to 7.5. Extrapolation of these values to subsurface conditions is difficult because it requires (1) a complete and reliable thermodynamic data base for the aqueous complexes; (2) complete and proper identification of the analyzed chemical constituents (including organic species) and, (3) knowledge of the volume of gases (CO_2 and H_2S) lost from a given volume of water. This last complication applies especially to wells producing oil and water because the solubility of CO_2 and H_2S in oil is variable. (See Kharaka and Barnes, 1973; and Kharaka and Mariner, 1977, for a more complete discussion of these problems.)

Computations of subsurface pH with SOLMNEQ II (Kharaka et al., in prep.) have been carried out for many years. Computations have shown that the activities of the species CH_3COOH^0, CH_3COO^-, $H_2CO_3{}^0$ and $HCO_3{}^-$ and the reactions

$$CH_3COOH^0 \rightleftarrows CH_3COO^- + H^+ \quad (17)$$

$$H_2CO_3{}^0 \rightleftarrows HCO_3{}^- + H^+ \quad (18)$$

are the main controls on the computed pH. Computed pH values in water dominated by acetate will be different than those dominated by carbonate because the dissociation constants (K) are different for these two weak acids (Fig. 8). Results have shown that computed pH values vary by as much as 2.0 pH units when all the alkalinity is attributed to carbonate species versus when values are obtained by using organic and inorganic species.

Computed pH values are more reliable in the case of geopressured-geothermal wells from the northern Gulf of Mexico basin because the amount of CO_2 and H_2S degassed from a given volume of water is known (Kharaka et al., 1979; Kraemer and Kharaka, (in prep.). The computed pH values for these waters range from 4.0 to 5.5 which are as much as 2.5 pH units lower than the pH values measured at the wellhead (Kharaka et al., 1980; Kraemer and Kharaka, (in prep.). The diagenetic minerals predicted from geochemical modeling using the computed pH values will be totally different from those using pH values measured at surface conditions. Kaiser and Richman (1981), for example, showed (Fig. 9) that kaolinite is the stable mineral (as shown also by examination of core samples) in the two aquifers tested at Pleasant Bayou, Texas, when the computed pH value (4.0) is used. Geochemical modeling indicates chlorite as the stable phase if the surface-measured pH of 6.5 and 6.2 are used (Fig. 9).

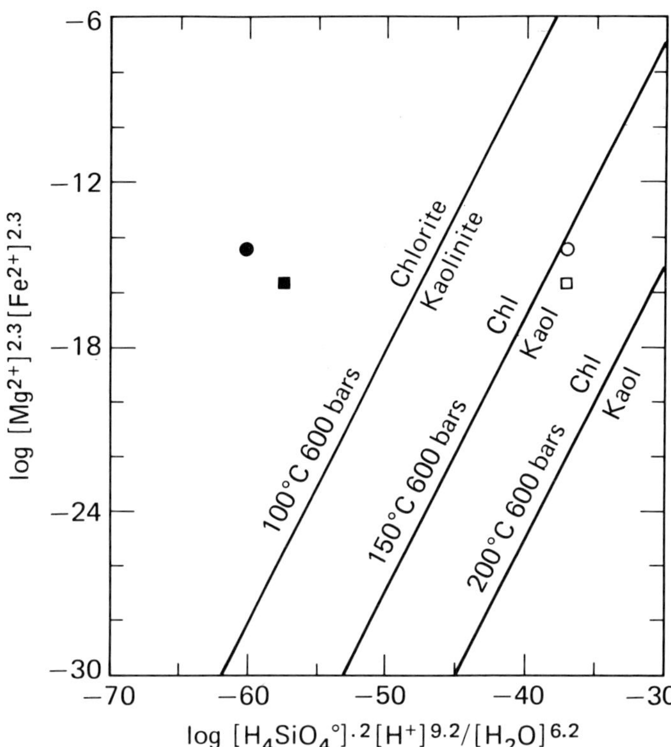

FIG. 9.—Stability diagram for the reaction 1.4 kaolinite + 2.3 Mg^{++} + 2.3 Fe^{++} + 6.2 $H_2O \rightleftarrows$ chlorite + $2H_4SiO_4^0$ + 9.2 H^+. Note that, at the *in situ* pH of 4.0 (open symbols), kaolinite is the stable mineral relative to chlorite. This is indicated also by petrographic examination of core samples (from Kaiser and Richman, 1981).

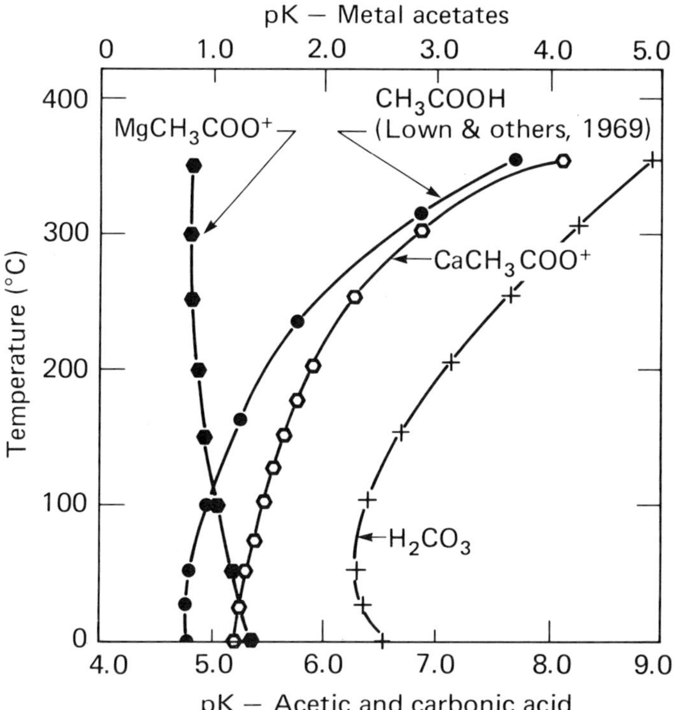

FIG. 8.—Effects of temperature on the stability constants of CH_3COOH^0, $H_2CO_3^0$, $CaCH_3COO^+$ and $MgCH_3COO^+$. The pH of subsurface waters is controlled mainly by the pK values of acetic and carbonic acids (data from Kharaka et al., in prep.).

The difference between the computed pH values at subsurface conditions and those measured at the wellhead is small (<0.5 pH units) where CO_2 and H_2S gases have not, or are assumed not to have, been lost from solution. The pH of sample 83-TX-12 measured at the wellhead (42°C) is 6.83. The computed value at reservoir temperature (98°C) is 6.52; however, a change of 0.5 pH units can lead to changes of as much as 3 kcal in the reaction state (ΔG_{diff}) of some minerals and may lead to reversal of saturation state (from dissolution to precipitation or vice versa) for many minerals.

Organic species as a source of CO_2.—

Secondary porosity results from dissolution of detrital and authigenic carbonate and silicate minerals. Precipitation of carbonate minerals, on the other hand, is an important cause of cementation in many reservoir rocks. Carbon dioxide is probably the most important agent in both of these processes (Curtis, 1978; Schmidt and McDonald, 1979; Hanor, 1978). Thermal and bacterial (temperatures < 80°C) decarboxylation of organic matter, including the dissolved species discussed in this paper, is generally believed to be the source of this CO_2 (Curtis, 1978; Schmidt and McDonald, 1979; Carothers and Kharaka, 1980; Lundegard et al., 1984). Direct evidence for the organic source of CO_2 comes from $\delta^{13}C$ values of CO_2 in natural gas, dissolved total carbonate and carbonate cements (Boles, 1978; Carothers and Kharaka, 1980; Lundegard et al., 1984).

The origin and role of carbon dioxide in secondary porosity have been discussed in detail (Lundegard et al., 1984; Lundegard and Land, this volume). We are in general agreement with the main conclusions of Lundegard and Land (this volume) including the one stating that CO_2 from decarboxylation of organic matter can produce a regional average of only a few (2–4%) percent secondary porosity. Lundegard et al. (1984) and Lundegard and Land (this volume), however, believe that organic matter must account for a regional average of 13% secondary porosity that is observed in cores from the Tertiary section of the Texas Gulf coast. It is possible that the problem lies in requiring such a high number (13%) of secondary porosity on a regional scale. The high number could result from sampling bias both in the siting of petroleum wells in favorable locations and the coring of promising zones. It is more probable that fluids charged with CO_2 and dissolved organic species are directed by faulting to favorable structures. Carbon dioxide from organic sources would be adequate to account for 10–15% secondary porosity in sandstones over about 25% of the area.

Effect on oxidation potential.—

Subsurface waters in sedimentary basins are highly reducing (Eh < -200 mV) and dissolved organic species are the main reducing species (Kharaka et al., 1980). Sulfate reduction is one of the earliest (reduction of oxygen to CO_2 is even earlier) reactions that takes place in sedimentary basins (Claypool and Kaplan, 1974; Carothers and Kharaka, 1980). The reaction represented by

$$SO_4^{--} + CH_3COO^- \rightleftarrows HS^- + 2HCO_3^- \quad (19)$$

is mediated by sulfate-reducing bacteria and commonly results in an almost complete reduction of sulfate to sulfide. The concentrations of sulfate (0.4 to 16 mg/l) in the waters from High Island field (Table 1) show that sulfate reduction is almost complete even in marine connate water from middle Pleistocene sediments.

Dissolved organic species have also been suggested (Surdam et al., 1984) as a reducing agent for ferric iron present in iron oxides and clays. The reaction, which can apply also to manganese species, may be represented by

$$2H_2O + CH_3COOH^0 + 8Fe^{+3} \rightarrow 8Fe^{++} + 2CO_2 + 8H^+. \quad (20)$$

The decarboxylation reaction alone is

$$CH_3COOH^0 \rightarrow CH_4 + CO_2. \quad (21)$$

Note that reaction (20) generates 2 moles of CO_2 per mole of acetic acid and one mole of proton per mole of Fe^{+3} reduced; decarboxylation results in one mole of CO_2 only. The increased CO_2 and the proton generated can dissolve additional minerals to produce more secondary porosity (Surdam et al., 1984). The high concentration of Fe^{++} going into solution (reaction 20), on the other hand, may eventually precipitate as siderite or ankerite (e.g., if pH is raised by the degassing of CO_2). The exact reactions involved in the reduction of ferric iron in sedimentary basins, however, are not known at this time and more field and laboratory studies are required for these important reactions.

Organic-inorganic complexes.—

Organometallic complexes have been suggested as the mechanism for the transport of many metals, including Pb and Zn in Mississippi Valley-type ore deposits (Giordano and Barnes, 1981) and aluminum in sedimentary basins (Surdam et al., 1982). Organic complexes are invoked because (1) inorganic complexes are believed to be inadequate for the transport of the required masses of the metals; (2) there is close association between liquid and gaseous organic species and these ore deposits; and (3) metals can form strong complexes with many naturally occurring organic species. (See Giordano and Barnes, 1981, for an excellent discussion and more references on this subject.)

SOLMNEQ II (Kharaka et al., in prep.) has been used to study organic-inorganic complexes in oil field brines from many sedimentary basins. Results show that complexes of Al, Zn, Pb, and U with mono- and dicarboxylic acid anions are minor (less than 1% of total metals). Using detailed chemical analysis and computed subsurface pH values, Kharaka et al. (1983b), showed that galena and sphalerite control the concentrations of Pb and Zn, but that as much as 10 mg/l Pb and 50 mg/l Zn could be transported in oil field brines mainly as chloride complexes. Kraemer and Kharaka (in prep.) conclude that geopressured-geothermal waters from the northern Gulf of Mexico basin are in equilibrium with both uraninite (UO_2) and coffinite ($USiO_4$), the dominant uranium species being UO_2^+, UO_2Cl^+ and $UO_2(CO_3)_2^{--}$. These organic acid anions, on the other hand, can complex a minor portion (as much as 10%) of Na, Ca and Mg in these waters.

Salicylate ($C_7H_4O_3^{--}$) forms very strong complexes with metals, and preliminary calculations by Giordano and Barnes (1981) showed that it could be responsible for the transport of Pb and Zn in an average Mississippi Valley-ore solution. The pK_1 and pK_2 values of Pb-salicylate are 7.5 and 13.7, respectively (Martell and Smith, 1977). No salicylate was found in the waters from High Island field (the procedures used would have identified them) or any other water; however, 150 mg/l (~0.1 mM) salicylate were added to the chemical composition of samples of oil field brines to study the complexes formed. SOLMNEQ II was used again after all the complexes of salicylate with inorganic species were added to the thermodynamic data base. Results showed that less than 1% of Pb and Zn would have been complexed with salicylate if it were present.

Detailed analyses oi the output from SOLMNEQ II showed the reasons for these unexpected results. Results showed that more than 99% of salicylate was complexed with H^+, Ca^{++}, Mg^{++} and Fe^{++}; Pb and Zn in these brines, on the other hand, were complexed with Cl forming mainly $PbCl_4^{--}$ and $ZnCl_4^{--}$ species. Complexes of aluminum with hydroxyl groups ($Al(OH)_4^-$, $Al(OH)_2^+$, $Al(OH)^{++}$) comprise more than 99% of total aluminum and the concentrations of aluminum-organic species are negligible.

The results indicate that organic-inorganic complexes may not be as important as previously believed in the transport of metals. Many organic species (e.g., monocarboxylic acid anions) form only weak complexes with metals. Other organic species (e.g., salicylate) that form very strong complexes with metals also form strong complexes with Na, Ca, Mg and Fe that are present in waters at concentrations that may be more than 4 orders of magnitude greater than that of the metals, thus more than compensating for the lower stabilities of their complexes. Organic species, we believe, indirectly control the concentrations of metals in subsurface waters by controlling the pH and Eh of these waters.

CONCLUSIONS

The concentrations of dissolved organic species in oil field waters can be very high (as much as 10,000 mg/l). Acetate and propionate are the dominant monocarboxylic acid anions; C_4 and C_5 dicarboxylic acid anions are the dominant species of this group and oxalate (C_2) is absent. The concentrations of other organic species are probably low.

Computations with geochemical models show that the organic acid anions control the pH and Eh of subsurface waters and thus strongly affect mineral diagenesis. These organic species can be decarboxylated to CO_2 and natural gas. The CO_2 generated may be responsible for the secondary porosity in many sedimentary basins. Metal complexes with known organic species do not seem to be quantitatively important. The number of possible organic species in oil field waters is immense and only a very small fraction of these have been identified and tested for metal-organic complexes. All the inorganic complexes of a newly identified organic species should be examined, however; consideration should not be limited to complexes with the metal of interest.

REFERENCES

BARNES, IVAN, AND CLARKE, F. E., 1969, Chemical properties of groundwater and their encrustation effects of wells: U.S. Geological Survey Professional Paper 498-D, 58 p.

BISCHOFF, J. L., AND DICKSON, F. W., 1975, Seawater-basalt interaction at 200°C and 500 bars: Implications for origin of sea-floor heavy-metal deposits and regulation of seawater chemistry: Earth and Planetary Science Letters, v. 25, p. 385–397.

BLATT, HARVEY, 1979, Diagenetic processes in sandstone, *in* Scholle, P. A., and Schluger, P. R., eds., Aspects of Diagenesis: Society of Economic Paleontologists and Mineralogists Special Publication 26, p. 141–157.

BOLES, J. R., 1978, Active ankerite cementation in the subsurface of Eocene of Southwest Texas: Contributions of Mineralogy and Petrology, v. 68, p. 13–22.

CAROTHERS, W. W., AND KHARAKA, Y. K., 1978, Aliphatic acid anions in oil field waters—Implications for origin of natural gas: American Association of Petroleum Geologists Bulletin, v. 62, p. 2441–2453.

———, ———, 1980, Stable carbon isotopes in oil-field waters and the origin of CO_2: Geochimica et Cosmochimica Acta, v. 44, p. 323–332.

CLAYPOOL, G. E., AND KAPLAN, I. R., 1974, The origin and distribution of methane in marine sediments, *in* Kaplan, I. R., ed., Natural Gases in Marine Sediments: New York, Plenum Press, p. 99–139.

COLLINS, A. G., 1975, Geochemistry of Oilfield Waters: New York, Elsevier Scientific Publishing Company, 496 p.

CORDELL, R. J., 1972, Depths of oil origin and primary migration: a review and critique: American Association of Petroleum Geologists Bulletin, v. 56, p. 2029–2067.

CURTIS, C. D., 1978, Possible links between sandstone diagenesis and depth-related geochemical reactions occurring in enclosing mudstones: Geological Society of London Journal, v. 135, p. 107–117.

DAVIS, J. B., 1967, Petroleum Microbiology: New York, Elsevier Scientific Publishing Company, 604 p.

DEGENS, E. T., HUNT, J. M., REUTER, J. H., AND REED, W. E., 1964, Data on the distribution of amino acids and oxygen isotopes in petroleum brine waters of various geologic ages: Sedimentology, v. 3, p. 199–225.

EISMA, E., AND JURG, J. W., 1967, Fundamental aspects of the diagenesis of organic matter and the formation of hydrocarbons: Proceedings, Seventh World Petroleum Congress, v. 2, p. 61–72.

GIORDANO, T. H., AND BARNES, H. L., 1981, Lead transport in Mississippi Valley-type ore solutions: Economic Geology, v. 76, p. 2200–2211.

GULLIKSON, D. M., CARAWAY, W., AND GATES, B. L., 1961, Chemical analyses and electrical resistivity of selected California oil-field waters: U.S. Bureau of Mines Report of Investigations 5736, p. 1–21.

HANOR, J. S., 1978, Precipitation of beachrock cements: mixing of marine and meteoric waters vs. CO_2 degassing: Journal of Sedimentary Petrology, v. 48, p. 489–502.

HASS, J. L., Jr., 1978, An empirical equation with tables of smoothed solubilities of methane in water and aqueous sodium cloride solutions up to 25 weight percent, 360°C, and 138 MPa: U.S. Geological Survey Open-File Report 78-1004, 41 p.

HOLDEMAN, L. V., CATO, E. P., AND MOORE, W. E. C., 1977, Anaerobe Laboratory Manual, 4th ed.: Virginia Polytechnic Institute and State, Blacksburg, VA, 130 p.

HUNT, J. H., 1979, Petroleum Geochemistry and Geology: San Francisco, Freeman and Company, 617 p.

KAISER, W. R., AND RICHMAN, D. L., 1981, Predicting diagenetic history and reservoir quality in the Frio Formation of Brazoria County, Texas and Pleasant Bayou test wells, *in* Bebot, D. G. and Bachman, A. L., eds., Proceedings, Fifth Conference on Geopressured-Geothermal Energy, p. 67–74.

KARTSEV, A. A., 1974, Hydrogeology of Oil and Gas Deposits: National Technical Information Service Report TT73-58022, 323 p.

KHARAKA, Y. K., AND BARNES, IVAN, 1973, SOLMNEQ: Solution-mineral equilibrium computations: NTIS Report PB 215-899, U.S. Department of Commerce, Springfield, VA., 81 p.

———, AND MARINER, R. H., 1977, Solution-mineral equilibrium in natural water-rock systems: Proceedings, Second International Symposium on Water-Rock Interaction, V. IV, Strasbourg, France, p. 66–75.

———, CAROTHERS, W. W., AND ROSENBAUER, R. J., 1983a, Thermal decarboxylation of acetic acid: Implications for origin of natural gas: Geochimica et Cosmochimica Acta, v. 47, p. 397–402.

———, LAW, L. M., AND SPECHT, D. J., 1983b, Hydrodynamics, geochemistry and metal transport of Gulf Coast brines: (Abs.), Proceedings, Geological Society of America Annual Meeting, Indianapolis, IN, p. 46.

———, LICO, M. S., AND CAROTHERS, W. W., 1980, Predicted corrosion and scale-formation properties of geopressured-geothermal waters from the northern Gulf of Mexico basin: Journal of Petroleum Technology, v. 32, p. 319–324.

———, ———, WRIGHT, V. A., AND CAROTHERS, W. W., 1979, Geochemistry of formation waters from Pleasant Bayou no. 2 well and adjacent areas in coastal Texas, *in* Dorfman, M. H., and Fisher, W. L., eds., Proceedings, Fourth United States Gulf Coast Geopressured-Geothermal Energy Conference: Research and Development: The University of Texas at Austin, p. 168–193.

KVENVOLDEN, K. A., AND WEISER, D., 1967, A mathematical model of a geochemical process: Normal paraffin formation from normal fatty acids: Geochimica et Cosmochimica Acta, v. 31, p. 1281–1309.

LEENHEER, J. A., MALCOLM, R. L., MCKINLEY, P. W., AND ECCLES, L. A., 1974, Occurrence of dissolved organic carbon in selected ground water samples in the United States: Journal of Research, U.S. Geological Survey, v. 2, p. 361–369.

LOCHTE, H. L., BURNAM, C. W., AND MEYER, H. W. H., 1949, Organic acids produced by high pressure gas wells: Petroleum Engineer, v. 21, p. C34-C40.

LOWN, D. A., THIRSK, H. R., AND WYNNE-JONES, L., 1970, Temperature and pressure dependence of the volume of ionization of acetic acid in water from 25 to 225°C and 1 to 3,000 bars: Transactions, Faraday Society, v. 66, p. 51–73.

———, ———, LUNDEGARD, P. D., LAND, L. S., AND GALLOWAY, W. E., 1984, Problem of secondary porosity: Frio Formation (Oligocene), Texas Gulf Coast: Geology, v. 12, p. 399–402.

MARTELL, A. E., AND SMITH, R. M., 1977, Critical Stability Constant: New York, Plenum, v. 3, 495 p.

MERINO, ENRIQUE, 1975, Diagenesis in Tertiary sandstones from Kettleman North Dome, California—I. Diagenetic mineralogy: Journal of Sedimentary Petrology, v. 45, p. 320–336.

MILLIKEN, K. L., LAND, L. S., AND LOUCKS, R. G., 1981, History of burial diagenesis determined from isotopic geochemistry, Frio Formation, Brazoria County, Texas: American Association of Petroleum Geologists Bulletin, v. 65, p. 1397–1413.

MOREL, F. M. M., 1983, Principles of Aquatic Chemistry: New York, John Wiley & Sons, 446 p.

PRICE, L. C., 1978, Crude oil and natural gas dissolved in deep, hot geothermal waters of petroleum basins. A possible significant new energy source, *in* Meriwether, J., ed., Proceedings, Third Conference on Geopressured-Geothermal Energy, v. 1, p. GI-167-249.

———, WENGER, L. M., GING, TOM, AND BLOUNT, C. W., 1983, Solubility of crude oil in methane as a function of pressure and temperature: Organic Geochemistry, v. 4, p. 201–221.

RANDOLPH, P., 1978, Natural gas content of geopressured aquifers, *in* Meriwether, J., ed., Proceedings, Third Geopressured-Geothermal Energy Conference, v. 1, p. ES-23-46.

RAPP, J. B., 1976, Amino acids and gases in some springs and an oil field in California: Journal of Research, U.S. Geological Survey, v. 4, p. 227–232.

SCHMIDT, V., AND MCDONALD, D. A., 1979, The role of secondary porosity in the course of sandstone diagenesis, *in* Scholle, P. A., and Schluger, P. R., eds., Aspects of Diagenesis: Society of Economic Paleontologists and Mineralogists Special Publication 26, p. 175–207.

SURDAM, R. C., BOESE, S., AND CROSSEY, L. J., 1982, Role of organic and inorganic reactions in development of secondary porosity in sandstones (abs.): American Association of Petroleum Geologists Annual Convention, Calgary, Canada, p. 116.

———, CROSSEY, L. S., AND LAHANN, R., 1984, Mineral oxidants and porosity enhancement (abs.): American Association of Petroleum Geologists Bulletin, v. 68, p. 532.

THURMAN, E. M., 1984, Humic Substances I. Geochemistry, Characterization and Isolation: New York, John Wiley & Sons, 380 p.

TISSOT, B. P., AND WELTE, D. H., 1978, Petroleum Formation and Occurrence: New York, Springer Verlag, 538 p.

TRUESDELL, A. H., AND SINGER, W., 1974, The calculation of aquifer chemistry in hot-water geothermal systems: Journal of Research, U.S. Geological Survey, v. 2, p. 271–278.

WILLEY, L. M., KHARAKA, Y. K., PRESSER, T. S., RAPP, J. B., AND BARNES, IVAN, 1975, Short-chain aliphatic acids in oil-field waters of Kettleman North Dome oil field, California: Geochimica et Cosmochimica Acta, v. 39, p. 1707–1710.

ZINGER, A. S., AND KRAVCHIK, T. E., 1973, Simpler organic acids in groundwater of the lower Volga region (genesis and possible use in prospecting for oil): Akademii Nauk Doklady, v. 202, p. 218–221.

ON THE REACTIVITY OF CARBONIC AND ORGANIC ACIDS AND GENERATION OF SECONDARY POROSITY

INDU D. MESHRI
Amoco Production Company, Tulsa, Oklahoma

ABSTRACT: Recently, an abundant concentration of short-chain aliphatic acids has been recognized in oil field waters. This has led to an interest in organic acids as a means of creating secondary porosity in sandstones through dissolution of carbonate and aluminosilicate minerals. This paper compares stoichiometric and thermodynamic reactivities of organic acids to those of carbonic acid. The stoichiometric reactivity is presented in the form of hydrogen ion donor capacity, and the enhanced dissociation in terms of solubility of the corresponding calcium salts in water. The hydrogen ion donor capacity of organic acids is calculated through the dissociation constant data and is shown to be 6 to 350 times higher when compared to that of carbonic acid.

Porosity preservation is discussed in terms of solubility of corresponding calcium salts of the organic acids. The solubility of calcium acetate is approximately three orders of magnitude higher than that of calcium carbonate. Thus, organic acids will keep more calcium ions in the solution, preventing cementation by calcium carbonate.

Gibbs Free Energy of dissolution of feldspars and calcite in the presence of carbonic and organic acids shows a better thermodynamic drive when the dissolution agent is an organic acid. This paper proposes dissolution-precipitation mechanisms involving organic and carbonic acids that can be tested through measurement of carbon isotopes of cements and the host rock, such as limestone and sandstone.

INTRODUCTION

Only recently have researchers viewed the presence of organic acids in oil field waters as a means of creating secondary porosity in sandstone. A considerable amount of literature exists on the experimental and theoretical aspects of dissolution of rock-forming minerals by organic acids during weathering. Petrologists, however, have only begun to recognize and apply this knowledge to the understanding of secondary porosity observed in the subsurface.

Organic acids in oil field brines were known early (Rogers, 1917), but were not identified as short-chain aliphatic acids until the work of Lochte et al. (1949). Wiley et al. (1975) were the first to use modern analytical techniques to quantify acid anions in subsurface samples. Collins (1961) has reported the presence of organic acids in oil field brines, but the most comprehensive study to date is by Carothers and Kharaka (1978). They measured the concentrations of acetate, propionate, butyrate, and valerate in 95 formation water samples from 15 oil and gas fields in California and Texas. They showed that below 80°C, aliphatic acid anions are in concentrations less than 60 mg/l and consist predominantly of propionate. Between 80–200°C, the concentrations are up to 4900 mg/l with acetate forming more than 90% of the total anions. No aliphatic acid anions were found above 200°C, presumably because they were converted to methane, ethane, propane, and CO_2 in the reservoir due to decarboxylation (Carothers and Kharaka, 1978).

The chief concern of previous investigators in determining the concentration of organic acids in oil field waters has been either to assess the role of organic acids as proximity indicators in petroleum exploration or to understand their role in the origin of natural gas.

The dissolution of minerals during weathering has been studied experimentally and/or theoretically by Correns (1963), Wollast (1967), Reesman and Keller (1968), and Huang and Keller (1970). The work by Huang and Keller points out that the solubilities of Si, Al, Fe, Ca, and Mg were higher in organic acids than in carbon dioxide-charged waters, and that the organic acids are capable of complexing aluminum and iron. These observations, although directed towards simulation of weathering, are applicable to the subsurface.

Researchers have now recognized that the organic acids generated during thermal maturation of organic matter may be producing secondary porosity in sandstones. Surdam et al. (1982) have shown experimentally that aliphatic acid solutions will dissolve carbonates and aluminosilicates. They were particularly concerned with the ability of the organic acids to complex and transport aluminum. Aluminum mobility has created a problem in understanding the diagenesis of sandstones. Siebert et al. (1984) have also investigated the role of organic acids in the dissolution of feldspars and complexing of aluminum.

Lundegard et al. (1984) have shown by mass balance calculations that CO_2 and organic acids are insufficient to explain the observed secondary porosity in the Frio Sandstone in the Gulf Coast. In these calculations, they assume complete dissociation for carbonic as well as organic acid such as acetic acid, which allows for a maximum hydrogen ion donor capacity from either acid. Maximum dissociation may never be achieved, however, and the true comparison can only be made by accounting for the actual dissociation at proper pressure-temperature and other pertinent controls.

The purpose of this paper is to show that organic acids, in theory, are more corrosive than carbonic acid based on their hydrogen ion donor capacity and solubility of their corresponding salts. Also, in presence of organic acids, dissolution reactions have a better thermodynamic drive.

STOICHIOMETRIC REACTIVITY OF ORGANIC ACIDS

The primary concern here is to show that hydrogen ion donor capacity can be easily calculated, as shown in Figure 1. Here, HA represents the undissociated acid, which dissociates into H^+ and an anion A^-, such as a bicarbonate, an acetate, a propionate, etc. The $[H^+]$ concentration of dissociated hydrogen ions in moles/liter is equal to the concentration of the corresponding anion, A^-. So, the numerator can be simply expressed as the square of the hydrogen ion concentration. Because the dissociated portion is often very small compared to the concentration of 1 mole/liter

Calculation Of Approximate pH

$$HA \longrightarrow H^{+} + A^{-}$$

$$K = \frac{[H^{+}][A^{-}]}{[HA]}$$

$$\approx \frac{[H^{+}]^{2}}{[HA]}$$

$$\approx [H^{+}]^{2} \quad \text{for } [HA] = 1$$

$$\therefore \quad [H^{+}] \approx \sqrt{K}$$

$$\text{and pH} = -\log [H^{+}]$$

FIG. 1.—Calculation of approximate pH as a square root of the dissociation constant of an acid provides a useful comparison of hydrogen ion donor capacity for a molal solution of an acid.

of the undissociated acid [HA], the concentration of the undissociated portion can be taken as equal to 1. Now the simplified form emerges, where the dissociation constant is simply the square of the hydrogen ion concentration.

Table 1 lists the calculated values of the pH of a few organic acids for which dissociation constant data are available. From this table, it can be seen that a mole of organic acid is capable of producing $[H^{+}]$ ranging from $10^{-1.44}$ moles/l for chloroacetic to $10^{-2.52}$ moles/l for trimethylacetic, whereas carbonic acid can produce $10^{-5.1}$ to $10^{-3.16}$ moles/l of $[H^{+}]$. Thus, the supply of $[H^{+}]$ from organic acids is 1–3 orders of magnitude higher when a stoichiometric comparison is made.

Dissociation constants at 25°C and the calculated pH values representing 6 to 350 times greater hydrogen ion donor capacity are shown in Table 2 for a few chosen organic acids that have been detected either experimentally by Surdam et al. (1982) or found in natural subsurface waters by Carothers and Kharaka (1978).

TABLE 1.—Dissociation constants and approximate pH values for some organic acids

Acid	Dissociation Constant		Approximate $[H^{+}]$	Approx. pH
	K_1	K_2		
Formic	$10^{-3.75}$		$10^{-1.88}$	1.88
Acetic	$10^{-4.75}$		$10^{-2.38}$	2.38
Propionic	$10^{-4.87}$		$10^{-2.49}$	2.44
n-Butyric	$10^{-4.82}$		$10^{-2.41}$	2.41
iso-Butyric	$10^{-4.82}$		$10^{-2.41}$	2.37
Trimethylacetic	$10^{-5.05}$		$10^{-2.52}$	2.52
Diethylacetic	$10^{-4.75}$		$10^{-2.37}$	2.37
Chloroacetic	$10^{-2.87}$		$10^{-1.44}$	1.44
Glycollic	$10^{-3.83}$		$10^{-1.92}$	1.92
Oxalic (Dibasic)	$10^{-1.3}$	$10^{-4.28}$	$10^{-0.65}$ $10^{-2.14}$	0.65 2.19
Trichdoro acetic acid				0.45
Carbonic	$10^{-6.35}$	$10^{-10.25}$	$10^{-3.16}$ $10^{-5.1}$	3.2 5.1

TABLE 2.—Efficiency of organic acids versus that of carbonic acid

Acid	Approximate pH	$[H^{+}]$	Efficiency
Carbonic Acid	≧3.2	6.3×10^{-4}	
Propionic Acid	2.44	3.6×10^{-3}	6×
Acetic Acid	2.38	4.2×10^{-3}	7×
Formic Acid	1.88	1.3×10^{-2}	20×
Oxalic Acid	0.65	2.2×10^{-1}	350×

SOLUBILITY OF THE CORRESPONDING SALTS

Another important factor that controls the stoichiometric efficiency of organic acids is the anion. When a solvent such as carbonic acid is expected to dissolve a carbonate cement, the carbonate ion concentration comes from the solvent (acid) as well as the solute (mineral). In the case where acetic acid is the solvent, at least half of the carbonate anion concentration that would be generated during dissolution is substituted by the acetate anion. Calcium acetate is also the byproduct of dissolution rather than calcium carbonate alone. Calcium acetate has a solubility in water that is 3 orders of magnitude higher (Fig. 2, Table 3) when compared to calcium carbonate, which very effectively will keep calcium ions in solution and prevent immediate re-precipitation of calcium as calcium carbonate. Thus, the higher solubility of the corresponding salt of an acid is an important factor in enhancing the dissociation of that acid and driving it to completion, whereas dissociation of carbonic acid is not driven to completion as assumed by previous workers (Lundegard et al., 1984).

THERMODYNAMIC REACTIVITY OF ORGANIC ACIDS

Petrologists have recognized that dissolution of calcite and several feldspars commonly creates secondary porosity (Stanton and McBride, 1976; McBride, 1977; Pittman, 1979; Larese et al., 1980; Al-Sheib and Shelton, 1981; and Curtis, 1983). In this section I have presented a few equations that represent these commonly observed dissolution reactions. Choice of reactions was based on two important criteria: 1) the transformation must represent a net reduction in the volume of solid phase products when compared to the volume of solid phase reactants; and 2) the transformation must represent a commonly observed phenomenon in the diagenetic realm.

Following the first criterion, molar volumes are used to calculate difference in volume of solid phase products and reactants. Second, Gibbs Free Energy function for the reaction, which in turn represents the thermodynamic drive for the reaction, is calculated in the presence of different solvents for the comparison of efficiency.

Calcite Dissolution Reactions

Calcite dissolution in presence of CO_2 or carbonic acid

$$CaCO_3 + CO_2 + H_2O \rightleftharpoons Ca^{++} + 2HCO_3^{-}$$

$$\Delta G_r^{\circ} = +28.37 \text{ kcal/mole}$$

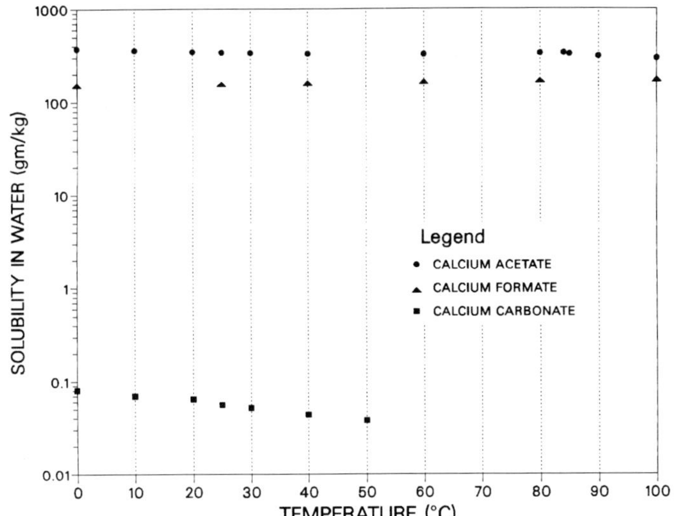

FIG. 2.—Solubility of corresponding calcium salts of formic and acetic acid is 3 to 4 orders of magnitude higher than the solubility of $CaCO_3$. Higher calcium concentration in solution resulting from higher solubility of calcium salts of organic acids would aid in preservation of secondary porosity.

TABLE 3.—Solubility* of calcium salts of corresponding acids (the solubilities are in GMS/liter)

Temp. (°C)	Calcium Acetate	Calcium Formate	Calcium Carbonate
0	374.0	161.5	0.081
10	360.0	—	0.070
20	347.0	—	0.065
25	342.0	166.0	0.056
30	338.5	—	0.042
40	332.0	170.0	0.044
60	327.0	175.0	0.038
80	335.0	179.5	—
84	338.0	—	—
85	329.0	—	—
90	311.0	—	—
100	297.0	184.0	—

*Solubility data given in this table are compiled from Seidell (1965).

Calcite dissolution in presence of acetic acid

$$CaCO_3 + CH_3COOH \rightleftharpoons Ca^{++} + HCO_3^- + CH_3COO^-$$
$$\Delta G_r^\circ = +11.20 \text{ kcal/mole}$$

Silicate Dissolution Reactions

Porosity increase in silicate conversions has also been documented in innumerable examples by petrologists. Recently, Siebert et al. (1984) have shown porosity increase due to feldspar grain dissolution through a following reaction:

$2NaA1Si_3O_8.CaAL_2Si_2O_8 + 4H^+ + 2H_2O$
Andesine Plagioclase

$2A1_2Si_2O_5\,(OH)_4 + 4SiO_2 + 2Na^+(aq) + Ca^{++}(aq)$
Kaolinite Quartz

Here they indicate "one volume of dissolved plagioclase could produce .66 volumes of Kaolinite and .30 volumes quartz." This indicates that volume is conserved over this transformation. How, then, can porosity generation occur? In reality it appears that they have erroneously quoted weight % as volume%. One mole of andesine plagioclase will produce 66 weight% kaolinite and 33 weight% quartz because in a chemical reaction mass must be conserved. Based on the molar volumes of solid phases in the above reaction, a net reduction in volume equal to 14.3% is obtained indicating secondary porosity generation. In the following section a few silicate conversions, their porosity generation potential, and the thermodynamic drive is given. Molar volumes and values of Gibbs Free Energy function for these reactions are obtained from thermodynamic data published by Robie et al. (1979). ΔV_s denotes the change in molar volumes of solid phases only.

Microcline to Illite Conversion

Microcline to illite conversion at 25°C (298°K) and 1 bar in H_2CO_3

$3KAlSi_3O_8 + 2H_2CO_3 + 12H_2O \rightarrow KAl_3Si_3O_{10}(OH)_2 +$
Microcline Carbonic Acid Illite (Beaver's bend)

$2K^+ + 6H_4SiO_4 + 2HCO_3^-$

$\Delta G_r^\circ = +102.6$ kcal/mole

$\Delta V_s = -49.0$ ml. or 15.3% volume reduction for solid phases is obtained.

Microcline to illite conversion at 25°C (298°K) and 1 bar in HCOOH

$3KAlSi_3O_8\ 2HCOOH\ 12H_2O \rightarrow KAl_3Si_3O_{10}(OH_2) \quad +$
Microcline Formic Acid Illite (Beaver's bend)

$2K^+ + 6H_4SiO_4 + 2HCOO^-$
Formate Anion

$\Delta G_r^\circ = +95.60$ kcal/mole at 298°K and 1 bar.

$\Delta V_s = 15.3\%$ volume reduction in solid phases at 298°K and 1 bar.

Microcline to illite conversion at 25°C (298°K) and 1 bar in CH_3COOH

$3KAlSi_3O_8 + CH_3COOH\ 12H_2O \rightarrow KAl_3Si_3O_{10}(OH)_2 +$
Microcline Acetic Acid Illite (Beaver's bend)

$2K^+ + 6H_4SiO_4 + 2CH_3COO^-$
Acetate Anion

$\Delta G_r^\circ = +23.73$ kcal/mole at 298°K and 1 bar.

$\Delta V_s = 15.3\%$ volume reduction in the solid phases at 298°K and 1 bar.

Similar calculations were made for muscovite-type illite but are not shown.

Microcline to Kaolinite

Microcline to kaolinite conversion at 25°C and 1 bar in H_2CO_3

$2KAlSi_3O_8$ (Microcline) + $2H_2CO_3$ (Carbonic acid) + $9H_2O \rightarrow Al_2Si_2O_5(OH)_4$ (Kaolinite) +

$2K^+ + 4H_4SiO_4 + 2HCO_3^-$

$\Delta G_r^\circ = +22.98$ kcal/mole at 298°K, 1 bar.

$\Delta V_s = -27.17$ ml. or 12.5%.

Microcline to kaolinite conversion at 25°C and 1 bar, in HCOOH

$2KAlSi_3O_8$ (Microcline) + $HCOOH$ (Formic Acid) + $9(H_2O) \rightarrow Al_2Si_2O_5(OH)_4$ (Kaolinite) +

$2K^+ + 4H_4SiO_4 + 2HCOO^-$

$\Delta G_r^\circ = +15.8$ kcal/mole at 298°K, 1 bar.

Microcline to kaolinite conversion at 25°C and 1 bar, in CH_3COOH

$2KAlSi_3O_8$ (Microcline) + $2CH_3COOH$ (Formic Acid) + $9(H_2O) \rightarrow Al_2Si_2O_5(OH)_4$ (Kaolinite) +

$2K^+ + 4H_4SiO_4 + CH_3COO^-$

$\Delta G_r^\circ = -4.28$ kcal/mole at 298°K, 1 bar.

Anorthite to Kaolinite Conversion

Since the data on free energies of solid solution of feldspars are composition dependent and therefore not always easily available, calculations were made for the end member anorthite.

Anorthite to kaolinite conversion at 25°C and 1 bar in H_2CO_3

$CaAl_2Si_2O_8$ (Anorthite) + H_2CO_3 (Carbonic Acid) + $H_2O \rightarrow Al_2Si_2O_5(OH)_4$ (Kaolinite) +

$Ca^{++} + CO_3$

$\Delta G_r^\circ = -15.0$ kcal/mole at 298°K, 1 bar.

$\Delta V_s = -1.26$ ml. or 1.26% at 298°K, 1 bar.

Anorthite to kaolinite conversion at 25°C and 1 bar in HCOOH

$CaAl_2Si_2O_8$ (Anorthite) + $2HCOOH$ (Formic Acid) + $H_2O \rightarrow Al_2Si_2O_5(OH)_4$ (Kaolinite) +

$2HCOO^-$ (Formate Anion) + Ca^{++}

$\Delta G_r^\circ = -16.5$ kcal/mole at 298°K, 1 bar.

Anorthite to kaolinite conversion at 25°C and 1 bar in CH_3COOH

$2CaAl_2Si_2O_8$ (Anorthite) + CH_3COOH (Acetic Acid) + $2H_2O \rightarrow 2Al_2Si_2O_5(OH)_4$ (Kaolinite) +

$2Ca^{++} \rightarrow$ Acetate Anion

$\Delta G_r = -36.9$ kcal/mole.

The last column on the right on Table 4 shows the free energy of reaction (ΔG_r°) for each conversion at 25°C. Comparison of ΔG_r° for any series of conversions involving the various converting agents shows that the ΔG_r° for the equation written with acetic acid is lower, which indicates a better thermodynamic drive.

MECHANISMS INDICATING ROLE OF ACIDS AS POROSITY-GENERATING AGENTS

It has been shown (Lundegard et al., 1984) that organically derived CO_2 is insufficient to explain observed secondary porosity in sandstones. This paper shows that activity of H^+ from organic diagenesis is stoichiometrically and thermodynamically more efficient for dissolution of carbonate as well as silicates. Can additional support be advanced, however, that H^+ from various sources is what actually does take part in dissolution and not CO_2? Support for this argument is advanced in terms of isotopic values of δC^{13} for a host limestone and carbonate cements. This example is a carbonate rock, but the logic also is applicable to carbonate cement in sandstone. Two possible pathways exist for dissolution. One pathway is based on the hypothesis of organically derived CO_2 and the other pathway on H^+ from organic acids.

Decarboxylation of Organic Acids

According to the popular hypothesis, CO_2 during organic diagenesis plays an important role in the creation of secondary porosity. If this is true, the following steps would be involved. A diagrammatic sketch is shown in Figure 3. Carbonic acid hypothesis was suggested by Schmidt and McDonald (1979) through decarboxylation mechanism. The first part of Figure 3 is based on their hypothesis of decarboxylation of organic matter in the source rock. Appropriate equations are written to show the transformations in an overlying sandstone and in the zone of cementation. The hypothesis suggested in this paper involves dissociation of organic acids during diagenesis of organic matter in the source rock. This hypothesis is shown in a parallel diagram (see Fig. 3). The two hypotheses are represented by the following steps.

Step 1A

With continued burial the organic matter in shales or in the organic-rich micritic zones would undergo organic diagenesis.

TABLE 4.—Comparison of (ΔG_r°) for carbonic, formic and acetic acid

Conversion	ΔV_s°	Converting Agent	ΔG_r° Kcal/mole
K-spar → Illite (Beaver's Bend Type)	−15.3%	H_2CO_3	+102.6
		HCOOH	+ 95.6
		CH_3COOH	+ 23.6
K-spar → Kaolinite	−12.5%	H_2CO_3	+ 23.0
		HCOOH	+ 15.8
		CH_3COOH	− 4.28
Anorthite → Kaolinite	−1.26%	H_2CO_3	− 15.0
		HCOOH	− 16.5
		CH_3COOH	− 36.9

ΔV° represents percent reduction in the volume of solid phase products when compared to solid phase reactants.

$$\text{Organic Matter} \xrightarrow[\text{Diagenesis}]{\text{P, T}} \text{Organic Acids for eq.}$$

$$C°H_3—C°(=O)—OH$$

where C° stands for isotopically lighter organic carbon.

$$\text{Organic Acids} \xrightarrow[\text{High T}]{\text{Decarboxylation}} C°H_4 + C°O_2 .$$

Step 2A

$C°O_2$ so evolved would produce carbonic acid, which upon dissociation produces H^+, $HC°O_3^-$ and $C°O_3^{2-}$.

$$C°O_2 + H_2O \rightarrow H_2C°O_3$$
$$H_2C°O_3 \rightarrow H^+ + HC°O_3^-$$
$$HC°O_3^- \rightarrow H^+ + C°O_3^{2-}$$

Step 3A

H^+ so evolved from dissociation of carbonic acid would dissolve the adjacent limestone or carbonate cement in sandstones.

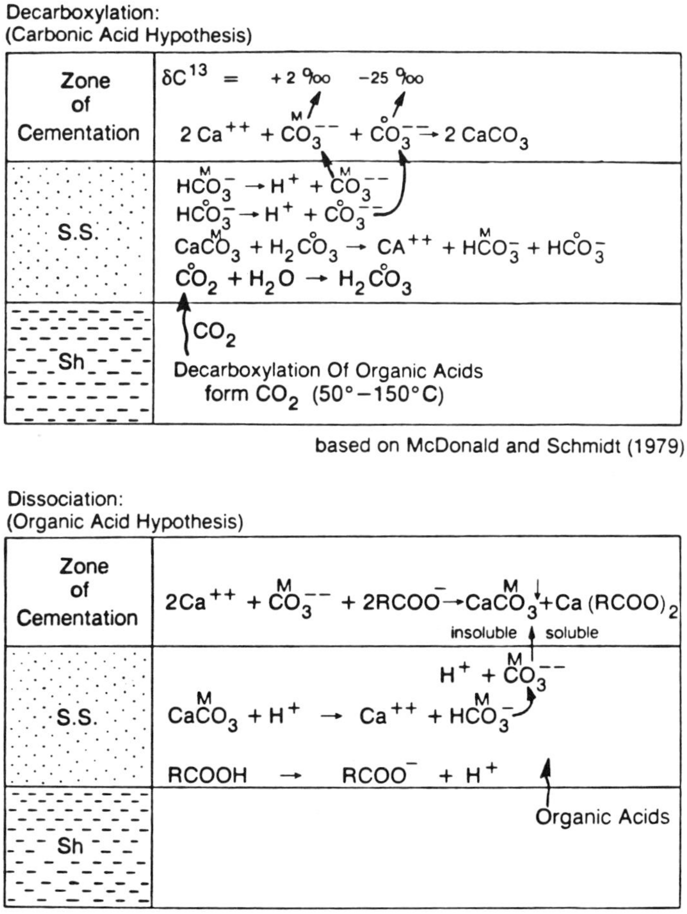

Decarboxylation:
(Carbonic Acid Hypothesis)

Dissociation:
(Organic Acid Hypothesis)

FIG 3.—Generation of secondary porosity due to carbonic or organic acids may be distinguished by the δC^{13} values of resulting $CaCO_3$ cements, as shown by decarboxylation versus dissociation mechanisms.

$$CaC^mO_3 + H^+ \rightarrow Ca^{2+} + HC^mO_3^-$$
$$HC^mO_3^- \rightarrow H^+ + C^mO_3^{2-}$$

C^m stands for isotopically heavy carbon, such as that from marine limestones.

Step 4A

Isotopically light $C°O_3^{2-}$ (Step 2A) and isotopically heavy $C^mO_3^{2-}$ (Step 3A) together would combine with Ca^{2+} in upward-moving formation waters precipitating carbonate cement.

$$Ca^{2+} + C°O_3^{2-} \text{ (from Step 2A)} \rightarrow CaC°O_3$$
$$Ca^{2+} + C^mO_3 \text{ (from Step 3A)} \rightarrow CaC^mO_3$$

Thus, the precipitated carbonate cement should be isotopically lighter than the host limestone or the original marine carbonate cement in a sandstone.

Dissociation of organic acids

A diagrammatic sketch showing this hypothesis is presented in Figure 3. In this hypothesis, H^+ comes from dissociation of organic acids generated during organic diagenesis which dissolves marine limestone, creating C^mO_2 with heavier carbon. The following steps are proposed in this hypothesis.

Step 1B

As the burial continues, the organic-rich micritic zones would undergo organic diagenesis at a suitable depth (temperature), evolving H^+ as suggested by the following mechanism.

$$\text{Organic Matter} \xrightarrow{\text{P, T}} \text{Organic Acids, } C°H_3{-}C°(=O){-}OH$$

$$C°H_3C°(=O)OH \xrightarrow{\text{dissociation}} [\,C°H_3{-}C°(=O){-}O\,]^- + H^+$$

Step2B

H^+ so evolved would produce acidic formation waters, which would move updip with compaction leading to dissolution of the overlying limestone. Organic acids would dissociate to give H^+ and acetate ions.

$$H^+ + CaC^mO_3 \rightarrow Ca^{2+} + HC^mO_3$$
$$HC^mO_3^- \rightarrow H^+ + C^mO_3^{2-}$$

Step 3B

The dissolution would continue until H^+ from organic diagenesis and H^+ from bicarbonate dissociation were exhausted.

Step 4B

Buffering action would occur, precipitating carbonate with carbon isotope values more akin to marine carbon.

$$C^mO_3^{2-} + Ca^{2+} \rightarrow CaC^mO_3 \text{ (isotopically heavy)}$$

If decarboxylation hypothesis is a dominant mechanism, then the carbonate-cemented zones will have lighter δC^{13} values when compared to the host limestone, and carbonate cements in sandstones will also have light δC^{13} values.

If dissociation hypothesis represents the dominant mechanism, then the dissolution will be controlled by generation of H^+, and the carbonate cements will have δC^{13} values akin to marine values. In addition, acetate ions would occur in the formation waters.

Using these criteria, the carbon isotope values can be interpreted as supporting decarboxylation versus the dissociation hypothesis to decipher late diagenesis mechanisms in addition to other known interpretations usually applied to early diagenesis.

CONCLUSIONS

1. Hydrogen ion donor capacity of organic acids can be 6 to 350 times greater than that of carbonic acid.
2. Approximate hydrogen ion concentration for various acids can be calculated from the dissociation constant data.
3. The solubility of calcium salts of organic acids is orders of magnitude greater than calcium carbonate and bicarbonate. This fact is responsible for prevention of porosity destruction by precipitation of insoluble carbonates.
4. Feldspar grain dissolution commonly observed represents 15% to 12% volume reduction for a mole of dissolved feldspar, thus supporting porosity enhancement observed by the petrologists during these conversions. The actual secondary porosity observed in a rock will depend on the amount and type of feldspar dissolved.
5. Feldspar grain dissolution in the presence of acetic acid typically requires 79 to 21 kcal/mole lower free energy than the free energies required in presence of carbonic acid, making porosity enhancement in the presence of organic acid thermodynamically more efficient.
6. Theoretical discussion on the thermodynamic and stoichiometric efficiency of organic acids presented in this paper, coupled with observations of other workers in terms of abundant concentration of organic acid anions, indicate that these acids play a major role in secondary porosity generation.

ACKNOWLEDGMENTS

I thank Amoco Production Company for the support of this research and for permission to publish this paper. I am also highly appreciative of discussion with colleagues and particularly for the reviews of the manuscript by Dr. E. D. Pittman and Dr. H. D. Winland.

REFERENCES

AL-SHEIB, Z. AND SHELTON, J., 1981, Migration of Hydrocarbons and secondary porosity in sandstones: American Association of Petroleum Geologists Bulletin, v. 65, p. 2433–2436.

CAROTHERS, W. W. AND KHARAKA, Y. K., 1978, Aliphatic acid anions in oil-field waters—Implications for origin of natural gas: American Association of Petroleum Geologists Bulletin, v. 62, p. 2441–2453.

COLLINS, G. W., 1961, Geochemistry of oil-field waters: Elsevier Publications.

CORRENS, C. W., 1963, Experiments on the decomposition of silicates and discussion of chemical weathering: Clays and Clay Minerals, v. 12, p. 443–460.

CURTIS, C. D., 1983, Link between aluminum mobility and destruction of secondary porosity: American Association of Petroleum Geologists Bulletin, v. 67, p. 380–383.

HUANG, W. H. AND KELLER, W. D., 1970, Dissolution of rock-forming silicate in organic acids: Simulated first-stage weathering of fresh mineral surfaces: American Mineralogist, v. 55, p. 2076–2094.

LARESE, R. E., THOMAS, J. B., PITTMAN, E. D., AND MESHRI, I.D., 1980, Development of secondary porosity in reservoir sandstones by dissolution of silicate mineral constituents (abs): American Association of Petroleum Geologists Bulletin, v. 64, p. 737.

LOCHTE, H. L., BURNAM, C. W., AND MEYER, H. W. H., 1949, Organic acids produced by high pressure gas wells: Petroleum Engineer, v. 21, p. C34–C40.

LUNDEGARD, P. D., LAND, L. S., AND GALLOWAY, W. E., 1984, Problem of secondary porosity: Frio Formation (Oligocene), Texas Gulf Coast: Geology, v. 12, p. 399–400.

MCBRIDE, E. F., 1977, Secondary porosity—Importance in sandstone reservoirs in Texas: Gulf Coast Association of Geological Societies, Proceedings, 27th Annual Meeting, p. 121–122.

PITTMAN, E. D., 1979, Recent advances in sandstone diagenesis: Annual Review of Earth and Planetary Sciences, v. 7, p. 39–62.

REESMAN, A. L. AND KELLER, W. D., 1968, Aqueous solubility studies of high-alumina and clay minerals: American Mineral, v. 53, p. 929–942.

ROBIE, R. A., HEMINGWAY, B. S., AND FISHER, J. R., 1979, Thermodynamic properties of minerals and related substances at 298.15°K and 1 bar pressure and at higher temperatures: U.S. Geological Survey Bulletin 1452.

ROGERS, G. S., 1917, Chemical relations of the oil-field waters in San Joaquin Valley, California: U.S. Geological Survey Bulletin 653, 119 p.

SCHMIDT, V., AND MCDONALD, D. A., 1979, The role of secondary porosity in the course of sandstone diagenesis: Society of Economic Paleontologists and Mineralogists Special Publication 26, p. 175–207.

SEIDELL, A., 1965, Solubilities: Inorganic and metal organic compounds, a compilation of solubility data from the periodic literature: American Chemical Society, v. 1., 1485 p.

SIEBERT, R. M., MONCURE, G. K., AND LAHANN, R. W., 1984, A theory of framework grain dissolution, in sandstones *in* McDonald, D. A. and Surdam R. C., eds., Clastic Diagenesis: American Association of Petroleum Geologists Memoir 37, p. 163–176.

STANTON, G. D. AND MCBRIDE, E. F., 1976, Factors influencing porosity and permeability of Lower Wilcox (Eocene) Sandstone, Karnes County, Texas (abs): American Association of Petroleum Geologists Bulletin, v. 60, p. 725–726.

SURDAM, R. C., BOESE, S. AND CROSSEY, L. J., 1982, Role of organic and inorganic reactions in development of secondary porosity in sandstones: American Association of Petroleum Geologists Annual Meeting, Calgary, Book of Abstracts, p. 116.

WILEY, L. M., KHARAKA, Y. K., PRESSURE, T. S., RAPP, J. B., AND BARNES, I., 1975, Short-chain aliphatic aids in oil-field waters and their contribution to the measured alkalinity: Geochemica et Cosmochimica Acta, v. 39, p. 1707–1710.

WOLLAST, R., 1967, Kinetics of the alteration of K-feldspar in buffered solutions at low temperature: Geochemica et Cosmochimica Acta, v. 31, p. 635–648.

CARBON DIOXIDE AND ORGANIC ACIDS: THEIR ROLE IN POROSITY ENHANCEMENT AND CEMENTATION, PALEOGENE OF THE TEXAS GULF COAST

PAUL D. LUNDEGARD[1] AND LYNTON S. LAND
Department of Geological Sciences, University of Texas at Austin, Austin, Texas 78712

ABSTRACT: Carbon dioxide produced by decarboxylation of organic matter is not a dominant factor in secondary porosity development. In typical sedimentary basins it may account for a regional average of a few volume percent secondary porosity, but it is insufficient to account for all of the secondary porosity in Texas Gulf Coast sandstones.

Petrographic data indicate dissolution of a minimum of 5 volume percent feldspar from Paleogene sandstones of the Texas Gulf Coast. The amount of acid required for feldspar dissolution depends on its mineralogy and the fate of aluminum. Feldspar dissolution in Texas Gulf Coast sandstones probably requires at least 1 mole of protons for every mole of feldspar dissolved.

Carbonate dissolution, a process of debatable importance in sandstones, was probably of major significance in many Gulf Coast shales. This is indicated by the strontium isotopic composition of carbonate cements in sandstones, and the decreasing weight percent calcite in shales in some wells. Carbonate cementation in sandstones probably occurred in response to lowered CO_2 solubility in upward-migrating formation water.

Considerations of the solubility of carbon dioxide and the distribution of organic acids with respect to temperature suggest that these species will be of great importance in porosity enhancement only if generated locally. Other potential sources of acid for dissolution reactions are reverse weathering reactions in shales and hydrous pyrolysis reactions between organic carbon and oxygen in H_2O (including synthesis of organic acids from kerogen).

INTRODUCTION

The CO_2—organic acid system is of great geologic interest as it is widely recognized as a major control on the pH of natural waters. For instance, variation in the partial pressure of carbon dioxide is believed to be an important variable in weathering reactions (Garrels and MacKenzie, 1967), the evolution of carbonate groundwater (Holland et al., 1962; Deines and Langmuir, 1974; Plummer et al., 1976), meteoric diagenesis of limestone (Hanor, 1978), and carbonate and silicate cementation and dissolution in clastic rocks (Curtis, 1978; Schmidt and McDonald, 1979). Organic acids are known to be important in weathering reactions (Ong et al., 1970; Stevenson, 1967; Tan, 1980), and may contribute significantly to the alkalinity of some oil field brines (Carothers and Kharaka, 1978). They may also be important in aqueous metal and trace metal transport because of their complexing properties (Baker, 1978; Barnes and Bourcier, 1983; Saxby, 1976; Surdam et al., 1984, p. 532). The goal of this paper is to evaluate the role that carbonic acid and organic acids played in porosity enhancement and cementation in Paleogene sandstones of the Texas Gulf Coast. In a subsequent paper we will present isotopic data on these and other carbon compounds and model their origin.

All dissolution, precipitation, and replacement reactions involving an acid will be affected by changes in the partial pressure of CO_2 and organic acid concentration. Such reactions may either increase or decrease the porosity of a rock and therefore its hydrocarbon reservoir quality, susceptibility to compaction, and bulk chemistry. For example, secondary porosity in sandstones, which in many basins is the dominant type of porosity (Schmidt and McDonald, 1979), may result either from the dissolution of silicate or carbonate minerals. This process is commonly attributed to the formation of CO_2 produced by decarboxylation of organic matter (Curtis, 1978; and Schmidt and McDonald, 1979). On the other hand, porosity reduction by carbonate cementation is commonly thought to be caused by the loss of dissolved CO_2 (Hanor, 1978; Blatt, 1979).

The Texas Gulf Coast Basin is an ideal place to study the interplay of subsurface carbon dioxide and organic acids with rock diagenesis for both geological and logistical reasons. The geologic framework of the Gulf Coast Basin has been studied in considerable detail. Rocks are presently at or near their maximum depth of burial (Murray, 1960) and most of the basin is still actively compacting (Jones, 1969). As a result, the hydrologic history, although complex, is simpler than that of basins that have undergone full compaction, uplift, and erosion. The large number of oil and gas wells in the region allow substantial selectivity in fluid sampling. A sizeable data base has been accumulated on aspects of rock diagenesis and pore water chemistry by geology students at the University of Texas at Austin, the Bureau of Economic Geology at the University of Texas at Austin, the U. S. Geological Survey, and others. In particular, several regional studies conducted by the Bureau of Economic Geology on both sandstone diagenesis and pore water chemistry have provided a very useful data base with which to integrate the results of this study. Loucks et al. (1984) point counted over 560 thin sections of Tertiary sandstones (mostly from the Wilcox and Frio Formations) from the Texas Gulf Coast. Morton et al. (1981) reported the analyses of major cations and anions in 78 Frio Formation waters from the Texas Gulf Coast. Galloway et al. (1982) determined the volumes of sandstone and shale in the Frio Formation throughout the Texas Gulf Coast. Fisher (1982) reported 37 analyses of Wilcox Formation pore water and quantitatively discussed their evolution. Finally, several studies conducted by the U.S. Geological Survey on dissolved organic acids, dissolved inorganic carbon, and other aspects of the pore water chemistry of the Frio Formation in the Houston and Corpus Christi areas have provided useful data and laid the groundwork for the present study (Carothers and Kharaka, 1978, 1980; Kharaka et al., 1977).

[1]Current address: Unocal Research Center, P.O. Box 76, Brea, California, 92621

GEOLOGIC SETTING

The Texas Gulf Coastal region is composed of a broad, gulfward-dipping wedge of Mesozoic and Cenozoic strata that accumulated almost continuously from the Jurassic to present (Fig. 1). Following rifting and formation of the continental margin, probably during Triassic time, Middle to earliest Late Jurassic redbeds and evaporites were deposited. The evaporites have subsequently been deformed into domes and diapirs that disturb both Mesozoic and Cenozoic deposits in the East Texas and Rio Grande Embayments. Considerable salt has also been removed by dissolution (Seni and Jackson, 1983).

Upper Jurassic and Cretaceous strata in the Texas Gulf Coast area are predominantly marine carbonate rocks that were deposited over broad shelves. During Middle Cretaceous (Aptian to Cenomanian) time, slow subsidence of the shelf and sporadic clastic input promoted growth of an extensive shelf-edge reef complex, called the Stuart City trend (Edwards and Sligo Formations) in the subsurface (Bebout and Loucks, 1974.). These rocks are presently buried to depths of 3300 to 5000 m, whereas shelfward deposits occur at shallower depths or are exposed in outcrop.

In early Cenozoic time the Gulf of Mexico basin began receiving large amounts of terrigenous sediment from northern and western sources in response to the Laramide orogeny. The supply of sediment generally exceeded subsidence rates, and the resulting progradation of fluvial-deltaic and strandplain facies produced a series of offlapping clastic wedges that buried the old carbonate shelf margin (Fig. 1). Rapid sediment loading of plastic muds caused large growth fault systems to form near the downdip edge of each wedge. Commonly, salt-withdrawal basins developed contemporaneously with deposition and further complicate the structure.

METHODS

Water samples were collected from onshore Cenozoic rocks (Wilcox Fm-Eocene; Frio Fm-Oligocene) from depths of 640 m to depths greater than 4200 m, where temperatures approach 200°C. Only wells perforated over narrow depth intervals and which had not been treated (e.g., flushed with acid) or water-flooded were sampled. Wells producing only a few barrels of water per day were avoided for water sampling but were acceptable for collection of gas samples. Samples were taken directly from the wellhead or from a connection as close to it as possible. Elsewhere, samples were collected from separators which had been well flushed with fresh formation water.

Total alkalinity was measured in the field by titration of 50 ml of freshly collected formation water to a pH of 2.8 or less with either 0.01639 N or 0.03278 N sulfuric acid. Samples for stable carbon isotopic analysis of dissolved inorganic carbon were collected in borosilicate glass bottles, one-third filled with formation water then filled completely

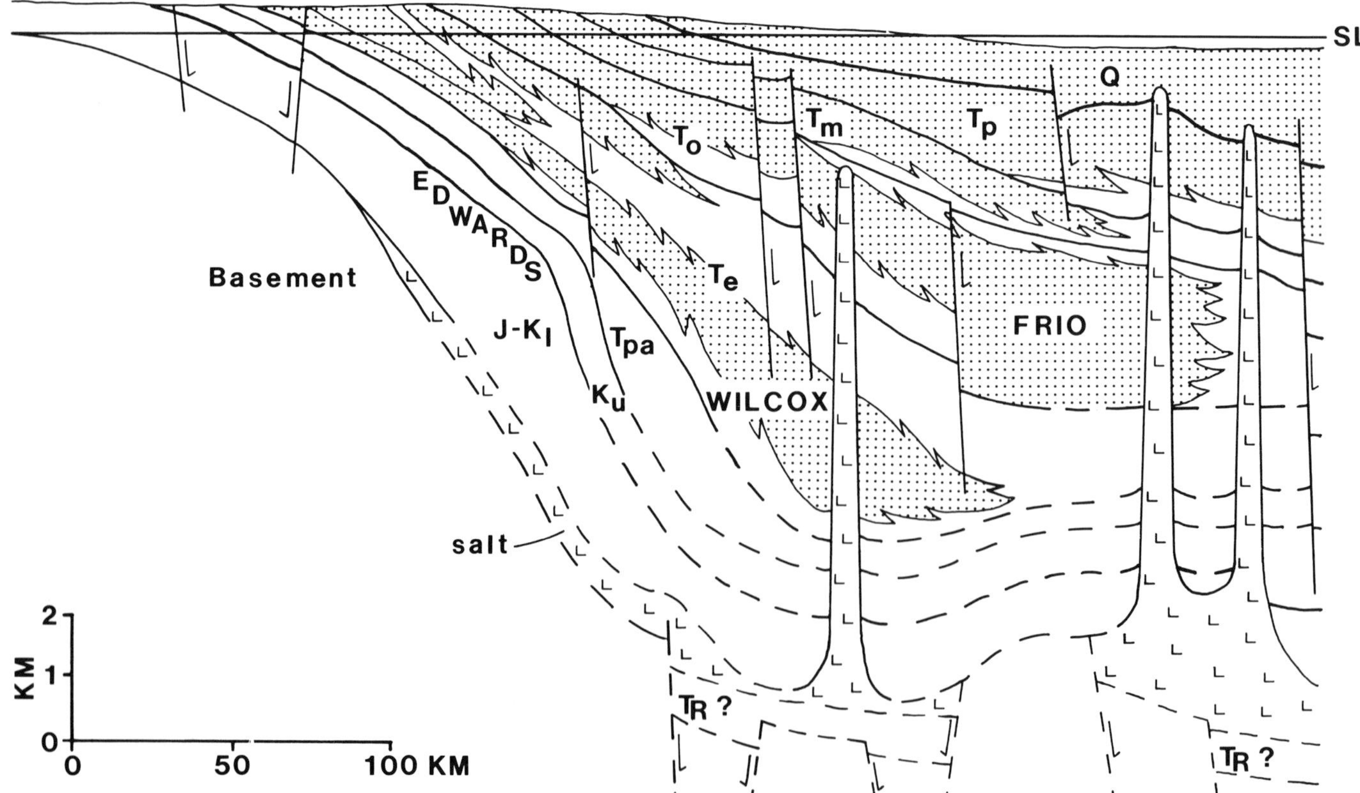

FIG. 1.—Schematic cross section of the Texas Gulf Coast. Note Gulfward dipping and thickening units and abundance of growth faults. As part of this study, data were collected from the Wilcox Formation, the Frio Formation, Miocene and Plio-Pliestocene reservoirs (offshore Louisiana), and the Edwards Limestone. Adapted from unpublished cross sections by W. E. Galloway (Bureau of Economic Geology, University of Texas at Austin) and R. Buffler (Institute for Geophysics, University of Texas at Austin).

with a saturated solution of barium hydroxide and capped. Water for dissolved organic carbon and volatile fatty acid analyses was collected in glass bottles and made slightly basic with 1 N NaOH. These samples were subsequently stored under refrigeration. Organic alkalinity was determined by the titrametric method of the American Public Health Association (1976, p. 527–529).

One hundred and eight wells in 31 fields and 17 counties were sampled for water and/or gas (Fig. 2). Of the 94 wells from which water was collected, 52 were judged to produce representative samples of formation water. The quality of water samples was evaluated by comparison of measured downhole temperatures and those calculated by silica and Na-Ca-K geothermometry (Fournier and Rowe, 1966; Fournier and Truesdell, 1973; Morey et al., 1962), and by the stable isotopic ratios of oxygen and hydrogen. Samples were discarded if the calculated temperatures differed from the measured ones by more than 15°C or if the isotopic ratios were extremely aberrant.

PETROGRAPHIC DATA

In order to assess the role carbon dioxide and organic acids had on the diagenesis of Tertiary sandstones, the petrographic data and observations of numerous investigators were used. The Bureau of Economic Geology, University of Texas at Austin, generously allowed us to study raw point count data on over 560 samples of Gulf Coast sandstones (mostly from the Wilcox and Frio Formations). These data were collected as part of a regional evaluation of prospects for geopressured-geothermal energy (Loucks et al., 1984). We also made use of data collected by Fisher (1982), Stanton (1977), and Boles and Franks (1979) on Wilcox sandstones, and Lindquist (1976) on Frio sandstones. Since diagenesis of Gulf Coast sandstones has been described in detail by these and other workers, only data relevant to the present study are discussed here.

Sandstones from the Frio and Wilcox Formations average approximately 8 and 5 volume percent carbonate cement, respectively (Stanton, 1977; Loucks et al., 1984; Fisher, 1982; Land and Fisher, in prep.). The cement is predominantly calcite in the Frio and ankerite in the Wilcox. Using oxygen isotopic data, Land (1984, p. 56) estimated that calcite in the sandstones of the Frio Formation precipitated at approximately 80°C. Similarly, Fisher (1982) estimated that calcite in the sandstones of the Wilcox Formation precipitated between 80 and 100°C and the ankerite precipitated between 95 and 120°C. The data of Loucks et al. (1984) show an average of 11 volume percent secondary porosity in sandstones of the Frio Formation and 7 volume percent in the sandstones of the Wilcox Formation in Texas (Fig. 3). Frio sandstones have consistently more secondary porosity than Wilcox sandstones throughout the Texas coastal region, except in east Texas (Fig. 3, area 6). In the central Texas coastal region, the area from which most of the fluid samples were collected (areas 3 and 4 in Fig. 3), Frio sandstones average 12 percent secondary porosity and Wilcox sandstones average 8 percent. According to Loucks et al. (1984), approximately half of this secondary porosity was produced by the dissolution of detrital and authigenic carbonates, the remainder by the dissolution of detrital feldspar, especially K-feldspar. Skeletal feldspar grains and oversized pores are abundant in these rocks. Some investigators have doubted that extensive carbonate dissolution has taken place (Moncure et al., 1984, p. 153; Siebert et al., 1984 p. 165), but the removal of feldspars is a process of unquestionable significance. Some of the kaolinite in Texas Gulf Coast sandstones clearly post-dates major dissolution since it can be found filling secondary pores in feldspar grains (Loucks et al. 1984; Milliken et al., 1981).

There are clearly several potential sources of bias or error in these data. There is some subjectitivity and operator bias in the identification of secondary porosity; but, while the origin of many intergranular pores is ambigous, few would argue that intragranular pores in feldspars (see cover of Au-

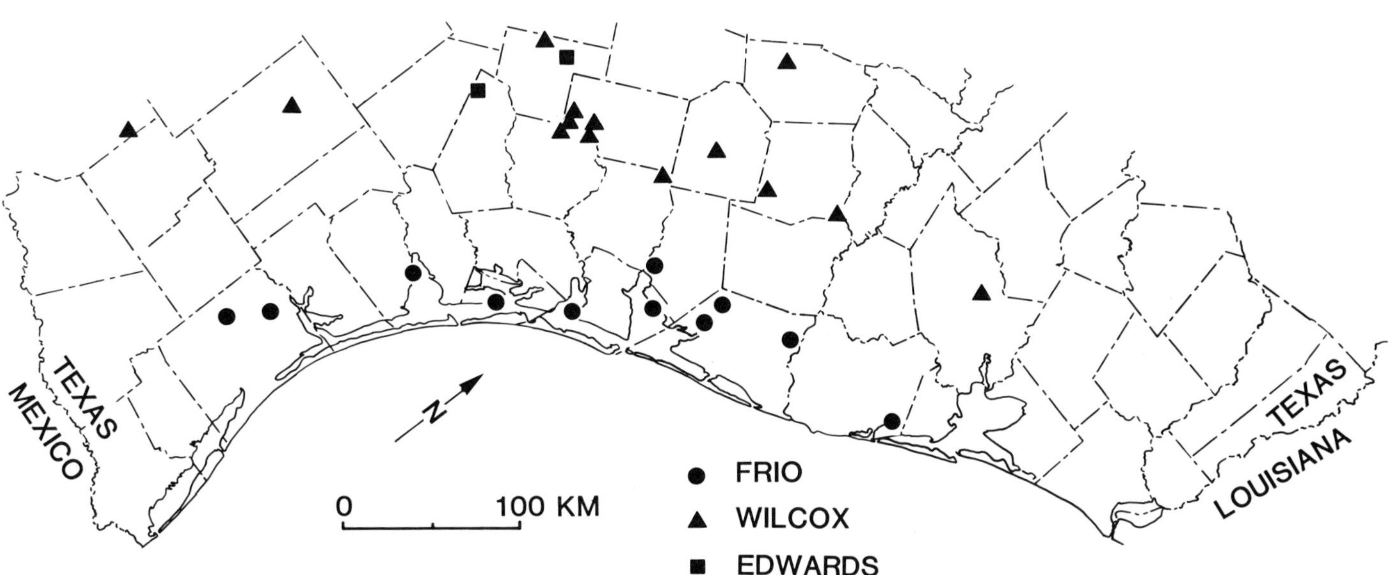

FIG. 2.—Location of the 31 oil and gas fields from which fluid samples were collected for this study. Some field locations are generalized to include two or more nearby fields.

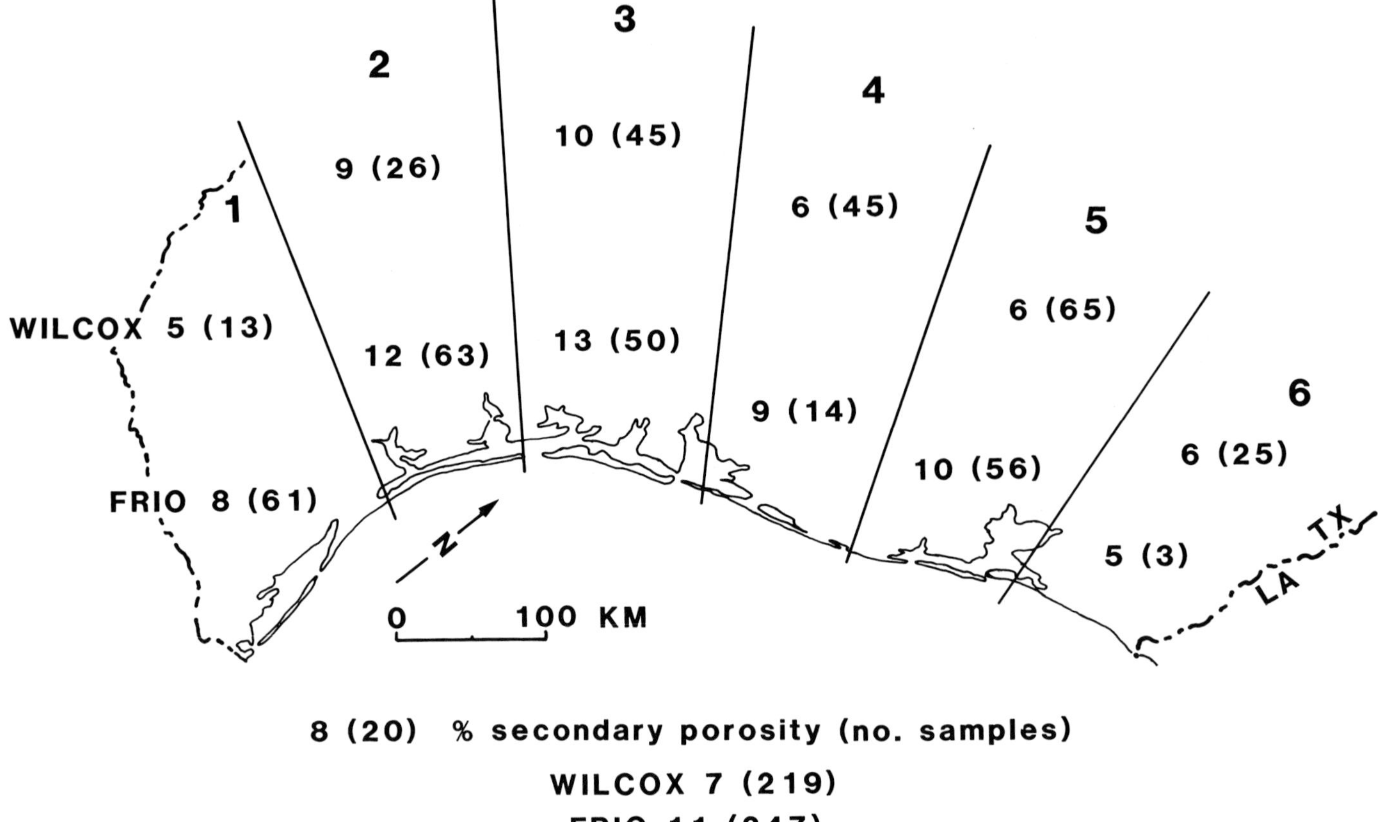

FIG. 3.—Map of the Texas Gulf Coast showing the reported average amounts of secondary porosity in the sandstones of the Wilcox (Eocene) and Frio (Oligocene) Formations for the six areas studied by Loucks et al. (1984). Note that the Frio has consistently more secondary porosity than the Wilcox.

gust 1984 issue of *Geology*), or pores over one grain diameter in width are caused by anything other than dissolution during burial. Also, because several petrographers did the point counts for Loucks et al. (1984), the effect on the total data set of a petrographer who is "pro-secondary" or "pro-primary" porosity would be lessened. No doubt some operator bias does exist, but its effect is difficult to assess. Because rock samples came largely from cores provided by oil companies, they are biased in favor of sandstones that were once considered potential reservoirs. The success ratio of explorationists, however, is far from perfect (as most are painfully aware) and many cores are taken in prospective reservoirs that turn out to be unproductive. Finally, there may have been some bias in the selection of thin section samples from the cores. In a test well for geopressured geothermal fluids in Brazoria County, Loucks et al. (1984) collected thin section samples on a random basis (every 0.6 m) through the sandier intervals. For the 118 samples with less than 25 percent clay matrix, the average reported secondary porosity was 11 volume percent (Fig. 4), which is the same as the average for the regional data (Fig. 3). If this well is typical of the others, then thin section sampling bias appears to be minimal. In this same well the petrographic data indicate an average of 2.5 volume percent framework grain dissolution porosity, mostly within plagioclase grains. The data of Moncure et al. (1984) indicate

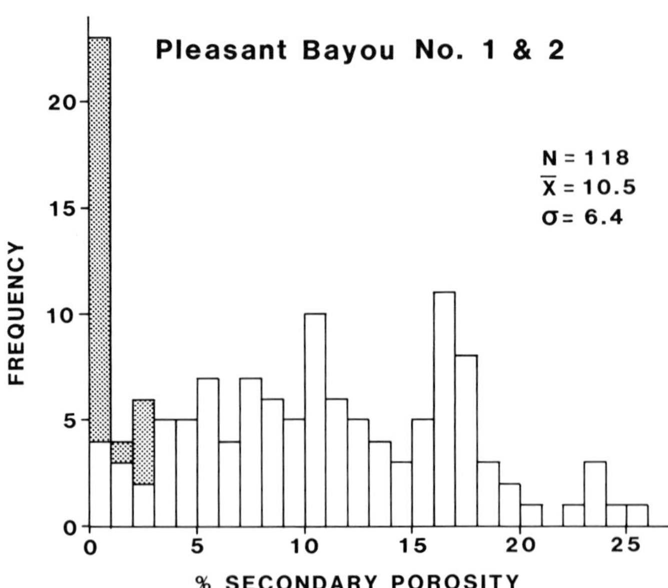

FIG. 4.—Volume percent secondary porosity in randomly selected sandstone samples from the Pleasant Bayou no. 1 and no. 2 wells. Stippled area is for samples with more than 25% clay matrix. Statistics are for those samples with less than 25% clay matrix. The average amount of secondary porosity is the same as the regional average (Fig. 3). Data from raw point counts of Loucks et al. (1984).

an average slightly greater than this in a short interval of the same core. We believe that 2.5 volume percent represents an absolute minimum amount of silicate dissolution. Intragranular pores are certainly secondary and the value 2.5 percent does not include secondary pores where grains have been completely dissolved (oversized pores). Other evidence suggests that much more feldspar was dissolved than is currently represented by intragranular pores. In sandstones from both the Wilcox and Frio Formations, the percent K-feldspar decreases with depth (Fig 5). This trend is not related to variations in grain size and indicates the probable dissolution of at least 5 percent K-feldspar. Most of this K-feldspar was dissolved completely and is not recognizable as intragranular pores in the rocks today. We believe these data collectively indicate that a minimum of 5 volume percent feldspar was dissolved from these Gulf Coast sandstones.

FELDSPAR DISSOLUTION

Petrographic data indicate a minimum of 5 volume percent feldspar dissolution. Here we consider the possible reactions by which this dissolution took place and the amount of acid that was consumed in the process, an essential step in assessing the mass balance of dissolution.

The amount of acid required for feldspar dissolution depends on several factors, one of which is mineralogy. Plagioclase in sandstones of the Frio and Wilcox Formations initially had a composition of approximately An_{20} and An_5, respectively (Land, 1984, Fig. 11; Land and Fisher, in prep.). As shown by reactions (1) and (2) below, less acid is required to dissolve a mole of alkali feldspar than a mole of plagioclase feldspar.

$$KAlSi_3O_8 + 4.5H_2O + H^+ \rightarrow K^+ + 0.5Al_2Si_2O_5(OH)_4 + 2H_4SiO_4 \quad (1)$$

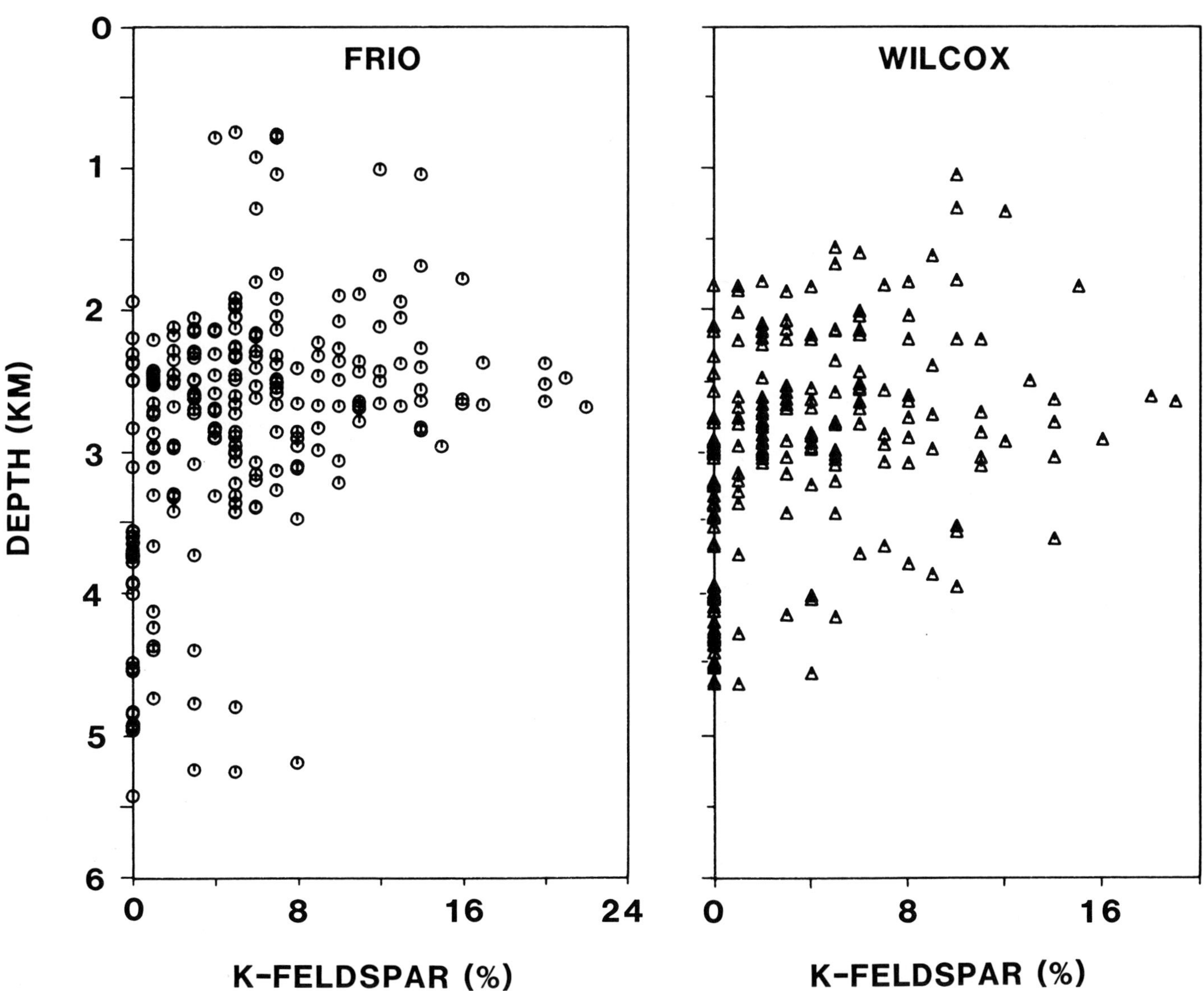

FIG. 5.—Volume percent K-feldspar versus depth for the sandstones of the Wilcox and Frio Formations in the Texas Gulf Coast. Note trend of decreasing K-feldspar with depth. Data from raw point counts of Loucks et al. (1984).

$$Na_{.8}Ca_{.2}Al_{1.2}Si_{2.8}O_8 + 3.8H_2O + 1.2H^+ \rightarrow$$
$$0.8Na^+ + 0.2Ca^{++} + 0.6Al_2Si_2O_5(OH)_4 + 1.6H_4SiO_4 \quad (2)$$

If kaolinite (or another clay mineral) does not consume the aluminum released by feldspar dissolution and the aluminum is instead released as an aqueous species, the acid required for feldspar dissolution depends on which aluminum ion pair dominates (Siebert et al., 1984), as shown by reactions (3) and (4) below.

$$Na_{.8}Ca_{.2}Al_{1.2}Si_{2.8}O_8 + 3.2H_2O + 4.8H^+ \rightarrow$$
$$0.8Na^+ + 0.2Ca^{++} + 1.2Al^{+3} + 2.8H_4SiO_4 \quad (3)$$

$$Na_{.8}Ca_{.2}Al_{1.2}Si_{2.8}O_8 + 8H_2O \rightarrow$$
$$0.8Na^+ + 0.2Ca^{++} + 1.2Al(OH)_4^- + 2.8H_4SiO_4 \quad (4)$$

At very low pH, dissolved aluminum may exist as Al^{+3} or, with increasing pH, as the $Al(OH)^{+2}$, $Al(OH)^+{}_2$, $Al(OH)_3^0$, or $Al(OH)_4^-$ ion pair with a parallel decrease in the amount of acid required per mole of feldspar dissolved. Kharaka et al. (1979, p. 176) calculated an *in situ* pH of 4.1 for a well in the Frio Formation in Brazoria County, Texas. Assuming *in situ* pH values for Gulf Coast pore fluids generally range from about 4 to 5, the thermodynamic data of SOLMNEQ (Kharaka and Barnes, 1973) indicate that the dominant aluminum hydroxide ion pair is $Al(OH)_2^+$ at 100°C. Accordingly, the dissolution of 1 mole of intermediate plagioclase by a reaction similar to reactions (3) and (4) would require about 2.4 moles of acid. It should be noted, however, that there is some disagreement among published dissociation constants for aluminum hydroxide ion pairs. In any case, very little feldspar dissolution can occur by reactions such as (3) and (4), because saturation with respect to clay minerals will quickly occur. As a result, aluminum concentrations will be buffered to very low values.

Several workers have recently suggested that feldspar dissolution involves, at least in part, aluminum complexation by organic ligands, in particular by acetate and oxalate (Surdam et al., 1984; Siebert et al., 1984). Such a process would help explain petrographic observations of partially dissolved feldspar grains next to well preserved carbonate fossil fragments. The effect of organic complexation on acid requirements for feldspar dissolution depends on the nature of the organic complex. According to Surdam et al. (1984), the aluminum-oxalate complex may have the following formula, $(AlC_2O_4 \cdot 4H_2O)^+$. If true, for a given feldspar composition an equal or greater amount of acid is required by this process [reactions (5) and (6)] than if aluminum is released as $Al(OH)_2^-$ or if it is consumed in clay mineral reaction products.

$$Na_{.8}Ca_{.2}Al_{1.2}Si_{2.8}O_8 + 1.2H_2C_2O_4 \text{ (oxalic acid)}$$
$$+ 8H_2O + 2.4H^+ \rightarrow 0.8Na^+ + 0.2Ca^{++} + 2.8H_4SiO_4$$
$$+ 1.2(AlC_2O_4 \cdot 4H_2O)^+ \quad (5)$$

$$KalSi_3O_8 + H_2C_2O_4 \text{ (oxalic acid)} + 8H_2O + 2H^+ \rightarrow$$
$$K^+ + 3H_4SiO_4 + (AlC_2O_4 \cdot 4H_2O)^+ \quad (6)$$

Aluminum-oxalate complexes with higher oxalate:aluminum ratios are also possible (Lind and Hem, 1975). The dominant form is controlled by the activity of oxalate in solution, but thermodynamic data for calculating speciation at elevated temperatures are lacking. While oxalate is a strong complexer of aluminum (Surdam et al., 1984), it has so far been found only in very low concentrations in natural formation waters (Kharaka et al., this volume). Acetate, on the other hand, is abundant in some formation waters but is a much weaker complexer of aluminum [reaction (7), Surdam et al., 1984].

$$Na_{.8}Ca_{.2}Al_{1.2}Si_{2.8}O_8 + 2.4H_4C_2O_2 \text{ (acetic acid)}$$
$$+ 8H_2O + 2.4H^+ \rightarrow 0.8Na^+ + 0.2Ca^{++}$$
$$+ 2.8H_4SiO_4 + 1.2[Al(H_3C_2O_2)_2 \cdot 4H_2O]^+ \quad (7)$$

Analyses of dissolved aluminum in Gulf Coast formation waters show that its concentration is nearly always less than 1.0 mg/1 (Morton et al., 1981, Appendix A; Kharaka et al., this volume), casting serious doubt on whether complexation of aluminum by organic ligands is a significant process.

Estimates of acid requirements for complexation reactions are confounded by uncertainty about the structure of the complexes at reservoir conditions and about how many hydroxyl ions substitute for coordinated water molecules in the complexes (Hem and Roberson, 1967). If all the aluminum in K-feldspar is consumed by an oxalate complex as in reaction (6), 2 volume percent K-feldspar dissolution would require that *all* the carboxyl groups in the organic matter in 16 shale volumes be converted to oxalate for complexation of aluminum from one sandstone volume (assuming Type III kerogen, the most oxygen-rich type, 6 weight percent of the kerogen is oxygen in carboxyl groups, shale has 0.5 weight percent organic matter and a solid density of 2.6 g/cc). Because abundant CO_2 and acetic acid are produced from kerogen during its thermal evolution (Tissot and Welte, 1978, p. 148–151; Carothers and Kharaka, 1978; Hunt, 1979, p. 166–169; Lundegard, 1985) and are possible proton sources for reactions (5), (6) and (7), it is apparent that only a small fraction of the carboxyl groups in the kerogen could actually produce oxalate, so the required shale:sandstone ratio must be much higher.

In summary, the mineralogy and composition of the feldspar and the fate of its aluminum are important controls on the amount of acid required for feldspar dissolution. Petrographic observations of authigenic kaolinite, some filling secondary pores, in Texas Gulf Coast sandstones (Loucks et al., 1984; Milliken et al., 1981), as well as occurrences of undissolved calcite adjacent to skeletal feldspars (Surdam et al. 1984; Seibert et al., 1984) indicate that feldspar dissolution by reactions (1) and (2), and reactions similar to (5), (6), and (7) took place. Although the exact amounts of acid and organic complexing agents cannot be specified, it is apparent that feldspar dissolution in the Gulf Coast subsurface requires at least 1 mole of protons for every mole of feldspar dissolved. This ratio will be used in all further calculations.

IMPLICATIONS OF CO_2 SOLUBILITY

Carbonate cementation and dissolution are sensitive to variations in the concentration of dissolved carbon dioxide. It is important, therefore, to consider how CO_2 solubility might vary in a sedimentary basin. CO_2 solubility is a function of three things, the ionic strength of the solution, temperature, and the fugacity of carbon dioxide. CO_2 solubility is commonly expressed in terms of the Henry's Law coefficient, $K_h = f_c/x = P_c \cdot \mu/x$, where f_c is the fugacity of carbon dioxide, x is the ratio of the moles of carbon dioxide to the sum of the moles of carbon dioxide and water in the solution, p_c is the partial pressure of carbon dioxide, and μ is the fugacity coefficient. Henry's Law coefficient varies with the temperature and ionic strength of the solution. For sedimentary conditions μ varies from unity by only a few percent (Ellis and Golding, 1963, Table 2), and other sources of error are at least this large (Michels, 1981, p. 4); therefore, for the purposes of discussion K_h will be considered equal to p_c/x. The partial pressure of CO_2,p_c, is equal to the total pressure times the mole fraction of CO_2(g).

Because CO_2 solubility is a function of several geologically variable parameters, it is not possible to compute a single profile for the profile of CO_2 (aq) versus depth for the Texas Gulf Coast. Insights into the implications of CO_2 solubility for diagenetic reactions can still be acheived, however. Figure 6 shows a calculated depth profile of CO_2 (aq) for a given mole fraction of CO_2 (g) and ionic strength, using a hydrostatic pressure gradient and a temperature gradient typical of the Frio Formation. The concentration of CO_2 (aq) increases continuously with increasing depth in spite of the well known fact that CO_2 solubility and temperature are inversely proportional to about 175°C (Ellis and Golding, 1963, Fig. 2). No geologically reasonable temperature gradient is great enough to overcome the effect that increasing total pressure with depth (and hence CO_2 partial pressure) has on increasing CO_2 solubility. Pressure gradients below 2 or 3 km in the Gulf Coast basin are typically greater than hydrostatic, indicating a more rapid increase in the concentration of CO_2 (aq) than shown in Figure 6. If the partial pressure of carbon dioxide is the dominant control on pH, this relationship indicates that formation water moving upward will tend to exsolve CO_2 and therefore to precipitate carbonate minerals. Under these conditions, only local sources of CO_2 are likely to be important in carbonate dissolution reactions. CO_2 generation in shales can be expected to cause dissolution of calcite within the shales but is unlikely to do the same for carbonate in sandstones far updip, unless the shales initially contained little carbonate and no $CaCO_3$ is encountered along the flow path.

In the previous example it was assumed that the mole fraction of CO_2(g) was constant with respect to depth. In fact, analyses of natural gas from the Texas Gulf Coast show that the mole fraction of CO_2 increases with depth (Fig. 7). Using the observed relationship of depth and mole percent CO_2(g), a revised profile of CO_2 solubility can be computed for the same hydrostatic pressure gradient and ionic strengths as before. Instead of assuming a constant mole fraction of CO_2(g), the average observed value for a given depth will be used. The results again show that the concentration of CO_2 (aq) increases rapidly and continuously with increasing depth (Fig. 8). In the vicinity of salt domes salinity variations will be an important control on the solubility of CO_2 and will modify the general trend toward increased solubility with increased depth.

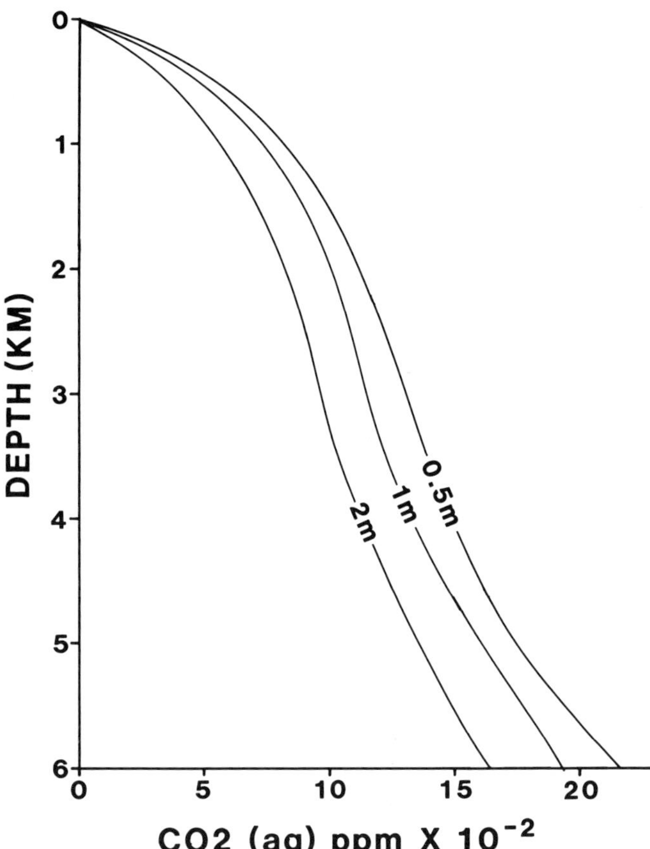

FIG. 6.—Concentration of CO_2 (aq) in 0.5, 1.0, and 2.0 m NaCl solutions versus depth for a constant mole fraction of CO_2 (g). Calculated from the Henry's Law coefficient data of Ellis and Golding (1963, Fig. 2), using a geothermal gradient of 30°C/km, a hydrostatic pressure gradient equal to 10.5 kPa/m (0.465 psi/ft), and the assumption that gas activity coefficients equal 1.0. Note that CO_2 (aq) concentration increases rapidly with depth.

Under some conditions the pH of formation water may be controlled by variables other than the partial pressure of carbon dioxide. Weak organic acids like acetic acid may contribute up to 100% of the alkalinity of formation water (Fig. 9; Carothers and Kharaka, 1978; Lundegard, 1985) and therefore have the potential to buffer the pH (Surdam et al., 1984). This potential will be evaluated later in this paper.

SECONDARY POROSITY

Carbon Dioxide from Decarboxylation

Thermally induced decarboxylation of organic matter has been shown to occur during burial and during simulated maturation (Robin and Rouxhet, 1978; Rouxhet et al., 1980; Johns, 1982; Kharaka et al., 1983). Given that decarbox-

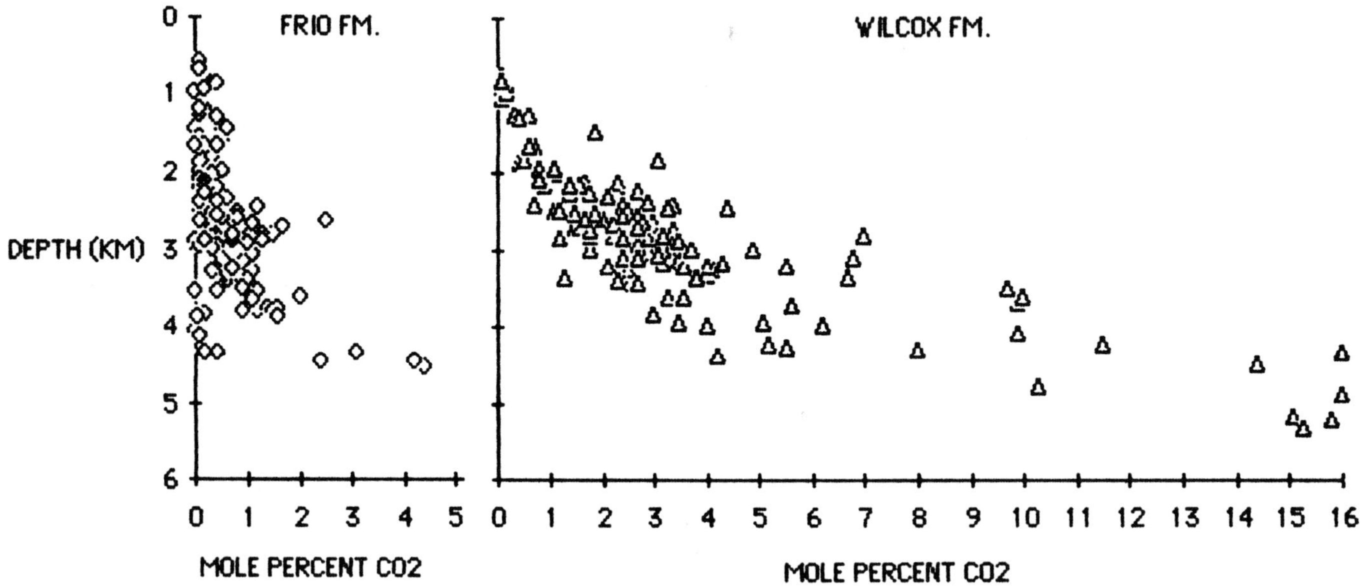

FIG. 7.—Mole percent CO_2 in the gas from the Wilcox and Frio Formations as a function of depth. The CO_2 content remains low until some critical depth is reached beyond which it increases rapidly. Note that for a given depth the CO_2 content of Wilcox gas is greater than that of Frio gas. Data from the present study and Moore (1976).

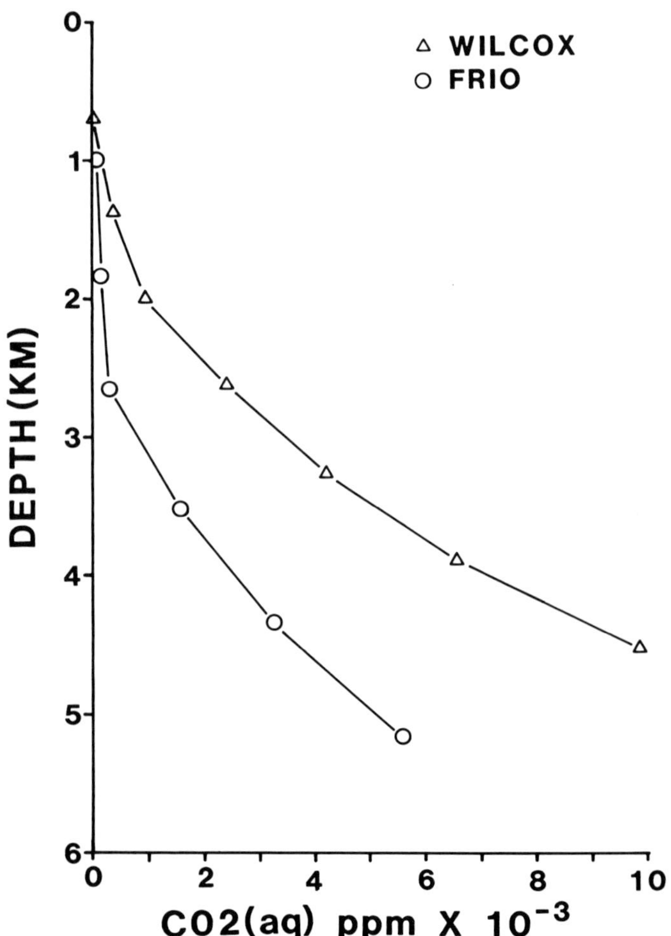

FIG. 8.—Concentration of CO_2 (aq) versus depth for observed values of mole fraction CO_2 (g) for the Wilcox and Frio Formations (Fig. 7). Calculated from the Henry's Law coefficient data of Ellis and Golding (1963, Fig. 2) for a 1 m NaCl solution, using a hydrostatic pressure gradient of 10.5 kPa/m (0.465 psi/ft), a 40° C/km thermal gradient for the Wilcox, a 30° C/km thermal gradient for the Frio, and the assumption that gas activity coefficients equal 1.0. Note that CO_2 (aq) concentration increases rapidly with depth.

ylation occurs, it is important to ask how much secondary porosity can be produced by the CO_2 it releases.

The amount of CO_2 that a given shale can produce by decarboxylation of its organic matter depends on the abundance of organic matter and its type, since the abundance of carboxyl groups is not the same in all types of organic matter (Rouxhet et al., 1980, Fig. 6.6). Tertiary shales of the Gulf Coast are typically low in organic carbon. Galloway et al. (1982, p. 679) reported an average of 0.28 weight percent for 140 analyses of shales from the Frio Formation. Our own analyses of 33 Gulf Coast shales averaged 0.39 weight percent organic carbon. Unpublished analyses by Amoco Production Company of 29 shales from the Wilcox and Frio Formations indicate similar concentrations. In our calculations we will assume an original value of 0.5 weight percent organic matter in order to account for the loss of volatiles during maturation and the fact that organic matter contains non-carbon elements.

Organic matter in Tertiary shales of the Gulf Coast is dominantly Type III and Type II, or woody and herbaceous (Galloway et al., 1982, p. 679; Amoco Production Company, unpublished data). Type III is the most CO_2-prone type of organic matter (Tissot and Welte, 1978). In order

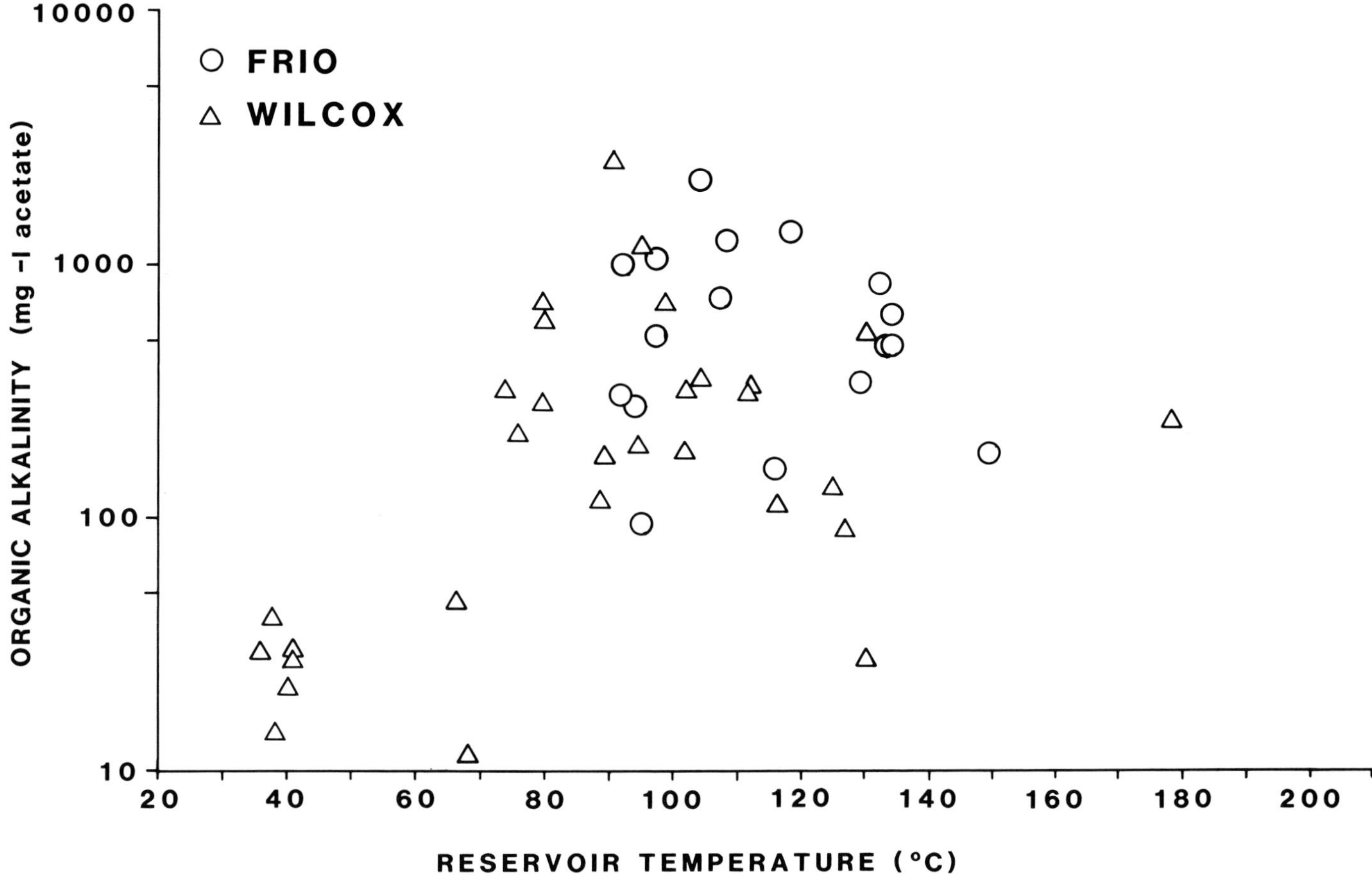

FIG. 9.—Semilogrithmic plot of organic alkalinity of waters from the Frio and Wilcox Formations versus temperature. Note that maximum concentrations occur at 80 to 120°C.

to make our calculations conservative, we will also assume that all the organic matter is of this type, although less CO_2-prone types may dominate in some areas. Thermally immature Type III kerogen could contain as much as 25 weight percent oxygen (Tissot and Welte, 1978, Table II, 4.11 and Fig. 11, 4.13). According to infrared absorption studies by Robin and Rouxhet (1978), approximately 25% of this oxygen may actually be present in carboxyl groups. We have therefore, chosen a value of 6.3 weight percent for the amount of oxygen in kerogen that could be converted to CO_2 by decarboxylation. These assumptions are the same as those used by Lundegard et al. (1984). Alternatively, one could use values for CO_2 generation from organic matter under experimental conditions (Harwood, 1977), which are not greatly different from the ones used here. We have adopted the approach described above, however, because it is based on observations of kerogen that has undergone maturation under natural conditions (Robin and Rouxhet, 1978; Rouxhet and others, 1980).

With the above assumptions, a maximum of 9.8×10^{-6} moles of CO_2 per gram of shale or 1.96×10^{-3} moles per gram of organic matter could be produced by decarboxylation of organic matter in Paleogene shales of the Texas Gulf Coast. If the shale has 10% porosity and a solid density of 2.6 g/cm, then each cubic centimeter of shale could produce 2.3×10^{-5} moles of CO_2 by decarboxylation.

The potential amount of secondary porosity produced by this CO_2 is dependent not only on the type and abundance of organic matter but also on the shale:sandstone ratio. Based on a large number of electric logs, Galloway et al. (1982, Table 2) estimated the volumes of sandstone and thermally mature shale (shale within the oil window) in the onshore Frio Formation. Using their data, the volumetric ratio of thermally mature shale to sandstone ranges from 0 to 13 and averages slightly less than three. The total shale-to-sandstone ratio averages only 4.4. Note that these ratios are considerably less than those assumed by some authors (Franks and Forester, 1984, p. 77; Siebert et al., 1984, p. 170). The effective shale:sandstone ratio could be greater only if deeper shales (below about 5 km) or shales in the offshore section contributed fluid to sandstones of the onshore Frio Formation. These scenarios require significantly greater distances of either vertical or horizontal fluid transport. If dissolution occurs by reactions (8) or (9) below, then on the average, Frio Formation shales can produce enough CO_2 by organic matter decarboxylation to make a maximum of 0.2 volume percent secondary porosity by calcite dissolution or 0.7 volume percent by feldspar dissolution.

$$CaCO_3 + CO_2 + H_2O \rightarrow Ca^{++} + 2HCO_3^- \quad (8)$$

$$2KAlSi_3O_8 + 2CO_2 + 11H_2O \rightarrow Al_2Si_2O_5(OH)_4 + 2K^+ + 4H_4SiO_4 + 2HCO_3^- \quad (9)$$

In some areas where the shale:sandstone ratio is high, Frio shales could generate enough CO_2 by decarboxylation to produce about 1% secondary porosity by calcite dissolution or about 3% by feldspar dissolution (Fig. 10). Comparison of Figures 3 and 10 shows that much more secondary porosity exists in Frio sandstones than can be explained by CO_2 produced by decarboxylation of organic matter in adjacent Frio shales.

Similar mass balance calculations for various shale:sand ratios and types and abundances of organic matter indicate that in most basins CO_2 from decarboxylation of organic matter can produce a regional average of only a few percent secondary porosity (Table 1; Bjorlykke, 1984). If limited by the amount of oxygen originally contained in carboxyl groups, Type III organic matter (the most oxygen-rich type) cannot yield more than about 1.96×10^{-3} moles CO_2 (or any mono-functional organic acid) per gram of organic matter (Lundegard et al., 1984, p. 399). The same limitations must also apply to any organic acids, such as acetic or oxalic, that might be generated from carboxyl groups in kerogen, and therefore must be taken into consideration when evaluating feldspar dissolution reactions where these acids act as presumed aluminum complexers [reactions (5), (6), and (7)].

Additional Sources of Acid

Other functional groups in kerogen.—

Organic matter contains oxygen in functional groups other than carboxyl, such as ester, ether, phenol, and carbonyl groups (Robin and Rouxhet, 1978; Rouxhet et al., 1980).

TABLE 1.—Secondary porosity from decarboxylation of kerogen

Type Organic Matter	Weight % Oxygen	Percent O In-COOH Groups	Volume % Calcite Dissolved	Volume % K-Spar Dissolved
III	25	25	1.6	4.4
II	15	13	0.5	1.4
I	10	7	0.2	0.5

Note: Calculations made using reactions (8) and (9) in the text, 1 weight percent organic matter, a shale porosity of 20, and a shale/sandstone ratio of 10. Amounts of dissolution shown are maximum values for either calcite or K-feldspar and are not additive.

Oxygen in these groups is generally liberated from kerogen at higher levels of thermal evolution, presumably in the form of CO_2 and H_2O, although the fate of these other functional groups during catagenesis is poorly known. A fraction of the total initial oxygen content remains in the kerogen to advanced stages of catagenesis (Fig. 11). Judging from the slope of its mean evolution path on a van Krevelen diagram, Type III kerogen loses about two thirds of its oxygen in the form of CO_2 and one third in the form of H_2O during catagenesis (Fig. 12). If it is assumed, therefore, that Type III kerogen, initially with 25 weight percent oxygen, produces CO_2 during catagenesis from 66% of this oxygen, approximately 2.6 times as much CO_2 is produced than by decarboxylation alone. This would be equivalent to the production of 5.17×10^{-3} moles of CO_2 per gram of organic matter. Calculations made by Laplante (1974) for Gulf Coast shales indicate that this is a reasonable value. Ignoring timing problems associated with the wider range of thermal evolution over which these other functional groups are lost from kerogen, the CO_2 released from shales in the Frio Formation could produce up to 0.5% secondary porosity by calcite dissolution or 1.8% by feldspar dissolution. This is still too little CO_2 to account for the secondary porosity in sandstones of the Frio Formation (Fig. 3).

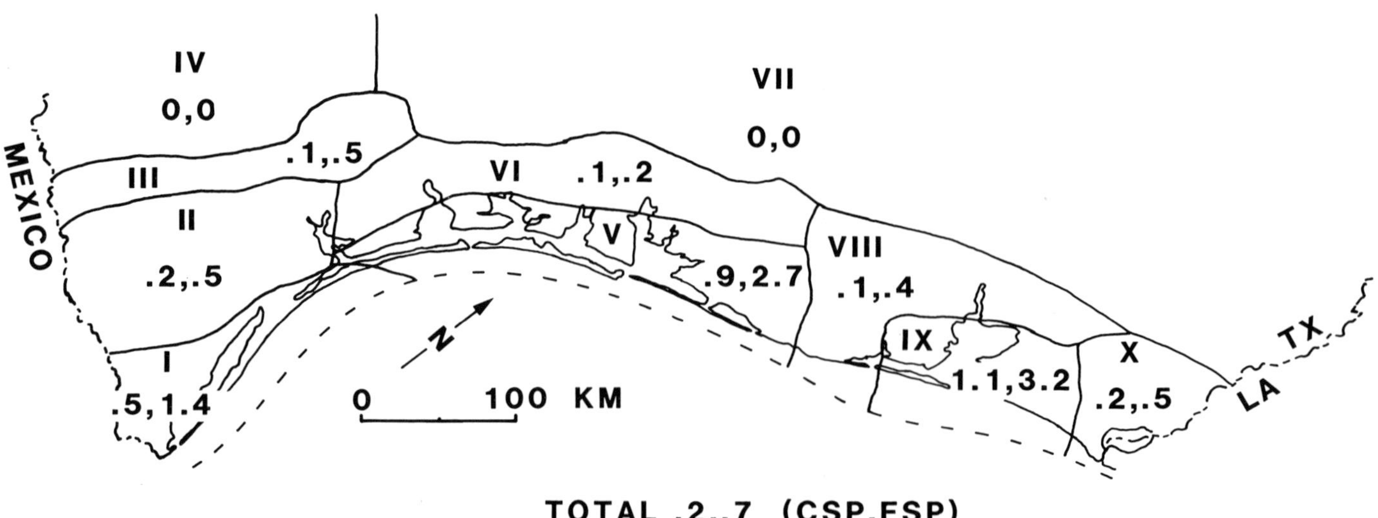

FIG. 10.—Volume percent calcite secondary porosity (CSP) or feldspar secondary porosity (FSP) in sandstones of the Frio Formation that can be produced by CO_2 from decarboxylation of organic matter in shales of the Frio Formation. Calculations based on the measured ratios of sandstone and thermally mature shale for areas I through X of Galloway et al. (1982, Table 2), reactions (8) and (9), and organic matter properties discussed in the text.

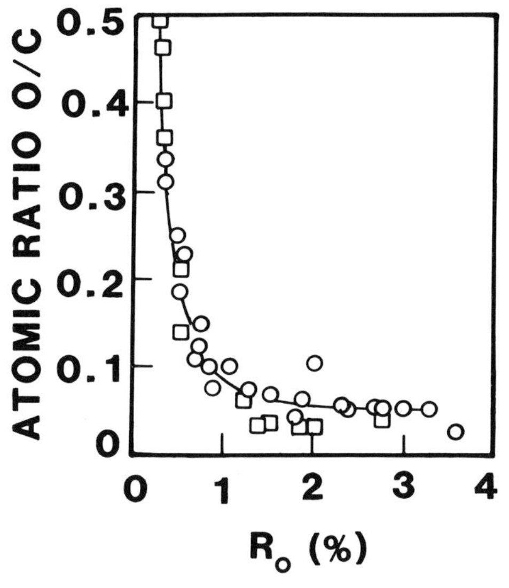

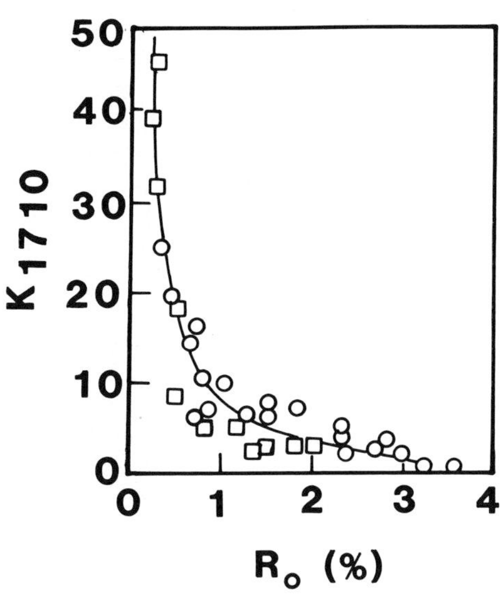

FIG. 11.—Plots of the atomic O/C ratio and the infrared adsorption peak characteristic of carboxyl and carbonyl groups in Type III kerogens (circles) and coals (squares) as a function of vitrinite reflectance. Modified from Rouxhet et al. (1980, Fig. 6.7).

Reverse Weathering Reactions

Reverse weathering reactions [equation (10)] are a possible means by

$$\text{silica} + \text{clay} + \text{cations} \rightarrow \text{Al-silicate} + \text{water} + H^+ \quad (10)$$

which additional acid could be generated during burial diagenesis (Mackenzie and Garrels, 1966). In particular, reactions which produce chlorite as a new alumino-silicate mineral will also produce protons because of the abundance of hydroxyl groups in the brucite layers of chlorite. In one of the more detailed studies of shale diagenesis, Hower et al. (1976) showed that in one Gulf Coast well the percent chlorite increased with depth at the expense of kaolinite. Based on these data, Boles and Franks (1979) suggested an acid-releasing reaction between these minerals. It should be pointed out that reverse weathering reactions require that shales be significant sinks for cations rather than sources as is commonly believed (Boles and Franks, 1979). Also, to assess the net acid release (or consumption), the effect of *all* the reactions in the shales must be weighed. This requires accurate chemical and mineralogical data. Unfortunately, such data are limited to the wells studied by Hower et al. (1976), so it is presently impossible to determine if regionally consistent patterns exist. Using the data of Hower et al. (1976), T. Jackson (pers. commun., 1984) found that the bulk shale system actually *consumed* protons during burial diagenesis. Thus, there is presently no evidence to suggest that reverse weathering reactions characterize clay diagenesis in the Gulf Coast.

Mineral Oxidants and Hydrous Pyrolysis Reactions

As has been mentioned, the abundance and distribution of oxygen functional groups in organic matter place clear limitations on the amount of CO_2 producible by simple elimination reactions. If carbon in organic matter can react with an external oxygen source, however, much more CO_2 can be produced than if the initial oxygen content of the kerogen is limiting. In the reducing environment of the deep subsurface, it is unclear what this source could be, although

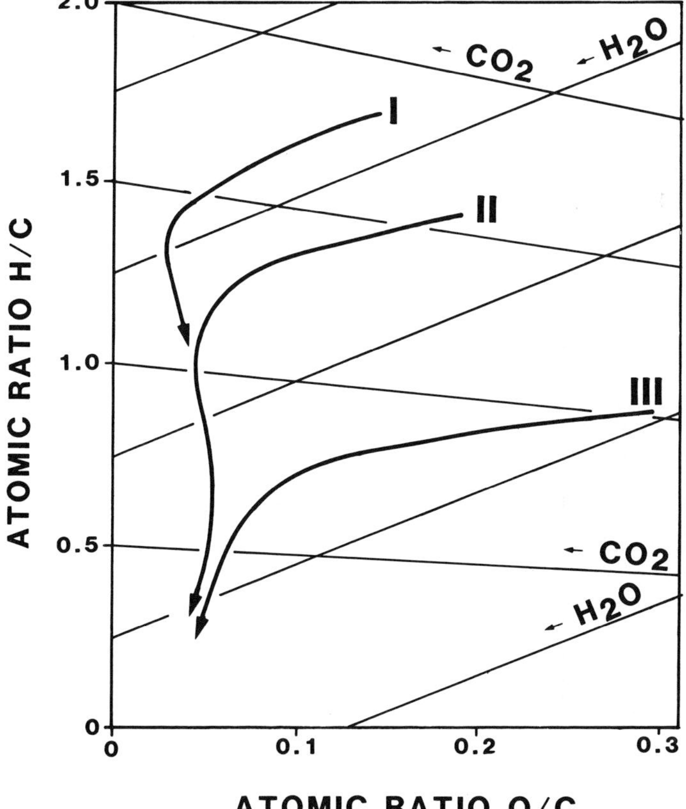

FIG. 12.—A van Krevelen diagram showing the mean evolution paths of Types I, II, and III kerogen, and evolution paths for the loss of CO_2 and H_2O. Note that Type III kerogen, the most abundant type in Gulf Coast Tertiary shales, has an evolution path that roughly bisects those for CO_2 and H_2O loss, indicating that approximately twice as much oxygen is lost in the form of CO_2. Modified from Tissot and Welte (1978, Fig. 11.5.1)

oxygen in H_2O is a possibility (Lundegard et al., 1984). There is presently no direct evidence that such "hydrous pyrolysis" reactions occur, although Hoering (1982) has demonstrated extensive exchange of hydrogen between organic matter and H_2O during simulated maturation experiments. Mineral oxidants such as ferric iron in clays or in ferric oxides have been suggested to have the potential to oxidize organic carbon to CO_2 (Eslinger et al., 1979; Surdam et al., 1984). Again, an external oxygen source must be available, as shown below.

$$CH_2O + 4Fe^{+++} + H_2O \rightarrow CO_2 + 4Fe^{++} + 4H^+ \quad (11)$$

Note also that according to reaction (11), 1 mole of protons is produced for every mole of iron reduced. Reactions of this type certainly have great potential for causing dissolution. A major uncertainty, however, is how much ferric iron is actually present in the detrital and authigenic clays and therefore how much was available to react with organic matter. Full evaluation of the role of mineral oxidants in acid generation must await the acquisition of regional data on iron content and its oxidation state in shales.

Sulfate or sulfur reduction is another possible mechanism of kerogen oxidation [reaction (12)]. Jenden (1983, p. 146), however, did not find any significant increase in the percent pyrite in shales of the Wilcox Formation in south Texas, as would be expected if sulfate reduction were active.

$$SO_4^{-2} + 2CH_2O + H^+ \rightarrow HS^- + 2H_2O + 2CO_2 \quad (12)$$

This, and the generally low concentrations of dissolved sulfate, sulfide, and H_2S in formation waters from the Gulf Coast Tertiary (except directly associated with salt domes; Morton et al., 1981; Fisher, 1982; Lundegard, 1985; Kharaka et al., 1977) suggest that sulfate reduction is not a volumetrically important reaction.

Relative Timing

The relative timing of acid-producing reactions and secondary porosity development is an important consideration (Lundegard et al., 1984, p. 401). In an actively compacting basin like the Gulf Coast, where the dominant fluid flow is up and out of the basin, a particular sandstone is most likely to have its porosity enhanced by acid production occurring downdip. Studies of the effect of catagenesis on oxygen-containing functional groups in kerogen show that carboxyl groups are the oxygen functions most easily lost (Rouxhet et al, 1980; Robin and Rouxhet, 1978; van Krevelen and Schuyer, 1957). Most of the oxygen and especially the carboxyl oxygen is lost by a vitrinite reflectance of 0.5% R_0 (Fig. 11). In the case of the Paleogene sandstones most kerogen decarboxylation probably occurred prior to major secondary porosity development in the sandstones and therefore could not have been of prime importance in producing the secondary porosity (Lundegard et al., 1984, p. 401).

Role of Meteoric Water

While in some basins an influx of meteoric water might have been important in causing dissolution and secondary porosity (Bjorlykke, 1984), this is highly improbable in the Paleogene of the Gulf Coast. Large volumes of meteoric water would be necessary to create the observed secondary porosity. Dissolution of 5 volume percent calcite according to reaction (8) would require approximately 1 liter of meteoric water equilibrated with a soil zone CO_2 pressure of $10^{-0.01}$ (mole fraction CO_2 = 0.97, 1 atm total pressure) per cc of rock. Such a flux of meteoric water (but at a lower CO_2 pressure) is not unreasonable for fully compacted basins with elevated recharge areas, such as the Alberta basin (Toth, 1978). The Gulf Coast basin is an actively compacting basin, however, where highly elevated recharge areas do not exist. Many of the sandstone reservoirs with abundant secondary porosity are currently geopressured, and the depth to geopressure was almost certainly shallower in the past. It is very unlikely that meteoric water can enter geopressured reservoirs against the hydraulic gradient. Oxygen isotopic data on the formation waters from below the top of geopressure indicate no mixing with meteoric water (Fig. 13).

Above the top of geopressure, in the hydropressured zone, oxygen isotopic data show that meteoric water has invaded the sandstones of the Wilcox Formation to depths of 3 km in some places (Fig. 13). Data on the spatial distribution of cements and their oxygen isotopic composition, however, indicate that cementation (and by analogy dissolution) both above and below present-day geopressure is similar and did not take place in the presence of meteoric water (Fisher, 1982; Land and Fisher, in prep.).

Long Distance Material Transport

Until this point the possibility of long distance, cross-formational fluid transport has been ignored in trying to account for the observed amount of secondary porosity in Paleogene sandstones of the Texas Gulf Coast. If the amount of CO_2 generated by decarboxylation of organic matter in shales is insufficient to produce the secondary porosity in the sandstones, what might be the role of acid-charged fluids moving upward from deeper in the basin? Recalling that sandstones of the Wilcox Formation, which partly underlie the Frio Formation sandstones, also have abundant secondary porosity (Figs. 1 and 3), it seems likely that such fluids must be derived from rocks older than the Wilcox so as to have affected both units. In fact, strontium isotopic data on carbonate cements in both the Wilcox and Frio sandstones indicate that fluids derived from the Mesozoic section have affected these units.

The $^{87}Sr/^{86}Sr$ ratio in seawater is well known through the Cenozoic (Burke et al., 1982) and increases more or less continuously from Late Jurassic to the present. This relationship can be used to help identify the source of diagenetic fluids. The $^{87}Sr/^{86}Sr$ ratio of carbonate cements in sandstones of the Frio and Wilcox Formations is compared with the strontium isotope age curve of Burke et al. (1982) in Figure 14. Note that some of the cements have isotopic ratios that intersect the age curve in the range expected for seawater of Frio and Wilcox age. The cements in these samples could either have precipitated from coeval seawater (a scenario that is incompatible with the oxygen iso-

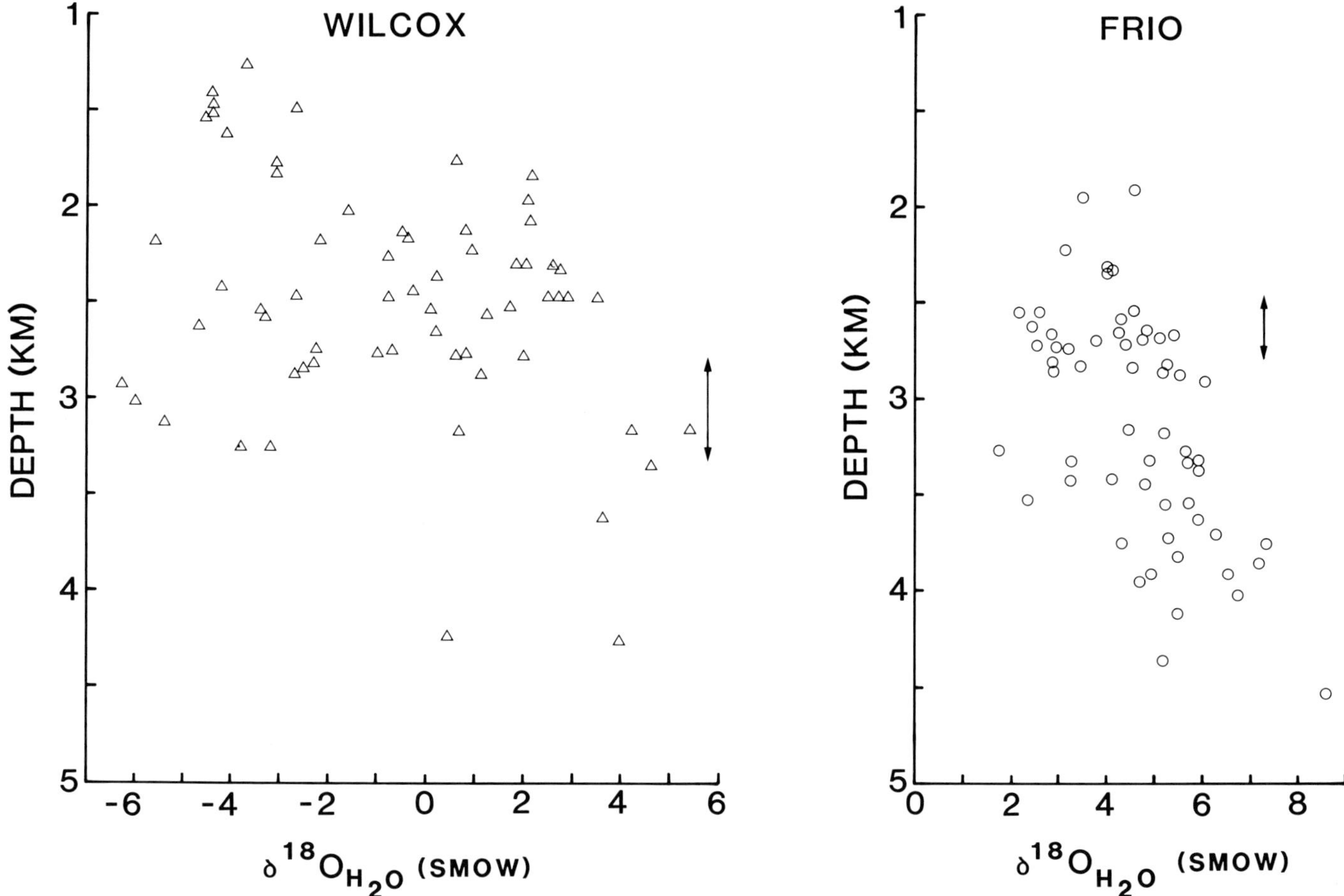

FIG. 13.—Oxygen isotopic composition of formation water from the Wilcox and Frio Formations as a function of depth. Approximate depth of the top of geopressure is shown by the arrows. Note that most of the Wilcox waters are more negative than the Frio waters, and are from above the top of geopressure. Mixing with meteoric water is the probable explanation. Frio data show no evidence of a meteoric component, however. Data from this study, Fisher (1982), and Land and Fisher (in prep.).

topic composition of the cements; Milliken, 1981; Fisher, 1982) or have been derived by remobilization of carbonate that was precipitated from coeval seawater, such as nannofossil calcite in shales. The latter scenario is compatible with the decreasing amount of calcite in shales in some wells (Fig. 15). Of special interest are the samples for which the strontium isotopic ratios are less radiogenic than expected for seawater of either Frio or Wilcox age. Two diagenetic processes might be expected to modify the $^{87}Sr/^{86}Sr$ of buried seawater. These are the transition of smectite to mixed-layer smectite/illite and the dissolution of feldspars. The former process *consumes* small amounts of strontium, according to Perry and Turekian (1974), and therefore will have only a minor impact on the isotopic ratio of pore fluids. Feldspars so far analyzed in the Texas Gulf Coast have $^{87}Sr/^{86}Sr$ ratios greater than 0.712 (Land and Fisher, in prep.), reflecting their predominantly plutonic and metamorphic source, so their dissolution could only result in *more* radiogenic ratios in the carbonate cements. Basaltic plagioclase with non-radiogenic "mantle" ratios (~0.704) has yet to be found in the Tertiary of the Gulf Coast. Barring dissolution of significant amounts of basaltic plagioclase, the only reasonable explanation for the $^{87}Sr/^{86}Sr$ ratios between 0.7068 and 0.7074 is that the carbonate for these cements was precipitated from pore water buffered by Sr-rich Mesozoic carbonate rocks. Because of the minimum in the isotopic ratio curve in the Mesozoic, it is not possible to determine whether these rocks were Cretaceous, Jurassic, or both. But these data do suggest that long distance, cross-formational fluid flow has occurred in the Gulf Coast Basin.

If fluids derived from the Mesozoic section are acidic or undersaturated with respect to calcite or feldspar for any reason, then their passage through the Tertiary sandstones could indeed cause additional dissolution porosity. If CO_2 is the acid involved, however, a major limitation to the amount of dissolution caused by cross-formational fluid flow exists, as pointed out earlier in the discussion of CO_2 solubility. Presumably any water in contact with the Mesozoic carbonates would be approximately saturated with respect to calcite. As this water moves upward (down the pressure gradient), it will exsolve CO_2, causing the pH to rise, thus favoring carbonate precipitation, not dissolution. This is, in fact, what the strontium isotopic data indicate has occurred. Carbonate dissolution is likely to occur only if a non-volatile acid is added to the water along the way. Pore water derived from Mesozoic rocks could be undersaturated

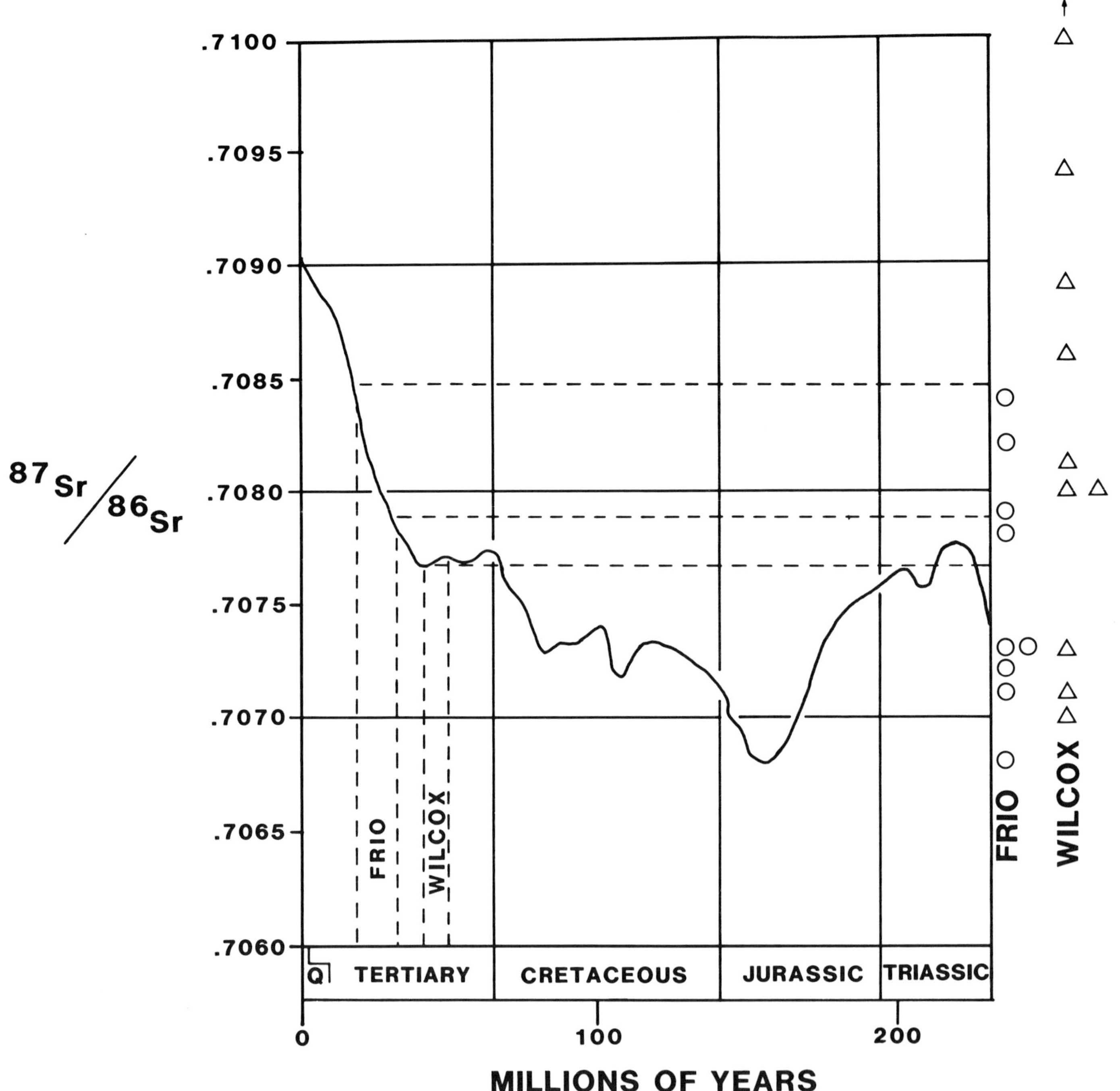

FIG. 14.—Strontium isotopic ratio for seawater from Burke et al. (1982). On the right are shown analyses of the strontium isotopic composition of carbonate cements in sandstones of the Frio and Wilcox Formations. The dashed lines indicate the range of values expected for carbonate precipitated from seawater of Wilcox and Frio age. Compositions lower than these ranges probably represent contributions from older rocks. After Land and Fisher (in prep.).

with respect to feldspar, however, since the Mesozoic section is in general silicate-poor. Rapid fluid flow along faults, where the rock:water ratio is low, could result in water reaching Tertiary sandstones still undersaturated with respect to feldspars.

The relationship of organic acid concentration to reservoir temperature indicates that the zone of maximum organic acid generation is between 80 and 120°C (Fig. 9; Carothers and Kharaka, 1978, Fig. 2). Organic acid concentration decreases rapidly with increasing temperature, presumably as a result of thermal decomposition (Kharaka et al., 1983). Therefore, water moving upward from depths where temperatures are much greater than 120°C will contain low concentrations of organic acids. As a result, fluids derived from deep in the basin will be inefficient at causing dissolution in sandstone reservoirs far updip. In

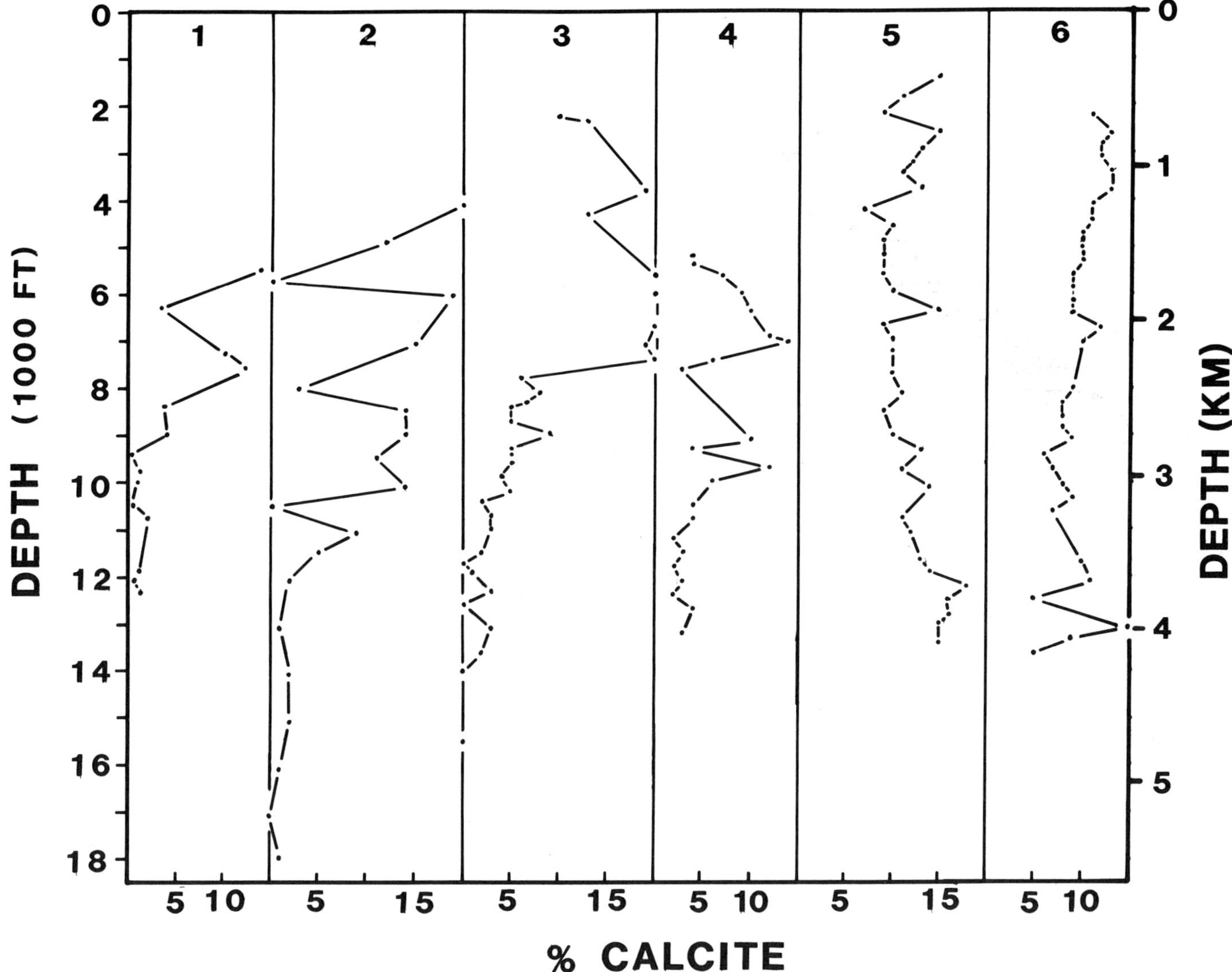

FIG. 15.—Weight percent calcite in Gulf Coast shales versus depth. Note that wells 1 through 4 show a decrease in the percent calcite between approximately 2.5 and 3.5 km. Wells 5 and 6, however, show no such trend. Location and source of data: 1—middle Texas coast, Wilcox Formation, Milliken and Land (1982); 2—upper Texas coast, Frio Formation to Miocene, Hower et al. (1976); 3—upper Texas coast, Frio Formation to Miocene, Freed (1981); 4—upper Texas coast, Frio Formation to Miocene, Freed (1981); 5 and 6—lower Texas coast, Vicksburg Formation to Miocene, Freed (1981).

order for organic acids to be efficient at causing dissolution in sandstones, they must be generated locally and escape neutralization in transit from shales to sandstones. Organic acid generation in carbonate-free shales will greatly increase the likelihood of the latter requirement.

CARBONATE CEMENTATION

Sandstones of the Frio Formation average 8 volume percent carbonate cement, which is predominantly calcite (Lindquist, 1976; Loucks et al., 1984). Sandstones of the Wilcox Formation average 5 volume percent carbonate cement, which is predominantly ankerite (Stanton, 1977; Loucks et al., 1984; Fisher, 1982). The petrography and the isotopic and trace element chemistry of these cements have been discussed in detail by various authors (Lindquist, 1976; Stanton, 1977; Boles and Franks, 1979; Loucks et al., 1984; Milliken, 1981; Fisher, 1982; Land, 1984; Land and Fisher, in prep.). Petrographic observations of sandstones of both the Frio and Wilcox Formations indicate that the bulk of the calcite cement precipitated after quartz cementation and before major secondary porosity development, whereas most of the ankerite and some of the calcite (ferroan) post-date dissolution.

Strontium isotopic ratios of calcite cements in both Frio and Wilcox sandstones indicate that some were derived from carbonate in coeval shales and others from underlying units (Fig. 14; Land, 1984; Land and Fisher, in prep.). Ankerite cement in the Wilcox is highly radiogenic, indicating that it probably contains strontium derived from the dissolution of feldspars. Temperatures inferred from oxygen isotopic data suggest emplacement of most of the calcite between

80 and 100°C in both formations, and ankerite emplacement in the Wilcox sandstones between 95 and 120°C (Land, 1984; Fisher, 1982; Land and Fisher, in prep.). The carbon isotopic composition of the cements averages −4.1 ‰ and −6.3 ‰ for the Frio and Wilcox Formations, respectively (Lundegard et al., 1984, Fig. 3). These values indicate some contribution of ^{13}C-depleted carbon, of probable organic derivation, but that little carbon with an isotopic composition as negative as −25 ‰ affected these cements (Lundegard et al., 1984; Lundegard, 1985). Carbon sources with an isotopic composition near zero per mil, such as skeletal carbon, must have dominated the isotopic composition of the cements in the sandstones.

Strontium isotopic data indicate that some of the carbonate cements were derived from coeval or older formations, and are supported by the observation that the carbonate content of some Gulf Coast shales decreases with increasing depth (Fig. 15). Mobilization of carbonate from shales to sandstones requires acid. For example, in order to mobilize sufficient carbonate to account for an averge of 8 volume percent calcite cement in the sandstones of the Frio Formation, approximately 6.91×10^{16} moles of CO_2 are required according to reaction (8). This is slightly more than the acid required to produce the secondary porosity in the sandstones by calcite and feldspar dissolution, a process which apparently post-dates most of the carbonate cementation. If the carbonate is mobilized by either CO_2 or acetic acid produced from carboxyl groups in organic matter, then a shale:sandstone ratio of approximately 86:1 is necessary. If two-thirds of all the oxygen in the organic matter is liberated as one of these acids (as discussed earlier), then a shale:sandstone ratio of approximately 33:1 is necessary. Because the shale:sandstone ratio in the onshore Frio Formation is only 4:4, the acid imbalance is clearly severe. The role of organic acids in mobilizing carbonate is probably greatest from 80 to 120°C, where their concentrations are greatest (Fig. 9; Carothers and Kharaka, 1978, Fig. 2).

It has been suggested that carbonate precipitation may occur when CO_2 is added to waters with high organic alkalinities, because the pH might be buffered by organic acid anions (Surdam et al., 1984). This hypothesis was tested using the computer algorithm PHREEQE (Parkhurst et al. 1982), to which thermodydamic data for acetic acid, and acetate complexes with sodium and calcium were added. A 0.5 m sodium chloride solution with various concentrations of acetic acid was equilibrated with calcite at 100°C and a specific pressure of carbon dioxide. Carbon dioxide pressure was then increased one order of magnitude, while holding acetate plus acetic acid constant but allowing the concentrations of acetic acid, acetate, and pH to vary. For concentrations of acetic acid as much as about 0.06 m, increased CO_2 pressure resulted in calcite undersaturation (Table 2). For concentrations of acetic acid greater than 0.06 m, increased CO_2 pressure resulted in calcite oversaturation. The highest organic acid concentration observed by us or by Carothers and Kharaka (1978, Fig. 2) in Gulf Coast formation water is approximately 2500 mg/1 acetate, or about 0.04 m. It may be argued that concentrations of organic acids in formation waters of Paleogene rocks may have been higher in the geologic past. We have also analyzed 30 formation waters from Plio-Pleistocene and Miocene reservoirs, however, and found that for equivalent temperatures the organic acid concentrations are no higher than in water from Paleogene reservoirs. Therefore, based on currently available data, we conclude that in the Gulf Coast Basin organic acid concentrations were never high enough to buffer the pH to the degree necessary for increased CO_2 pressure to cause calcite precipitation. In basins where much higher organic acid concentrations have been observed (Carothers and Kharaka, 1978), more efficient pH buffering by organic acid anions might to be expected.

TABLE 2.—Calcite equilibria and ph buffering by acetate

Σ HAc + Ac$^-$ (moles/1)	Calcite dissolved (+), precipitated (−)* (mmoles/1)
0.00	+1.83
0.02	+1.15
0.04	+0.55
0.06	−0.03
0.08	−0.64
0.10	−1.29

* For a one order of magnitude increase in the partial pressure of carbon dioxide (10^{-1} to 10^0), 100°C, and 0.5 m NaCl. See text for discussion.

The decreased solubility of CO_2 in upward-migrating formation water that occurs as a result of decreased pressure is a more reasonable explanation for carbonate cementation. This process can be modelled using the curves in Figure 8. Taking the calculated concentration of $CO_2(aq)$ in Wilcox Formation water at 175°C (6682 ppm), one may ask how CO_2 solubility will change if this water moves upward to 100°C. If the water simply equilibrates with the lower temperature and pressure of CO_2, the solubility of $CO_2(aq)$ will decrease by approximately 5701 ppm. This could cause the precipitation of up to 4.8 cc of calcite per liter of water. While oversimplified, this exercise illustrates the tremendous potential upward-migrating formation waters can have for precipitating carbonate minerals. Cements with strontium isotopic compositions indicating either local derivation or derivation from much deeper, older rocks could clearly form in this way.

CONCLUSIONS

The role of CO_2 produced by decarboxylation of organic matter in secondary porosity development has been over estimated in past studies. In typical sedimentary basins it may account for regional averages of a few volume percent secondary porosity but is insufficient to account for all of the secondary porosity in Paleogene sandstones of the Texas Gulf Coast. Petrographic data indicate dissolution of a minimum of 5 volume percent feldspar from the sandstones of the Wilcox and Frio Formations. The acid requirements for feldspar dissolution depend on feldspar mineralogy and the fate of aluminum. In these two formations feldspar dissolution probably required at least 1 mole of protons for every mole of feldspar dissolved.

Considerations of CO_2 solubility and the distribution of organic acids with respect to temperature suggest that deep, distant sources of these species are not likely to be of great importance in porosity enhancement. In pore waters migrating upward (down the pressure gradient), the solubility of CO_2 decreases rapidly, favoring carbonate precipitation, not dissolution. Dissolved organic acids, which could cause both carbonate and silicate dissolution, decrease rapidly in concentration at temperatures above 120°C. Consequently, pore waters migrating upward from temperatures much greater than this will contain few organic acids and be inefficient at causing dissolution.

Other potential sources of acid for dissolution reactions have been identified and are worthy of further study. These include hydrous pyrolysis reactions between organic carbon, and oxygen in H_2O to yield additional CO_2 (perhaps enhanced by mineral oxidants, Surdam et al., 1984), and reverse weathering reactions in shales. Any consideration of acid-producing reactions should involve a careful evaluation of the relative timing of events and basin hydrology.

Carbonate dissolution, a process of debatable importance in the sandstones, was probably of major significance in many Gulf Coast shales. This is indicated by the strontium isotopic composition of carbonate cements in sandstones and the decreasing weight percent calcite in shales in some wells. Strontium isotopic data indicate that some carbonate cement in the sandstones was derived from coeval shales and some was derived from older, deeper rocks. The acid required to mobilize carbonate from shales to sandstones for cementation is of the same order of magnitude as, and at least partly independent from, that required to form secondary porosity. This fact greatly worsens the problem of accounting for sufficient acid for dissolution. The major control on carbonate cementation was probably the lowered solubility of CO_2 in upward-migrating formation water. Carbonate precipitation as a result of the addition of CO_2 to acetate-buffered pore water (Surdam et al., 1984 p. 142) is an unlikely mechanism for the origin of carbonate cements in the Gulf Coast Basin.

ACKNOWLEDGMENTS

This study was funded by National Science Foundation Grant EAR-8121009 and by the Geology Foundation of the University of Texas at Austin. We thank numerous oil and gas companies for permission to sample and for assistance in collecting water and gas. The Bureau of Economic Geology, University of Texas at Austin, generously allowed us to study their unpublished point count data. Wendy McPherson prepared the rough draft of Figure 1. We thank the following people for their helpful comments on early drafts of the manuscript: P. Blanchard, W. E. Galloway, P. B. Gold, R. L. Kugler, K. L. Milliken, E. F. McBride, P. L. Parker, and J. M. Wolff.

REFERENCES

American Public Health Association, 1976, Standard methods for the examination of water and wastewater, 14th ed.: American Public Health Association, Washington, D.C., 1193 p.

Baker, W. E., 1978, The role of humic acid in the transport of gold: Geochimica et Cosmochimica Acta, v. 42, p. 645–649.

Barnes, H. L., and Bourcier, W. L., 1983, Role of organic and inorganic complexes in ore metal transport at low temperatures (abs.): Geological Society of America Abstracts with Programs, v. 15, p. 521.

Bebout, D. G., and Loucks, R. G., 1974, Stuart City trend lower Cretaceous, South Texas: a carbonate shelf-margin model for hydrocarbon exploration: The University of Texas at Austin, Bureau of Economic Geology, Report of Investigations No. 78, 80 p.

Bjorlykke, K., 1984, Formation of secondary porosity: How important is it?: American Association of Petroleum Geologists Memoir 37, p. 277–286.

Blatt, H., 1979, Diagenetic processes in sandstone: Society of Economic Paleontologists and Mineralogists Special Publication 26, p. 141–157.

Boles, J. R., and Franks, S. G., 1979, Clay diagenesis in Wilcox sandstones of southwest Texas: Implications of smectite diagenesis on sandstone cementation: Journal of Sedimentary Petrology, v. 49, p. 55–70.

Burke, W. H., Denison, R. E., Hetherington, E. A., Koepnick, R. B., Nelson, H. F., and Otto, J. B., 1982, Variation of seawater $^{87}Sr/^{86}Sr$ throughout Phanerozoic time: Geology, v. 10, p. 516–519.

Carothers, W. W., AND Kharaka, Y. K., 1978, Aliphatic acid anions in oilfield waters: Implications for origin of natural gas: American Association of Petroleum Geologists Bulletin, v. 62, p. 2441–2453.

———, and ———, 1980, Stable carbon isotopes of HCO_3^- in oil-field waters-Implications for the origin of CO_2: Geochimica et Cosmochimica Acta, v. 44, p. 323–332.

Curtis, C. D., 1978, Possible links between sandstone diagenesis and depth-related geochemical reactions occurring in enclosing mudstones: Geological Society of London Journal, v. 135, p. 107–117.

Deines, P., and Langmuir, D., 1975, Stable carbon isotope ratios and the existence of a gas phase in the evolution of carbonate ground waters: Geochimica et Cosmochimica Acta, v. 38, p. 1147–1164.

Ellis, A. J., and Golding, R. M., 1963, The solubility of carbon dioxide above 100°C in water and in sodium chloride solutions: American Journal of Science, v. 261, p. 47–60.

Eslinger, E., Highsmith, P., Albers, D., and DeMayo, B., 1979, Role of iron reduction in the conversion of smectite to illite in bentonites in the Disturbed Belt, Montana: Clays and Clay Minerals, v. 27, p. 327–338.

Fisher, R. S., 1982, Diagenetic history of Eocene Wilcox sandstones and associated formation waters, south-central Texas: Ph.D. Dissertation, The University of Texas at Austin, Austin, Texas, 185 p.

Fournier, R. O., and Rowe, J. J., 1966, Estimation of underground temperatures from the silica content of water from hot springs and wet-steam wells: American Journal of Science, v. 264, p. 685–697.

———, R. O., and Truesdell, A. H., 1973, An empirical Na-K-Ca geothermometer for natural waters: Geochimica et Cosmochimica Acta, v. 37, p. 1255–1275.

Franks, S. G., and Forester, R. W., 1984, Relationships among secondary porosity, pore-fluid chemistry and carbon dioxide, Texas Gulf Coast: American Association of Petroleum Geologists Memoir 37, p. 63–79.

Freed, R. L., 1981, Shale mineralogy and burial diagenesis of Frio and Vicksburg Formations in two geopressured wells, McAllen Ranch area, Hidalgo County, Texas: Gulf Coast Association of Geological Societies Transactions, v. 31, p. 289–294.

Galloway, W. E., Hobday, D. K., and Magara, K., 1982, Frio Formationof Texas Gulf Coastal Plain: Depositional systems, structural framework, and hydrocarbon distribution: American Association of Petroleum Geologists Bulletin, v. 66, p. 649–688.

Garrels, R. M., and MacKenzie, F. T., 1967, Origin on the chemical composition of some springs and lakes, *in* Stumm, W., ed., Equilibrium Concepts in Natural Water Systems: American Chemical Society, Advances in Chemistry Series 67, p. 222–242.

Hanor, J. S., 1978, Precipitation of beachrock cements: Mixing of marine and meteoric waters vs. CO_2 degassing: Journal of Sedimentary Petrology, v. 48, p. 489–502.

Harwood, R. J., 1977, Oil and gas generation by laboratory pyrolysis of kerogen: American Association of Petroleum Geologists Bulletin, v. 61, p. 2082–2102.

Hem, J. D., and Roberson, C. E., 1967, Formation and stability of aluminum hydroxide complexes: U. S. Geological Survey Water-Supply Paper 1827-A, 55 p.

Hoering, T. C., 1982, Thermal reactions of kerogen with added water, heavy water, and pure organic substances: Carnegie Institution, Director of Geophysical Laboratory Annual Report, 1981–1982, p. 397–402.

HOLLAND, H. D., KIRSIPU, T. V., HUEBNER, J. S., AND OXBURGH, U. M., 1962, On some aspects on the chemical evolution of cave waters: Journal of Geology, v. 72, p. 36–37.

HOWER, J., ESLINGER, E. V., HOWER, M. E., AND PERRY, E. A., 1976, Mechanism of burial metamorphism of argillaceous sediment: 1. Mineralogical and chemical evidence: Geological Society of America Bulletin, v. 87, p. 725–737.

HUNT, J. M., 1979, Petroleum geochemistry and geology: W. H. Freeman and Co., San Francisco, 617 p.

JENDEN, P. D., 1983, Maturation of organic matter in the Paleocene-Eocene Wilcox Group, south Texas: Relationship to clay diagenesis and sandstone cementation: Ph.D. Dissertation, University of California, Los Angeles, California, 274 p.

JOHNS, W. D., 1982, The role of the clay mineral matrix in petroleum generation during burial diagenesis, *in* Van Olphen, H., and Veniale, F., eds., International Clay Conference 1981: Developments in Sedimentology 35: Elsevier, New York, p. 655–664.

JONES, P. H., 1969, Hydrology of Neogene deposits in the northern Gulf of Mexico Basin: Louisiana Water Resources Research Institute, Bulletin No. GT-2, 105 p.

KHARAKA, Y. K., AND BARNES, I, 1973, SOLMNEQ: Solution-mineral equilibrium computations: U. S. Geological Survey Computer Contributions, National Technology Information Report PB-215-899, 82 p.

———, Y. K., CALLENDER, E., AND WALLACE, R. H., Jr., 1977, Geochemistry of geopressured geothermal waters from the Frio Clay in the Gulf Coast region of Texas: Geology, v. 5, p. 241–244.

———, Y. K., CAROTHERS, W. W., AND ROSENBAUER, R. J., 1983, Thermal decarboxylation of acetic acid: Implications for origin of natural gas: Geochimica et Cosmochimica Acta, v. 47, p. 397–402.

———, Y. K., LICO, M. S., WRIGHT, V. A., AND CAROTHERS, W. W., 1979, Geochemistry of formation waters from Pleasant Bayou No. 2 well and adjacent areas in coastal Texas, *in* Dorfman, M. H., and Fisher, W. L., eds., Proceedings: Fourth United States Gulf Coast Geopressured-Geothermal Energy Conference: Research and Development: The University of Texas at Austin, Austin, Texas, p. 168–193.

LAND, L. S., 1984, Frio sandstone diagenesis, Texas Gulf Coast: A regional isotopic study: American Association of Petroleum Geologists Memoir 37, p. 47–62.

LAPLANTE, R. E., 1974, Hydrocarbon generation in Gulf Coast Tertiary sediments: American Association of Petroleum Geologists Bulletin, v. 58, p. 1281–1289.

LIND, C. J., AND HEM, J. D., 1975, Effects of organic complexation on chemical reaction of aluminum: U. S. Geological Survey Water-Supply Paper 1827-G, 83 p.

LINDQUIST, S. J., 1976, Sandstone diagenesis and reservoir quality, Frio Formation (Oligocene), south Texas: M.A. Thesis, The University of Texas at Austin, Austin, Texas, 147 p.

LOUCKS, R. G., DODGE, M. M., AND GALLOWAY, W. E., 1984, Regional controls on diagenesis and reservoir quality in Lower Tertiary sandstones along the Texas Gulf Coast: American Association of Petroleum Geologists Memoir 37, p. 15–45.

LUNDEGARD, P. D., 1985, Carbon dioxide and organic acids: Origin and role in diagenesis (Texas Gulf Coast Tertiary): Ph.D. Dissertation, The University of Texas at Austin, Austin, Texas, 145 p.

———, Land, L. S., AND Galloway, W. E., 1984, Problem of secondary porosity: Frio Formation (Oligocene), Texas Gulf Coast: Geology, v. 12, p. 399–402.

MACKENZIE, F. T., AND GARRELS, R. M., 1966, Silica-bicarbonate balance in the ocean and early diagenesis: Journal of Sedimentary Petrology, v. 36, p. 1075–1084.

MICHELS, D. E., 1981, CO_2 and carbonate chemistry applied to geothermal engineering: Lawrence Berkeley Laboratory, Geothermal Reservoir Engineering Management Program No. 15, UC-66a, 27 p.

MILLIKEN, K. L., AND LAND, L. S., 1982, Fluid dynamics for cap-rock formation in Gulf Coast: Discussion: American Association of Petroleum Geologists Bulletin, v. 66, p. 2685–2687.

———, ———, and LOUCKS, R. G., 1981, History of burial diagenesis determined from isotopic geochemistry, Frio Formation, Brazoria County, Texas: American Association of Petroleum Geologists Bulletin, v. 65, p. 1397–1413.

MONCURE, G. K., LAHANN, R. W., AND SIEBERT, R. M., 1984, Origin of secondary porosity and cement distribution in a sandstone/shale sequence from the Frio Formation (Oligocene): American Association of Petroleum Geologists Memoir 37, p. 151–161.

MOORE, B. J., 1976, Analyses of natural gases, 1917–74: U. S. Department of Interior, Bureau of Mines, Computer Printout 1–76, 889 p.

MOREY, G. W., FOURNIER, R. O., AND ROWE, J. J., 1962, The solubility of quartz in water in the temperature interval from 25° to 300°C: Geochimica et Cosmochimica Acta, v. 26, p. 1029–1043.

MORTON R. A., GARRET, C. M., Jr., POSEY, J. S., HAN, J. H., AND JIRIK, L. A., 1981, Salinity variations and chemical compositions of waters in the Frio Formation, Texas Gulf Coast: The University of Texas at Austin, Bureau of Economic Geology, Report to the Department of Energy under Contract no. DE-AC08-79ET27111, 96 p.

MURRAY, G. E., 1960, Geologic framework of Gulf Coastal Province of the United States, *in* Shepard, F. P., Phleger, F. B., van Andel, T. H., eds., Recent Sediments, Northwestern Gulf of Mexico: American Association of Petroleum Geologists Publishing Co., p. 5–33.

ONG, H. L., SWANSON, V. E., AND BISQUE, R. E., 1970, Natural organic acids as agents of chemical weathering: U. S. Geological Survey Professional Paper 700C, C130–137.

PARKHURST, D. L., THORSTENSON, D. C., AND PLUMMER, L. N., 1982, PHREEQE - A computer program for geochemical calculations: U. S. Geological Survey Water Resources Investigations 80–96, 210 p.

PERRY, E. A., Jr., AND TUREKIAN, K. K., 1974, The effects of diagenesis on the redistribution of strontium isotopes in shales: Geochimica et Cosmochimica Acta, v. 38, p. 929–935.

PLUMMER, L. N., VACHER, H. L., MACKENZIE, F. T., BRICKERr, O. P., AND LAND, L. S., 1976, Hydrogeochemistry of Bermuda - a case history of ground water diagenesis of biocalcarenites: Geological Society of America Bulletin, v. 87, p. 1301–1316.

ROBIN, P. L., AND ROUXHET, P. G., 1978, Characterization of kerogens and study of their evolution by infrared spectroscopy: Carbonyl and carboxyl groups: Geochimica et Cosmochimica Acta, v. 42, p. 1341–1349.

ROUXHET, P. G., ROBIN, P. L., AND NICAISE, G., 1980, Characterization of kerogens and of their evolution by infrared spectroscopy, *in* Durand, B., ed., Kerogen: Editions Technip, Paris, p. 163–190.

SAXBY, J. D., 1976, The significance of organic matter in ore genesis, *in* Wolf, K. H., ed., Handbook of Strata-Bound and Stratiform Ore Deposits, v. 2, Geochemical Studies: Elsevier, New York, p. 111–133.

SCHMIDT, V., AND McDONALD, D. A., 1979, The role of secondary porosity in the course of sandstone diagenesis: Society of Economic Paleontologists and Mineralogists Special Publication 26, p. 175–207.

SENI, S. J., AND JACKSON, M. P. A., 1983, Evolution of salt structures, East Texas diapir province, part 2: Patterns and rates of halokinesis: American Association of Petroleum Geologists Bulletin, v. 67 p. 1245–1274.

SIEBERT, R. M., MONCURE, G. K., AND LAHANN, R. W., 1984, A theory of framework grain dissolution in sandstones: American Association of Petroleum Geologists Memoir 37, p. 163–175.

STANTON, G. D., 1977, Factors influencing porosity and permeability, Wilcox Group (Eocene), Karnes County, Texas: M.A. Thesis, The University of Texas at Austin, Austin, Texas, 156 p.

STEVENSON, F. J., 1967, Organic acids in soil, *in* McClaren, A. D., and Peterson, G. H., eds., Soil Biochemistry: Marcel Dekker, Inc., New York, p. 119–146.

SURDAM, R. C., BOESE, S. W., AND CROSSEY, L. J., 1984, The chemistry of secondary porosity: American Association of Petroleum Geologists Memoir 37, p. 127–149.

TAN, K. H., 1980, The release of silicon, aluminum, and potassium during decomposition of soil minerals by humic acid: Soil Science, v. 129, p. 5–11.

TISSOT, B. P., AND WELTE, D. H., 1978, Petroleum formation and occurrence: Springer-Verlag, Berlin, 538 p.

TOTH, J., 1978, Gravity-induced cross-formational flow of formation fluids, Red Earth region, Alberta, Canada: Analysis, patterns, and evolution: Water Resources Research, v. 14, p. 805–843.

van KREVELEN, D. W., AND SCHUYER, J., 1957, Coal Science: Elsevier, New York, 352 p.

APPLICATION OF ORGANIC/INORGANIC DIAGENESIS TO POROSITY PREDICTION

LAURA J. CROSSEY, RONALD C. SURDAM
Department of Geology/Geophysics, University of Wyoming, Laramie, Wyoming 82071

AND

RICHARD LAHANN
CONOCO Research, Ponca City, Oklahoma

ABSTRACT: Mineral oxidants are examined as possible agents for the production of water soluble organic solvents from kerogen in the course of progressive burial of sedimentary rocks. In particular, the diagenetic conversion of smectite to illite is quantitatively evaluated as an oxidation agent using data from the Texas Gulf Coast as well as experimental results.

The reduction of mineral oxidants and consequent oxidation of organic matter may be as effective a mechanism in releasing peripheral difunctional carboxylic acid groups from the kerogen as thermal degradation in the natural system. Porosity enhancement in the subsurface is often limited by the ability of the diagenetic fluids to transport aluminum. Difunctional carboxylic acids (as well as other organic solvents such as phenols) complex aluminum, effectively increasing the solubility of aluminosilicate minerals. The solubility of carbonate minerals is also increased by these organic solvents. The coincidence in time, temperature, and space of mixed-layer smectite/illite transformations with peak concentrations of organic acids in oil field brines suggests a possible mechanism for the generation of difunctional carboxylic acids. This mechanism would allow highly soluble difunctional organic acids to pass through adjacent sandstones just prior to hydrocarbon generation. These acids are ideal solvents for dissolving carbonates and/or aluminosilicates out of pores and pore throats, thereby enhancing porosity and permeability in hydrocarbon reservoirs.

INTRODUCTION

Porosity enhancement in the form of carbonate grain, carbonate cement, and aluminosilicate framework grain dissolution is the dominant type of porosity in many sandstone hydrocarbon reservoirs. As primary porosity may be, to an extent, predicted by sedimentologic trends, the explorationist must use geochemical trends to predict the occurrence and extent of porosity enhancement. The prediction of secondary porosity must take into account the reactions within and between the source and reservoir rocks in the sedimentary prism.

Past attempts to explain observed porosity enhancement with the action of carbonic acid produced from thermal decarboxylation of organic material have been largely unsuccessful and point to another mechanism (Lundegard et al., 1984). The presence of organic acids in oil field brines has been known for some time (Carothers and Kharaka, 1978), but only recently has information on the effect of these organic acids on aluminosilicate and carbonate minerals been demonstrated (Surdam et al., 1984). The elucidation of the mechanism by which porosity enhancement occurs is essential for the prediction of porosity trends.

Objective.—

In the present study, mineral oxidants are examined as possible agents for the production of difunctional organic acids from kerogenous material in the course of the progressive burial of sedimentary rocks. In particular, the diagenetic conversion of smectite to illite is quantitatively evaluated as an oxidative agent using Gulf Coast petrologic and chemical data, as well as experimental results.

The Origin and Role of Organic Acids during Diagenesis.—

Carboxylic acids have been observed in many oil field brines in significant quantity (Carothers and Kharaka, 1978). Most abundant are the monofunctional acetic and propionic acids (concentrations of as much as 10,000 ppm). Carothers and Kharaka note that, in many cases, the alkalinity of a brine is dominated by these short-chain acids. Analyses of four oil field brines from the North Coles Levee field of the southern San Joaquin Valley are presented in Table 1. These brines show very little carbonate/bicarbonate, yet have significant titratable alkalinity due to the presence of the organic acids. Thus, the pH of oil field brines is more likely to be buffered by acetate than the carbonate system. Less abundant in brines are the difunctional organic acids, such as oxalic acid (Surdam et al., 1984). Experiments on the solubility of aluminosilicate minerals (feldspars, zeolites, and clays) indicate that aluminum concentrations can be greatly increased in solutions containing difunctional organic acids (Surdam et al., 1984). The magnitude of this effect is pH-dependent with a maximum enhancement of solubility at pH = 5. Porosity enhancement in the course of the progressive burial of sediments is limited by the ability of the pore fluids to transport aluminum. In the absence of complexing agents such as organic acids, aluminum concentrations are exceedingly low (about 1 ppm) in the pH range of 4 to 8. Thus, any material dissolved from a mineral grain would be reprecipitated in the immediate vicinity of the grain, and porosity reduction rather than porosity enhancement would result. The increase in aluminum mobility provided by the presence of difunctional acids allows the mass transport of aluminosilicate, as well as carbonate, material to occur. Other organic solvents may also be active during diagenesis in organic/inorganic systems. Often a discrepency exists between the total dissolved carbon in oil field brines and the sum of the carbon in the form of bicarbonate alkalinity and carboxylic acids. For example, the analysis of coal gasification condensates (Mason, 1983) shows that 85 to 90% of the dissolved carbon is in the form of phenols; acetic acid constitutes the remainder. If any analogy exists between large-scale coal gasification experiments and the natural kerogen maturation process, a high potential for the generation of phenols during the diagenesis of certain kerogen types is present. Phenols have also been observed in oil field brines. Preliminary analysis of brines

Table 1.—NORTH COLES LEVEE OIL FIELD WATER ANALYSES (STEVENS)

	25–31	18–30	14–31	12–31
Lab pH	6.75	7.00	6.94	7.05
ppm Ca^{++}	830	889	889	624
ppm Mg^{++}	35.6	33.6	17.4	33.6
ppm Na^{+}	8724	10100	9253	7559
ppm K^{+}	258	233	169	245
meq Ca^{++}	41.	44.	44.	31.
meq Mg^{++}	2.93	2.76	1.43	2.76
meq Na^{+}	379.	439.	402.	328.
meq K^{+}	6.60	5.96	4.32	6.27
ppm Cl^{-}	11056	13871	12639	10528
ppm F^{-}	306	324	352	317
ppm $PO_4^{=}$	8	23	31	23
ppm Br^{-}	129	134	120	102
ppm $SO_4^{=}$	5	25	35	29
meq Cl^{-}	311.	391.	356.	296.
meq F^{-}	16.	17.	18.	16.
meq $PO_4^{=}$	0.17	0.48	0.65	0.48
Meq Br^{-}	1.61	1.68	1.50	1.28
meq $SO_4^{=}$	0.10	0.52	0.73	0.60
meq titratable alkalinity	79.00	74.00	80.00	54.50
Balance:	21.57	7.42	−5.32	−1.54
Total meq	839.	977.	910.	739.
% excess	5	2	−1	0
T.CO_2 (mmol/l)	4.21	4.90	3.40	5.35
@25° C:				
HCO_3^- (mmol/l)	3.01	4.00	2.70	4.46
HCO_3^- (meq/l)	3.01	4.00	2.70	4.46
H_2CO_3 (mmol/l)	1.20	0.90	0.70	0.89
log PCO_2 (atm)	−1.45	−1.58	−1.69	−1.58
TDS (mg/l)	21352	25633	23505	19461
HCO_3^-/Tot. Alk(%)	3.81	5.41	3.38	8.18
Organic Acids				
ppm acetate	3700	3750	3730	2670
ppm propionate	980	862	896	600

from the southern San Joaquin Valley, California (Stevens Sandstone) indicates that phenols constitute 5 to 10% of the dissolved organic carbon. Our recent experimental dissolution studies have shown that some phenols (hydroquinone and catechol) are as effective as difunctional carboxylic acids in complexing aluminum (Fig. 1).

Further evidence for the production of organic acids (carboxylic and phenolic) from kerogen comes from the analysis of the kerogen itself. Experiments involving low temperature (100°C) aqueous thermal degradation of kerogen yield small amounts of difunctional acids, depending on the kerogen type. Type III kerogen, with a high oxygen content, is the greatest producer of the oxygen-containing carboxylic acids. It is speculated that the oxidation of kerogen, coupled with thermal degradation, may produce greater quantities of difunctional acids. Oxidative degradation of kerogen is a commonly used laboratory technique for the study of kerogen structure (Vitorovic, 1980). Four kerogens were selected for a pervasive oxidation study: Green River Fm (Eocene-WY, Type I), Monterey Fm (Miocene-CA, Type II), Woodford Fm (Devonian-OK, Type II), and a Mesozoic brown coal (Type III). Samples were oxidized with $KMnO_4$ for two days according to the technique of Bone et al. (1926). Results of the pervasive oxidation experiments are presented in Table 2. All samples are similar in terms of difunctional/monofunctional product ratios (normalized to the organic carbon content of the sample) with the exception of the brown coal. The coal produced more difunctional acid than monofunctional acid. The study of preoil generation pyrolysates from hydrous pyrolysis experiments similar to those presented by M. Lewan (pers. commun., 1985) may prove to be another technique for the direct analysis of kerogen changes during maturation. This possibility is currently being evaluated.

The relatively recent application of Nuclear Magnetic Resonance (NMR) techniques to the study of kerogens (Miknis et al., 1982) has provided a direct method of examining structural changes in the kerogen as it matures. Figure 2 shows two Nuclear Magnetic Resonance (NMR) spectra, both of a brown coal. One is a near surface specimen and the other is the same coal at 9041 ft (2755 m). The loss of the smaller peaks representing carbonyl and phenol groups indicates that these groups have been released and expelled by the source rock at depth.

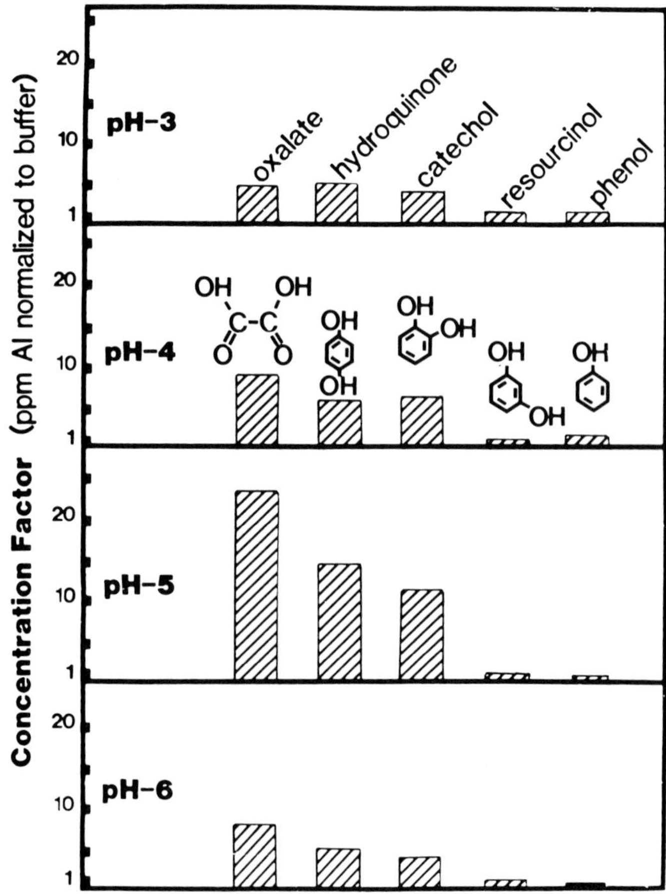

FIG. 1.—Aluminum solubility determined from andesine dissolution experiments. Aqueous phases contained 1000 ppm phenols or carboxylic acids. The pH in the aqueous phase was buffered by a sodium-acetate buffer. Runs were made for two weeks at 100°C. Aluminum values have been normalized to the acetate buffer (from Surdam and Crossey, 1985).

TABLE 2.—KEROGEN OXIDATION RESULTS

Formation	Oxalate* (ppm)	Acetate* (ppm)	Propionate* (ppm)	Formate* (ppm)
Green River Fm	191	43	393	173
Monterey Fm	144	34	283	254
Woodford Fm	203	29	276	369
Australian brown coal	12,000	1,850	1,640	1,176

*Fluid/solid ratio equal to 100/1.

Formation	% ORGANIC CARBON AS: Difunctional carboxylic acid	Monofunction carboxylic acid	Carboxylic acid
Green River Fm	7	33	40
Monterey Fm	8	47	55
Woodford Fm	11	47	58
Australian brown coal	41	23	64

Mineral Oxidants.—

Mineral oxidants have been suggested as interacting with organic matter in sediments by Curtis (1978), Hoffman and Hower (1979), and others. Eslinger et al. (1979) discuss the role of iron reduction in the conversion of smectite to illite. Ferric iron has been considered for its oxidizing capacity in early diagenesis as well (Berner, 1980). Several mineral phases with the capacity to act as oxidizing agents are present in sedimentary rocks. These include clay minerals, oxide coatings on grain surfaces, and polysulfides. Several lines of evidence point to the interaction of clay minerals and organic material during diagenesis. We have examined the potential of clay diagenesis as a mineral oxidant, but do not mean to suggest that the other possibilities are unimportant in some cases.

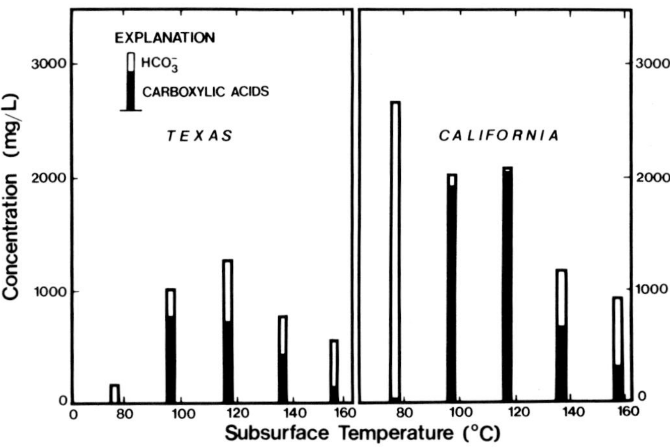

FIG. 3.—Histograms of alkalinity (concentration versus temperature) for the oil field waters from the Texas Gulf Coast and Central Valley, California. Note that the alkalinity is dominated by carboxylic acids over the temperature interval 80° to 140°C. Samples at temperatures greater than 80°C were averaged over 20°C intervals. Modified from Carothers and Kharaka (1978).

Clay Diagenesis.—

Carothers and Kharaka (1978) have examined brines from a variety of locations, including the Gulf Coast region and California (Central Valley). When the abundance of organic acids is plotted against the *in situ* temperature, a distribution pattern emerges with a maximum between 100° and 120°C in both locations (Fig. 3). The low temperature limit of the presence of organic acids is due to the action of bacteria; as temperature increases, the bacteria are eliminated due either to temperature or the release of poisonous

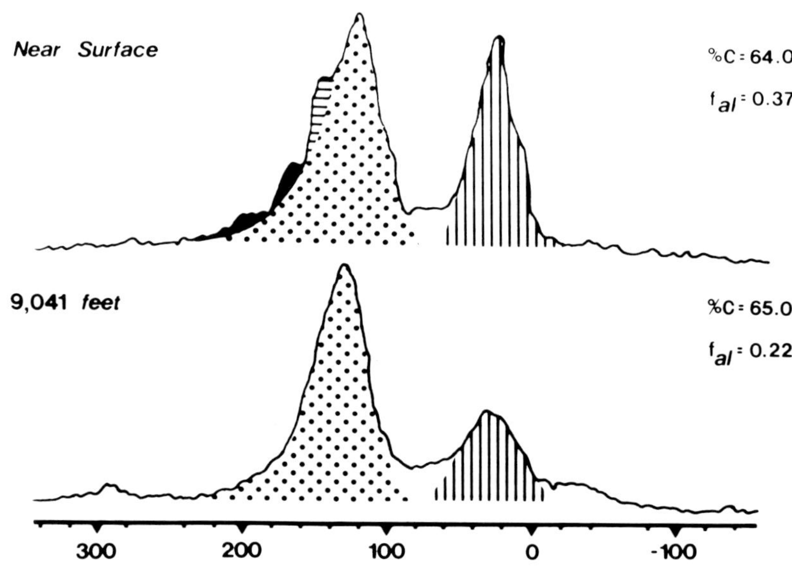

FIG. 2.—NMR spectra showing the successive changes in kerogen with depth in a Mesozoic brown coal (from Surdam and Crossey, 1985).

phenols from the organic material. The high temperature limit to the presence of organic acids is due to thermal decarboxylation. In general, the short-chain acids are not stable above 200°C. The peak abundance of the organic acids occurs at the same temperature as the ordering of mixed-layer smectite/illite (Hower, 1981). Upon conversion from random to allevardite-type ordering, there is a net release of iron from octahedral sites (Hower et al., 1976; Boles and Franks, 1979). Whether the iron is released and then reduced, or reduced within the clay structure itself and released as ferrous iron (Almon, 1974), an electron transfer must occur. The oxidation of organic matter dispersed in the shales is an ideal donor. Chemical analyses of clays from the Texas Gulf Coast show a decrease in iron content (Fig. 4). One would expect to find, as a complementary trend, diagenetic minerals containing ferrous iron. Chlorite and ankerite are two commonly observed ferrous minerals. Moncure et al. (1984) describe a 5 ft (1.5 m) sand interval overlying a thicker shale sequence from the Gulf Coast. The mineralogical and porosity trends are summarized in Figure 5. Porosity increases toward the sand-shale contact, as does chlorite abundance. This is the trend one would expect if the shales were the source of both the iron and organic solvent. Higher upsection in the sand is a kaolinitic zone, possibly representing the limit of transport of the aluminum from framework grain dissolution.

Applications of Clay-Organic Reactions to the Gulf Coast Porosity Balance.—

A calculation utilizing the Gulf Coast data of Hower et al. (1976), modifications of Boles and Franks' (1979) recalculations, typical kerogen compositions, and our experimental data on feldspar solubility is described below. First, the amount of iron available for reduction is depen-

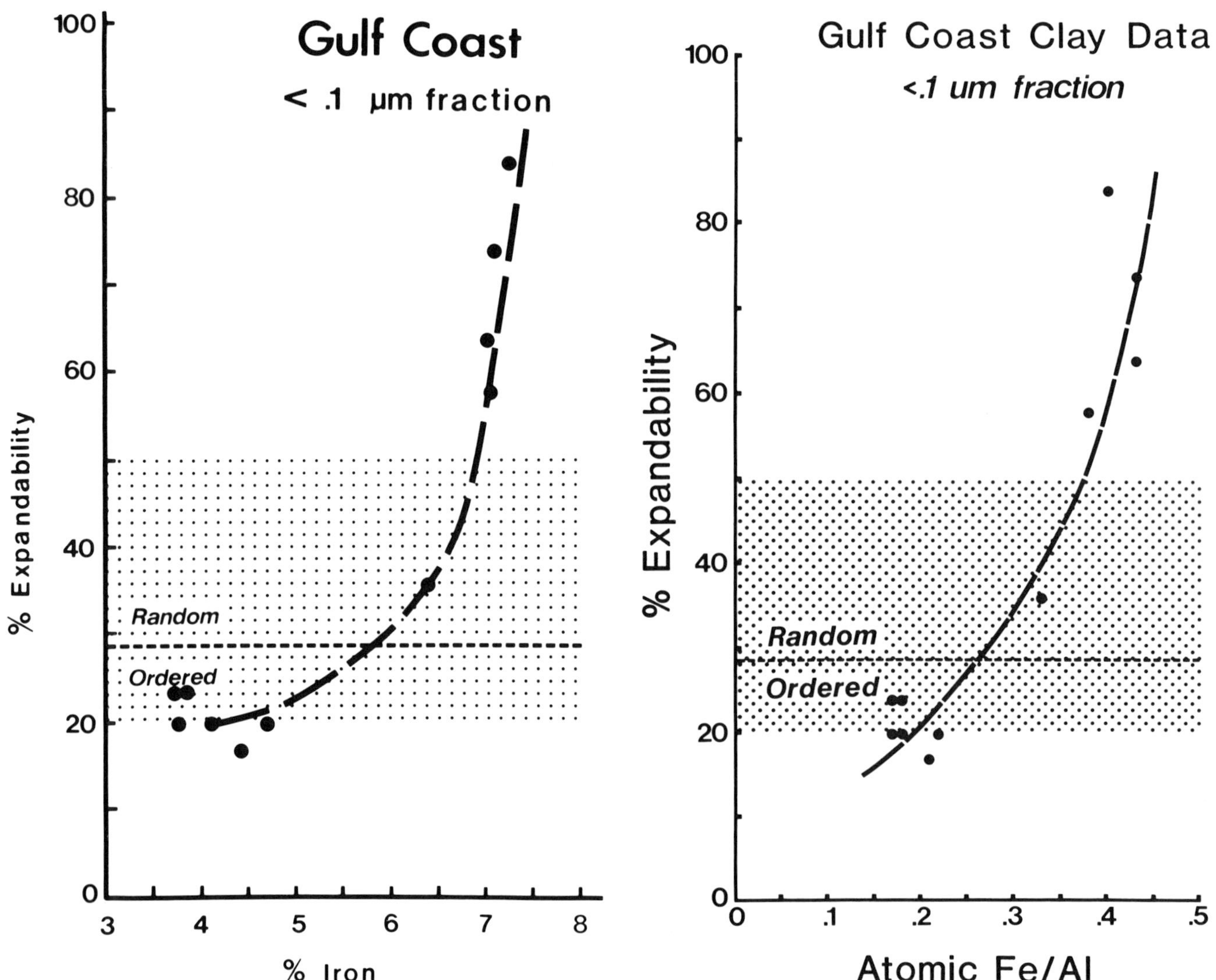

FIG. 4.—Diagram showing the variation in total iron in the mixed-layer clays with percent expandability in the Gulf Coast Tertiary (a). Data are from Hower et al. (1976). Note the apparent decrease in iron content over the stratigraphic interval from 80 to 20% expandability. The same trend results when atomic Fe/Al is plotted versus percent expandability (b). According to Hower et al. (1976), the clays change from random to ordered structures at approximately 28% expandability. Modified from Surdam and Crossey (1985).

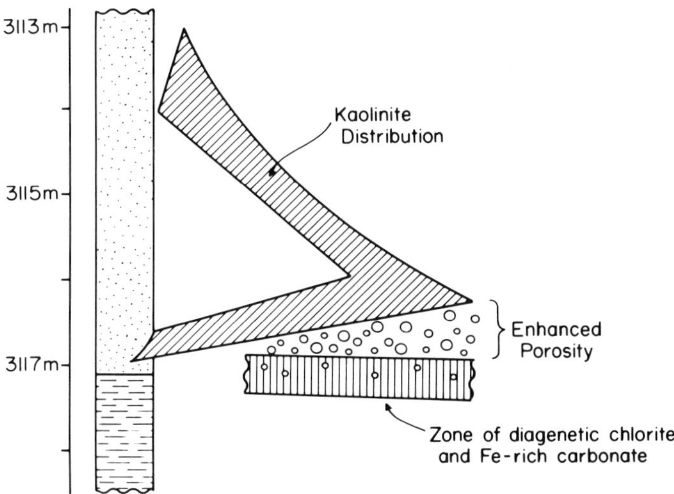

FIG. 5.—Schematic reconstruction of petrologic data from Moncure et al. (1984). Chlorite abundance increases toward the sand/shale contact. Immediately above is a zone of enhanced porosity. A zone enriched in kaolinite overlies the zone of enhanced porosity.

dent on the amount and composition of the mixed-layer clay fraction. Boles and Franks (1979) suggest, based upon reexamination of chemical and mineralogical data presented by Hower et al. (1976), that an appropriate composition for the initial smectite (100% expandable) is:

$$(K_{.2}Na_{.2}Ca_{.4})^{+1.2}\ (Mg_{.8}Fe_{.8}Al_{2.4})^{-.8}\ (Al_{.4}Si_{7.6})^{-.4}\ O_{20}(OH)_4. \quad [1]$$

Boles and Franks also calculate the following illite product composition:

$$(K_{1.1}Mg_{.4}Fe_{.3}Al_{3.4})^{-.1}\ (Al_{1.0}Si_{7.0})^{-1.0}\ O_{20}(OH)_4. \quad [2]$$

These compositions are based on mineral reconstructions from linear regression fits to expandability/composition data taken from Hower et al. (1976). The smectite composition has an interlayer charge of 1.2 equivalents per $[0_{20}(OH)_4]$ unit. This layer charge is appropriate for a vermiculite, and too high a charge to permit swelling with ethylene glycol (generally 1.0 eq. per $[0_{20}(OH)_4]$ unit is considered to be the division between smectite and vermiculite). The 1.2 layer charge is 50% greater than that determined by cation exchange capacities (CEC) [Hower and Mowat (1969)]. The illite composition calculated by Boles and Franks has similar problems: the layer charge of 1.1. eq./$[O_{20}(OH)_4]$ unit is far less than the 1.7 eq. thought to be required for dehydration (fixing) of potassium.

These inconsistencies in the calculated compositions of smectite and illite result from two sources. One, as noted by Boles and Franks (1979), is that there are two reactions operating: smectite to randomly ordered smectite/illite, and randomly ordered smectite/illite to ordered smectite/illite. The application of linear regression calculations is not appropriate for the overall reaction series. The second source is that the fine clay fraction upon which the composition calculations are based contain impurities that must be accounted for.

If the compositional data from Hower et al. (1976) for the finest size fraction (the same size fraction Boles and Franks used) are corrected for kaolinite abundance, normalized, and plotted against expandability (Fig. 6), several interesting patterns emerge. For the elements silicon, magnesium, iron, and potassium, distinct breaks in the slope occur between 36 and 24% expandability; based on X-ray diffraction analysis, Hower et al. (1976) interpreted 28%

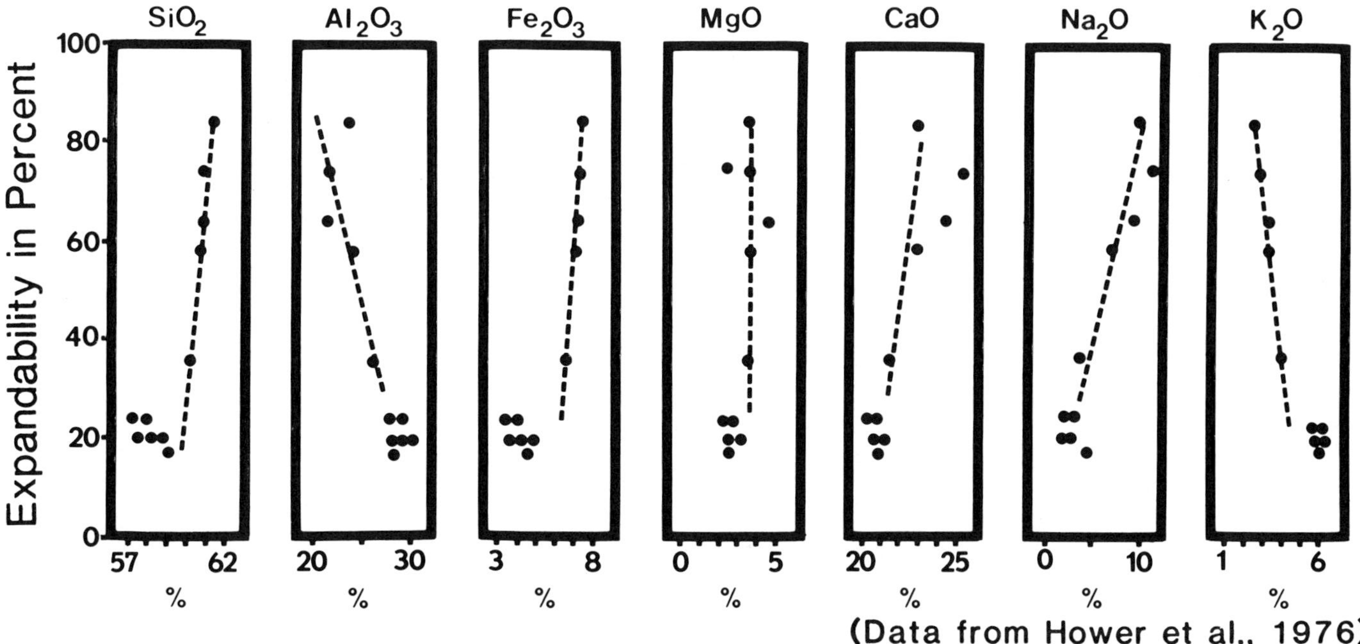

FIG. 6.—Percent oxide in the clay mineral fraction of the Gulf Coast Tertiary plotted against percent expandability in the mixed-layer clay. Note decreases in SiO_2 and Fe_2O_3, and increasing K_2O. Data from Hower et al. (1976).

TABLE 3.—KEROGEN CONVERSION REACTIONS

	KEROGEN → ACETATE
TYPE I	$CH_{1.72}O_{0.12} + 0.74Fe_2O_3 + 0.14H_2O \rightarrow 0.5CH_3COOH + 1.48FeO$
TYPE II	$CH_{1.28}O_{0.18} + 0.46Fe_2O_3 + 0.36H_2O \rightarrow 0.5CH_3COOH + 0.92FeO$
TYPE III	$CH_{0.88}O_{0.20} + 0.24Fe_2O_3 + 0.56H_2O \rightarrow 0.5CH_3COOH + 0.48FeO$
	KEROGEN → CO_2
TYPE III	$CH_{0.88}O_{0.20} + 2.24Fe_2O_3 \rightarrow CO_2 + 0.44H_2O + 4.48\ FeO$
	KEROGEN → OXALATE
TYPE I	$CH_{1.72}O_{0.12} + 2.24Fe_2O_3 \rightarrow 0.5H_2C_2O_4 + 4.48FeO + 0.36H_2O$
TYPE II	$CH_{1.28}O_{0.18} + 1.96Fe_2O_3 \rightarrow 0.5H_2C_2O_4 + 3.92FeO + 0.14H_2O$
TYPE III	$CH_{0.88}O_{0.20} + 1.74Fe_2O_3 + 0.06H_2O \rightarrow 0.5H_2C_2O_4 + 3.48FeO$
	KEROGEN → HYDROQUINONE
Type I	$CH_{1.72}O_{0.12} + 0.91Fe_2O_3 \rightarrow 0.17H_2C_6O_2 + 1.82FeO + 0.7H_2O$
TYPE II	$CH_{1.28}O_{0.18} + 0.63Fe_2O_3 \rightarrow 0.17H_2C_6O_2 + 1.26FeO + 0.48H_2O$
TYPE III	$CH_{0.88}O_{0.20} + 0.41Fe_2O_3 \rightarrow 0.17H_2C_6O_2 + 0.82FeO + 0.28H_2O$

The use of Fe_2O_3 does not imply discrete iron oxide but is more convenient than using the smectite compositions.

expandibility as the onset of allevardite ordering. In addition to constructing regression lines for only the random portions of the collapse curves, two data points have been deleted because of anomolously high calcium concentrations (the points with 74 and 64% expandability). The rationale used for the omission of these two points is that both sodium and calcium exist in the clay structure as exchange ions. When Na_2O is plotted against expandability, a regression line passing nearly through the origin (−0.04% Na_2O) at 0% expandability is generated. When the two high calcium values are included in the CaO-expandability regression, the regression coefficient is 0.87, and the line indicates −0.5% CaO at 0% expandability. If the two high calcium values are omitted, the regression coefficient improves to 0.92, and an intercept of 0.04% CaO at 0% expandability is predicted. About 1% of calcite by weight would account for the excess CaO. The data of Hower et al. (1976) suggest that calcite is present in the coarser size fraction in this portion of the section.

The smectite composition obtained by the extrapolation of the regression lines to 100% expandability was adjusted by 'removal' of CaO and normalization to the CEC (100 meq/100 g) found by Hower and Mowatt (1969). The resulting montmorillonite structural formula (below) is consistent with generally accepted ideas of montmorillonite composition: (1) layer charge of less than 1.0 eq./$[O_{20}(OH)_4]$ unit, (2) the charge deficiency is primarily in the octahedral layer, (3) Na^+ as the most abundant exchange ion.

$$(Ca_{.11}Na_{.29}K_{.24})^{+.75}\,(Al_{2.66}Fe_{.74}Mg_{.66})^{-.45}\,(Al_{.30}Si_{7.70})^{-.30}\,O_{20}(OH)_4 \quad [3]$$

This structural formula suggests that 0.74 moles of Fe^{+3} may be released for every 761 g (molecular weight) of initial smectite. As many natural smectite-illite systems cease reacting at 20% expandability, 40% of the iron would remain in the mixed-layer clay structure (Fig. 4). As described above, this iron would be released and incorporated into such late authigenic phases as chlorite, ankerite, or pyrite.

Assuming, as outlined above, that the iron in the mixed-layer illite/smectite clay is reduced and that organic matter is oxidized, the effect of these reactions may be quantitatively evaluated. As a first-order approximation, it will be assumed that the shale contains 50% mixed-layer illite/smectite clay of the initial composition [3] described above. Furthermore, it is assumed that 60% of the iron in the clay is released as it passes from 100 to 20% expandable

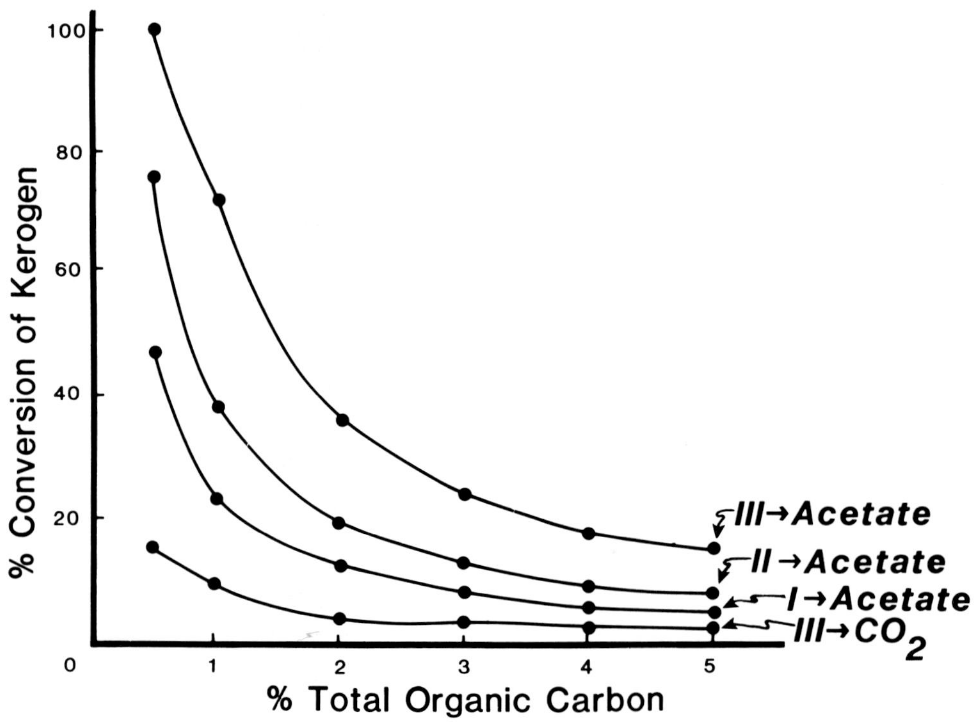

FIG. 7.—Percent conversion of kerogen versus percent total organic carbon for production of acetate and CO_2 by the reactions shown in Table 3.

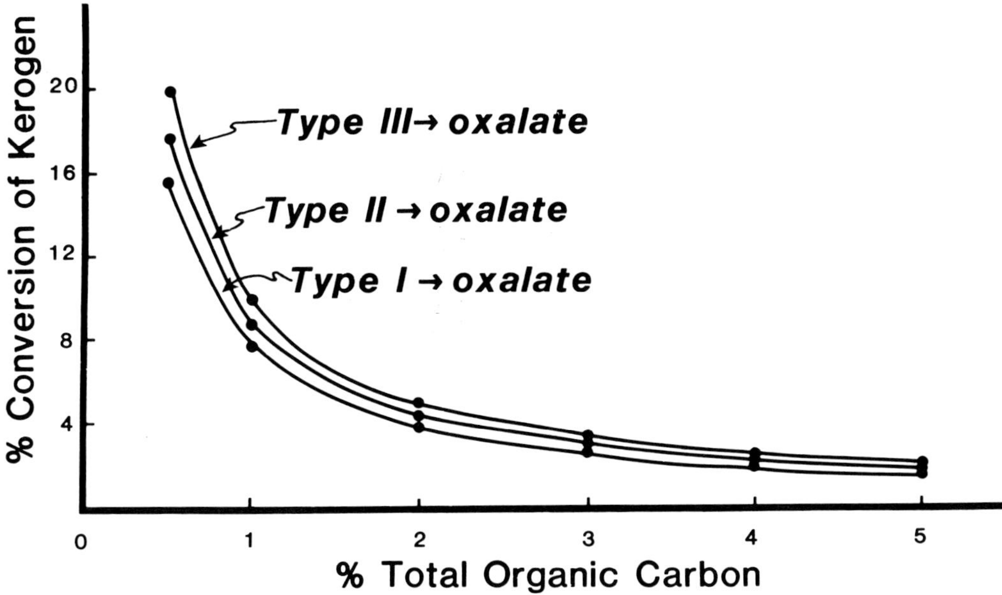

FIG. 8.—Percent conversion of kerogen to oxalate by reactions shown in Table 3.

(Fig. 4). Thus, 0.44 moles of Fe^{+3} are available for every 1522 g of shale. To evaluate the effect of this amount of iron, reactions of the form:

$$CH_2O + 2Fe_2O_3 \rightarrow CO_2 + H_2O + 4FeO$$

will be used. Estimates of the compositions for kerogen of various types are taken from Tissot and Welte (1978). These compositions are as follows:

Type I	$CH_{1.72}O_{.12}$
Type II	$CH_{1.28}O_{.18}$
Type III	$CH_{0.88}O_{.20}$

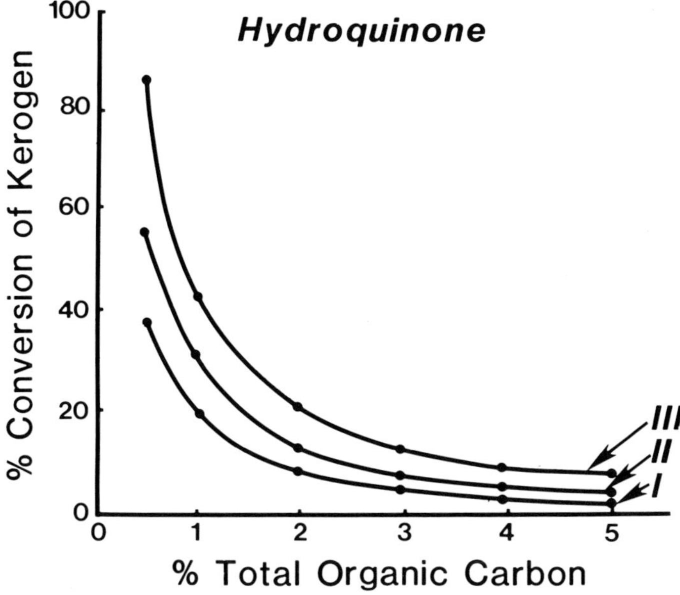

FIG. 9.—Percent conversion of kerogen to hydroquinone by reactions shown in Table 3.

For each kerogen type, an expression converting the kerogen to acetate, oxalate, and hydroquinone is given (Table 3). For Type III kerogen, a reaction is also written for conversion to CO_2. For simplicity, ferric iron is expressed as Fe_2O_3 rather than the smectite composition. Using the assumptions described above and the reactions in Table 3, the fraction of kerogen converted to the specific organic product has been determined for various initial concentrations of organic matter (see Figs. 7, 8, and 9).

Assuming a density of 2.3 g/cc for the shale and using the 20% conversion factor obtained for Type III kerogen from Figure 8, about 0.1 moles of oxalate would be formed from each liter of shale. From our experimental work (Surdam et al., 1984) at pH 5.10, 640 ppm oxalic acid will carry 72 ppm aluminum. If the released oxalate is diluted into 1 liter of water, 0.035 moles of aluminum could be transported. This would represent 6.3 grams of andesine, or 2.4 cc of feldspar removed per liter of shale. At a shale-to-sand ratio of 10 to 1, 2.4% whole rock porosity would result from the process (Table 4). Using hydroquinone as the organic solvent, a similar calculation results in 2.8% whole rock volume % porosity. These data are in excellent agreement with petrographic observations of the Frio Sandstone of Texas (Boles, 1983; see Figure 10).

Factors leading to a possible overestimation of the organic acid production from kerogen by the previously described mechanism are as follows:

TABLE 4.—POROSITY ENHANCEMENT DUE TO FELDSPAR DISSOLUTION

Solvent	Kerogen Type	Shale/Sand	Volume % Porosity Due to Andesine Dissolution
Oxalate	III	10/1	2.6
Oxalate	III	20/1	5.2
Hydroquinone	III	10/1	2.8
Hydroquinone	III	20/1	5.6

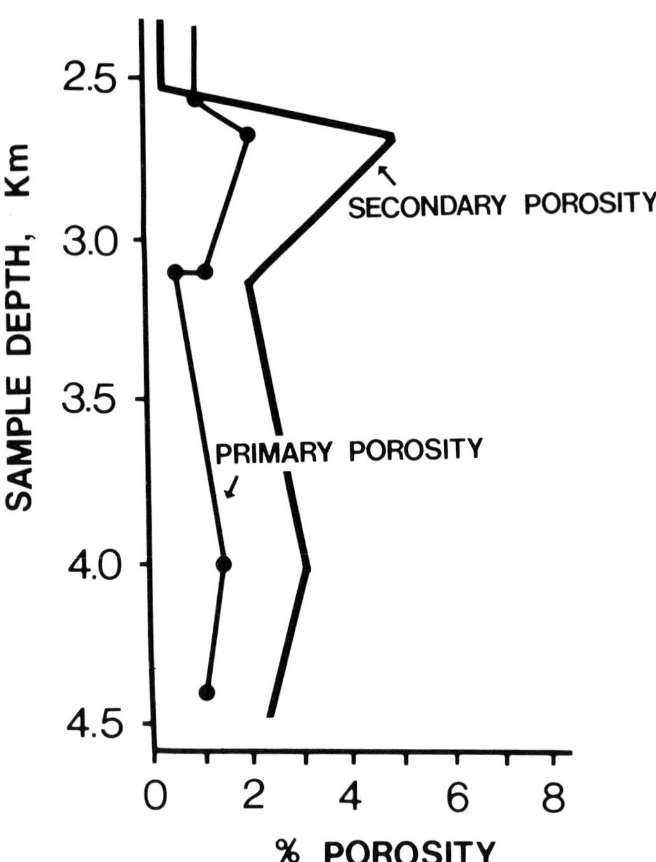

FIG. 10.—Observed porosity in the Frio Sandstone of the Texas Gulf Coast. Whole-rock porosity is the sum of the two curves. Modified from Boles (1983).

1) the assumption that all of the reacted kerogen forms oxalate;
2) the assumption that the iron reduction process is oxidizing only kerogen.

Further considerations suggest, however, that the estimate of the porosity enhancement due to this process is in fact conservative:

1) the assumption that iron from mixed-layer clay diagenesis is the only mineral oxidant;
2) the assumption that only half of the shale consists of mixed-layer smectite/illite;
3) the assumption that the kerogen molecule is homogenous and reacts uniformly;
4) the assumption that experimental data represent optimum aluminosilicate solubility;
5) the assumption that the pore fluids flow homogenously through the sandstones;
6) the assumption that there is no thermal degradation of kerogen leading to the production of organic acids.

SUMMARY

The role of clay diagenesis in kerogen maturation has long been an interesting, but unsolved problem. The coincidence in time, temperature and space of mixed-layer clay reactions and kerogen maturation in source rocks circumstantially has suggested some kind of genetic relationship. Almon (1974) suggested that this genetic relationship involves electron transfer from the kerogen to clay acceptor sites (i.e., Fe^{+3}) to produce two free radicals from the kerogen. These acceptor sites could be Fe^{+3} within the clay and the free radicals could include carboxylic and phenolic groups. If there is an oxidation-reduction mechanism linking clay diagenesis and kerogen maturation in the natural system, it would have a profound effect. The reduction of any mineral oxidant and simultaneous oxidation of kerogen would effectively increase the concentration of difunctional organic solvents. These difunctional organic solvents increase the solubility of aluminosilicates, such as plagioclase, by several orders of magnitude (Surdam et al., 1984). In this paper it has been shown that in the Gulf Coast, the reduction of Fe^{+3} in mixed-layer clays contained in Tertiary shales and concomitant oxidation of kerogen could account for all the observed aluminosilicate framework grain dissolution observed in adjacent sandstones.

Not only does the mixed-layer clay diagenesis offer the potential for oxidation/reduction reactions, but it also results in a significant increase in fluid flux. If transfer or redistribution of material is to ocur, some type of fluid flux is necessary to transport the water soluble organic acids and their dissolved or chelated load.

The prediction of secondary porosity requires an understanding of the mechanisms by which porosity enhancement occurs. The interaction of clay minerals and kerogen warrants further investigation, particularly with regard to the extent of the interaction, and the role that both kerogen type and clay mineral composition may play in the interaction. Whether one evaluates carbonic acid or organic acids as the solvent responsible for porosity enhancement, the contribution of mineral oxidants to organic/inorganic reactions should not be overlooked.

REFERENCES

ALMON, W. R., 1974, Petroleum-forming reactions: Clay catalyzed fatty acid decarboxylation: Ph.D. Thesis, University of Missouri, Columbia, Missouri. 117 p.

BERNER, B. A., 1980, Early Diagenesis; A Theoretical Approach: Princeton University Press, Princeton, N.J., 237 p.

BOLES, J. R., 1983, Physical-chemical controls on diagenesis: American Association of Petroleum Geologists Clastic Diagenesis School, Notes, p. 1–81.

——— AND FRANKS, S. G., 1979, Reply; Cementation of sandstones: Journal of Sedimentary Petrology, v. 49, p. 1362.

BONE, W. A., HORTON, L., AND WARD, S. G., 1926, Researches on the chemistry of coal, Part VI: Royal Society Proceedings, A, v. 110, p. 537.

CAROTHERS, W. W. AND KHARAKA, Y. K., 1978, Aliphatic acid anions in oil-field waters—Implications for origin of natural gas: American Association of Petroleum Geologists Bulletin, v. 62, p. 2441–2453.

CURTIS, C. D., 1978, Possible links between sandstone diagenesis and depth related geochemical reactions occurring in enclosing mudstones: Journal of the Geological Society of London, v. 135, p. 107–117.

ESLINGER, E., HIGHSMITH, P., ALBERS, D., AND DEMAYO, B., 1979, Role of iron reduction in the conversion of smectite to illite in bentonites in the disturbed belt, Montana: Clay and Clay Minerals, v. 27, p. 327–338.

HOFFMAN, J. AND HOWER, J., 1979, Clay mineral assemblages as low grade metamorphic geothermometers: Applications to the thrust faulted disturbed belt of Montana, U.S.A.: Society of Economic Paleontologists and Mineralogists Special Publication 26, p. 55–79.

HOWER, J., 1981, Shale diagenesis, *in* Longstaffe, F. J., ed., Short Course in Clays and the Resource Geologist: Mineralogical Association of Canada, Calgary, Canada, Chapter 4, p. 60–80.

——— AND MOWATT, 1969, The mineralogy of illite and mixed-layer illite/montmorillonite: The American Mineralogist, v. 51, p. 825–854.

———, ESLINGER, E., AND HOWER, M., 1976, Mechanisms of burial metamorphism of argillaceous sediments. 1. Mineralogical and chemical evidence: Geological Society of America Bulletin, v. 87, p. 725–737.

LUNDEGARD, P. D., LAND, L. S., AND GALLOWAY, W. E., 1984, Problem of secondary porosity: Frio Formation (Oligocene), Texas Gulf Coast: Geology, v. 12, p. 399–402.

MASON, J. M., 1983, Hydrophilic organic compounds produced during underground coal gasification: M.S. Thesis, University of Wyoming, Laramie, Wyoming, 73 p.

MIKINIS, F. P., SMITH, J. W., MAUGHAN, E. K., AND MACIEL, G. E., 1982, Nuclear Magnetic Resonance: A technique for direct nondestructive evaluation of source-rock potential: American Association of Petroleum Geologists Bulletin, v. 66, p. 1396–1401.

MONCURE, G., LAHANN, R., AND SIEBERT, R., 1984, Origin of secondary porosity and cement distribution in a sandstone/shale sequence from the Frio Formation (Oligocene): American Association of Petroleum Geologists Memoir 37, p. 151–163.

SURDAM, R., BOESE, S., AND CROSSEY, L., 1984, The chemistry of secondary porosity: American Association of Petroleum Geologists Memoir 37, p. 127–151.

SURDAM, R. C., AND CROSSEY, L. J., 1985, Organic-inorganic reactions during progressive burial: Key to porosity and permeability enhancement and preservation: Philosophical Transactions of the Royal Society of London, Series A 315, p. 135–156.

TISSOT, B. P. AND WELTE, D. H., 1978, Petroleum Formation and Occurrence: New York, Springer-Verlag, 638 p.

VITOROVIC, D., 1980, Structure elucidation of kerogen by chemical methods, *in* Durand, B., ed., Kerogen: Paris, Technip, p. 301–338.

DIAGENESIS OF THE BRENT SANDSTONE IN THE STATFJORD FIELD, NORTH SEA

K. BJØRLYKKE[1] AND A. BRENDSDAL[2]
Department of Geology
University of Bergen

ABSTRACT: The sandstones of the Middle Jurassic Brent Group are major reservoir rocks in the North Sea. The overall reservoir quality of the sandstones is generally good, but porosity and permeability vary greatly within the reservoirs. Mica, particularly biotite, has been the most reactive mineral component during diagenesis of these sandstones. The primary distribution of mica and its degree of alteration to illite and kaolinite control, to a large extent, the porosity and especially the permeability. Detailed mineralogical analyses of cores from the Statfjord field show that most diagenetic kaolinite formed from alteration of mica; feldspar was less important as a precursor. The alteration of muscovite to kaolinite is accompanied by an increase in volume that significantly reduces permeability. Abundant siderite has been precipitated between sheets of biotite. Microprobe analyses of the iron carbonate show that it has a distinctive composition which is influenced by the geochemical microenvironment within the biotite. Carbonate-cemented layers or nodular lenses may be important as permeability barriers. There is some evidence supporting leaching of carbonate cement but none suggesting that large parts of the sandstones of the Brent Group were ever carbonate-cemented and subsequently leached. Secondary porosity, formed by leaching of feldspar, is commonly observed but does not contribute much to the overall porosity. Plagioclase and albite lamellae in perthites are selectively dissolved, whereas potassium feldspars have secondary overgrowths. The main leaching phase is interpreted to have been relatively early and probably caused by flushing of meteoric water. Later (post-mid-Cretaceous) diagenesis was essentially isochemical, but some leaching may have occurred during deep burial. Alteration of unstable mica buffered pore water with respect to potassium, which protected the K-feldspar in relation to Na-feldspar. Micas also have an important effect on compaction and seems to enhance pressure dissolution against quartz and feldspar.

INTRODUCTION

The Middle Jurassic Brent Group sandstones are one of the most important reservoir rocks in the northern North Sea. The Brent Group is the major producer in the Statfjord field and adjacent oil fields in the Norwegian and British sectors. Mid-Jurassic sandstones are also the main exploration target in the mid-Norwegian and north Norwegian shelf. The porosity and permeability of the sandstones of the Brent Group in the Viking Graben have been generally good, although quite variable, depending upon sedimentary facies and diagenetic alteration. The prediction of reservoir quality, both regionally and within a reservoir, requires a detailed understanding of the processes that cause reduced porosity. The Brent Sandstone of the Statfjord field was one of the examples described by Schmidt and McDonald (1979a, 1979b), in which the porosity was interpreted to be mostly secondary due to leaching of feldspar and carbonate cement. We consider the porosity in the Brent Sandstone to be mostly primary and that leaching of feldspar and carbonate cement has contributed only 2–3% additional porosity. The formation of secondary porosity is commonly compensated for by precipitation of authigenic minerals so that the total porosity is not increased. We interpret the cause of leaching to be flushing by relatively early meteoric water rather than dissolution during deep burial due to CO_2 released through the maturation of kerogen, as claimed by Schmidt and McDonald (1979a, 1979b).

Surdam et al. (1984, p. 142) point out that elevated P CO_2 cannot be the cause of dissolution of aluminosilicates. They suggest, on the basis of laboratory work, that carboxylic acid (acetic acid) produced during the maturation of kerogen is capable of complexing aluminium and thereby making feldspar more soluble. Lundegard et al. (1984), however, point out that the oxygen content of the kerogen may also be a limiting factor for the formation of carboxylic acids from kerogen. It is very important to establish the mechanism that causes leaching of feldspar and carbonate cement, because it provides the basis for predicting the distribution of secondary porosity in the subsurface. The meteoric water-leaching model predicts that leaching occurs relatively early but may repeat itself during tectonic uplift. The most proximal facies (i.e., in deltas) will be more intensely leached than the distal marine facies and turbidites. Meteoric water may extend to a depth of 2 km or more below sea level (Galloway, 1984). If the leaching is caused by the formation of carbon dioxide or carboxylic acids from maturing kerogen, one would expect leaching to occur at greater burial depth and that the most intense leaching would be found in sandstones in close proximity to source rocks.

The reservoir quality of the Brent Sandstone is generally good, but it varies greatly both vertically and laterally. To be able to predict the three-dimensional distribution of reservoir properties, it is important to understand both the primary depositional environment and the diagenetic processes that modified the primary porosity and permeability. The present paper will briefly review the stratigraphy and depositional environment of the Brent Group in the Statfjord area. Mineralogical analyses done by one of the authors (A. B.) on one of the cores (33/9-A23) will be used as an example to illustrate processes that we consider to be valid for most of the Statfjord area and beyond (Fig. 1).

STRATIGRAPHY AND SEDIMENTOLOGY

General introductions to the stratigraphy and regional geology of the North Sea have been published by Deegan and Schull (1977) and Ziegler (1981). Papers dealing more specifically with Middle Jurassic sedimentation in the northern North Sea include Chauvin and Valachi (1980); Skarpnes et al. (1980); Hamar et al. (1980); Kirk (1980); Eynon (1980); Proctor (1980); Hancock and Fisher (1981); Hallett (1981);

[1]Present Address: Department of Geology, University of Oslo, P.O. Box 1047, Blindern, 0316 Oslo 3
[2]Present Address: Statoil Stavanger.

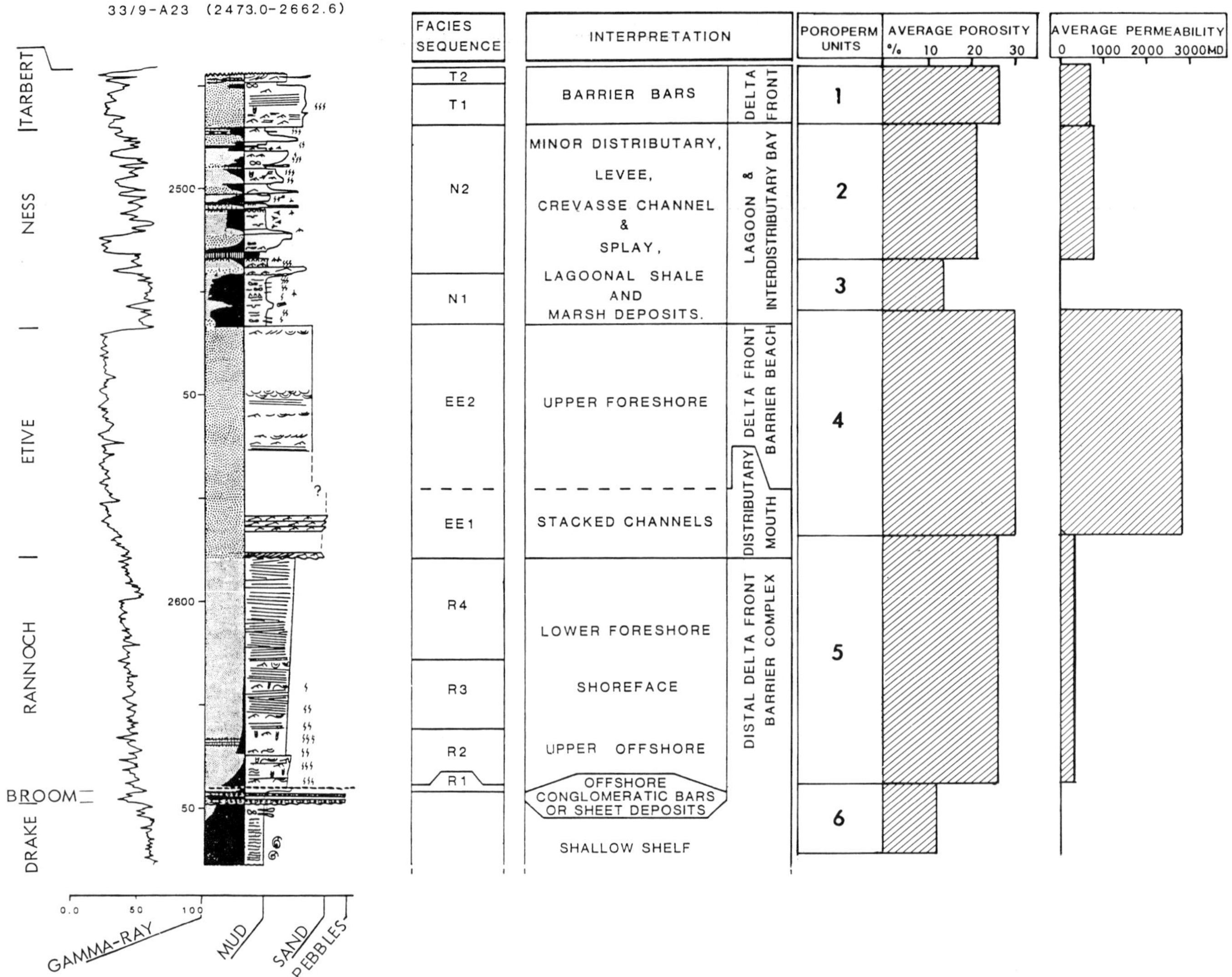

FIG. 1.—Lithologic log of a well in the Statfjord field (33/9-A23) with interpretation of the depositional environment (after Dag Bergliens). Average porosity and permeability data from Mobil Exploration Norway, Inc.

Budding and Inglin (1981); Gray and Barnes (1981); and Nagy et al. (1984).

The Statfjord Formation, which is of Late Triassic to Early Jurassic (Rhaetian-Sinemurian) age, consists of fluvial sediment and of marine sandstones in the upper part (Skarpnes et al., 1980) and is succeeded by the Dunlin Shale. The basal part of the overlying Brent Group (Broom Formation) consists of a thin, poorly sorted clayey sandstone and granule conglomerate with chamosite ooids and calcite and siderite cement. The Rannoch Formation represents a coarsening upward cycle which includes the overlying Etive Formation (Fig. 1). This cycle can be interpreted as a progradation of the lower shoreface of the Rannoch Formation to the foreshore facies of the Etive Formation. It may be cut by distributary channels (Bergslien, 1983) as also observed in the Cormorant field (Budding and Inglin, 1981). The thickness of the Rannoch-Etive barrier sequence is rather variable and may be related to its progradations across a shallow shelf. Assuming that the shelf subsidence is nearly constant, this sequence will be thicker during low progradation rates (Budding and Inglin, 1981). Differences in rates of subsidence, however, may be due to synsedimentary faulting, possibly as early movements along the faults that were activated later during tectonic rotation of blocks during the Kimmerian phases (Hamar et al., 1980).

The Ness Formation consists of shales, coal beds and thinner sandstones representing shoals and lagoonal shale, marsh deposits, and crevasse splay sandstones deposited in a more protected environment behind a barrier (Budding and Inglin, 1981; Bergslien, 1983). The Tarbert Formation represents a progradation of a marine shoreface facies with reworked distributory channels and may represent a delta front facies (Bergslien, 1983). The Tarbert Formation is commonly followed by an unconformity at the base of the overlying Heather Formation. The overlying low-permeability Cretaceous shales had an important effect in draping over the Upper Jurassic rotated blocks and sealing the underlying sandstones, probably already at that time building up an abnormal pore pressure (Bjørlykke, 1983).

DIAGENETIC ALTERATION OF THE SANDSTONES OF THE BRENT GROUP

Several petrographic descriptions of the Brent Sandstone and interpretations of its diagenetic history have been published, including Hancock and Taylor (1978); Blanche and Whitaker (1978); Bjørlykke et al. (1979); and Irwin and Hurst (1983). The Brent Sandstone is characterized by a relatively high feldspar content corresponding to an arkosic to subarkosic composition. The Rannoch Formation and parts of the Ness Formation have rather high mica contents. This composition reflects tectonic uplifts associated with mid-Jurassic rifting, which exposed fresh basement rocks on the platform and highs around the graben. The high content of muscovite and biotite results in characteristic diagenetic reactions and produces a mineralogy and a pore geometry which may cause formation damage during production.

Most authors agree that porosity and permeability in the sandstones have been reduced by the formation of diagenetic kaolinite and illite and the precipitation of carbonate cement. Some new (secondary) porosity has been created by the leaching of feldspar and carbonate cement. There has been some disagreement, however, concerning the timing of these diagenetic reactions and the importance of leaching during deeper burial (2–4 km). Schmidt and McDonald (1979a) interpreted the present porosity in the Brent Sandstone to be essentially secondary, formed by leaching of carbonate and feldspar during deep burial as a consequence of CO_2 produced by decarboxylation of kerogen. Bjørlykke et al. (1979) and Bjørlykke (1980, 1984) emphasized the effects of leaching of these sandstones by early meteoric water and considered this to be more important than late-stage dissolution by carbonic acids. Most of the previous publications have been petrographic descriptions of samples of the Brent Sandstone, using data from X-ray diffractometry (XRD), scanning electron microscopy (SEM) and thin sections. Relatively few systematic quantitative data on the stratigraphic and regional distribution of clastic and diagenetic minerals have been published.

In the present paper diagenetic reactions and their effect on porosity and permeability are discussed using one complete core (33/9-A23) through the Brent Group as an example. Seventy-five samples of this core have been analyzed by XRD and 23 modal analyses from thin sections were carried out (Fig. 2). X-ray diffractometry data are presented as relative X-ray diffractogram peak intensities (Figs. 3a, b). These data are representative of relative mineral concentrations. Quantitative determinations of mineral composition were not attempted because of the great uncertainty of such calculations.

The concentration of feldspar is highly variable through the Brent Group (Fig. 3a), and it is difficult to find any significant trend, except that the Ness Formation may have a lower total feldspar content than the other formations. K-feldspar is more abundant than plagioclase, which is a relatively pure albite. Perthitic feldspars are common and display selective leaching of the albite lamellae (Fig. 4). Feldspar overgrowths on clastic feldspars are very common, but authigenic feldspars make up less than 1 % of the total volume of the rocks. The overgrowths often show a sawtooth-like pattern of feldspar crystals. This is an early stage of feldspar overgrowth (Waugh, 1978), and it seems that continued growth of authigenic feldspar leads to a coalescence of the small crystals into smooth crystal surfaces. Microprobe analyses of the overgrowths of K-feldspar show very pure K-feldspar with a much lower Na content than in the clastic feldspars (see analyses by Bjørlykke et al., 1979). Authigenic Na-feldspars are very rare and have only been identified in the Broom Formation and the lower Rannoch Formation. Textural relationships here suggest that it forms as pore-filling cement, as clastic "host" grains are not observed. Many feldspar grains show evidence of leaching, but the leaching is generally not extensive except in the Broom Formation. Leaching of feldspar varies greatly between neighboring grains, and incipient leaching during weathering prior to deposition seems to have controlled the degree of dissolution. Secondary porosity due to feldspar leaching does not, according to our estimates, add more than 1–3% porosity, and significant porosity is often lost due to precipitation of authigenic aluminous silicate minerals such as kaolinite. Feldspar overgrowth postdates the main leaching phase, but in a few cases there is evidence of leaching later than the overgrowth.

The dissolution of mica seems to have "buffered" the pore water with respect to potassium so that potash feldspar was more stable. Selective leaching of albite has occurred because there has been no corresponding supply of sodium to the pore water. This differs from the Gulf Coast, where K-feldspar is selectively leached and authigenic albite is

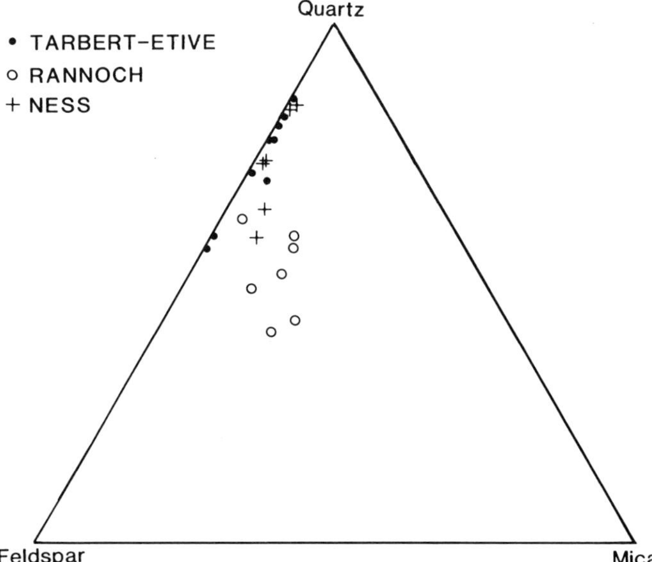

FIG. 2.—Modal analyses of 22 samples of the Brent Formation from well A-23.

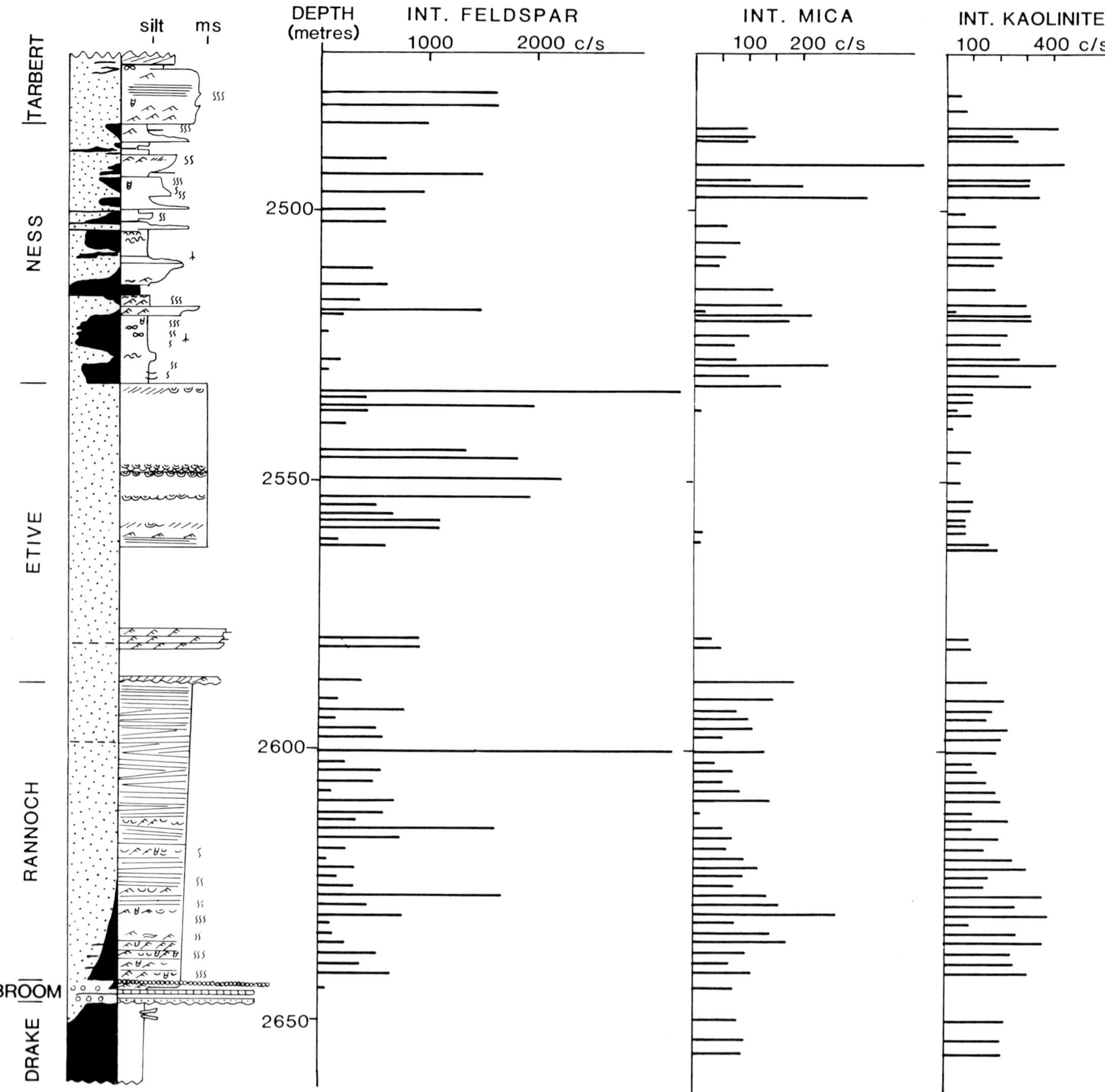

FIG. 3a.—Vertical distribution of feldspar, mica and kaolinite based on unweighted X-ray diffraction (XRD) data.

formed due to supply of Na^+ from smectite in shales or from evaporites (Boles, 1982) (Table 2).

CLASTIC MICA AND ITS DIAGENETIC ALTERATION PRODUCT

Clastic mica is an important and characteristic component in the Brent Group, particularly in the Rannoch and Ness Formations. In the Broom, Etive and Tarbert Formations mica is a minor component. Biotite is always more altered diagenetically than muscovite. In the Ness and Rannoch Formations biotite has expanded and split into thin layers of altered and potassium-depleted biotite. Authigenic carbonate, mainly siderite, and feldspar, occur between the sheets of altered biotite. The siderite crystals in biotite show up particularly well in electron backscatter pictures. They are probably formed from iron released during the breakdown of biotite (Fig. 5). No siderite is found associated with muscovite. There is very little kaolinite between the

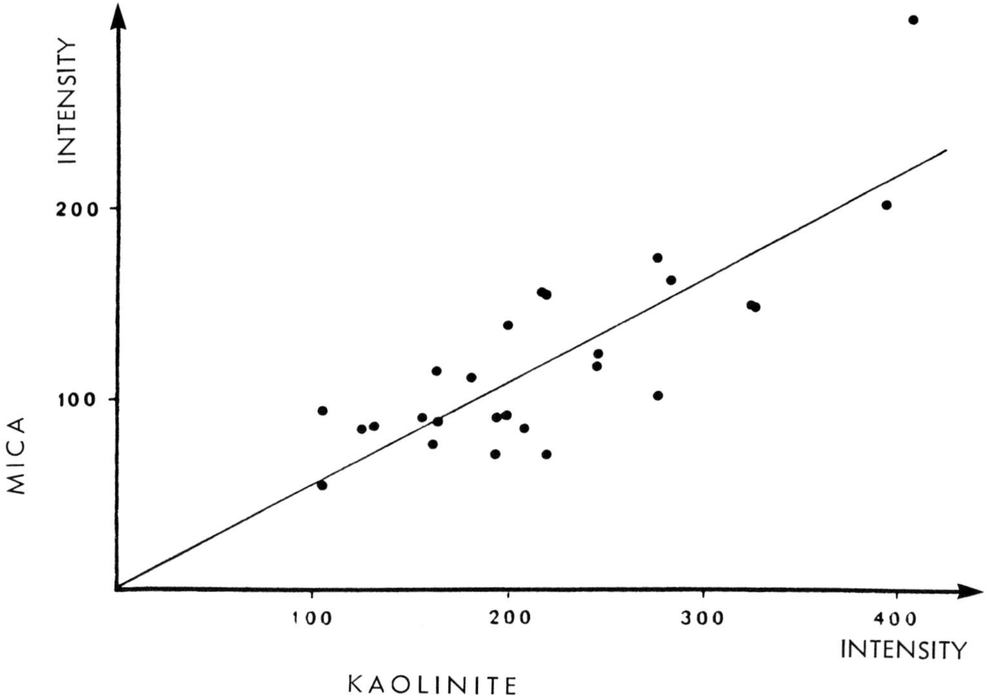

FIG. 3b.—Correlation between mica and kaolinite content. Correlation coefficient 0.8.

biotite sheets, but authigenic kaolinite occurs in pore spaces around biotite. In the Tarbert and Etive Formations practically no biotite survived leaching and only ghosts of biotite are left.

Muscovite is also altered to variable degrees. In the Rannoch and Ness Formations all transitions exist between unaltered fresh muscovite and almost completely kaolinized muscovite (for microprobe analyses see Bjørlykke et al., 1979). Muscovite squeezed between quartz sand grains is commonly unaltered, but the same sheet of mica alters to kaolinite where it is not confined and can expand into available pore space (Figs. 6a, b). Kaolinite also may form between thin laminae of nearly unaltered muscovite (Fig. 5). Grains of quartz and feldspar are often dissolved against mica, resulting in dissolution of about half the original clastic grain (Fig. 6b). This represents a type of pressure dissolution which contributes significantly to compaction and which may have been underestimated in earlier studies. It

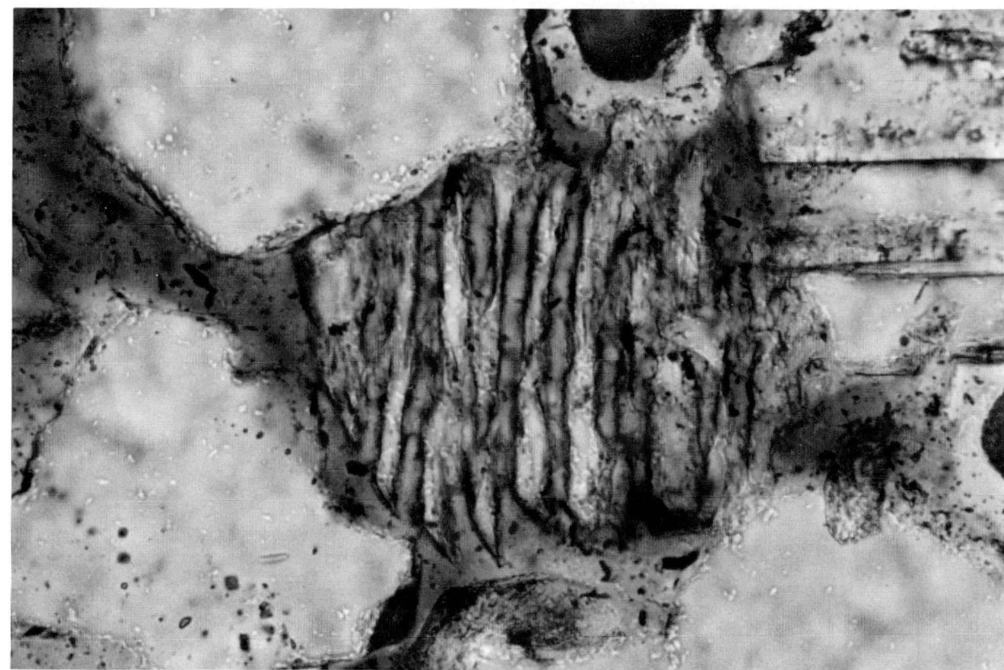

FIG. 4.—Photomicrograph showing selective leaching of albite lamellae in perthites A, B, in a feldspar grain from the Etive Formation, Field of view is 0.4 mm.

TABLE 1.—Microprobe Analyses of Carbonate Minerals in the Rannoch Formation

		Mg	Ca	Mn	Fe	C
Ankerite	1.	2.02	4.25	0.16	1.47	8.04
Ankerite	2.	2.26	3.99	0.16	1.48	8.05
Ankerite	3.	2.41	4.06	0.10	1.38	8.03
Siderite in low Mg calcite	4.	0.26	0.67	0.31	6.52	8.12
Siderite in Biotite	5.	1.13	0.85	0.17	5.63	8.11
Siderite in Biotite	6.	1.25	0.72	0.14	5.47	8.21
Siderite in Biotite	7.	1.09	0.89	0.18	5.75	8.04

The analyses are based on 24 oxygen in the formulas.

seems that mica has a catalytic effect on quartz and feldspar dissolution, as discussed by Gjeldsvik and Bjørkum (1984). In the Tarbert and Etive Formations kaolinization of muscovite is generally more advanced than in the Ness and Rannoch Formations, probably due to its higher permeability (Fig. 1).

Both thin sections and SEM analyses show that kaolinite occurs as two types: 1) pore-filling kaolinite of pseudohexagonal crystals in accordion- or spiral-shaped aggregates; and 2) kaolinite replacements of mica. The pore-filling kaolinite is formed as a result of alteration of feldspar, biotite and other unstable silicate minerals. Pore-filling kaolinite associated with authigenic anatase is observed in the Etive Formation. Authigenic titanium minerals suggest that biotite is the precursor for this kaolinite. Clusters that can be interpreted as pseudomorphs after feldspar are rare, however. Kaolinite is also precipitated in the space left by the dissolution of albite lamellae in perthites (see Fig. 4). Pore-filling kaolinite is quantitatively less important than the kaolinite replacements of mica. In the micaceous part of the Brent Sandstones, mica is clearly a more important precursor for kaolinite than feldspar, and there is a good correlation between mica and kaolinite content (Fig. 3b).

TABLE 2.—A Comparison Between Types of Feldspar Leaching and Precipitation in the Gulf Coast (Frio Formation) and the Brent Formation of the North Sea

GULF COAST (FRIO FORMATION)	NORTH SEA (BRENT FORMATION)
Leaching of K-feldspar	Selective leaching of albite
Albitization of plagioclase (Boles, 1982)	Precipitation of authigenic K-feldspar
Na^+-saturation from smectite-illite transformation or from evaporites	K^+-saturation from transformation of mica (incl. biotite)
Typical of passive margins with smectitic shales and/or evaporites	Typical of rift basins with exposed basement highs producing immature sediments, including biotite

CARBONATE CEMENT

The sandstones of the Brent Group are generally low in carbonate cement, but the cement is important in certain horizons. The distribution of carbonate-cemented layers may be critical to the reservoir properties. The cemented layers form permeability barriers to vertical flow and thereby control the flow of hydrocarbons during production. The most important carbonate minerals are low-Mg calcite (LMC) and siderite. Small amounts of ankerite are also found.

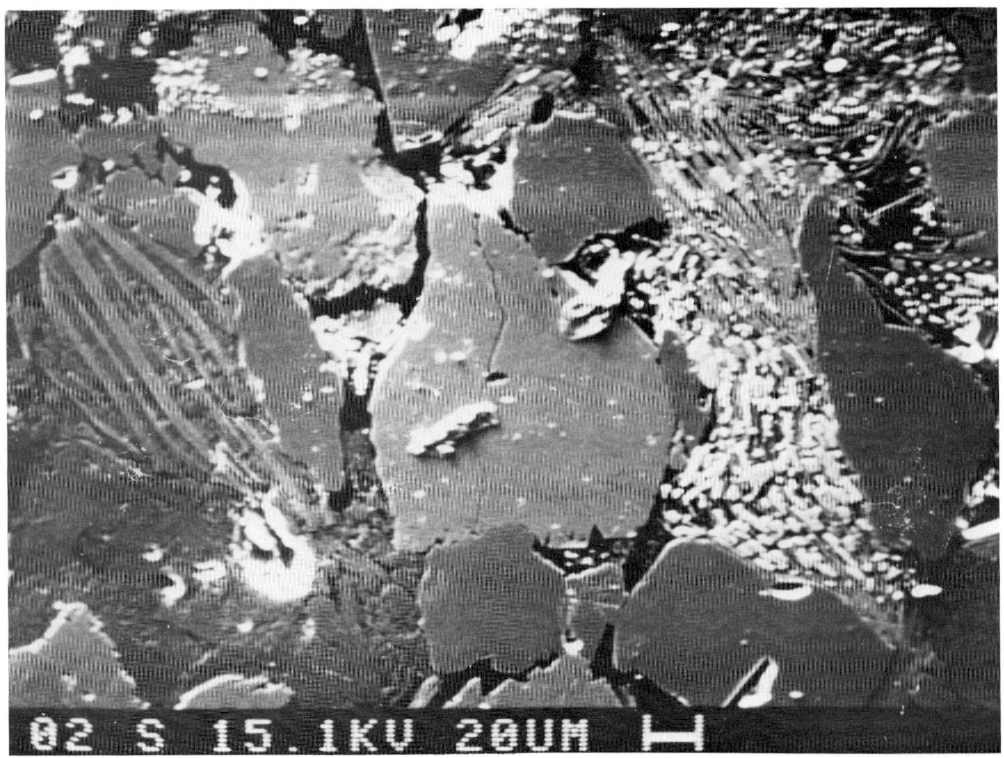

FIG. 5.—Electron backscatter photographs showing siderite crystals between sheets of degraded biotite and kaolinite between sheets of degraded muscovite. Siderite is showing higher backscatter intensities (lighter because of higher atomic number). Kaolinite is precipitated around the degraded biotites and not between the mica sheets. For description of method, see White et al. (1984).

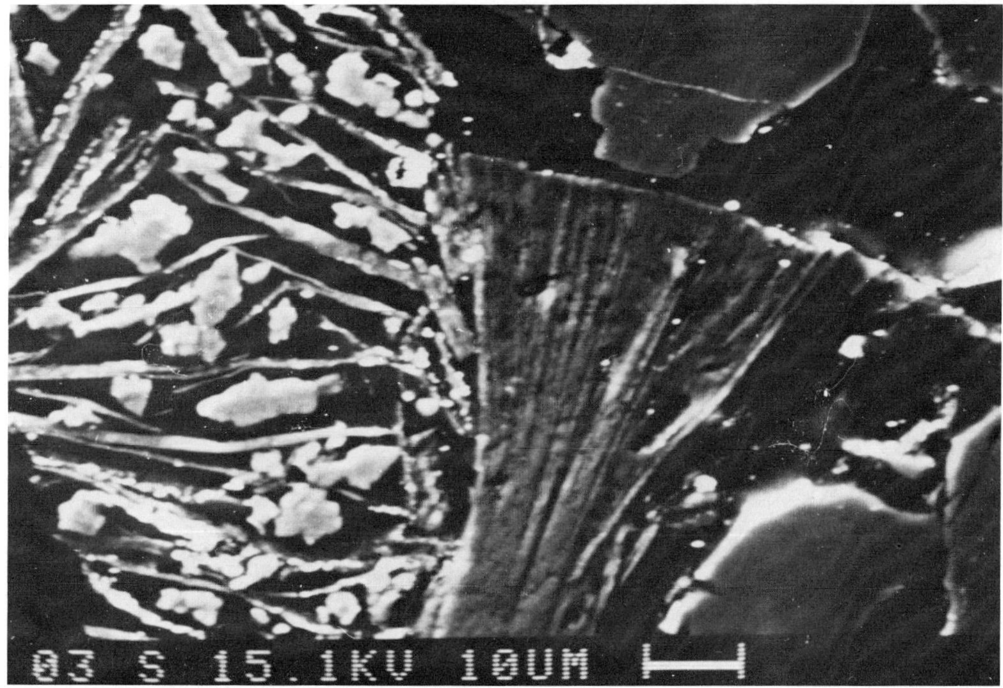

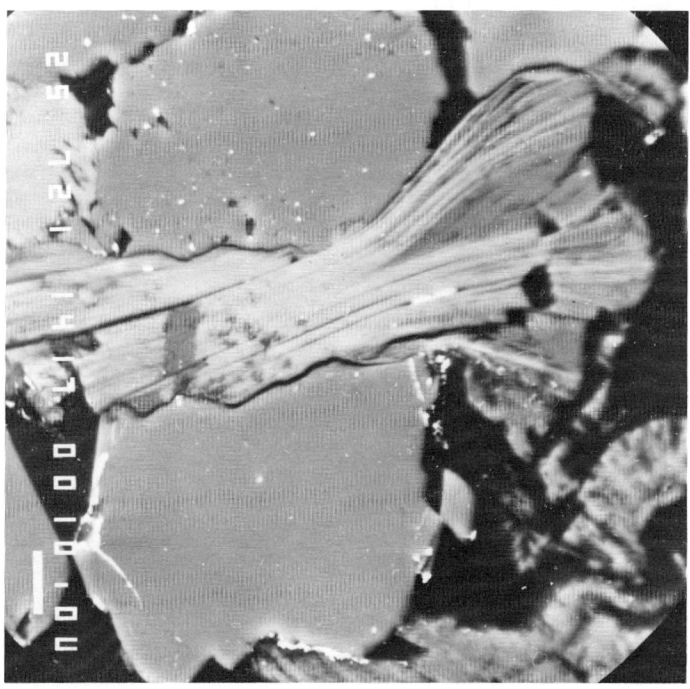

FIG. 6.—Alteration of muscovite to kaolinite and expansion into available pore space, destroying porosity and permeability (Rannoch Formation). A. Electron backscatter picture of muscovite with biotite with siderite on the left side. B. Backscatter photomicrograph of mica squeezed between two quartz grains. Note very significant dissolution of quartz against the mica sheet.

The conglomeratic facies of the Broom Formation are partly cemented with low-Mg calcite, whereas the uppermost few centimeters of the formation are cemented with siderite. The low-Mg calcite unit is a very coarse-grained poikilotopic cement. Each crystal of cement encloses several grains of authigenic kaolinite, pyrite, feldspar and quartz overgrowths. Atomic absorption analyses of this LMC cement show a Sr content of 110 ppm. The relatively low Sr content combined with textural evidence suggests that this LMC represents a late diagenetic precipitation of calcite. It may have formed from dissolution of earlier calcite cement.

In the Rannoch Formation, on the other hand, low-Mg calcite has a Sr content of 600 ppm. Calcite cement with Sr content of this level must represent first generation calcite cement formed directly in pore water during dissolution of aragonite (Kinsman, 1969; Pingitore, 1976, 1978). The high Sr content in the calcite cement suggests relatively local reprecipitation of the dissolved aragonite, i.e., a moderately closed system (Brand and Veizer, 1980). A semi-closed system during aragonite dissolution and calcite precipitation was probably a result of low permeability in the mica-rich Rannoch Formation. During freshwater flushing, the pore-water would flow slowly so that the dissolving aragonite would maintain high Sr/Ca ratios. Assuming a partition coefficient of 0.05, dissolving aragonite with 5000 ppm Sr would precipitate calcite with 250 ppm Sr (Veizer, 1983). Calcite with 600 ppm Sr therefore suggests a semi-closed system.

Leaching of mica contributes to a basic reaction in pore water, particularly in the film near mineral surfaces and between mica sheets. A pH gradient will be established from the mica surface as K^+ is released into the pore water and replaced by H^+ (Bjørlykke et al., 1979; Boles and Johnsen, 1984). The slight increase in pH between the mica sheets may explain the preferred precipitation of calcite (in muscovite) and siderite (in biotite). Siderite formed between sheets of biotite has a distinct composition, particularly with respect to Mg content relative to siderite associated with calcite (Table 1). This shows that the siderite is strongly influenced by the local geochemical environment created by the dissolution of biotite. Ankerite has a rather uniform composition and is interpreted to have formed during deepest burial, probably near present depth.

The cleaner sandstones of the Etive and the Tarbert Formations and parts of the Ness Formation have a much higher permeability than sandstones of the Rannoch Formation, although porosity is similar. This difference most probably existed during early diagenesis, and fluid flow through these sandstones was probably considerably faster in this well than in the Rannoch Formation. These formations do contain siderite, however. They also show evidence of much stronger leaching of muscovite. Biotite is almost completely absent, probably due to leaching.

POROSITY IN THE BRENT SANDSTONES—PRIMARY OR SECONDARY?

The Brent Sandstone of the Statfjord field was cited as an example of sandstone reservoirs with essentially secondary porosity (Schmidt and McDonald, 1979b). Their interpretation implied that the Brent Sandstone was, for the most part, fully carbonate-cemented and that most of the porosity was formed by dissolution of carbonate cement and feldspar. This conclusion was based on observed evidence of carbonate leaching and some leaching of feldspar. There is no evidence, however, to suggest that carbonate cements, which now occur only in very thin beds or nodules, should ever have been pervasive. Leaching of feldspar is common, but later feldspar overgrowths and late carbonate cements such as ankerite usually show no sign of etching. Petrographic evidence described above suggests that leaching was relatively early and was caused by flushing of meteoric pore water. Late-stage leaching was very limited. We suggest, therefore, that most of the leaching was early and was related to freshwater flushing and that the pore water during deeper burial was essentially in equilibrium with feldspar and was buffered by carbonate, both in the sandstones and in the source rocks. There is some evidence of leaching of carbonate cement, however, and also of precipitation of calcite and siderite which may have occurred during deeper burial. This may suggest local dissolution and reprecipitation of carbonate cement without creating much net increase in increased porosity.

Leaching of feldspar and mica is stronger in the more permeable sandstone formations (Etive and Tarbert). This is a pattern expected from meteoric water leaching. Late diagenetic leaching, on the other hand, may be expected to affect the sandstones associated with shale and coal beds more strongly. Sandstones, which must have released significant amounts of CO_2 and which directly overlie coal beds in the Ness Formation, appear not to be particularly strongly leached. This is contrary to what would be expected if CO_2 or carboxylic acids released from coal were important for the dissolution of feldspar.

Dissolution of feldspar and carbonate cement is, in our opinion, quantitatively less important than has been claimed by Schmidt and McDonald (1979b). The CO_2 released by kerogen, particularly sapropelic kerogen in oil source rocks, is insufficient to cause large-scale leaching (Bjørlykke, 1980, 1984; Lundegard et al., 1984). The source rock in the Viking Graben (the Kimmeridge shale) contains dominantly sapropelic kerogen (Geoff, 1983) and would therefore not have released very large quantities of CO_2 (Hunt, 1979). Calculations show that even humic source rocks will release insufficient CO_2 to account for significant amounts of secondary porosity in sedimentary basins (Bjørlykke, 1980, 1984). In the Viking Graben the CO_2 released by maturation of kerogen has probably been neutralized by the carbonate and feldspar in the source rock and thus buffered the pore water before it entered the reservoir rock. Evidence of late-stage leaching may be observed, but it is quantitatively not important.

CONCLUSIONS

Petrography of reservoirs such as the Brent Sandstone describes the end products of reactions that have taken place at temperatures, pressures and pore water compositions very different from the present. The timing of diagenetic reactions is often difficult to interpret, and textural relationships may not always be conclusive. Several diagenetic processes may occur at the same time in near equilibrium with the pore water.

Pore-filling cement such as kaolinite in the Brent Sandstone is enclosed in quartz overgrowth and is therefore earlier or at least contemporaneous with the main phase of quartz cementation. Both the leaching of feldspar and the formation of kaolinite are interpreted to have been caused by the flow of meteoric water into the basin during relatively moderate burial depth. The diagenetic history of the Brent Sandstone is summarized in Figure 7.

By mid-Cretaceous time the Viking Graben was covered by a blanket of low permeability shales, and overpressuring developed during rapid subsidence. This resulted in near stationary pore water and more isochemical diagenesis.

Petrographic analyses of the Brent Sandstone show that mica is the most important factor controlling the distribution of porosity and permeability. High mica content in the lower energy environment of the Rannoch Formation (lower

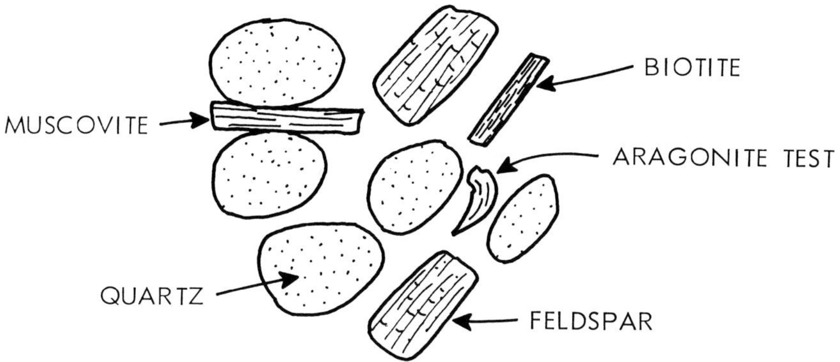

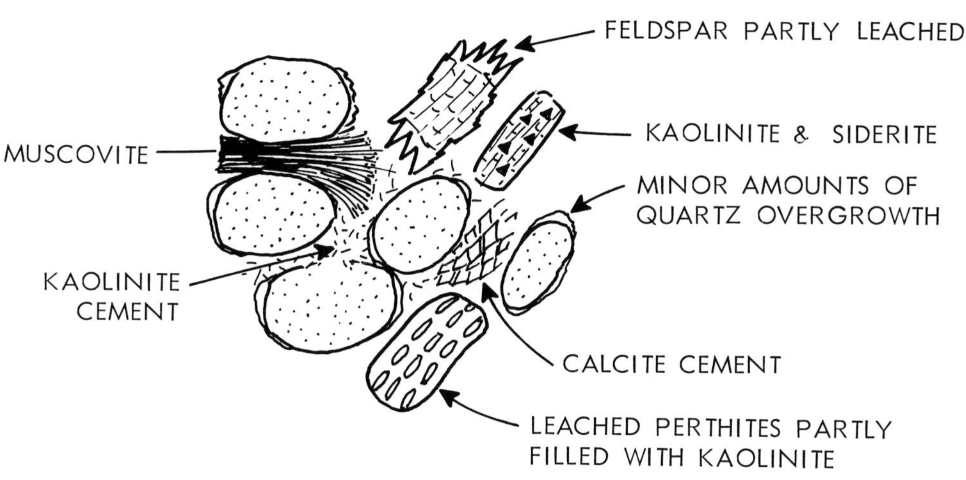

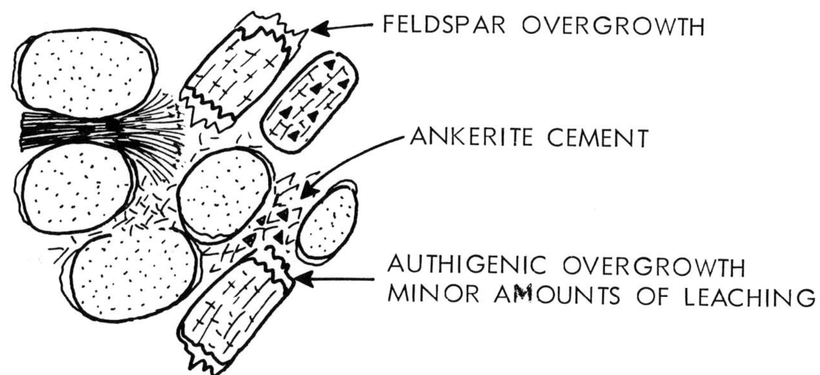

FIG. 7.—Summary of burial diagenesis in the Brent Formation in the Statfjord field.

foreshore) and Ness Formation (lagoonal) exerts a major control on the primary porosity and permeability and on diagenetic reactions. The present study indicates that most of the diagenetic kaolinite in these sandstones is formed by the alteration of mica and that feldspar is a less important precursor. Muscovite altered into kaolinite expands and flares into the pores, causing serious reduction in permeability. Biotite is also often associated with aggregates of siderite precipitated in the local geochemical environment within and around the degraded biotite. Mica may have reduced effective packing, and early compaction quartz and feldspar grains are often partly dissolved by pressure dissolution along mica grains (Fig. 6b).

The presence of unstable mica also contributed to a potassium buffering of the pore water, which enhanced selective leaching of sodium feldspar and subsequent K-feldspar overgrowth. The high content of mica, particularly biotite, is typical of sandstones derived from exposed basement highs in rift basins. This primary composition has a pronounced effect on the diagenetic evolution. Sandstones of this study are thus very different from the sandstones along the Gulf Coast, in which potassium feldspar and plagioclase are selectively dissolved and authigenic albite is precipitated (Boles, 1982) (Table 2). In our study most of the dissolution of feldspar, mica and carbonate can be related to flushing by meteoric water early in the burial history. The selective leaching that is observed in the most permeable beds of the Brent Groups is consistent with what might be expected from meteoric water-flushing in Upper Jurassic and lowermost Cretaceous time. After deposition of the overlying Lower Cretaceous mudstones, the sandstones were probably sealed off from major pore water-flushing except for a slow upward movement of compactional water. The pore water was then probably in near equilibrium with the minerals present and a diagenetic semi-closed system prevailed.

ACKNOWLEDGMENTS

The Statfjord unit owners are thanked for making core and data available for this study. The manuscript has benefited from critical comments by Snorre Olaussen, Statoil Stavanger, and Henning Dypvik, University of Oslo. The authors are grateful to Dag Bergslien for the use of data from his unpublished thesis. Financial support from the Norwegian Research Council for Scientific and Applied Research is gratefully acknowledged.

REFERENCES

Bergslien, D., 1983, Brentgruppen på Blokk 33/9 i Nordre Nordsjøen. Sedimentasjon og Modell for Avsettingsmiljø: unpublished Thesis, University of Bergen.

Bjørlykke, K., 1980, Clastic diagenesis and basin evolution: Revista del Instituto de Investigaciones Geologicas, Universidad de Barcelona, v. 34, p. 21–44.

———, 1983, Diagenetic reactions in sandstones, *in* Parker, A. and Sellwood, B. W., eds., Sediment Diagenesis: Reidel Publishing Company, p. 169–213.

———, 1984, Formation of secondary porosity: How important is it?, *in* McDonald, D. A. and Surdam, R. C., eds.: Clastic Diagenesis: American Association of Petroleum Geologists Memoir 37, p. 277–286.

———, Elverhøy, A. and Malm, A. O., 1979, Diagenesis in Mesozoic sandstones from Spitsbergen and the North Sea—A comparison: Geologische Rundschau, v. 68, p. 1152–1171.

Blanche, J. B. and Whitaker, J. H., McD., 1978, Diagenesis of part of the Brent Sand Formation (Middle Jurassic) of the northern North Sea basin: Journal of Geological Society of London, v. 135, p. 73–82.

Boles, J. R., 1982, Active albitization of plagioclase, Gulf Coast Tertiary: American Journal of Science, v. 282, p. 165–180.

Boles, J. R. and Johnsen, K. S., 1984, Influence of mica surfaces on pore water pH: Chemical Geology, v. 43, p. 303–317.

Brand, U. and Veizer, J., 1980, Chemical diagenesis of a multicomponent carbonate system: Trace elements: Journal of Sedimentary Petrology, v. 50, no. 4, p. 1219–1236.

Budding, M. C. and Inglin, H. G., 1981, A reservoir geological model of the Brent Sands in southern Coremorant, *in* Illing, L. V. and Hobson, G. D., eds., Petroleum Geology of the Continental Shelf of North-West Europe: Heyden & Sons Ltd., London, p. 326–334.

Chauvin, A. L., and Valachi, L. Z., 1980, Sedimentology of the Brent and Statfjord Formations at Statfjord Field *in* The Sedimentation of the North Sea Reservoir Rocks: Norwegian Petroleum Society, Geilo meeting, 1980, chap. XVI, p. 1–17.

Deegan, C. E. and Schull, B. J., 1977, A proposed standard nomenclature for the central and northern North Sea: Report of Institute of Geological Sciences No. 77/25, Bulletin Norwegian Petroleum Directorate No. 1.

Eynon, G., 1980, Basin development and sedimentation in the Middle Jurassic in the northern North Sea, *in* The Sedimentation of the North Sea Reservoir Rocks: Norwegian Petroleum Society, Geilo meeting, 1980, chap. IV, p. 1–19.

Galloway, W. E., 1984, Hydrological regime of sandstone diagenesis, *in* McDonald, D. A. and Surdam, R. C., eds., Clastic Diagenesis: American Association of Petroleum Geologists Memoir 37, p. 3–13.

Geoff, J. C., 1983, Hydrocarbon generation and migration from Jurassic source rocks in the E. Shetland and Viking Graben of the northern North Sea: Journal of Geological Society of London, v. 140, p. 445–474.

Gjeldsvik, N. and Bjørkum, P. A., 1984, The influence of mica on quartz during diagenesis: An example from the North Sea (abs.): Clay Mineralogical Society, 33rd Annual Clay Mineral Conference, Oct. 1984, Baton Rouge, Louisiana.

Gray, W. D. T. and Barnes, G., 1981, The Heather oil field, *in* Illing, L. V. and Hobson, G. D., eds., Petroleum Geology of the Continental Shelf of North-West Europe: Heyden & Sons Ltd., London, p. 335–341.

Hallet, D., 1981, Refinement of the geological model of the Thisth Field, *in* Illing, L. V. and Hobson, G. D., eds., Petroleum Geology of the Continental Shelf of North-West Europe: Heyden & Sons Ltd., London, p. 315–325.

Hamar, G. P., Jacobsen, K. H., Ormaasen, D. E. and Skarpnes, O., 1980, Tectonic development of the North Sea north of the central highs, *in* The Sedimentation of the North Sea Reservoir Rocks: Norwegian Petroleum Society, Geilo meeting, 1980, Chap. III, p. 1–11.

Hancock, N. J., and Fisher, M. J., 1981, Middle Jurassic North Sea deltas with particular reference to Yorkshire, *in* Illing, L. V. and Hobson, G. D., eds., Petroleum Geology of the Continental Shelf of North-West Europe: Heyden & Sons Ltd., London, p. 186–195.

———and Taylor, A., 1978, Clay mineral diagenesis and oil migration in the Middle Jurassic Brent Formation: Journal of Geological Society of London, v. 135, p. 69–72.

Hunt, J. M., 1979, Petroleum Geochemistry and Geology: Freeman & Co., San Francisco, 716 pp.

Irwin, H. and Hurst, A., 1983, Application of geochemistry to sandstone reservoir studies, *in* Brooks, J., ed., Petroleum Geochemistry and Exploration of Europe: Geological Society of London, Blackwell, London, p. 127–146.

Kinsman, D. J. J., 1969, Interpretation of Sr^2 concentrations in carbonate minerals and rocks: Journal of Sedimentary Petrology, v. 39, p. 486–505.

Kirk, R. M., 1980, Statfjord Field—a North Sea giant: American Association of Petroleum Geologists Bulletin, v. 64, p. 95–116.

LUNDEGARD, P. D., LAND, L. S. AND GALLOWAY, W. E., 1984, Problems of secondary porosity: Frio Formation (Oligocene), Texas Gulf Coast: Geology, v. 12, 399–402.

NAGY, J., DYPVIK, H. AND BJAERKE, T., 1984, Sedimentological and paleontological analyses of Jurassic North Sea deposits from deltaic environments: Journal of Petroleum Geology, v. 7, p. 169–188.

PINGITORE, N. E. JR., 1976, Vadose and phreatic diagenesis: Processes, products and their recognition in corals: Journal of Sedimentary Petrology, v. 46, p. 985–1006.

———, 1978, The Behavior of Zn^{2+} and Mn^{2+} during carbonate diagenesis: Theory and applications: Journal of Sedimentary Petrology, v. 48, p. 799–814.

PROCTOR, C. V., 1980, Distribution of Middle Jurassic facies in the east Shetland Basin and their control of reservoir capability, *in* The Sedimentation of the North Sea Reservoir Rocks: Norwegian Petroleum Society, Geilo meeting, 1980, chap. XV, p. 1–22.

SCHMIDT, V. AND MCDONALD, D. A., 1979a, Texture and recognition of secondary porosity in sandstones: Society of Economic Paleontologists and Mineralogists Special Publication 26, p. 209–225.

———AND———, 1979b, The role of secondary porosity in the course of sandstone diagenesis: Society of Economic Paleontologists and Mineralogists Special Publication 26, p. 157–207.

SKARPNES, O., HAMAR, G. P., JACOBSEN, K. H. AND ORMAASEN, D. E., 1980, Regional Jurassic setting of the North Sea north of the central highs, *in* The Sedimentation of the North Sea Reservoir Rocks: Norwegian Petroleum Society, Geilo meeting, 1980, p. XIII, 1–8.

SURDAM, R. C., BOESE, S. W. AND CROSSEY, L. J., 1984, The chemistry of secondary porosity, *in* McDonald, D. A. and Surdam, R. C., eds., Clastic Diagenesis: American Association of Petroleum Geologists Memoir, 37, p. 127–150.

VEIZER, J., 1983, Chemical diagenesis of carbonates—Theory and application of trace element technique: American Association of Petroleum Geologists, Short Course No. 10, p. 1–100.

WAUGH, G., 1978, Authigenic K-feldspar in British Permo-Triassic sandstones: Journal of Geological Society of London, v. 135, p. 51–56.

WHITE, S. H., SHAW, H. F. AND HUGGET, J. M., 1984, The use of backscattered electron imaging for petrographic studies of sandstones and shales: Journal of Sedimentary Petrology, v. 54, p. 487–492.

ZIEGLER, P. A., 1981, Evolution of sedimentary basins in north-west Europe, *in* Illing, L. V. and Hobson, G. D., eds., Petroleum Geology of the Continental Shelf of North-West Europe: Heyden & Sons Ltd., London, p. 3–39.

THERMAL MASS TRANSFER IN SYSTEMS CONTAINING QUARTZ AND CALCITE

JAMES R. WOOD
COFRC La Habra, Ca. 90631

ABSTRACT: Advective diagenesis predicts that large-scale mineral migration will occur when saturated pore fluids move across isotherms (or isobars). A general picture of the change in a given porosity field can be calculated if the fluid flow and temperature (or pressure) fields are known. For more detailed analysis it is necessary to know the details of the chemical equilibria involved, their temperature and pressure dependence, the host rock mineralogy and, in particular, the thermal mass transfer coefficients for the phases involved. In this paper thermal mass transfer coefficients for quartz and calcite have been calculated in detail over the temperature range 0–200°C. Simple approximations (involving only equilibrium constants) have been obtained for the concentrations of the aqueous species as well as for the solid mass precipitated or dissolved.

Quartz solubility is prograde over the entire temperature range of interest, and the mass transfer coefficients for quartz range from 0.1 ppm/C° at 0°C to about 4 ppm/C° at 200°C.

In a closed system calculation shows that calcite solubility is prograde up to 125°C, and then becomes retrograde, while in open systems calcite solubility is retrograde over this temperature range. The mass transfer coefficients for calcite range from +0.1 ppm/°C in a closed system at 0°C to greater than −60 ppm/°C at 0°C under 62 atm of CO_2 pressure. At higher temperatures these coefficients seem to trend toward a common asymptotic value of 0(?) ppm/°C at temperatures greater than 200°C.

These calculations suggest that *low* temperatures and *high* CO_2 pressures are most effective in mobilizing calcite but that high temperatures are more effective in mobilizing quartz. The calculations support recent suggestions that decarboxylation of organic matter is involved in the development of secondary porosity.

INTRODUCTION

Advective diagenesis has been defined as the component of diagenesis due to, or arising from, the movement of pore fluids (Wood and Hewett, 1982; this volume; Davis et al., 1985). Previous contributions to the theory of advective diagenesis have concentrated on the spatial distribution of the diagenetic products and the role of geometric factors.

The purpose of this paper is to show how the models for advective diagenesis described by Davis et al. (1985) and by Wood and Hewett (this volume) can be used to address the problem of quartz and calcite migration in the deep subsurface. This paper will detail the calculation of the thermal mass transfer coefficients for quartz and calcite, since such calculations have not been previously published.

STATEMENT OF THE PROBLEM

The integrated porosity equation for a monomineralic system consisting of pore fluid and a quartz matrix can be written (Wood and Hewett, this volume),

$$\phi = \int_t \{(\rho_f/\rho_s)\, \alpha_{T(q)}\, \underline{V} \cdot \text{grad } T\} dt + \phi_o \qquad (1)$$

where ϕ is the porosity field, ρ_f is the fluid density, ρ_s is the solid (quartz) density, $\underline{V}$ is the fluid flow field, T is the temperature field, t is time and ϕ_o is the original or non-advective porosity field. In contrast to the *fields*, $\alpha_{T(q)}$, the thermal mass transfer coefficient for quartz, is a *scalar* quantity and represents the mass of the *phase* quartz transferred per unit temperature change (units of ppm/°C).

Equation (1) can be regarded as a product of a

$$\{\text{chemical term}\} = (\rho_f/\rho_q)\, \alpha_{T(q)} \qquad (2)$$

and a

$$\{\text{field term}\} = \underline{V} \cdot \text{grad } T, \qquad (3)$$

which are, to a first approximation, independent of each other. The field, or geometric, term has been analyzed (Wood and Hewett, this volume) and the principal consideration here will be the scalar chemical term.

Thermal Mass Transfer Coefficient: Definition

The thermal mass transfer coefficient for a mineral is defined as the mass of that mineral which must be dissolved or precipitated in order that an aqueous fluid can maintain thermodynamic equilibrium with that mineral over a unit temperature change.

The quantity α_T can be computed from thermodynamic data describing the equilibrium between quartz and aqueous fluids or taken directly from solubility measurements of quartz as a function of temperature, provided that the measurements refer to a system in which the solution composition is the same as in the system of interest. In simple cases, the system quartz − H_2O for example, the mass transfer coefficient can be equated directly to the temperature coefficient for quartz solubility,

$$\alpha_{T(\text{qtz})} = (dC_{s(\text{qtz})}/dT)_P \qquad (4)$$

where $C_{s(\text{qtz})}$ is the concentration (in ppm) of $SiO_{2(aq)}$ in a fluid in equilibrium with quartz at temperature T. Since $C_{s(\text{qtz})} = K_q$, α_T is simply the temperature derivative of the equilibrium constant for the reaction $\text{Quartz}_{(s)} = SiO_{2(aq)}$ and can be related to the standard state enthalpy of reaction. In general, however, the interpretation of α_T is not so straightforward.

Two-Mineral Systems

If other silica-bearing minerals are present or other solid-fluid reactions are occurring, then the problem becomes more involved since the mass of SiO_2 that precipitates as quartz or dissolves from quartz becomes a function of these variables as well. The interphase mass transfer becomes a complex function of simultaneous and in some cases competing equilibria.

When two minerals, say quartz and calcite, are present

in a rock and both are in equilibrium with the pore fluid, the integrated porosity equation (1) is written

$$\phi = (\rho_f [\alpha_{T(q)}/\rho_q + \alpha_{T(c)}/\rho_c] \underline{V} \cdot \text{grad } T)t + \phi_o. \quad (5)$$

The term in brackets is the chemical term and contains all the chemistry-dependent parameters in the advective diagenesis model. Since the mineral densities are essentially constant with temperature, the only significant variation is in the α_T parameters.

Organization of Paper

A step-by-step approach will be followed in the analysis below to insure a complete analysis and to facilitate development of algorithms. This paper is organized as follows. For any general system four steps are involved:

1. define a chemical model,
2. build a thermodynamic model for that chemical model,
3. construct a mathematical model by combining the chemical and thermodynamic models, and
4. solve the mathematical model.

The following sections are concerned with the details of solubility calculations for the systems quartz-water and calcite-water over a 200°C temperature range. The paper concludes with a discussion of the problem of transporting large quantities of calcite in the subsurface and the approach to that problem using the advective model.

QUARTZ SOLUBILITY

The chemical and thermodynamic properties of the system SiO_2–H_2O are well known over a wide range of temperatures and pressures. The solubility of quartz in water has been reviewed by Holland and Malinin (1979). In addition, Fleming and Crerar (1982) have discussed the ionization of silicic acid, while experimental work on quartz solubility was done by Kennedy (1950), Morey (1962), Anderson and Burnham (1965), and Crerar and Anderson (1971), among others.

The Chemical Model

The chemical model for the quartz-solution system consists of solid SiO_2 (quartz), the aqueous molecules $H_4SiO_4^0$, H_2O, and the ions, H^+, $H_3SiO_4^-$ and OH^-. The pertinent chemical reactions in this system are:

$$H_2O = H^+ + OH^- \quad (6)$$

$$H_4SiO_4^0 = H_3SiO_4^- + H^+ \quad (7)$$

and

$$SiO_{2(qtz)} + H_2O_{(l)} = H_4SiO_4^0 \quad (8)$$

The corresponding mass action expressions for these reactions are:

$$K_w = [H^+][OH^-] \quad (9)$$

$$K_a = [H_3SiO_4^-][H^+]/[H_4SiO_4^0] \quad (10)$$

and

$$K_q = [H_4SiO_4^0] \quad (11)$$

where the brackets indicate concentrations (in moles/kg_{H_2O}). In this paper, activity/concentration corrections are ignored. We can either regard the solutions as being dilute enough that the activity coefficients are close to unity or we can regard the equilibrium constants as referring to the solution composition of interest.

The Thermodynamic Model

The thermodynamic model for this system consists of the equilibrium constants for the above reactions over the temperature range 0–300°C. The pK ($-\log_{10} K$) values are plotted as a function of temperature in Figure 1. Values for pK_a and pK_q were computed from expressions given by Fleming and Crerar (1982) using K_w values from Olofsson and Hepler (1975).

The Mathematical Model

In this system the quantities that are considered unknowns are the concentrations of the aqueous species, $[H^+]$, $[OH^-]$, $[H_3SiO_4^-]$ and $[H_4SiO_4^0]$. At any temperature the silicic acid concentration $[H_4SiO_4^0]$ is given directly by the equilibrium constant (equation 11) and the ion concentrations are obtained by combining the electrical-neutrality equation

$$[H^+] = [H_3SiO_4^-] + [OH^-] \quad (12)$$

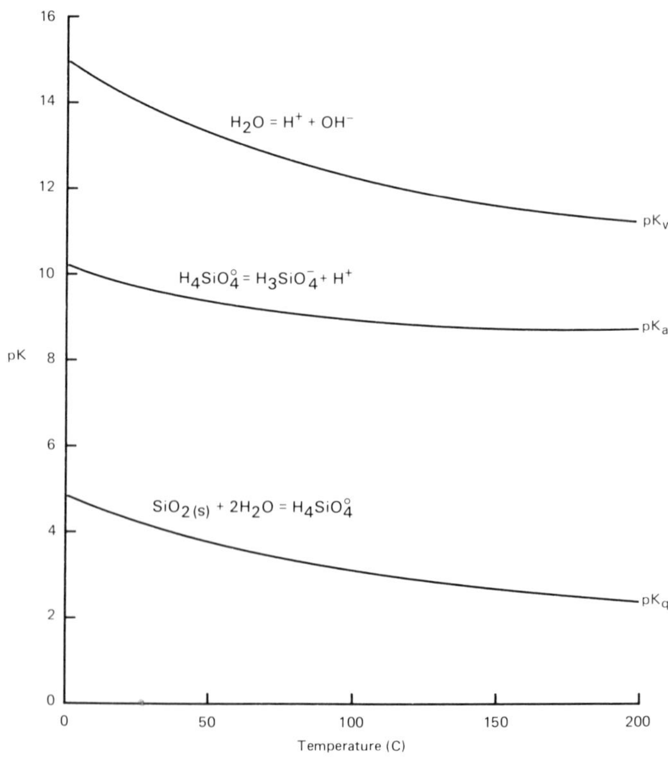

FIG. 1.—Thermodynamic model of the system SiO_2–H_2O over the temperature range 0–200°C. Reactions indicated on curves.

with the mass action equations (9) and (10) to obtain

$$[H^+] = K_a[H_4SiO_4^0]/[H^+] + K_w/[H^+] \quad (13)$$

or

$$[H^+] = (K_aK_q + K_w)^{1/2} \quad (14)$$

If we now define a quantity, S_q, as the mass of quartz that must dissolve to saturate an aqueous fluid at some temperature T, it follows that S_q is essentially the total dissolved silica

$$S_q = [H_4SiO_4^0] + [H_3SiO_4^-] \quad (15)$$

$$= K_q(1 + K_a/[H^+] \quad (16)$$

and is usually identified with the solubility of the phase. Curves showing the variation of S_q with temperature are presented in Figure 2 for two cases, quartz in pure water and in 0.001 m NaOH. It is apparent that quartz is much more soluble at high pH levels, but the thermal mass transfer coefficient, $\alpha_{T(q)}$, is not necessarily severely affected.

The Thermal Mass Transfer Coefficient for Quartz

Since the thermal mass transfer coefficient for quartz, $\alpha_{T(q)}$, is defined as the amount of quartz that must be precipitated or dissolved to maintain equilibrium over a *unit* temperature change, and S_q is defined as the mass required to saturate a fluid at T, the relationship between the two parameters must be

$$\alpha_{T(q)}, = d(S_q)/dT \times (\text{At. Wt.} \times 10^3), \quad (17)$$

where S_q has been converted from molality to ppm to conform with the units of ppm/C. $\alpha_{T(q)}$ can be obtained by taking the derivative of S_q with respect to temperature using equation (14) to estimate $d[H^+]/dT$. Because a sum of terms is involved in equation (13), it is difficult to reduce the expression without approximation. Assuming that $[H_4SiO_4^0] >> [H_3SiO_4^-]$, however, which will be true for all pH $< pK_a$, then

$$S_q = K_q \quad (18)$$

and

$$\alpha_{T(q)} = d\,K_q/dT \quad (19)$$

$$= K_q\,\Delta H_q^0/RT^2, \quad (20)$$

where ΔH_q^0 is the standard state enthalpy for reaction (8) at T and the relation

$$d \ln K/dT = \Delta H/RT^2 \quad (21)$$

has been used.

Equation (20) shows that $\alpha_{T(q)}$ is proportional to the enthalpy of reaction for the case where $H_4SiO_4^0$ is the dominant silica species. For the case $[H_3SiO_4^-] >> [H_4SiO_4^0]$, the result is somewhat different. Now we have

$$S_q \doteq [H_3SiO_4^-] \quad (22)$$

$$= K_aK_q/[H^+] \quad (23)$$

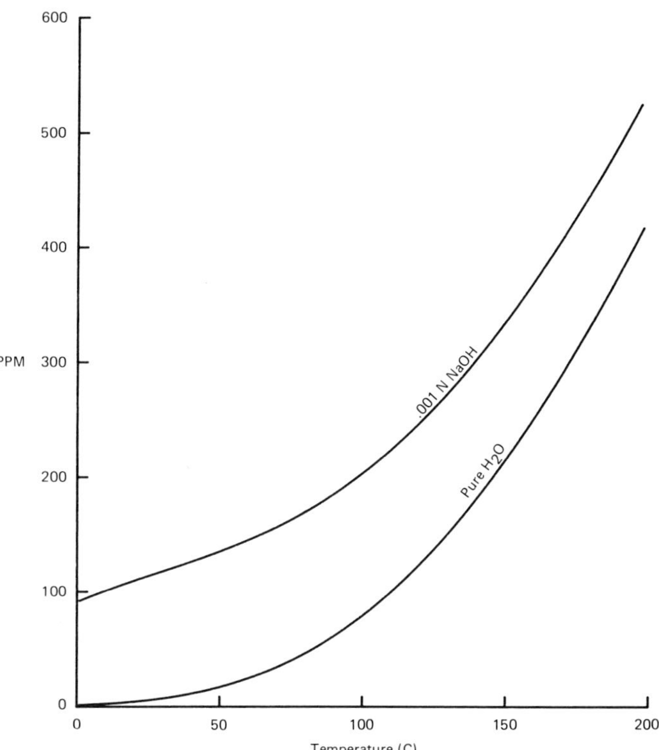

FIG. 2.—Solubility of quartz in pure H_2O and in 0.001 N NaOH as a function of temperature.

and

$$d(pS_q)/dT = d(pK_q)/dT + d(pK_a)/dT - d(\text{pH})/dT. \quad (24)$$

The term $d(\text{pH})/dT$ may be evaluated using equation (14). Note two limiting cases:

(i) when $[H_3SiO_4^-] >> [OH^-]$,

$$\text{pH} = (pK_q + pK_a)/2$$

$$d(pS_q)/dT = [d(pK_q)/dT + d(pK_a/dT)]/2,$$

and,

(ii) when $[OH^-] >> [H_3SiO_4^-]$

$$\text{pH} = pK_w/2$$

$$d(pS_q)/dT = d(pK_q)/dT + d(pK_a)/dT - [d(pK_w)/dT]/2.$$

Since $d(pK)/dT = -2.303d$, $(\ln K)/dTd(pS_q/dT)$ can be expressed in terms of the relevant enthalpies of reaction. Because of the approximations involved, however, these expressions are more useful in illuminating the thermodynamic connection between $\alpha_{T(q)}$ and S_q than in actually evaluating the thermal mass transfer coefficients.

A more practical, and general, approach is to use equations (14) and (16) to calculate the solution composition and quartz solubility over a temperature range centered on the temperature of interest and then obtain $\alpha_{T(q)}$ by numerical differentiation. A suitable formula (Scheid, 1968, pp. 245) for the derivative is

$$\alpha_{T(i)} = (-2y_{k-2} - y_{k-1} + y_{k+1} + 2y_{k+2})/10h, \quad (25)$$

where $y_k = S_q$ at temperature T_k and h is the temperature interval. (To calculate α_T at 100°C in a solution where $H_4SiO_4^0$ is the dominant species for example, use values of $[H_4SiO_4^0]$ at 98, 99, 101 and 102°C and set $h = 1$.)

Values of $\alpha_{T(q)}$ computed in this fashion over the temperature range 0–200°C are plotted in Figure 3 for the two cases quartz-water and quartz in 0.001 m NaOH. $\alpha_{T(q)}$ lies in the range 0–5 ppm/°C, even in the NaOH solutions, and from 50°C on, the two curves are practically coincident.

The curves in Figure 3 lead to an interesting conclusion, namely that the mass of quartz transferred by advection will not be greatly influenced by pH, even for pH values greater than pK_a. What difference exists tends to become less as temperature increases. (These results strictly refer to the system quartz-water and should not be applied to more complex systems, particularly those containing other silica phases, such as feldspars and clays.)

CALCITE SOLUBILITY—CLOSED SYSTEMS

The solubility of calcite in water has been reviewed by Holland and Malinin (1979) and also by Plummer and Busenberg (1982). Experimental work has been done by Ellis (1959, 1963), Malinin (1963) and Holland and Borcsik (1965) among others, whereas detailed solubility calculations under a variety of conditions have been made by Garrels and Christ (1965) and Stumm and Morgan (1970).

The calcite-water system is similar to the quartz-water system in that only one solid phase is involved, but the dissociation of the solid into ions, the formation of the complexes HCO_3^- and $H_2CO_3^0$, and the presence of $CO_{2(g)}$ change the analysis significantly and lead to a wider range of behaviors. In the following discussion it will be assumed that the system consists only of $CaCO_3$, H_2O and salts. In particular, the partial pressure of $CO_{2(gas)}$ is not fixed, so that, at equilibrium for any temperature and fluid composition, the moles of total dissolved calcium, T_{Ca}, are equal to the moles of total dissolved carbon, T_C.

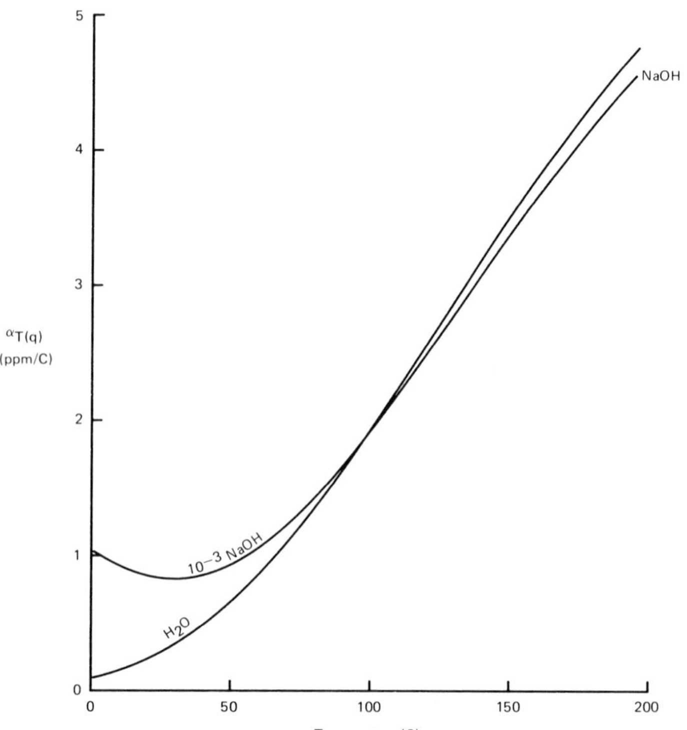

FIG. 3.—Thermal mass transfer coefficients for quartz in pure H_2O and in 0.001 N NaOH as a function of temperature.

The Chemical Model

The chemical model for the calcite-water system consists of the ions H^+, Ca^{++}, OH^-, HCO_3^-, $CO_3^=$, the neutral species H_2O, $H_2CO_3^0$ and $CO_{2(g)}$ and the solid $CaCO_3$. The chemical reactions to be considered include the water hydrolysis equation (7), the heterogeneous reactions

$$CaCO_{3(aq)} = Ca^{++} + CO_3^= \tag{26}$$

$$H_2CO_3^0 = H_2O + CO_{2(g)} \tag{27}$$

and the homogeneous reactions

$$H_2CO_3^0 = H^+ + HCO_3^- \tag{28}$$

$$HCO_3^- = H^+ + CO_3^=. \tag{29}$$

The mass action equations for these reactions are:

$$K_c = [Ca^{++}]\,[CO_3^=] \tag{30}$$

$$K_h = [H_2CO_3^0]/P_{CO2} \tag{31}$$

$$K_1 = [H^+]\,[HCO_3^-]/[H_2CO_3{}^0] \tag{32}$$

$$K_2 = [H^+]\,[CO_3^=]/[HCO_3^-]. \tag{33}$$

The Thermodynamic Model

The thermodynamic model consists of the set of equilibrium constants for the above reactions. These functions are plotted over the temperature range 0–200°C in Figure 4 in terms of pK values. There is significant variation in all of the pK values over this temperature range, particularly pK_w. The pK values for the carbonate dissociation reactions, K_1 and K_2, initially decrease with increasing temperature, but pass through a minimum at about 50°C and then increase with increasing temperature. pK_c for calcite increases with temperature over the entire range.

The pK_1 and pK_2 curves in Figure 4 divide the temperature range into regions which can be interpreted as domains for the various aqueous carbonate species. According to equation (33), when $[Co_3^=] = [HCO_3^-]$, pH $= pK_2$, and, from equation (32), when $[HCO_3^-] = [H_2CO_3^0]$, pH $= pK_1$. Thus, a pK-T plot is, in effect, a pH-T plot when referring to the fluid phase and defines dominant-species fields in the fluid phase as a function of temperature.[1]

The Mathematical Model

The mass balance on total dissolved carbon,

$$T_C = [H_2CO_3^0] + [HCO_3^-] + [CO_3^=] \tag{34}$$

[1]Plots of this type are particularly useful when it is necessary to make approximations to simplify the solubility expressions and will be used in the calculations below. pK_a serves a similar function for the quartz-water system; in fluids with pH values greater than pK_a, $[H_3SiO_4^-]$ is greater than $[H_4SiO_4^0]$.

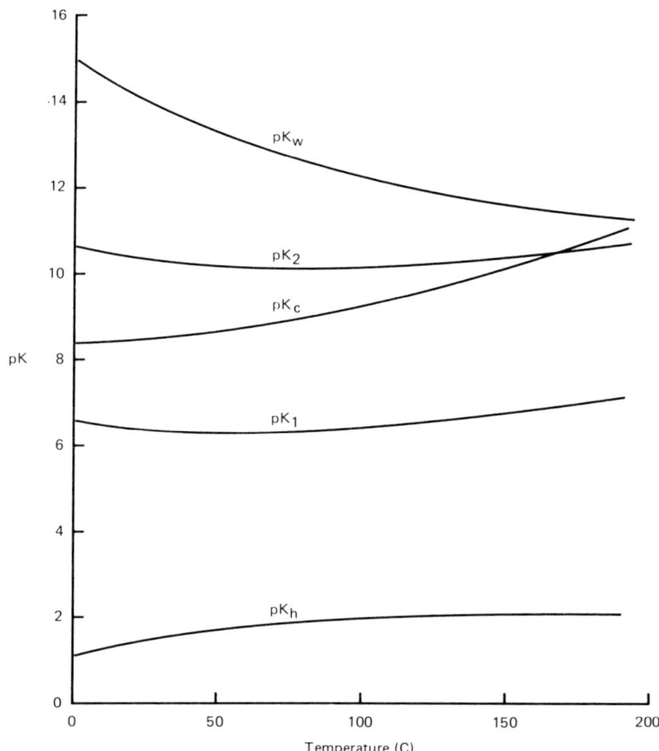

FIG. 4.—Thermodynamic model of the system $CaCO_3$–H_2O over the temperature range 0–200°C. See text for explanation of subscripts on *pK* titles.

holds for both open and closed systems, and, using equations (32) and (33), may be written

$$T_C = \beta[H_2CO_3^0]. \tag{35}$$

Where the distribution coefficient, β is defined as

$$\beta = 1 + K_1/[H^+] + K_1K_2/[H^+]^2. \tag{36}$$

It is convenient to set the model up to include solutions containing salts with common ions. For this system two types of salts are sufficient, one that dissolves to produce calcium ions and one that produces carbonate ions. As far as the mathematical model for this system is concerned, the salts

$$m_1 = [CaCl_2] \tag{37}$$

and

$$m_2 = [Na_2CO_3] \tag{38}$$

can be used as proxies for any number of salts with common ions.

A variable, S_c, is defined similar to S_q, such that at any particular temperature S_c is the mass of calcite that must dissolve in order to saturate the solution with calcite. With this definition, the following mass balances

$$T_{Ca} = S_c + m_1 = [Ca^{++}] \quad \text{(total dissolved calcium)} \tag{39}$$

$$T_C = S_c + m_2 = \beta[H_2CO_3^0] \quad \text{(total dissolved carbon)} \tag{40}$$

can be written for a closed system. Since

$$K_1K_2 = [H^+]^2[CO_3^=]/[H_2CO_3^0], \tag{41}$$

we find, from equations (32), (33) and (40), that equation (41) leads to

$$[CO_3^=] = (S_c + m_2)K_1K_2/\beta[H^+]^2 \tag{42}$$

and

$$[HCO_3^-] = (S_c + m_2)K_1/\beta[H^+]. \tag{43}$$

Substituting reactions (39) and (42) into the mass action equation (30), yields

$$K_c = (S_c + m_1)(S_c + m_2)K_1K_2/\beta[H^+]^2, \tag{44}$$

or, expanding and rearranging,

$$S_c^2 + (m_1 + m_2)S_c + \{m_1m_3 - \beta[H^+]^2K_c/K_1K_2\} = 0. \tag{45}$$

Equation (45) can be solved exactly using the quadratic formula to yield,

$$S_c = [-B + (B^2 - 4AC)^{1/2}]/2A, \tag{46}$$

where

$$A = 1 \tag{47}$$

$$B = (m_1 + m_2) \tag{48}$$

$$C = \{m_1m_2 - \beta[H^+]^2K_c/K_1K_2\}. \tag{49}$$

Equation (46) expresses S_c in terms of the hydrogen ion and known quantities. The remaining step is to substitute S_c into the electrical-neutrality equation

$$[Na^+] + 2[Ca^{++}] + [H^+] = [HCO_3^-] + 2[CO_3^=] + [OH^-] + [Cl^-], \tag{50}$$

eliminate $[OH^-]$ using equation (10), $[HCO_3^-]$ using equation (43), and $[CO_3^=]$ using equation (42), so that equation (50) becomes

$$2m_2 + 2(S_c + m_1) + [H^+] = K_1(S_c + m_2)/[H^+] + 2K_1K_2(S_c + m_2)/\beta[H^+]^2 + KK_w/[H^+] + 2m_1. \tag{51}$$

Equation (51) may be solved by scanning over the pH range 0–14 looking for the value of $[H^+]$ which satisfies (51). This is a somewhat involved calculation (see example 1 "$CaCO_3$ in Pure Water" in Garrels and Christ, 1965, pp. 77), and is usually carried out numerically. The concentrations of salts, m_1 and m_2, are embedded in equation (51) in such a way that no easy algebraic reduction is possible.

When m_1 and m_2 are zero,

$$S_c = (\beta[H^+]_2K_c/K_1K_2)^{1/2}, \tag{52}$$

and reaction (51) becomes

$$2S_c + [H^+] = S_c K_1 \beta [H^+] (1 + 2K_2/[H^+]) + K_w/[H^+]. \quad (53)$$

The solution compositions calculated numerically from equation (53) for calcite-saturated solutions are plotted as a function of temperature in a speciation diagram (Fig. 5). It is apparent that the bicarbonate ion is the dominant carbon species over much of the temperature range. At low temperatures the carbonate ion concentration is nearly equal to that of the bicarbonate ion, but $[CO_3^=]$ decreases rapidly with temperature. The concentration of carbonic acid increases steadily with temperature and apparently will replace $[HCO_3^-]$ as the principal carbon species at temperatures above 200°C.

The relations in Figure 5 suggest that the electrical-neutrality equation is dominated by three ions, calcium, bicarbonate and hydroxide, and that equation (50) can be approximated by

$$2[Ca^{++}] = [HCO_3^-] + [OH^-]. \quad (54)$$

This form of the electrical-neutrality equation permits algebraic reduction to a complete analytic solution. Solving equation (54) analytically for $[H^+]$ and then back-substituting for the rest of the aqueous species yields

$$[H^+] = (K_w^2 K_2/K_c)^{1/3} \quad (55)$$

$$[Ca^{++}] = S_c = (K_c K_w/K_2)^{1/3} \quad (56)$$

$$[CO_3^=] = (K_c^2 K_2/K_w)^{1/3} \quad (57)$$

$$[HCO_3^-] = (K_c K_w/K_2)^{1/3} \quad (58)$$

$$[OH] = (K_c K_w/K_2)^{1/3} \quad (59)$$

$$[H_2CO_3^0] = K_w/K_1 \quad (60)$$

$$P_{CO2} = K_w/K_h K_1 \quad (61)$$

as primitive expressions for the ion concentrations and P_{CO2}. It is interesting that the expressions for calcium, bicarbonate and hydroxide ions are identical, and that carbonic acid concentration is fixed at each temperature by the hydrolysis constants for reactions (6) and (28).

These approximate expressions do, in fact, describe the solubility of calcite in pure water fairly well over the temperature range 0–200°C and agree with other, more exact, calculations. For example, calculating the pH at 25°C using the pK values given by Garrels and Christ (1965, pp. 77–78) for this temperature ($pK_w = 14$, $pK_2 = 10.2$, $pK_c = 8.3$), equation (55) yields pH = 10.0 compared to their value of 9.95. Using equation (56) as an approximate expression for calcite solubility, $pS = 4.0$ compared with their estimate of 3.9 (for $[Ca^{++}]$).

Discussions: Calcite Solubility in Closed Systems

The solubility of calcite calculated from the exact equations is shown as a function of temperature in Figure 6. This plot, which corresponds to the $[Ca^{++}]$ curve in Figure 5, shows that the solubility of calcite is initially prograde (increases with increasing T), passes through a maximum at about 125°C and then becomes retrograde (decreases with increasing temperature).

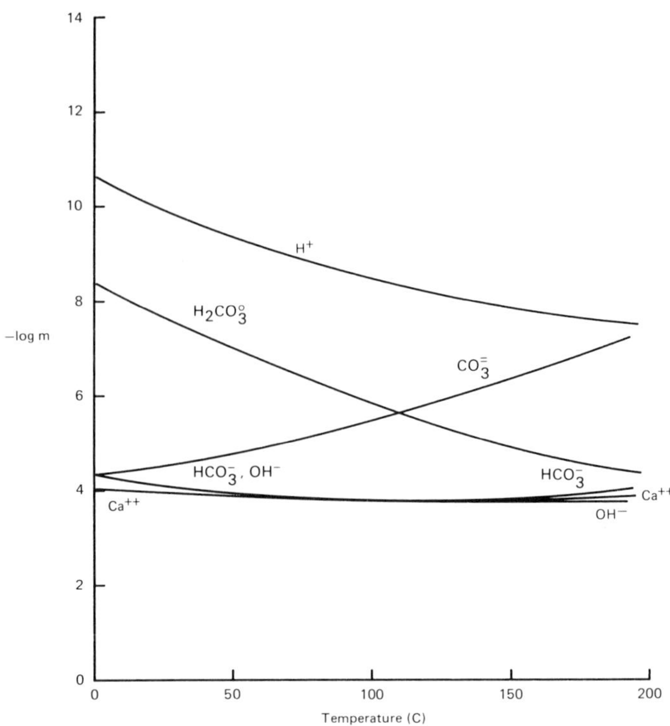

FIG. 5.—Speciation diagram for the system calcite-water over the temperature range 0–200°C. Curves calculated from exact electrical-neutrality equation. See text for explanation.

A plot of pH versus temperature for these solutions (Fig. 7) shows that the pH decreases monotonically from a value of 10.6 at 0°C to about 7.5 at 200°C. Superimposing plots of pH = pK_1 and pH = pK_2 on this figure demonstrates that $[HCO_3^-]$ is the dominant dissolved carbonate species from 0°C to about 200°C. The "intrinsic" partial pressure of CO_2 in equilibrium with these solutions calculated from equation (61) varies from less than 10^{-7} atm to more than 10^{-3} atm over this temperature range (Fig. 8).

Calculations thus suggest that calcite solubility may not always be retrograde but becomes slightly prograde at low temperatures in closed systems and under conditions of intrinsic P_{CO2}. No single reason for the prograde behavior can be given, but since it occurs in bicarbonate-dominated solutions, analysis of the approximate solutions from equations (55) through (61) may provide some insights.

Thermal Mass Transfer Coefficient for Calcite in Closed Systems

For calcite, $\alpha_{T(c)}$ is the temperature derivative of S_c,

$$\alpha_{T(c)} = dS_c/dT \cdot 10^5, \quad (62)$$

where the 10^5 factor converts from molality to ppm. Figure 8 is a plot of $\alpha_{T(c)}$ versus temperature. The magnitudes of $\alpha_{T(c)}$ are much lower than the quartz values at all temperatures, ranging from +0.07 ppm/°C at about 70°C to −0.07 ppm/°C to 180°C.

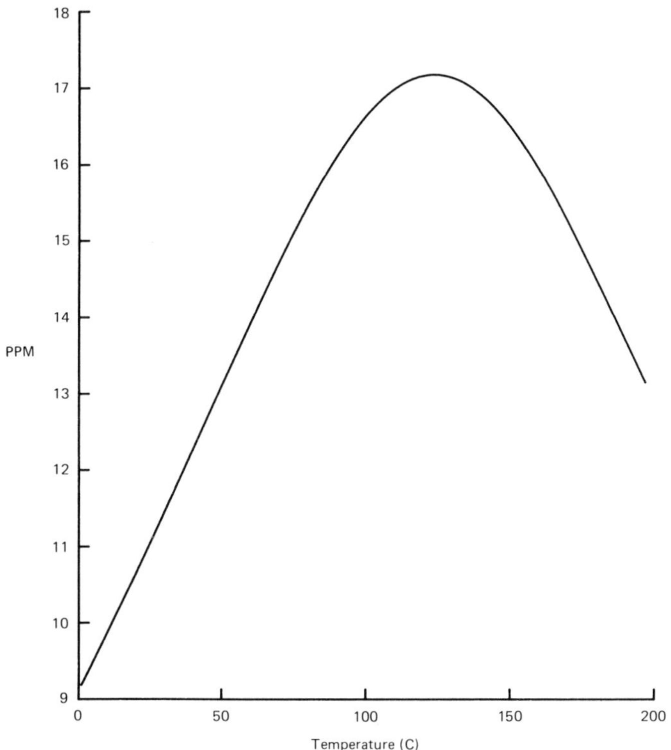

FIG. 6A.—Solubility of calcite in pure H_2O over the temperature range 0–200°C. Closed system, partial pressure of CO_2 variable determined by equilibria in system.

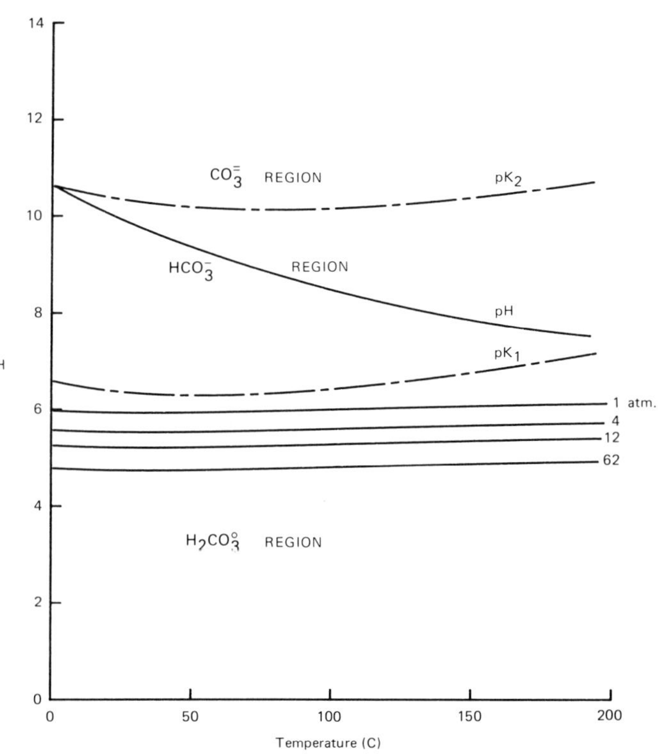

FIG. 7.—Plot of pH versus temperature for both closed (upper curve labeled "pH") and open (lower curves labeled "1" through "62") systems. The number on the lower pH curves refers to the partial pressure of CO_2 imposed on the system. The dashed curves refer to the first and second dissociation constants of carbonic acid and delineate the dominant ion fields for the dissolved carbon species (see below).

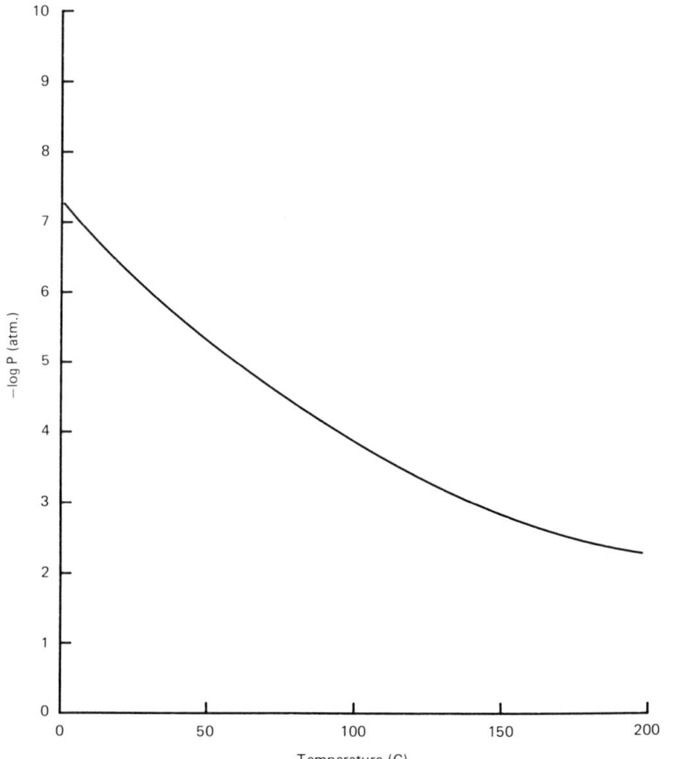

FIG. 6B.—Intrinsic partial pressure of CO_2 in equilibrium with calcite in the system $CaCO_3$–H_2O over the temperature range 0–200°C.

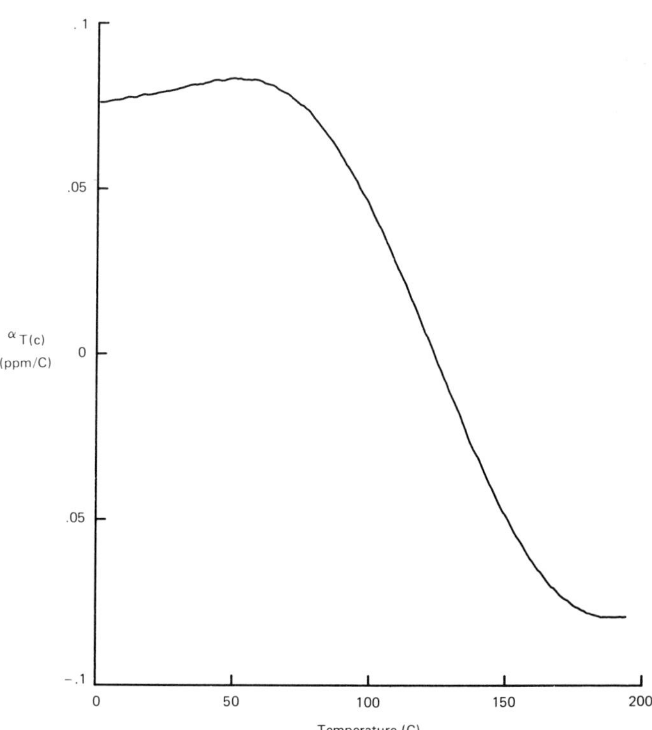

FIG. 8.—Plot of the thermal mass transfer coefficient for the closed system $CaCO_3$–H_2O as a function of temperature from 0 to 200°C.

CALCITE SOLUBILITY—OPEN SYSTEMS: THE EFFECT OF $CO_{2(g)}$

The calculations presented so far refer to a closed system, meaning that the partial pressure of $CO_{2(g)}$ is fixed externally. The CO_2 partial pressure can also be specified as an initial condition, however, an "open" system, and the whole problem reworked under that condition. The following discussion refers to open systems in which the partial pressure of $CO_{2(g)}$ is somehow fixed or imposed externally.

The Chemical and Thermodynamic Models for an Open System

The chemical and thermodynamic models for an open calcite-water system are the same as for the closed system. The only changes occur in the mathematical model. Consequently, the previous discussion for these two models applies here.

The Mathematical Model

Equation (32) shows that specifying P_{CO2} fixes $[H_2CO_3^0]$ and makes $[HCO_3^-]$ and $[CO_3^=]$ functions of $[H^+]$ only, so that equations (32), (33) and (34) can be rewritten as

$$[H_2CO_3^0] = K_h\, P_{CO2} \tag{63}$$

$$[HCO_3^-] = K_hK_1P_{CO2}/[H^+] \tag{64}$$

$$[CO_3^=] = K_hK_1K_2P_{CO2}/[H^+]^2. \tag{65}$$

Equations (34–35) still apply, but since total carbon is no longer conserved in the system, equation (40) is no longer valid. The balance on total dissolved calcium, equation (39), remains unchanged, but the balance on total dissolved carbon, equation (34), becomes

$$T_C = \beta K_hP_{CO2}. \tag{66}$$

Note that the total dissolved carbon is a function of the imposed CO_2 fugacity and pH and is not a function of either m_2, the salt with the common anion, or dissolved calcite. This change in the total carbon balance, along with the changes in equations (31), (32) and (33), are the primary differences between the open and closed system.

Substituting equations (39) and (65) into the mass action equation (30) for calcite now yields

$$K_c = (S_c + m_1)\,(K_hK_1K_2P_{CO2}/[H^+]^2), \tag{67}$$

or,

$$S_c = K_c[H^+]^2/(K_hK_1K_2P_{CO2}) - m_1. \tag{68}$$

If $m_1 = 0$ and the electrical-neutrality equation is approximated as before,

$$2[Ca^{++}] = [HCO_3^-] + [OH^-], \tag{69}$$

then the primitive solutions for the open system are:

Low P_{CO2}.—

$$[H^+] = [K_hK_1K_2P_{CO2}(K_hK_1P_{CO2} + K_w)/2K_c]^{1/3} \tag{70}$$

$$[Ca^{++}] = S_c = [K_c(K_hK_1P_{CO2} + K_w)^2/(4K_hK_1K_2P_{CO2})]^{1/3} \tag{71}$$

$$[CO_3^=] = [K_c^2K_hK_1K_2P_{CO2}/4(K_hK_1P_{CO2} + K_w)^2]^{1/3} \tag{72}$$

$$[HCO_3^-] = [2K_c(K_hK_1P_{CO2})/K_2)K_hK_1P_{CO2} + K_w)] \tag{73}$$

$$[H_2CO_3^0] = K_hP_{CO2} \tag{74}$$

$$P_{CO2} = (\text{fixed}). \tag{75}$$

These equations apply to the temperature range 0–200°C for P_{CO2} values similar to those generated by the closed system.

High P_{CO2}.—

For P_{CO2} values greater than 0.01 atm, the approximate electrical-neutrality equation

$$2[Ca^{++}] = [HCO_3^-] \tag{76}$$

can be invoked, because as the partial pressure of $CO_{2(g)}$ increases, the pH decreases and $[OH^-]$ can be neglected compared with $[HCO_3^-]$, leading to a simpler set of primitive solutions,

$$[H^+] = [K_h^2K_1^2K_2P_{CO2}^2/2K_c]^{1/3} \tag{77}$$

$$[Ca^{++}] = S_c = [K_cK_hK_1P_{CO2}/4K_2]^{1/3} \tag{78}$$

$$[CO_3^=] = [K_c^2K_2/(4K_hK_1P_{CO2})]^{1/3} \tag{79}$$

$$(HCO_3^-] = [2K_cK_hK_1P_{CO2}/K_2]^{1/3}, \tag{80}$$

where $[H_2CO_3^0]$ and P_{CO2} are given by equations (74) and (75). These solutions are easily derived from equations (70) through (73) by setting the CO_2 pressures high enough that $K_hK_1P_{CO2} >> K_w$, in effect setting the CO_2 partial pressure high enough that the hydoxyl ion can be neglected compared to the bicarbonate ion.

Discussion: Calcite Solubility in Open Systems

The solubility of calcite calculated from the high P_{CO2} equation (78) for CO_2 pressures of 1, 4, 12 and 64 atm increases markedly from the closed system values (Fig. 9). These solubility curves are qualitatively similar to those in Holland and Malinin (1979), but displaced to lower values, probably reflecting smaller values for K_c.

It is clear that the calcite solubility is retrograde over the temperature range 0–200°C when these relatively high P_{CO2} values are imposed on the system. Equations (77) and (78) show that both $[H^+]$ and calcite solubility are proportional to $(P_{CO2})^{1/3}$.

Thermal Mass Transfer Coefficients for Calcite in Open Systems

The $\alpha_{T(c)}$ coefficients for open systems are given by equation (63) using equations (71) or (78) for S_c. Plots of $\alpha_{T(c)}$ versus temperature are shown in Figure 10A. The inset, Figure 10B, shows that at P_{CO2} values of less than approximately 10^{-2} atm, the $\alpha_{T(c)}$ temperature variation starts to reflect closed system behavior. At still lower P_{CO2} values,

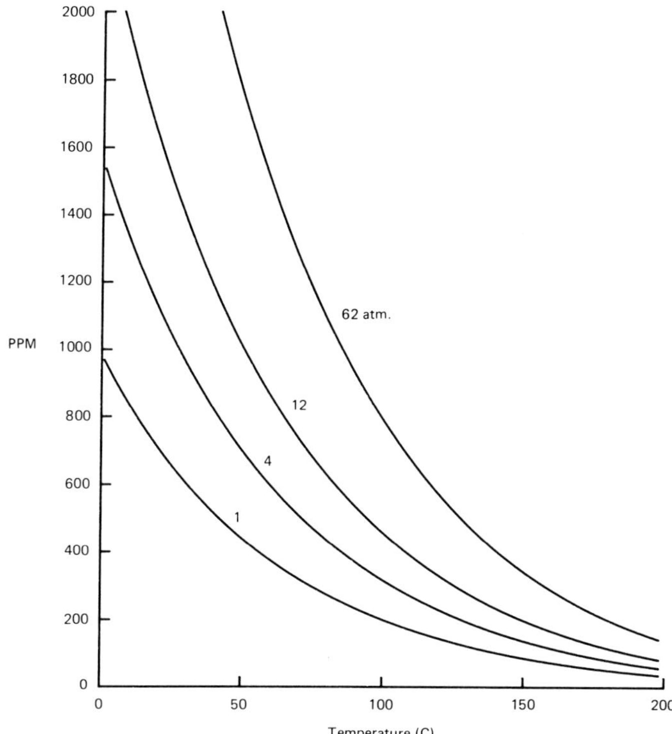

FIG. 9.—Solubility of calcite in open systems as a function of temperature and the partial pressure of CO_2. Numbers on curves refer to partial pressure of CO_2 (in atmospheres) imposed on the system.

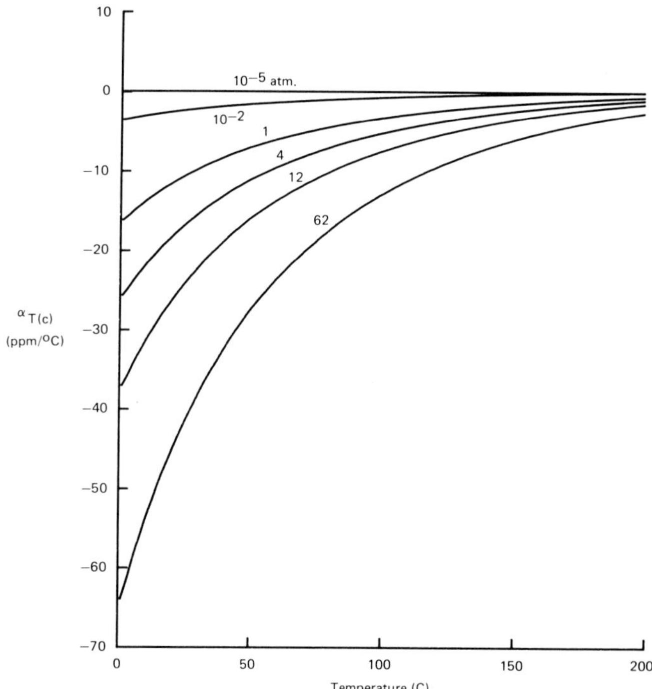

FIG. 10A.—Thermal mass transfer coefficients for calcite in the open system $CaCO_3$–H_2O–CO_2 as a function of temperature and CO_2 pressure.

on the order of 10^{-3} atm, both the $\alpha_{T(c)}$ functions and the solubility become positive over part of the temperature range as they approach the P_{CO2} values of the closed system.

APPLICATION TO DIAGENESIS

The preceding calculations of quartz and calcite solubilities under various conditions are primarily intended to show that the temperature coefficients for the chemical term in the advective diagenesis model can be obtained from solubility calculations. The diagenetic consequences of using the parameters calculated above in the advective model can be evaluated by analyzing a system in which a fluid flows through a rock containing both quartz and calcite.

Equation (5) shows that the chemical term for the porosity field, $[\alpha_{T(q)}/\rho_q + \alpha_{T(c)}/\rho_c]$, depends on the algebraic signs of $\alpha_{T(q)}$ and $\alpha_{T(c)}$ and their relative magnitudes. The porosity field depends on the sense of fluid movement with respect to the temperature field, however, and it is the combination of the algebraic sign of the α_T's with the sense of fluid movement which determines the source/sink relationships for any particular mineral. The following discussion elaborates on this point.

Some Important Cases

Case 1.—

$\alpha_{T(q)}$ and $\alpha_{T(c)}$ both positive or both negative.

When $\alpha_{T(q)}$ and $\alpha_{T(c)}$ both have the same sign, quartz and calcite either dissolve or precipitate together, depending on the sign of $\underline{V} \cdot$ grad T. In the system quartz-calcite-fluid under diagenetic conditions, $\alpha_{T(q)}$ is always positive (Fig. 3). Similarly, $\alpha_{T(c)}$ is positive for closed systems at temperatures less than 125°C, or for some open systems with very low CO_2 fugacities (Fig. 8). The advective model would

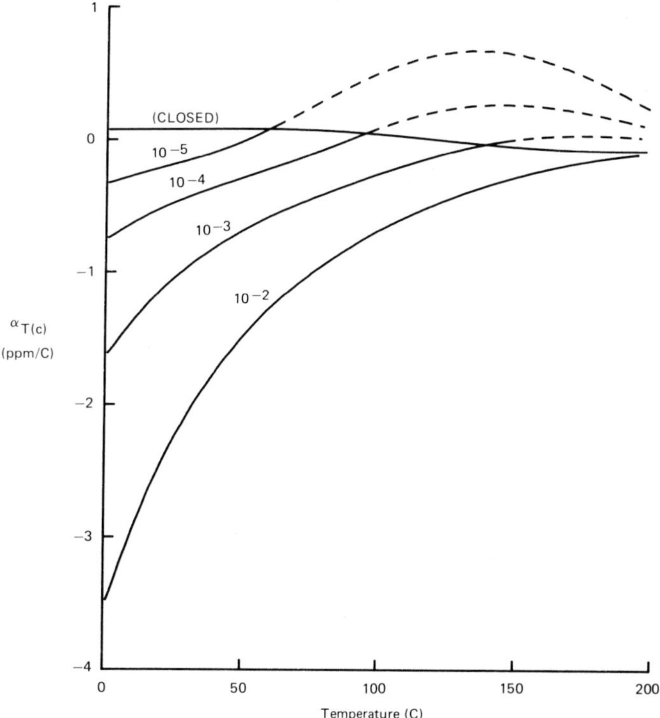

FIG. 10B.—Thermal mass transfer coefficients for calcite in the open system $CaCO_3$–H_2O–CO_2 as a function of temperature for low CO_2 pressures. The CO_2 pressures shown are in the range of the "intrinsic" CO_2 pressures for the closed system.

predict that quartz overgrowths and calcite cements occur at low temperatures in closed systems when the fluid is moving upward (cooling), whereas both quartz and calcite dissolve when the fluid is descending (heating).

Case 2.—

$\alpha_{T(q)}$ and $\alpha_{T(c)}$ have opposite signs.

When $\alpha_{T(q)}$ and $\alpha_{T(c)}$ have opposite signs, one mineral dissolves while the other precipitates. Which mineral precipitates and which mineral dissolves at a particular location depends on the algebraic signs of α_T relative to the sense of fluid movement determined by $\underline{V} \cdot$ grad T. This case applies to the replacement of one mineral for another, as for instance, in the process of isomorphic replacement.

Case 3.—

$\alpha_{T(q)} = -\alpha_{T(c)}$.

Whenever $\alpha_{T(q)}$ and $\alpha_{T(c)}$ have values such that the chemical term vanishes altogether, there is no net porosity change. The calcite removal just balances the quartz precipitation or vice versa, and the result is *volume for volume* replacement. This may be the required condition for isomorphic replacement and suggests that that process can occur during advective diagenesis when the chemical terms for two minerals are equal and opposite in sign.

Case 4

$\alpha_{T(q)} >> \alpha_{T(c)}$ or $\alpha_{T(q)} << \alpha_{T(c)}$.

When the *magnitudes* of the chemical terms are grossly out of balance, $\alpha_{T(i)} >> \alpha_{T(j)}$, then conditions are appropriate for complete cementation with the one phase. When $\alpha_{T(q)} >> \alpha_{T(c)}$, for example, the rock pores will eventually fill with quartz in regions where the fluid flow is directed upward. In regions where the flow is directed downward, quartz will appear to dissolve more readily or rapidly than calcite. In the quartz-calcite-fluid system, these conditions are likely to occur when the system is closed, since the thermal mass transfer coefficients for calcite are much less than for quartz (0.07 ppm/°C versus 1–3 ppm/°C) in the temperature range 50–150°C.

When systems containing calcite are open and subjected to high CO_2 fugacities, conditions are appropriate for calcite to dominate the mass transfer. When P_{CO2} values exceed, say 10^{-2} atm, $\alpha_{T(c)}$ can have large negative values, exceeding 100 ppm/°C at temperatures less than 100°C. Under such conditions, calcite will precipitate in regions where the fluid flow is directed downward and will be leached in regions of upward fluid movement.

Secondary Porosity: The Case For Calcite Leaching by Advection

It has been suggested that large-scale movement of minerals, particularly calcite, occurs in the subsurface (Land and Dutton, 1978; Land, 1984; Bjorlykke, 1982; Boles and Franks, 1979). The mass transfer involved is often sufficient to create secondary porosity comparable to the original depositional porosity, and some oil fields (e.g., Rangeley gas and oil field in Colorado) may owe their existence to secondary porosity.

Whereas there is more or less general agreement that large masses of material have moved, there is less agreement on the transfer mechanism. One argument calls for a large supply of acid, usually carbonic acid, to dissolve the calcite on a mole-for-mole basis. Calculations (Lundegard and Land, 1984; Bjorlykke, 1982, and others), however, show that there is not enough carbon present in sediments to account for the quantities of carbonic acid required, even if all the available organic carbon in the nearby rocks was converted to CO_2. If so, these arguments would rule out any postulated mechanism which calls on molar acid-to-solid (CO_2 to calcite) ratios on the order of unity.

For the advective model considered here, it is easy to compute the relative concentrations of organic and inorganic carbon in the fluid.

For an open system with high CO_2 fugacity, equations (77)–(80) show that the total dissolved carbon is approximately $[H_2CO_3^0] + [H\bar{C}O3] = K_h P_{CO2} + 2S_c$, while the total inorganic carbon is equal to the amount of dissolved calcite, S_c. It follows that the total dissolved organic carbon is $K_h P_{CO2} + S_c$.

The dissolved carbon from organic sources will thus be larger than the carbon obtained from calcite by an amount $K_h P_{CO2}$. The effect of increasing the partial pressure of carbon dioxide is primarily to increase the dissolved organic carbon (principally as $H_2CO_3^0$), but the amount of dissolved calcite also increases with increasing P_{CO2}. This is in agreement with previous calculations (Bjorlykke, 1982), which show that the conversion of large amounts of organic carbon to acid is necessary to dissolve significant amounts of calcite.

These results also suggest that very high CO_2 pressures only increase the concentration of the $H_2CO_3^0$ molecule in solution and only marginally increase the calcite solubility. From a mass transfer aspect, however, it is the *rate* of change of calcite solubility with increasing CO_2 pressure that is important. For any given temperature, the thermal mass transfer coefficient increases proportionally to $(P_{CO2})^{1/3}$,

$$\alpha_T = k(T)\,[P_{CO2}]^{1/3}$$

where

$$k(T) = d[K_c K_h K_1 / 4K_2]^{1/3} / dT.$$

Several conclusions can be drawn from this relation, depending on the type of flow system involved.

Convective Flow Fields

When a convective flow field is involved, the fluid streamlines form closed loops and a fluid parcel eventually returns to its original position with its original chemical composition. This recirculating type of flow affects net transfer of the solid matrix by virtue of flowing through a temperature field (see Wood and Hewett, 1982). If that matrix is calcite, and the CO_2 pressure is high enough, then calcite slowly migrates along the streamlines.

This type of a system derives its mass transfer potential

chiefly by virtue of fluid recirculation. Circulation of pure water would eventually effect a substantial transfer of solid matrix, but the rate of the transfer can be enhanced by a number of factors, including increasing carbon dioxide pressure in the case of a carbonate matrix.

The effect of increasing CO_2 pressure under these conditions is an increase in the *rate* of matrix transfer. More solid is transferred in a given time for a given flow rate, or alternatively, the flow rate can be reduced and still be effective.

Figure 10A shows that the rate of thermal mass transfer can change from nearly zero for a closed system to values on the order of -10 ppm/°C for a CO_2 pressure of 1 atm. Consequently, one is tempted to conclude that advective transfer of calcite is very ineffective in the absence of a source of carbon to generate CO_2.

Forced Fluid Flow

When the fluid flow is forced, the streamlines do not form closed loops (see Wood and Hewett, this volume) and the mass transfer due to the higher CO_2 pressures becomes less of a factor. This occurs because the organically-generated carbon moves out of the system with the fluid, forcing the CO_2 source(s) to continually supply carbon dioxide to the incoming fluid; and, as calculated above, the mass of CO_2 required to saturate the fluid increases in direct proportion to the CO_2 pressure.

The effect of a non-recirculating flow is thus to reduce the efficiency of carbonate mass transfer. It can be anticipated that more detailed calculations on specific models will show that they are significantly less efficient in transferring carbonates, perhaps by several orders of magnitude.

It is too early to conclude that the development of secondary porosity in carbonate-bearing rocks is itself evidence of recirculating fluids, but theory supports the conclusion that convecting systems in conjuction with elevated levels of carbon dioxide favor development of secondary porosity in such rocks.

The fact that the transfer of calcite is enhanced at lower temperatures also fits the organic maturation model (Tissot and Welte, 1978) in which CO_2 is an early reaction product. Franks and Forester (1984) have argued this case for the development of secondary porosity in the Texas Gulf Coast.

Along these lines it is known that high CO_2 pressures commonly exist in the subsurface (see Hunt, 1979, pp. 150–185), but the only other high pressure gases found in large quantities in the subsurface are hydrocarbon gases (methane, ethane, etc.), N_2 and H_2S (Hunt, 1979). These gases will not directly affect the thermal mass transfer coefficients for the common diagenetic minerals. (They may have some influence on the percentage of organic carbon converted to CO_2 and on total pressure.)

SUMMARY

In summary, theoretical calculation of the thermal mass transfer coefficients for calcite support arguments and observations that calcite should be preferentially mobilized in the subsurface in response to increasing carbon dioxide concentrations. Lower temperatures would appear to be favored over higher temperatures for the most intense transfer, but higher CO_2 pressures can compensate for temperature up to a point. The advective model for calcite mobilization would appear to dovetail nicely with the current thinking in kerogen maturation.

Some caution needs to be exercised in generalizing the results presented here. In particular, it must be kept in mind that the mineralogy treated is very simple, and that the presence of other carbonates (e.g., dolomite, ankerite, etc.) can affect the thermal mass transfer coefficients for calcite. Fluid composition can also have an effect, as implied in equation (44).

Application of the advective model to deep burial diagenesis offers a new approach to the problem of generation of secondary porosity in calcite-bearing rocks, particularly by emphasizing the distinction between systems with circulating fluids and those with non-recirculating fluids. Future work can be directed toward more specific cases and more complex mineralogies.

REFERENCES

ANDERSON, G. M. AND BURNHAM, C. W., 1965, The solubility of quartz in supercritical water: American Journal Science, v. 263, p. 494–511.

BJORLYKKE, K. O., 1982, Clastic diagenesis and basin evolution: Revista del Instituto de Investiagaciones Geologica, v. 34, p. 21–44.

BOLES, J. R. AND FRANKS, S. G., 1979, Clay diagenesis in Wilcox sandstones of southwest Texas: Implications of smectite diagenesis on sandstone cementation: Journal Sedimentary Petrology, v. 49, p. 55–70.

CERAR, D. A. AND ANDERSON, G. M., 1971, Solubility and solvation reactions of quartz in dilute hydrothermal solutions: Chemical Geology v. 8, p. 107–122.

DAVIS, S. H., ROSENBLAT, S., WOOD, J. R., AND HEWETT, T. A., 1985, Convective fluid flow and diagenetic patterns in domed sheets: American Journal of Science, p. 207–223.

ELLIS, A. J., 1959, The solubility of calcite in carbon dioxide solutions: American Journal of Science, v. 257, p. 354–365.

———, 1963, The solubility of calcite in sodium chloride solutions at high temperatures: American Journal of Science, v. 261, p. 259–267.

FLEMING, B. A. AND CRERAR, D. A., 1982, Silicic acid ionization and calculation of silica solubility at elevated temperature and pH: Geothermics, v. 11, p. 15–29.

FRANKS, S. G. AND FORESTER, R. W., 1984, Relationships among secondary porosity, pore-fluid chemistry and carbon dioxide, Texas Gulf Coast, *in* McDonald, D. A. and Surdam, R. C., eds., Clastic Diagenesis: American Association of Petroleum Geologists, Tulsa.

GARRELS, R. M. AND CHRIST, C. L., 1965, Solutions, Minerals, and Equilibria: Harper & Row, N. Y., p. 77–83.

HOLLAND, H. D., AND BORCSIK, M., 1965, On the solution and deposition of calcite in hydrothermal systems: Symposium, Problems of Postmagmatic Ore Deposition, v. 2, Prague, p. 364–374.

——— AND MALININ, S. D., 1979, the solubility and occurrence of nonore minerals , *in* Barnes, H. L., ed., Geochemistry of Hydrothermal Ore Deposits, 2nd ed.: Wiley-Interscience, New York.

HUNT, J. M., 1979, Petroleum Geochemistry and Geology: W. H. Freeman and Company, San Francisco.

KENNEDY, G. C., 1950, a portion of the system silica-water: Economic Geology, v. 45, p. 629–653.

LAND, L. S., 1984, Frio Sandstone Diagenesis, Texas Gulf Coast: A Regional Isotopic Study, *in* McDonald, D. A., and Surdam, R. C., eds., Clastic Diagenesis: American Association of Petroleum Geologists, Tulsa.

———, AND DUTTON, S. P., 1978, Cementation of a Pennsylvanian deltaic sandstone: Isotopic data: Journal of Sedimentary Petrology, v. 48, p. 1167–1176.

LUNDEGARD, P. D., AND LAND, L. S., 1984, Problem of secondary porosity: Frio Formation (Oligocene), Texas Gulf Coast: Geology, v. 12, p. 399–402.

MALININ, S. D., 1963, An experimental investigation of the solubility of calcite and witherite under hydrothermal conditions: Geokhimiya, v. 7, p. 631–646.

MOREY, G. W., 1962, The action of water on calcite, magnesite, and dolomite: American Mineralogist, v. 47, p. 1456–1460.

OLOFSSON, G., AND HEPLER, L. G., 1975, Thermodynamics of ionization of water over wide ranges of temperature and pressure: Journal of Solution Chemistry, v. 4, p. 127–144.

PLUMMER, L. N. AND BUSENBERG, E., 1982, The solubilities of calcite, aragonite and vaterite in CO_2-H_2O solutions between 0 and 90°C, and an evaluation of the aqueous model for the system $CaCO_3$-CO_2-H_2O: Geochimica et Cosmochimica Acta, v. 46, p. 1011–1040.

SCHEID, F., 1968, Theory and Problems of Numerical Analysis: McGraw-Hill, N. Y

STUMM, W., AND MORGAN, J. J., 1970, Aquatic Chemistry: Wiley-Interscience, N. Y.

TISSOT, B. P., AND WELTE, D. H., 1978, Petroleum Formation and Occurrence: Springer-Verlag, New York.

WOOD, J. R., AND HEWETT, T. A., 1982, Fluid convection and mass transfer in porous sandstones—a theoretical model. Geochimica et Cosmochimica Acta, v. 46, p. 1707–1713.

APPENDIX

A RELATIONSHIP BETWEEN CALCITE SOLUBILITIES IN OPEN AND CLOSED SYSTEMS

There is an interesting relationship between the calcite solubility curves in an open system and those in a closed system. If a solubility curve is calculated for each P_{CO2} value corresponding to an intrinsic P_{CO2} value, then a family of curves is produced, each one of which is tangent to the closed system solubility curve at one point.

Figure 11 shows a series of solubility curves for open calcite-water systems plotted against temperature over the range of CO_2 partial pressures corresponding to closed calcite-water systems. These curves are initially retrograde but pass through a minimum and become prograde. Superimposing the solubility curve for the closed system on Figure 11 shows that open system solubility curves are tangent[2] to the closed system curve, and thus that the solubility curve for calcite in a closed system is defined by a family of open system solubility curves.

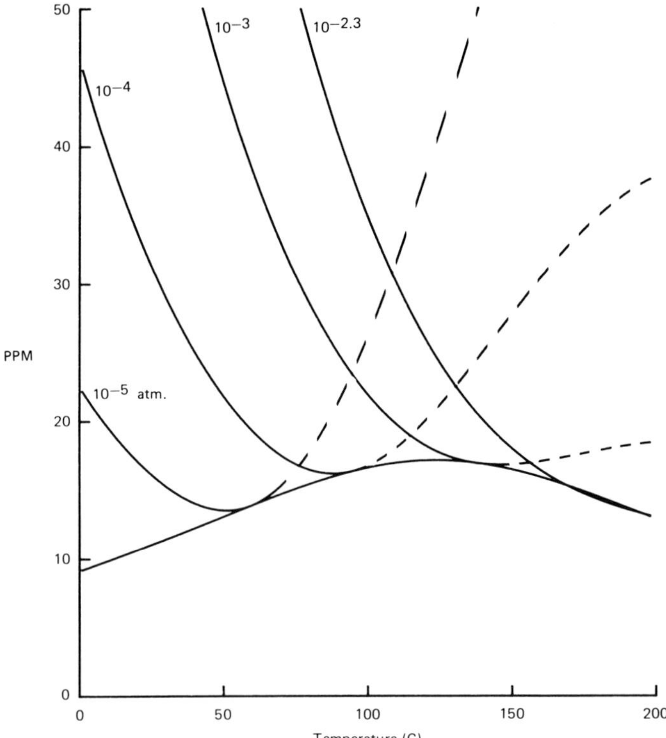

FIG. 11.—Plot of calcite solubility in both open and closed systems. The full curve on the bottom refers to the closed system. The upper curves refer to open systems under the CO_2 pressure indicated. The dashed portions of the upper curves refer to physically unattainable solution compositions. See text for explanation.

[2]This tangent point does not correspond to the minimum in the open system curves, but occurs where $T_C = T_{Ca}$ in the open system and can be demonstrated by showing that the solution compositions for the open and closed systems are the same at the point of tangency.

FORCED FLUID FLOW AND DIAGENESIS IN POROUS RESERVOIRS—CONTROLS ON THE SPATIAL DISTRIBUTION

J. R. WOOD AND T. A. HEWETT
Chevron Oil Field Research Company, Box 446, La Habra, CA 90631

ABSTRACT: When pore fluids flow across isotherms and maintain chemical equilibrium with the mineral assemblage, a certain amount of matrix mass transfer must occur due to the temperature dependence of the equilibrium constants involved. The fluid flow field V interacts with the temperature field T to produce a spatially varying component of the total diagenetic field (the advective diagenetic field), which is proportional to $V \cdot$ grad T where grad T is the gradient of the temperature field. The spatial distribution of the advective field depends on the geometry of the system (e.g., domes, anticlines, etc.) and to a first approximation is independent of the *authigenetic* mineral assemblage. The intensity of the alteration, or equivalently, the time required to reach a certain degree of alteration, is proportional to the permeability and thermal conductivities of the rocks involved. For cases when simple forced flow dominates, the alteration is relatively homogeneous vertically. The lateral variation, however, can be intricate, even for simple geometries, since the distribution of the altered assemblage is proportional to the directional derivative of the function(s) describing the structures.

The porosity isopleths (points of equal porosity change) migrate through the rock, diverging from regions of net mass removal and converging on regions of net accumulation. The movement of these isopleths provides a sense of the net mass flows in the system.

INTRODUCTION

The purpose of this paper is to discuss the role of a *forced* fluid flow in the diagenetic modification of a porous reservoir, particularly with respect to the spatial distribution of the diagenetic assemblage. The dynamics of convective fluid flow and the attendant diagenesis have been discussed previously (Wood and Hewett, 1982, 1984; Davis et al., 1985) and will not be treated here. The diagenesis associated with forced fluid flow has not been treated, however, even though it is in some aspects simpler to treat than convective flow. This paper will develop the necessary theory to include forced flow in the diagenetic calculations.

We start with the assumption that pore fluids are not static and that the circulation of fluids is intimately related to the spatial distribution of the diagenetic mineral assemblage. The rock alteration arising as a result of pore fluid flow will be referred to as "advective" or "fluid" diagenesis, partly to distinguish it from diagenetic alteration that is not related to pore fluid flow and partly to relate it generically to fluid flow.

Free and forced flow.—

Fluid diagenesis can be discussed in terms of either forced flow or free flow. Forced flow results from application of a force to a boundary of the fluid, whereas free flow occurs when unbalanced body forces act on every particle of fluid. Flow driven by a hydrodynamic head is an example of forced flow, as is flow driven by compaction of sediments or by the dewatering of hydrous minerals, such as illite and montmorillonite. Convective flow, driven by an unstable density gradient in the fluid, is an example of free flow and can occur when a non-uniform temperature field induces a lateral change in the fluid density which cannot be balanced by the normal hydrostatic gradient. Convective fluid flow can also occur when a change in pore fluid density occurs locally by dissolving salt or gases. In general, we note that the conditions for a static pore fluid are that the pressure at all points in the fluid be hydrostatic and that there be no lateral density gradients.

ASSUMPTIONS AND THE PHYSICAL MODEL

Local chemical equilibrium.—

One key assumption underlies much of the treatment of advective diagenesis to date, namely that the pore fluid maintains a state of chemical equilibrium with the solid matrix. This is not a mandatory assumption, but it greatly facilitates the theoretical treatment and serves as a point of departure for analysis of systems in which the pore fluids and solids are not in chemical equilibrium. There are good reasons to believe that subsurface fluids do approach equilibrium with their host rocks, however. These include the observations that:

(1) diagenetic mineral assemblages tend to be composed of minerals which are thermodynamically compatible;
(2) these assemblages can coexist in the presence of a pore fluid; and
(3) the compositions of many pore fluids plot on or near the invariant points on the relevant phase diagrams.

In addition, textural evidence often suggests a state of equilibrium was approached since authigenic minerals are often well crystallized and, in many cases, euhedral.

Physical model.—

If a pore fluid is moving slowly enough to maintain chemical equilibrium with the host rock, then there must be a net movement of solid material when the fluid moves from a state defined by one set of equilibrium conditions to another. For example, the (thermodynamic) equilibrium state is a function of temperature T, and if a fluid parcel moves across isotherms, the concentrations of dissolved material must change as the fluid moves. If we consider a system consisting of a solid matrix of quartz and a pore fluid consisting of pure water, then the concentration of dissolved SiO_2 changes as the fluid moves from temperature T_1 to T_2 either by precipitating or dissolving quartz along the fluid path.

THE DIAGENETIC EQUATION

Mass balance.—

If we take a mass balance on a fluid element moving at a constant velocity V_o, then we find

$$\partial\phi/\phi t = V_o(\rho_f/\rho_s)\alpha_T\,\partial T/\partial s, \tag{1a}$$

where ϕ is porosity, t is time, ρ_f is the fluid density, ρ_s is the solid density, α_T is the temperature coefficient for the solubility of a mineral, T is temperature and s is distance along the flow path. Equation (1) basically states that the porosity change, and hence the diagenesis, is proportional to the product of the fluid velocity and the rate of change of temperature along the flow path. The term $V_o\,\partial T/\partial s$ is equivalent to the vector equation (Kaplan and Lewis, 1971).

$$V_o\,\partial T/\partial s = V\cdot \text{grad}\,T, \tag{2}$$

where V is the fluid flow field and grad T is the gradient of the temperature field. Thus, equation (1) can be written

$$\partial\phi/\partial t = (\rho_f/\rho_s)\alpha_T\,V\cdot \text{grad}\,T. \tag{1b}$$

Written in vector notation, it becomes apparent that the diagenetic alteration can be regarded as a scalar field, similar to a temperature or pressure field, and can be manipulated as such. In particular, we expect that contours of the diagenetic field will map as surfaces similar to isotherms and isobars.

This discussion applies to the case of a fluid moving in a temperature field, but the theory can be readily extended to other thermodynamic potentials, pressure, for example. For the present purposes, however, only temperature effects will be considered. Note also that variables such as permeability and fluid viscosity are not explicit in the diagenetic equation (1b). This is because the fluid flow has been specified to be V_o and has not been derived from the equations of motion.

FORCED FLOW THROUGH A WAVY POROUS LAYER

The geologic model.—

Equation (1b) can be applied to a diagenetic setting by considering the following example involving forced flow. Assume that a porous sand body, consisting only of quartz grains, is embedded between two impermeable shales and that the pore fluid is moving as indicated in Figure 1. Assume also that the sand layer has a thickness h and is folded into a gentle sinusoidal form. Under these conditions, the upper and lower sand/shale contacts will be reasonably parallel. We will attempt to calculate the expected distribution of authigenic quartz after the fluid has been flowing in the body for some time. To a first approximation, none of the variables on the RHS of equation (1b) are functions of time and thus the integral is simply

$$\phi = [(\rho_f/\rho_s)\alpha_T\,V\cdot \text{grad}\,T]\,t + \phi_o, \tag{3}$$

where ϕ_o is the original porosity. The new porosity field, and hence the diagenetic field, can be calculated from equation (3) once the temperature and flow fields have been specified. For the model considered here, these fields can be obtained as follows.

The sand/shale contact.—

Let the sand/shale contact be described by a function of the form $z_c = f(x)$, where z is depth (positive down) and x is horizontal distance taken to be zero at some convenient point. Since there are two sand/shale contacts to be considered, designate the upper (*top*) contact as z_t and the lower (*bottom*) contact as z_b. Also, let the upper and lower con-

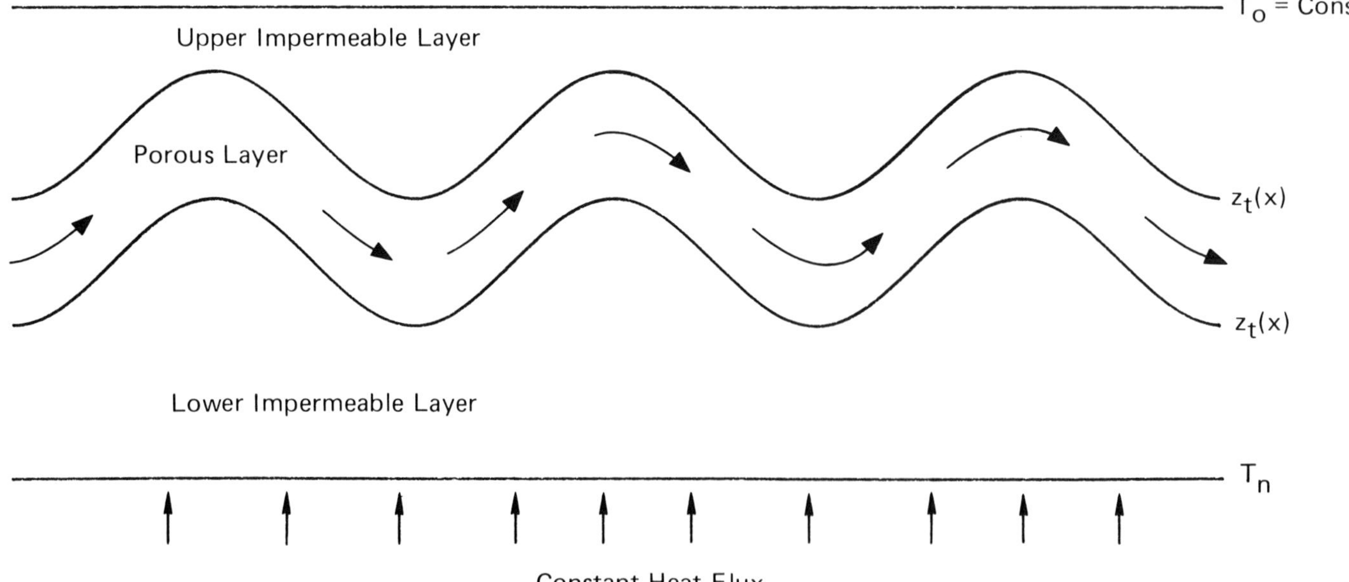

FIG. 1.—Schematic diagram of system: a wavy, porous sand layer embedded between two impermeable shales. The sand/shale contacts are described by the functions $Z_t(x)$ and $Z_b(x)$, which are simple cosine functions here. The temperature field is fixed by the constant heat flux from below and the fixed surface temperature.

tacts be everywhere parallel such that

$$z_t = f(x) \tag{4a}$$

$$z_b = f(x) + h, \tag{4b}$$

where h is a constant greater than zero.

The fluid flow field.—

For forced flow, the (impermeable) sand/shale contact is a streamline, and thus the stream function must be of the form

$$\Psi(x, z) = V_o[z_b - z], \tag{5}$$

so that for $z = z_b$, $\Psi = 0$ and for $z = z_t$, $\Psi = V_{oh}$.

The x and z components of the fluid flow field are then given by the usual derivatives

$$u = -\,\partial\Psi/\partial z \tag{6a}$$
$$= V_o$$

$$w = \partial\Psi/\partial x \tag{6b}$$
$$= V_o\, dz_b/dx,$$

where u is the horizontal component of the flow field and w is the vertical component. Thus, for parallel sand/shale contacts, the horizontal component of the fluid velocity field is constant while the vertical component varies as the horizontal (x) derivative of the function describing the sand/shale contact. The complete flow field is

$$V = u\, i + w\, k \tag{7a}$$
$$= V_o\,[i + dz_b\,(x)/dx\; k], \tag{7b}$$

where i and k are unit vectors in the x and z directions, respectively.

The temperature field.—

Assume that the heat transfer is purely conductive everywhere. Then, the temperature field can be approximated by a set of horizontal isotherms defined by

$$T = (dT/dz)_o z + T_o) \tag{8}$$

where $(dT/dz)_o$ represents the geothermal gradient and T_o is a prescribed surface temperature. Figure 2 (a–d) shows a series of temperature fields for thermal conductivity contrasts ranging from 0.25 to 4. The extreme values show how the isotherms tend to parallel bedding (Fig. 2d) or become perpendicular to bedding (Fig. 2a). The unperturbed temperature field in Figure 2c depicts the temperature field described in equation (8). The gradient of this temperature field is

$$\text{grad}\, T = (dT/dx)i + (dT/dy)j + (dT/dz)k \tag{9}$$
$$= (dT/dz)_o k,$$

where i, j and k are unit vectors in the x, y and z directions,

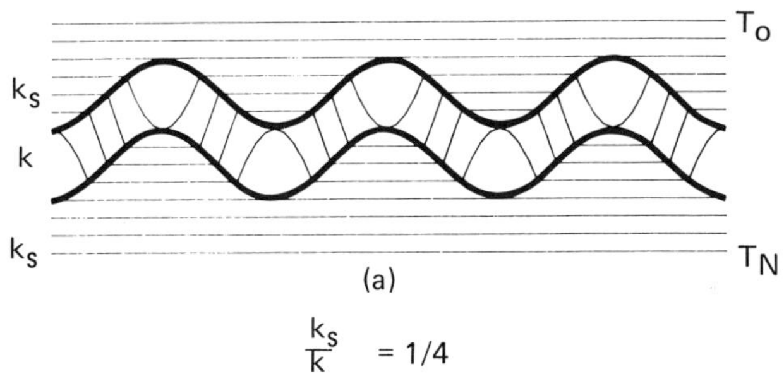

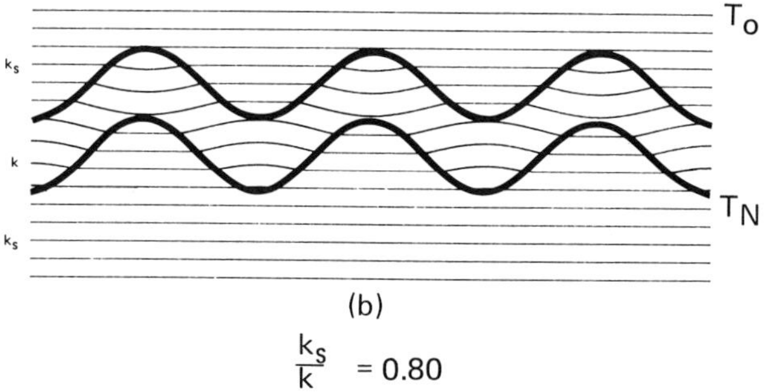

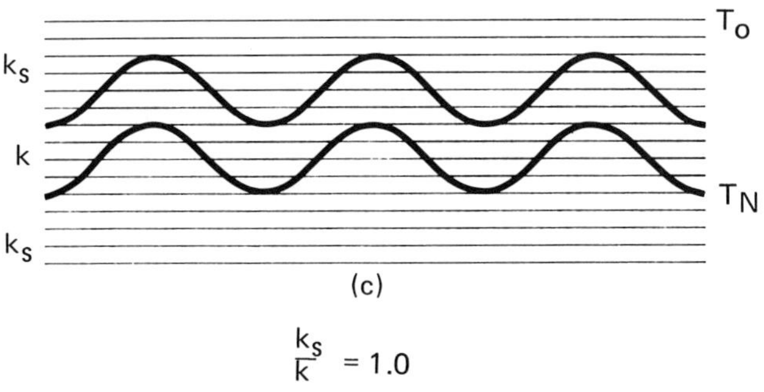

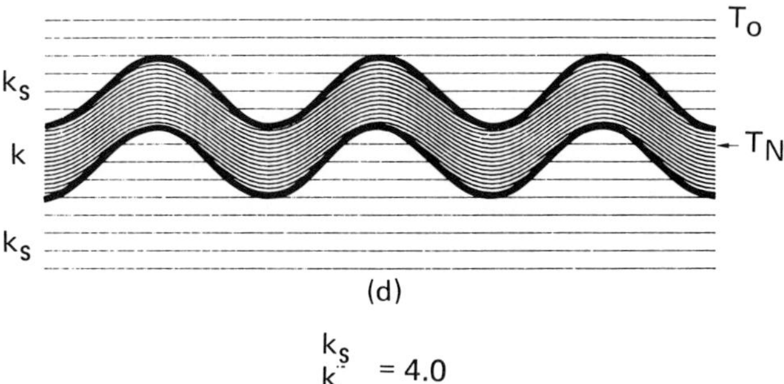

FIG. 2.—Temperature fields in a composite system as a function of the thermal conductivity ratio. Lower case k refers to the thermal conductivity of the middle wavy layer, while k_s is the thermal conductivity of the bounding layers. For small values of k_s/k, the isotherms tend to become perpendicular to the layer boundaries (a) (middle layer acts as a conductor). For large k_s/k ratios, the isotherms gather in the middle layer and tend to conform to the layer boundaries (d) (middle layer acts as an insulater). When the ratio is unity (c), the isotherms are indifferent to the presence of the inner layer. For most geologic conditions, values of the k_s/k ratios would probably lie between 0.8 (b) or 1–25.

respectively. From equations (7) and (9), then

$$V \cdot \nabla T = V_o\, dz_t(x)/dx \left(\frac{dT}{dz}\right)_o . \qquad (10)$$

The "cosine" fold.—

Since the actual form of the sand/shale contacts is arbitrary, let

$$z_t = A \cos(2\pi\, x/1), \qquad (11)$$

which describes the upper sand/shale contact where A is the amplitude and 1 is the wavelength. Figure 2b shows the distribution of isotherms for the temperature field described in equation (8), where the surface temperature $T^\circ = 25^\circ$ C and $(dT/dz)_o = 25^\circ$ C/km. The sand layer has arbitrarily been placed 1 km below the surface.

Using equations (11) in (10) and substituting into (3), we have

$$\phi = A' \sin(2\pi\, x/1)t + \phi_o, \qquad (12)$$

where the constants have been collected into

$$A' = [2\,\pi\, A/1]\,[V_o(\rho_f/\rho_s)\alpha_T(dT/dz)_o] .$$

Equation (12) shows that the diagenetic field is shifted 90° relative to the structure described by the function (equation11). This is a particular result for the contact function chosen here and is not a general result. Nevertheless, it does show how the diagenetic field depends on the geometry of the rocks.

For this particular geometry the maximum in the diagenetic field occurs at $x = \pi/4$ and goes to zero at $x = 0$ and $x = \pi/2$ (Fig. 3). Also, the intensity of the alteration, A', depends on the amplitude of the contact function and thus, more tightly folded beds will require less time to attain a given degree of alteration.

Since we have assumed that only quartz is present in the porous layer, and since α_T for quartz is positive, quartz will precipitate in the limb of the fold in which the fluid is ascending but will dissolve in the limb where the fluid is descending. There will be no porosity change at either the crest or trough of the structure. This diagenetic pattern is characteristic of diagenesis involving forced flow when the isotherms are essentially horizontal. Since the fluid flow parallels the sand/shale contact and because the isotherms are (approximately) horizontal over the crest of the structure, it follows that the fluid flow path will never cut sharply across isotherms in the vicinity of extrema in the structure.

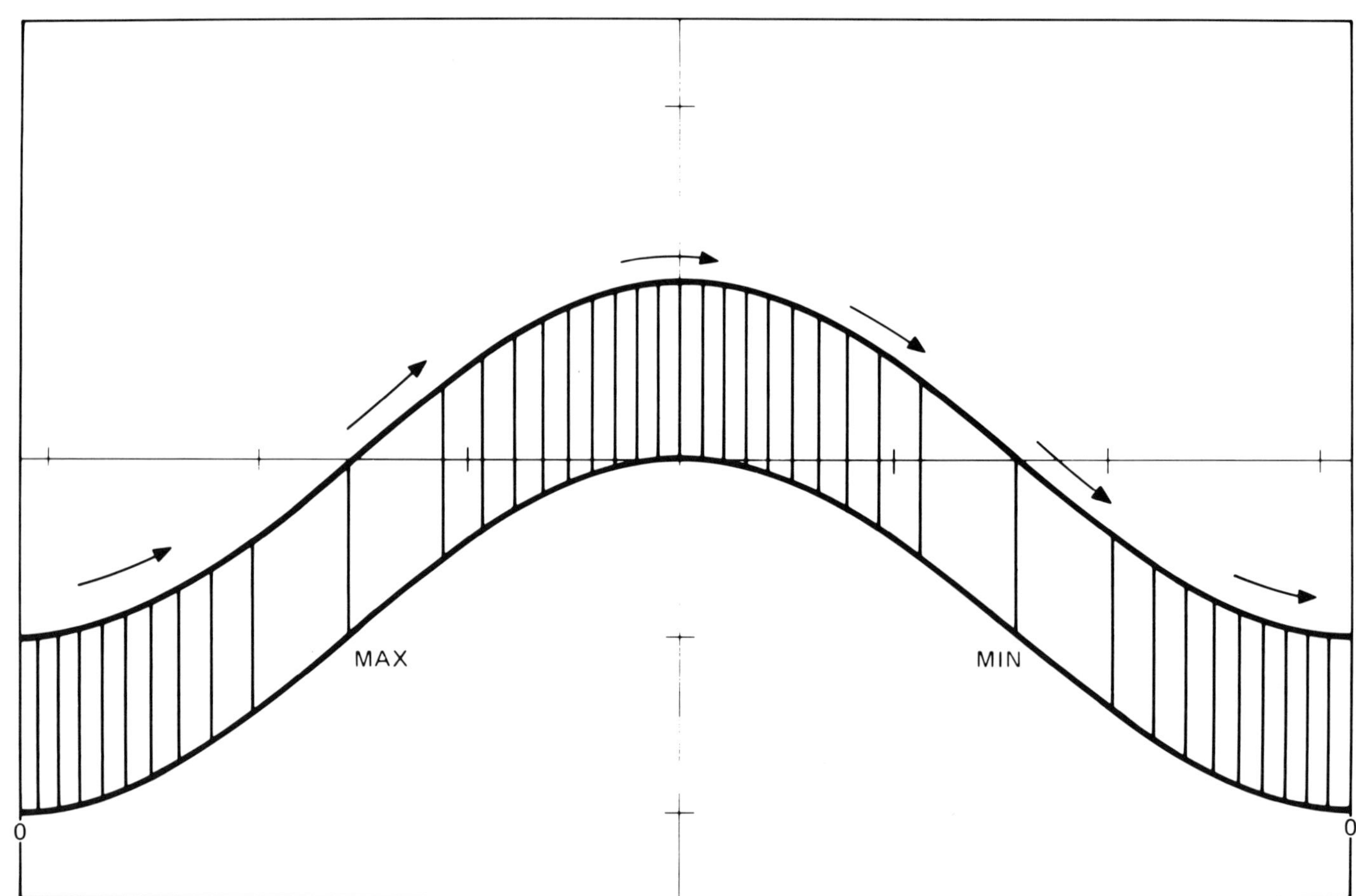

FIG. 3.—Contours of diagenetic intensity in a cosine layer. The diagenesis is uniform vertically but varies laterally, reaching extreme values (MAX and MIN) at $\pm\pi/4$ and going to zero at the crests and troughs ($\phi = 0, \pm\pi/2$). Note that the contours change most rapidly approaching the crests and troughs of the structure as a result of the rapidly changing curvature of the cosine function.

The important conclusion from a diagenetic standpoint is that forced flow alone generally cannot account for the transport of diagenetic minerals into the crests or troughs of structures. If diagenetic materials are to accumulate in the crests or troughs of structures, a flow field with a non-zero vertical component must exist at or near these points. A convective flow field, for example, will have non-zero vertical components near the anticlinal crests and could cause alteration in that region (Davis et al., 1985).

A more realistic temperature field.—

The temperature field described by equation (8) is somewhat simplified even for purely conductive heat flow. A conductive temperature field, which includes the effects of different thermal conductivities in different lithologies, would be:

$$T = z\left(\frac{dT}{dz}\right)_o + T^\circ \qquad \text{(upper shale)(13a)}$$

$$T = \{z - (1 - k_s/k)[z_b(x) - z_t(x)]\} (dT/dz)_o + T^\circ \qquad \text{(lower shale) (13b)}$$

$$T = \{(k_s/k)z + (1 - k_s/k)z_t(x)\} (dT/dz)_o + T^\circ, \qquad \text{(sand) (13c)}$$

where k is the thermal conductivity of the sand layer (+ pore fluid) and k_s is the thermal conductivity of the shales (+ pore fluid) (Davis et al., 1985). $(dT/dz)_o$ refers to the geothermal gradient in the upper shale. These equations assume that the upper and lower shales have the same thermal conductivity, $z_b(x)$ and $z_t(x)$ are as described before. Note that if $k = k_s$, equations (13a,c) reduce to equation (8).

The effect of different thermal conductivities in the sand and shales on the temperature field is to introduce a lateral perturbation in the temperature field in the sand layer. This perturbation shows up in the non-linear behavior of the temperature field in the x-direction (Fig. 2b). Equation (13c) shows that the perturbation will propagate into the lower shale if z_b differs from z_t by more than an additive constant. These equations show that, as the thermal conductivity ratio changes, the isotherms can either rotate in the sand unit until they are nearly perpendicular to bedding (Fig. 2a), or stack up in the sand and tend to become subparallel to z_t (Fig. 2d).

Taking the gradient of equation (13c), we obtain

$$\text{grad } T = [(k_s/k)\, k + (1 - k_s/k)(dz_t/dx)i](dT/dz)_o. \qquad (14)$$

Comparing equation (14) with (9), we see that the effect of correcting the temperature field for the thermal conductivities is to introduce an additional component in grad T. This component vanishes for $k = k_s$ or for $dz_t/dx = 0$ and becomes small as k approaches k_s. In spite of this difference in the temperature fields, the diagenetic field obtained by taking the vector product of equations (7b) and (14) is identical to equation (10), the expression obtained for the uncorrected temperature field.[1]

DIAGENETIC PATTERNS IN THREE DIMENSIONS

The preceding discussion was confined to diagenetic alteration in two dimensions (the x–z plane). The diagenetic equation itself is inherently three-dimensional, but in the general case, it is difficult to portray graphically the alteration in three-dimensions. For the case of uniform forced flow, however, the diagenesis is essentially uniform in the vertical direction, and it is possible to depict this case's three-dimensions relatively easily.

The analog to the cosine function (equation 11) describing the sand/shale contact is the surface

$$Z(x,y) = A \cos(2\, N_x\, \pi\, x/1_x) \cos(2\, N_y\, \pi\, y/1_y) \quad (0 < x < 1_x), (0 < y < 1_y) \qquad (15)$$

in two dimensions, where 1_x is the wave length of the first harmonic in the x-direction, 1_y is the wave length of the first harmonic in the y-direction, N_x (= 1, 2, . . .) is the Nth harmonic in the x-direction and N_y is the Nth harmonic in the y-direction. Assuming that the upper and lower sand/shale contacts are parallel, then

$$z_b = Z(x,y) \qquad (16)$$

and

$$z_t = Z(x,y) + c. \qquad (17)$$

Planar forced flow.—

Consider a flow field which is always parallel to an arbitrarily oriented vertical plane. As long as the flow does not have a component perpendicular to that plane, it is a two-dimensional flow and can be analyzed as before in terms of a stream function. In fact, equation (10) will describe such flow with proper definition of $Z_T(x)$. We note that dZ_t/dx is a special case of the directional derivative (cf. Kaplan and Lewis, 1970, v. II, p. 913), which is defined as

$$\nabla_S\, Z_T = \nabla Z_T \cdot S, \qquad (18)$$

where S is a unit tangent vector along the path in the chosen direction. That is, S is a unit vector tangent to the curve defined by the intersection of $Z_T(x,y)$ and the arbitrary vertical plane, so that

$$S = \cos\beta\, i + \sin\beta\, j, \qquad (19)$$

where β is the angle between the positive x-axis and S. The directional derivative is thus a generalization of the partial derivative and gives the rate of change of $Z(x,y)$ in the di-

[1]This is a special result, however, that occurs because of the form of equations (4a) and (4b). Since z_b only differs from z_t by a constant, the vertical component of the flow field (w) is simplified, and this carries over into the velocity function equation (7b). For more general cases, the diagenetic field will be a function of the thermal conductivity ratio.

rection β, which is the quantity required in equation (10). Writing

$$\nabla_S Z_T = Z_x \cos\beta + Z_y \sin\beta, \tag{20}$$

where Z_x and Z_y are partial derivatives in the x and y directions, respectively, and substituting in equation (10), we have

$$\phi = A' [Z_x \cos\beta + Z_y \sin\beta]\, t + \phi_o \tag{21}$$

as a general expression for the integrated diagenetic equation for planar forced flow. Note that equation (21) is identical to (10) for $\beta = 0$.

Diagenesis in three dimensions.—

Equation (21) shows that the three-dimensional diagenetic field for planar flow can be mapped in two dimensions by plotting the function $\nabla_S Z(x,y)$ and scaling for the amplitude A'. The diagenetic field depends on both the flow direction relative to the contact surface and the derivatives of $Z(x,y)$. Several examples illustrating the dependence of the diagenetic pattern on the direction of fluid flow are shown in Figures 4b–d for a sin x sin y surface. (This is effectively the same as a cos x cos y surface but the surface is zero for $x = y = 0$ and hence the cells are centered in the plot.) In these figures, the diagenetic field is represented by contouring the porosity on the contact surface. Contours representing equal values of porosity will be referred to as "isopors." Note that the contours refer to *diagenetic* porosity and not *total* porosity.[2]

In Figure 4b the fluid flows across the structure as indicated. In regions where it flows upward (over the positive structures), it is cooling and hence a prograde mineral such as quartz would precipitate. In regions where the fluid is descending and heating, quartz would dissolve (Figs. 4b–d). A retrograde mineral would have the opposite behavior. As the fluid flows over the crests of the structure, the fluid begins to descend and cool, hence the solubility/dissolution patterns change as indicated by the different patterns. The descending fluid reaches its minimum along the same line that the ascending fluid reaches the crest in this example, so the pattern of dissolution-precipitation forms a network of rectangular cells.

A similar analysis holds for the cell pattern in Figure 4d, where the fluid flow direction is at right angles to that in Figure 4b. Comparing Figure 4b with Figure 4d shows that changing the flow direction dramatically shifts the diagenetic field. The cellular pattern is preserved, but the cells have all shifted by one-half a wave length down and to the right.

For fluid flows oblique to the principal axes, the cellular patterns of the diagenetic field exhibited in Figures 4b and 4c shift into a more linear mode (Fig. 4c), where a lineation becomes more pronounced until, for a flow at 45° to that in Figures 4b or 4c, the field would consist of straight lines running diagonally. Figure 4c shows the development of a linear pattern for a flow at 30° clockwise from the flow in Figure 4b. Note that this diagenetic field (4c) shows little resemblance to the geometric field (Fig. 4a) which controls it.

These examples clearly show the influence of the orientation of the flow field on the diagenetic field. The examples discussed are too simplified to apply directly to real structure(s), but extension to real geologies simply requires redefinition of Z(x,y). The simple geometry used in Figures 4a–d does convey the main point that it is the fluid flow direction across a structure that is all important in establishing the advective diagenetic field. The same approach can be applied to more complex structures.

Note that the isopors are not stationary but migrate across the diagenetic surface with time. If a given contour has a value ϕ (or $\phi - \phi_o$) at time t_o, at a time Δt later, mass will have been removed or added at points along that contour, resulting in a new value for the contour. The contour can be regarded as migrating toward lower values when there is a net mass accumulation or as migrating toward larger values when there is a net mass removal.

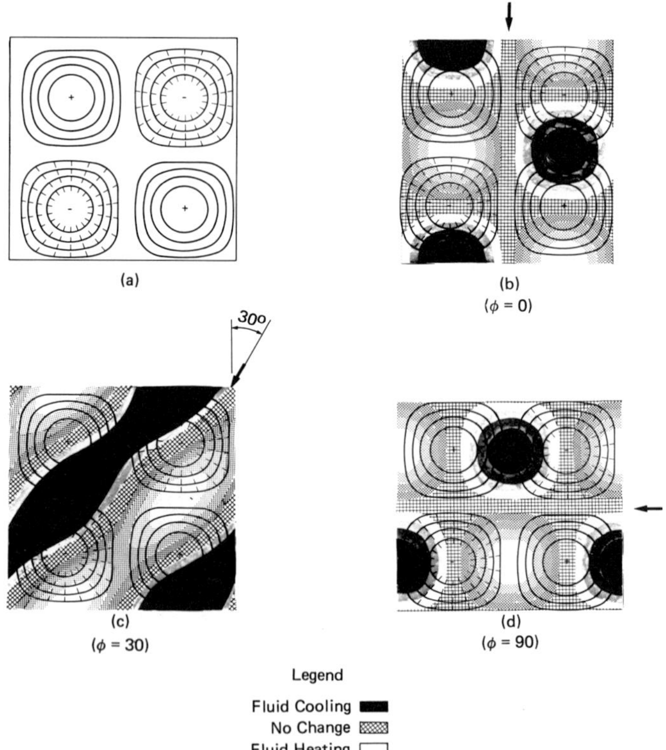

FIG. 4.—Plan view of structural and diagenetic contours for a sin(x)sin(y) or cos(x)cos(y) structure. (a) Base map shows the elevation contours, where the "+" sign indicates a dome and the "−" sign indicates an antidome. The base map is overlaid on the diagenetic contours indicated by different patterns (b–d) to show the spatial distribution of the diagenesis relative to a fixed structure. (b) shows the distribution of the diagenetic contours relative to the structure for a fluid flowing parallel to the x axis (arrow). (d) is the analogous case for a flow parallel to the y axis. The shifts in diagenetic pattern relative to (b) are obvious. (c) shows the diagenetic pattern for a flow at 30° clockwise from the x axis. The diagenetic centers coalesce and form continuous bands or "fairways" of porosity enhancement or porosity loss.

[2]Diagenetic porosity refers to the change in the porosity field due to diagenesis. As developed here, the diagenetic porosity is only the component arising from movement of pore fluid across isotherms.

A particular isopleth moves in a direction normal to its contour. Since the gradient of a function is normal to level surfaces for that function, we can obtain a function describing the trajectory of the contours as they migrate across the surface $Z(x,y)$ by taking the gradient of the function describing the porosity contours. Since mass flows in the opposite direction to porosity, we can equally regard the trajectories as mass trajectories or paths of mass migration simply by reversing the sense of direction.

DISCUSSION

The spatial distribution of the diagenetic assemblage is controlled primarily by the "topography" or geometry of the boundary between the permeable and impermeable layers. Although a simple geometry was used as an example, the general equations (12) and (21), which are of the form

$$\phi - \phi_o = \{\text{scale factor}\} \cdot \{\text{chemical term}\} \cdot \{\text{geometric term}\}t,$$

where

$$\{\text{scale factor}\} = V_o\,(\rho_f/\rho_s)\,[(1 - k_s/k)]/c(dT/dz)_o$$

$$\{\text{chemical term}\} = \alpha_T$$

and

$$\{\text{geometric term}\} = [Z_x \cos B + Z_y \sin B],$$

also show this. The theory thus separates the purely physical terms from the purely chemical terms. This provides certain computational advantages, particularly in the separation of the chemical term.

Isolating the chemical term α_T allows us to make certain generalizations about the diagenetic process. First, we note that all of the chemical parameters are contained in this term and that no other parameters are related to diagenetic reactions. Consequently, all questions regarding chemical processes such as cementation, dolomitization, isomorphic replacement, etc., have to be implicit in the α_T parameters. This is a general result and applies to diagenesis produced from convective flows as well. The chemical terms determine *what* phases precipitate and *what* phases dissolve, but do[es] not influence the position in space where the alteration occurs. The geometric term determines *where* the alteration occurs, but is indifferent to the actual diagenetic mineralogy involved. To a first approximation, the equations describing chemical diagenesis can be uncoupled from the equations describing the spatial distribution of the diagenetic products. Or, stated differently, since all of the chemistry is contained in the α_T terms, the spatial distribution of a diagenetic assemblage is independent of the diagenetic phase(s) involved.

Finally, we can see the role that certain parameters, such as the geothermal gradient, the thermal conductivity, the fluid velocity, etc., play in diagenesis. These parameters combine to determine a "scale factor" which influences *how much* alteration occurs, how much time is required for a given process to reach a certain level of intensity in a given location, and how growth of a structure influences diagenesis.

REFERENCES

DAVIS, S. H., ROSENVLAT, S., WOOD, J. R., AND HEWETT, T. A., 1985, Convective Fluid Flow and Diagenetic Patterns in Domed Sheets: American Journal of Science v. 285. p. 207–223.

KAPLAN, W. AND LEWIS, D. J., 1970, Calculus and Linear Algebra: John Wiley & Sons, Inc., New York.

WOOD, J. R. AND HEWETT, T. A., 1982, Fluid convection and mass transfer in porous sandstones—a theoretical model. Geochemica et Cosmochimica Acta, v. 46 [10], p. 1707–1713.

——AND, 1984, Reservoir Diagenesis and Convective Fluid Flow, *in* McDonald D. A. and Surdam R. C., eds., Clastic Diagenesis: AAPG, Tulsa, OK.

FLUID INCLUSION, STABLE ISOTOPE, AND VITRINITE REFLECTANCE EVIDENCE FOR THE THERMAL HISTORY OF THE BONE SPRING LIMESTONE, SOUTHERN GUADALUPE MOUNTAINS, TEXAS

CHARLES E. BARKER AND ROBERT B. HALLEY
U.S. Geological Survey, Box 25046, Denver, CO 80225 U.S.A.

ABSTRACT: The Permian (Leonardian) Bone Spring Limestone is a significant oil reservoir rock in the Delaware Basin of west Texas and southern New Mexico. Analysis of fluid inclusions trapped in fracture-filling cements, oxygen isotope composition of the cements, and thermal maturity measurements of host rock provide a detailed record of the thermal and fluid history for this area of the western Delaware Basin. Although these cements have petrographically distinct multiple zones, homogenization temperatures (T_h) of oil fluid inclusions and $\delta^{18}O$ data divide these cements into two (early and late) groups. Early cements have $\delta^{18}O$ near O‰ and contain dull-blue fluorescent oil inclusions whose T_h is mainly between 55°C and 90°C. The late cements occur in fractures that crosscut the early cements and are strongly oriented subparallel to the local structures formed during the Laramide orogeny and Basin and Range uplift. The late cements have $\delta^{18}O$ between −8‰ and −9‰ and contain sparse oil inclusions similar to those in the early cement. Primary fluid inclusions in the late cements have a T_h mainly between 90°C and 135°C. Petrographic observations suggest that overlap in T_h between early and late cements is due to: (1) incorporation of fragments of early cements into crosscutting late cements; (2) local neomorphism of early calcite along margins of late calcite; and (3) formation of secondary inclusions (some distinguished by yellow fluorescence) during uplift and exposure.

When interpreted through burial history reconstruction, T_h and $\delta^{18}O$ data suggest that oil migrated relatively early while this rock was about 55°C to 75°C. These temperatures were attained at near maximum burial depth of 1–2 km (estimated 30°C/km paleogradient) at the end of the Permian. The average T_h values of primary inclusions of about 110°C are consistent with maximum temperatures of ca. 120°C interpreted from a mean vitrinite reflectance (R_m) of 0.7%, suggesting that the late cements record the maximum temperature attained in the rock. The change in $\delta^{18}O$ of about 8‰ from the early to late cement is nearly that expected from the difference in T_h between these cements. Although maximum burial depth was reached near the end of the Permian, maximum temperature was not recorded until geothermal gradients apparently increased in the Tertiary. The evidence for this higher geothermal gradient is recorded in cements from fractures formed during early Tertiary deformation.

Calculations from calcite/water fractionation relations and T_h indicate that early calcite cement precipitated from water that had $\delta^{18}O$ of about +9‰ and later cements precipitated from water of about +7.5‰. These values are typical for waters remaining after intense evaporation or for deep basin brines. Temperature determinations are associated with sufficient error that both cement generations may have been precipitated from the same water, and cement isotopic differences may have resulted solely from temperature changes.

INTRODUCTION

The Permian (Leonardian) Bone Spring Limestone is an organic carbon-rich basinal limestone and shale facies in the Delaware Basin of west Texas and southern New Mexico. In the study area the Bone Spring has been fractured and cemented by several generations of calcite. The fractures appear to have formed early in the history of the rock by a combination of selective cementation and soft sediment deformation when some Bone Spring sediment was still in a semiplastic state. Parts of voids in these fractures are still open today, and potentially a complete diagenetic record could be preserved in fracture-filling cements. Fluid inclusion and isotopic analyses of zoned calcite cement that fills early and later fractures provide a partial record of the thermal and fluid history of the Bone Spring Limestone.

Our study applies mineral- and kerogen-based geothermometers (Appendix) to determine the thermal history of the Bone Spring Limestone by the relative chronology of paleotemperatures measured in individual cement zones and the enclosing host rock. Geothermometers based on mineral and kerogen diagenesis record different types of temperature information. Fluid inclusions record the temperature and composition of the fluid present when an authigenic phase forms. Measurement of fluid inclusion homogenization temperatures (T_h) from superimposed or crosscutting mineral layers indicates the temperature and relative time of thermal events in the rock's history.

The change of oxygen isotope ratios ($\delta^{18}O$) between generations of calcite cement can be used to calculate the change in mineral precipitation temperature (Epstein et al., 1953) if the water that deposited these cements had a constant oxygen isotope composition. Isotopic analysis of calcite cements is a bulk method that requires relatively large samples compared to the volume and size of a fluid inclusion. Whereas the closure of a single fluid inclusion records the temperature and fluid composition over an infinitesimally small length of geologic time, paleotemperatures computed from $\delta^{18}O$ data indicate the ambient temperature and fluid conditions over a wider range of cement deposition and consequently over a larger portion of geologic time.

Vitrinite reflectance (R_m) records the maximum temperature reached at the place from which the sample was taken. Maximum paleotemperature at any geologic time in the burial history of this sampled place can be computed using an empirical calibration curve (Barker and Pawlewicz, in prep.). A series of R_m measurements taken at various levels in a sedimentary section can be used to compute a line consisting of maximum paleotemperature points. This line is equal to maximum paleogeothermal gradient if R_m resulted from a peak thermal event that affected the entire sedimentary section. We believe that this condition is fulfilled in most sedimentary basins, because the basins normally provide close to steady-state thermal conditions.

The maximum paleogeothermal gradient is a crucial reference line to compare with the modern geothermal gradient and paleogeothermal gradients calculated from mineral-based geothermometers. The thermal history of the sedimentary section is indicated by comparing mineral-based geothermometric data and present-day temperatures to the maximum temperature indicated by R_m. Fluid inclusions and stable isotopes in successive cement layers or in minerals precipitated in crosscutting fractures record details of part of that rock's thermal history. The thermal history recorded by these geothermometers is limited to times of mineral deposition or recrystallization, and these diagenetic events must be fixed in the burial history to interpret the paleotemperature data. These details are added to the thermal history by reconstructing the burial history for a given sediment package and converting burial depth through geologic time to a temperature-absolute time path using apparent paleogeothermal gradients (Waples, 1981). The temperature-absolute time curve allows a thermal event that leaves a record of its temperature to be interpreted in terms of paleodepth and geologic time.

FRACTURE HISTORY AND PETROLOGY

The Bone Spring Limestone is exposed along the western edge of the southern Guadalupe Mountains (Fig. 1) and consists of black, thin-bedded fossiliferous mudstone and wackestone interpreted to be deposited in quiet, relatively deep water (King, 1948). Debris flows consisting of shelf-derived intraclasts occur locally within these basin mudstones (Pray and Stehli, 1963). The upper portion of the Bone Spring Limestone is the lowest formation exposed in the fault blocks that constitute the southern terminus of the Guadalupe Mountains.

The mudstones show extensive soft sediment deformation from downslope slumping and creep (King, 1948; Newell et al., 1953; Figs. 2 and 3). This deformation fractured some partially lithified horizons and these fractures subsequently were filled with sediment and calcite cement (Figs. 3–6). These fractures have several features that indicate they are essentially synsedimentary voids formed when a part of the enclosing sediment was still in a plastic state (Table 1; Figs. 3 and 5). The entire rock must have lithified after slumping because these fractures remained open during considerable burial; some parts of the fractures remain open today. The first-generation cements are "early" in a relative chronology because they are crosscut by a near vertical, northwesterly-trending, calcite-filled fracture set (Fig. 6).

The two major groups of calcite-filled fractures in the Bone Spring Limestone have distinctly different origins—early soft sediment deformation and later tectonic influences (King, 1948). The relationship between these fracture sets is clearly discerned in the field (Fig. 6A). The early fractures are short, apparently randomly oriented irregular voids filled with a dark calcite that is commonly zoned (symmetrically light and dark banded; Fig. 6B). Early calcite cements (fracture-filling) are discontinuous irregular vein-like to stellate vuggy forms, commonly with uniform zonation but in some instances chaotically zoned due to repeated uneven fracturing. The early cements are limited in distribution, not strongly oriented, partially controlled by sedimentary structure, and cut by all later fractures. Locally, the zoned early calcite has been cut and recemented by numerous zoned veinlets, making a confused mass of crosscutting veins. These chaotically zoned cements were not used in our study because of the difficulty in deciphering the cementation history. Crosscutting the early fractures are long, subparallel, more regular, near vertical fractures filled with calcite that is typically white in outcrop. Both these cements and individual zones within them may be separated under the microscope by their distinct textures (Fig. 6B). We found that only Bone Spring Limestone exposures in Guadalupe Canyon, and the "pipeline section," an exposure of the Bone Spring about one mile to the north-northwest along an El Paso Natural Gas Pipeline, contained both early and late cement. The type section to the north in Bone Canyon, and in Black Canyon, a well-exposed section to the south, contained only late cement and were not studied further.

The late cement generally appears massive but may be locally zoned in transmitted light and commonly is complexly zoned under cathodoluminescence. The late cement fills fractures that are strongly oriented (north-northwest and near vertical), crosscut all earlier fractures or sedimentary structures and are widespread throughout the Bone Spring Limestone of the southern Guadalupe Mountains.

A very fine-grained cement intermediate in age between the early and late cements occurs locally in the Bone Spring Limestone. This cement fills the void remaining after the symmetrically zoned early cement was deposited. In thin section this cement is micritic and because of its fine-grained character, fluid inclusions are difficult to find and observe. This cement was not studied further.

TABLE 1.—EVIDENCE THAT DARK FRACTURES IN THE BONE SPRING LIMESTONE WERE FORMED EARLY IN THE BURIAL HISTORY.

(1) Some fractures in this set are partially filled with micrite that appears to have been squeezed into the void.

(2) These fractures are commonly associated with contorted beds, are randomly oriented, and terminate in short distances, usually less than 20 cm.

(3) Fracture surfaces are sometimes bounded by bedding planes.

(4) Some fractures terminate at bedding planes and/or slump planes between sediment packages of slightly different lithology (and different competency?).

(5) This fracture set does not crosscut other sets and is crosscut by all later sets.

(6) The fractures have a limited distribution in outcrop and they lack the strong directional alignment and relationship to the trend of local structures that would be expected with fractures of tectonic origin.

Fracturing and neomorphism.—

Neomorphism of early cement and host limestone commonly makes it difficult to discern the origin of the cement and fluid inclusions. During fracturing, pieces of early cement and host rock may lodge in fractures and become incorporated in later cements. As cementation proceeds, there may be some neomorphism of early cements and rock fragments along contacts with later cement (Figs. 7 and 8). Neomorphism of early cement along petrographically obscure reentrants and fractures may produce fluid inclusions whose T_h is that of the late cement in a zone close to the contact between the cements. Late cement may also enclose what appear to be recrystallizated host limestone fragments, without resetting the T_h of their indigenous fluid inclusions. These fragments may be distinguished by their low-tem-

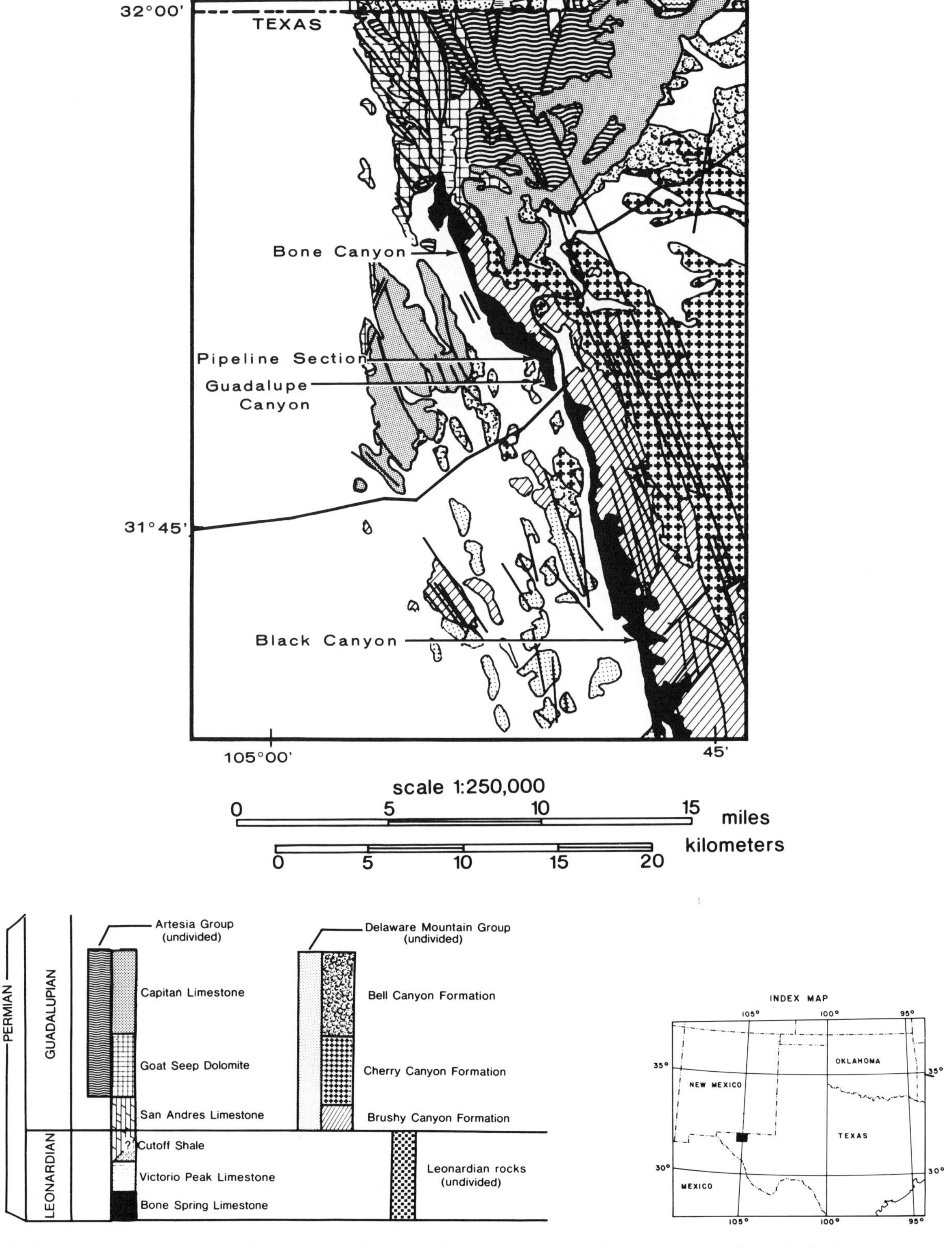

FIG. 1.—Geologic map and stratigraphy of the southern Guadalupe Mountains region with emphasis on the Bone Spring Limestone exposures and the study areas mentioned in the text. Stratigraphic nomenclature from King (1948). Capitan Limestone and Goat Seep Dolomite are equivalent to, but not part of, the Artesia Group. Adapted from Scholle and Halley (1980).

FIG. 2.—Soft sediment deformation, upper Bone Spring Limestone, Black Canyon area, southern Guadalupe Mountains, Texas. Early-formed fractures (Figs. 3, 4, and 5) are associated with this type of local soft sediment deformation. These features are caused by sediment slumping down the paleoslope (Newell et al., 1953). Early fractures filled with zoned calcite cement were found in Guadalupe Canyon and the pipeline section but not in Bone and Black Canyons (Fig. 1)—although all these sections show slump features.

perature fluid inclusions and mottled brownish color in transmitted light (Fig. 7). Some host rock fragments appear to recrystallize and take on the optical character of the late cement, except for a brownish stain and a low T_h. Late cement may also have brownish stains along fracture planes and cleavage partings, making it difficult to eliminate those inclusions derived from the host sediment based solely on the color of the enclosing mineral.

Secondary, high-temperature inclusions may be introduced into the early cement by fracturing and subsequent cementation. These later high-temperature cements are sometimes detectable because of a corresponding color change (loss of brown color) in affected cements.

Secondary fluid inclusions in the low T_h range found along planes cutting the late cement must contribute to the overlap in T_h between the early and late cements. These inclusions apparently were deposited during cooling by uplift and erosion in the Tertiary. Although this generation of inclusions is obvious in the relatively inclusion-free late cement, they could be easily misidentified in the inclusion-rich and low-temperature early cement. These later secondary planes, in some cases, are identified by oil fluid

FIG. 3.—Early cement filling fractures (arrow) related to folds formed by soft sediment deformation, upper Bone Spring Limestone, Guadalupe Canyon, southern Guadalupe Mountains, Texas. Exposed in roadcut along U.S. Highway 62, 2.5 km east of intersection with Texas State Highway 54.

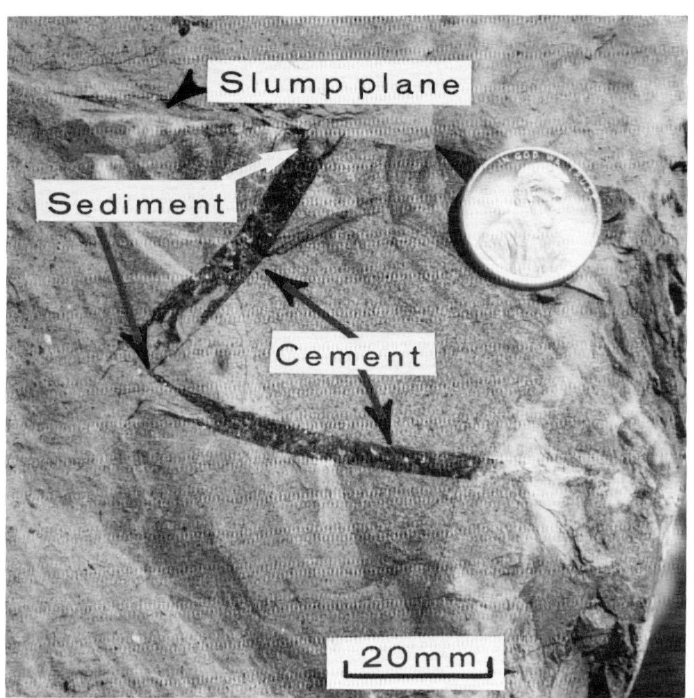

FIG. 4.—Evidence that sediment was still plastic when early fractures were open: randomly oriented fractures with sediment infill at open end (dark arrows) that truncates at slump(?) plane (white arrow) Bone Spring Limestone, Southern Guadalupe Mountains, Texas.

inclusions that fluoresce yellow under ultraviolet light (366 nm), in sharp contrast to the typical dull to bright blue fluoresence of inclusions in the early and late cements.

Neomorphism and fracturing result in the incorporation of older cement and host rock in young cement and the introduction of new cement in older cement. Neomorphic alteration along contacts between cement generations sometimes prevents clear assignment of inclusions to early or late cements, but in most cases the assignment can be made. Nevertheless, the origin of any given inclusion may remain uncertain and only a few inclusions were found to be obviously primary. We feel that the inclusion of old material in new cements and the introduction of new fluids and cements along fractures or twin planes in older cements account for much of the overlap between the T_h of the early and late cements in the Bone Spring Limestone. As will be shown later, this overlap is not apparent in the oxygen isotope analysis of these cements.

FLUID INCLUSIONS

Both aqueous and oil inclusions occur in the fracture-filling calcite. Oil inclusions are much more common than aqueous inclusions in both early and late cements. In the present work, analyses have been predominantly from oil inclusions. These inclusions may be less susceptible to stretching and other forms of reequilibration that can affect aqueous inclusions (Appendix).

Oil fluid inclusions.—

Oil fluid inclusions are common in early cement but rare in late cement (Figs. 9, 10, and 11). The oil in the primary inclusions is a clear fluid that fluoresces a dull to bright light blue under excitation by ultraviolet (366 nm) light. Light blue fluorescence suggests that this oil has an API gravity of at least 45° (Anderson, 1975). These inclusions are small, usually 5 to 15 μm in the longest dimension, and irregular tubular or rectangular to triangular in form. Primary fluid inclusions in the early cement occur as inclusion-rich clusters or bands parallel to cathodoluminescent zones. Primary fluid inclusions in the late cement occur as isolated irregular-shaped single inclusions, inclusions along

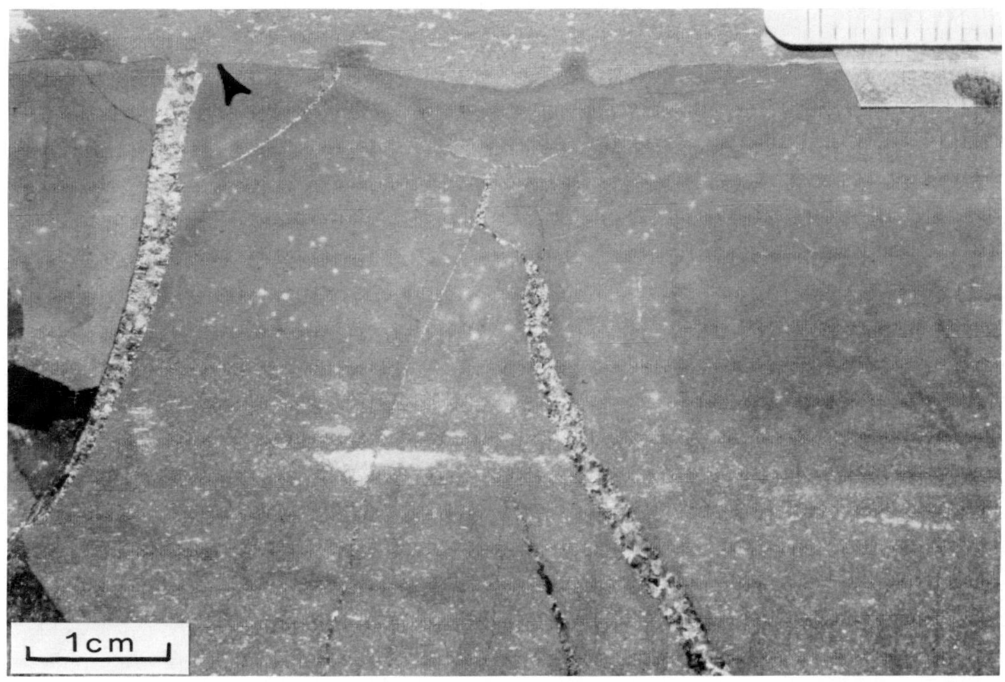

FIG. 5.—Fracture filled with early cement truncating on a bedding plane (see arrow), suggesting an origin related to sediment characteristics rather than a tectonic overprint, Bone Spring Limestone, Guadalupe Canyon, Texas.

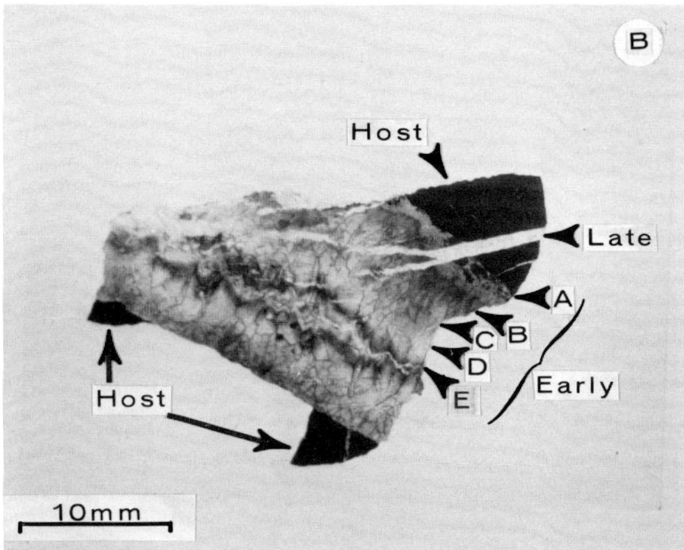

FIG. 6.—(A) Crosscutting relationship between the early and late cement. Late cement fills a fracture set that has a near vertical orientation and a strike averaging N 23° W (n = 40), Bone Spring Limestone, Guadalupe Canyon, Texas. (B) Thin section showing crosscutting relationship between early and late cement. Five petrographically distinct zones (A to E) are found in the early cements; however, we found no significant difference in T_h or $\delta^{18}O$ between zones in the early cement.

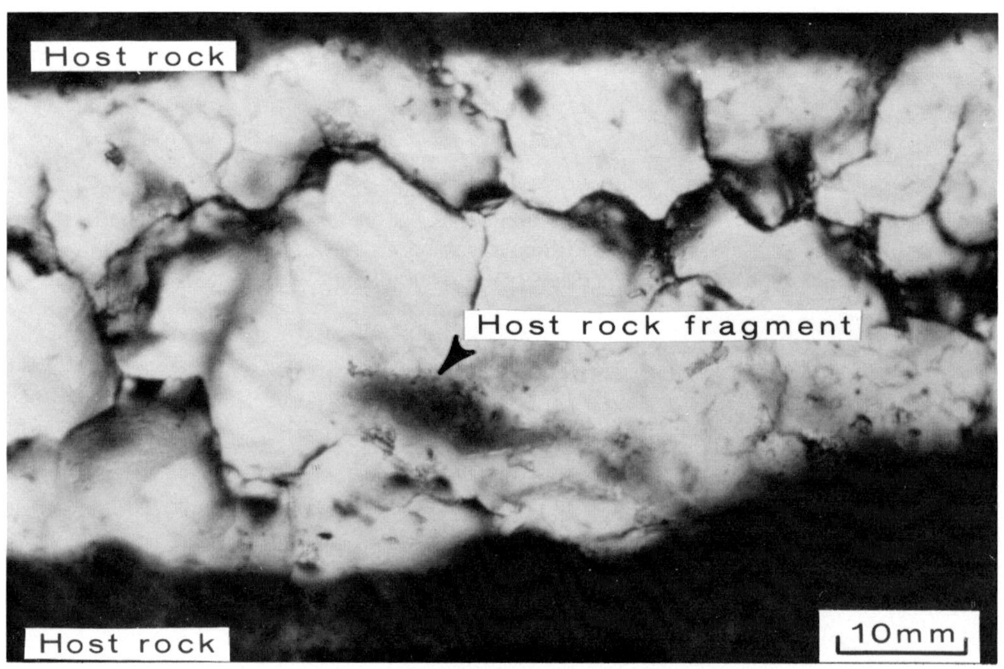

FIG. 7.—Fragment of early cement (see arrow), containing fluid inclusions that have a "low" homogenization temperature (T_h) of 55°C to 75°C, embedded in the late cement containing fluid inclusions that have a T_h of about 90°C to 135°C. Bone Spring Limestone, Guadalupe Canyon, Texas.

growth-zoned crystals (Figs. 9 and 10), or as inclusions along cathodoluminescent zones. Easily recognized secondary inclusions in the early cement were typically found in healed fracture planes that crosscut cathodoluminescent zones or crystallographic features, or in three-dimensional arrays of inclusions along cleavage or twinning planes. Secondary fluid inclusions along the fracture planes may have a yellow fluorescence, suggesting that they contain an oil with an API gravity of less than 45° (Anderson, 1975). Secondary fluid inclusions in the late cement are rare, but a few planes of inclusions along healed fracture planes were observed. Some of these secondary inclusions have a low T_h and may have formed during uplift of the basin.

The homogenization temperatures of inclusions in early and late cements overlap, in part, because neomorphic effects obscure cement relations. Primary inclusions related to crystal growth or those occurring as isolated clusters away from potentially neomorphosed cements indicate distinct thermal events. The T_h data from primary inclusions indicate the early cement precipitated at about 55° to 75°C (average 70°C) and the late cement at about 90° to 135°C (average 110°C; Fig. 12). Although we could petrographically

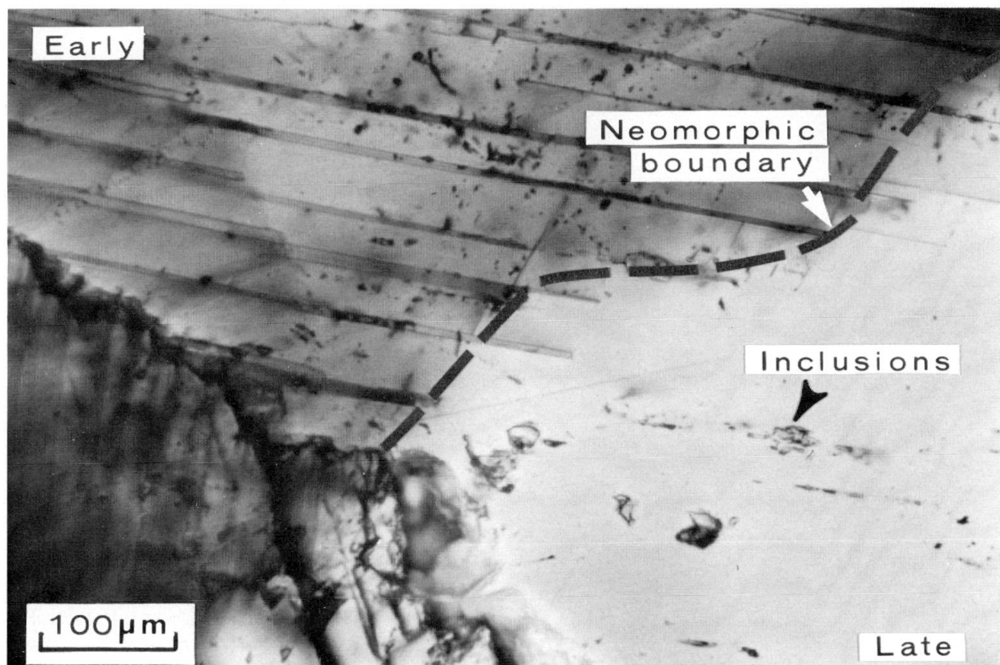

FIG. 8.—Neomorphism of early cement by late cement is indicated by: (1) the continuation of twin planes from the early cement into the crosscutting late cement; (2) remnant fluid inclusions commonly in fragments of early cement that have T_h like that of the early cement but that physically exist in the higher temperature late cement (Fig. 11). Note that the early cement is petrographically distinguished from the late cement by the presence of twinning. Bone Spring Limestone, Guadalupe Canyon, Texas.

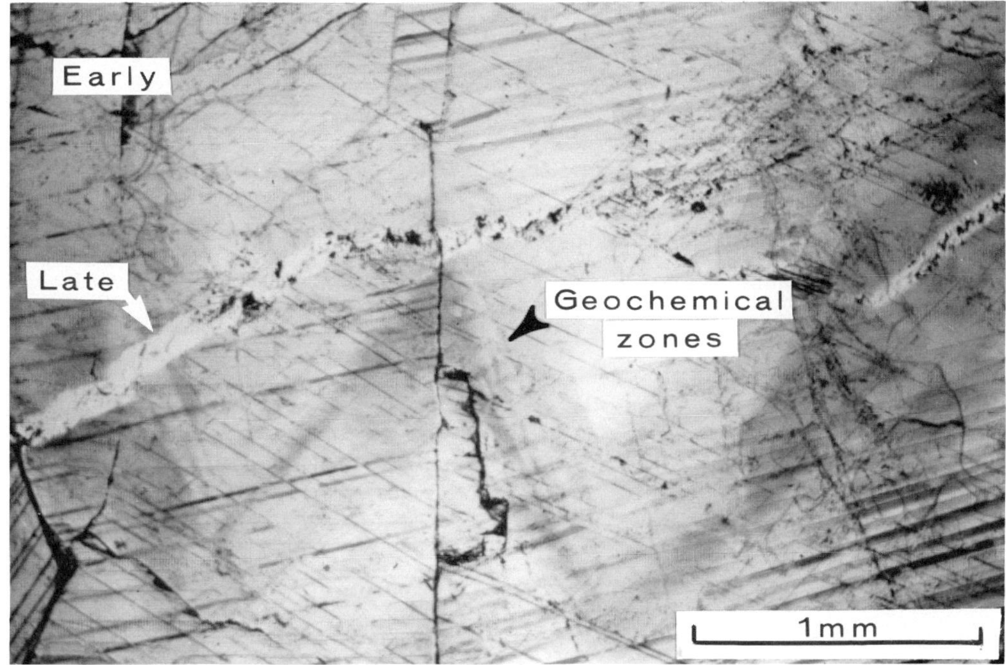

FIG. 9.—Growth zones (see dark arrow) in the early cement. Fluid inclusions in this cement homogenize at about 55° to 75°C. Early cement is crosscut by late cement (light arrow) that contains fluid inclusions that have a T_h of about 90° to 135°C. Many inclusions in both cements are considered primary because they lie in bands parallel to luminescent bands and/or crystal growth zones. Bone Spring Limestone, Guadalupe Canyon, Texas.

identify five zones in the early cement, the T_h data showed approximately the same range for all zones and we chose to lump the data into early and late cement groups.

Aqueous fluid inclusions.—

Aqueous fluid inclusions are rare in these cements. The aqueous inclusions, 10 to 20 μm in size, were detected by their non-fluorescent character and distinct freezing behavior. We did not homogenize the aqueous fluid inclusions because reequilibration or stretching during heating is apparently a significant problem in calcite hosts. The aqueous inclusions are difficult to freeze because of their high salinity. The few we measured would not allow conclusive inferences concerning the salinity of the fluid present during calcite precipitation. The inclusions that were measured did not initially freeze upon cooling to −196°C, although the vapor bubble did expand slightly. Heating the supercooled inclusions to the range −80°C to −100°C nucleated a solid phase accompanied by deformation of the vapor bubble. Further heating showed an indistinct eutectic tem-

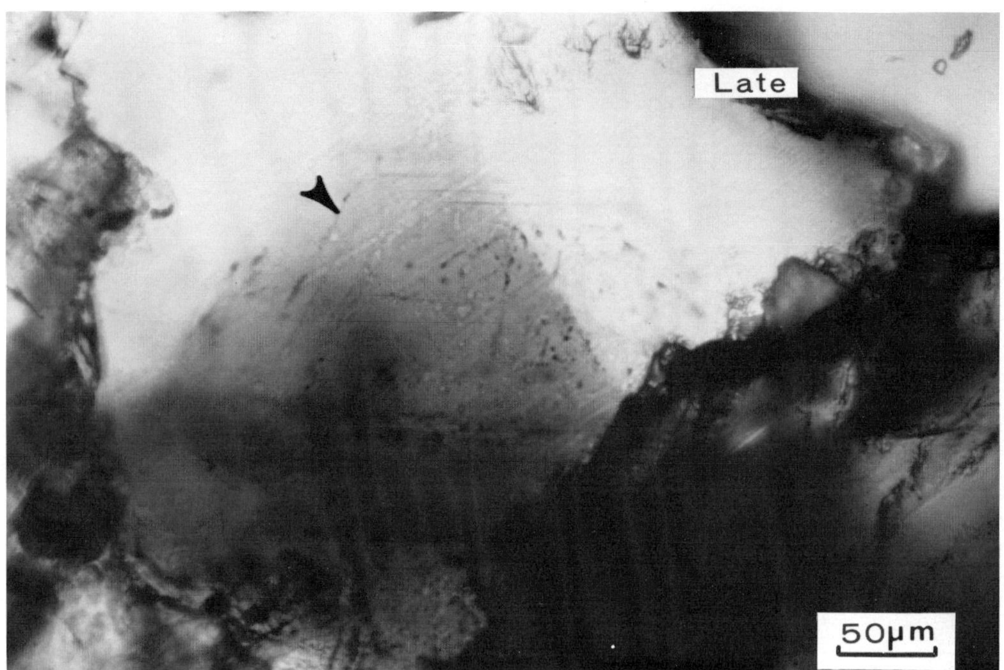

FIG. 10.—Growth-zoned (arrow) single calcite crystal in late cement. Bone Spring Limestone, southern Guadalupe Mountains, Texas. Oil fluid inclusions trapped along these growth zones are considered to be primary because they are related to precipitation of the enclosing mineral. Partial shading imparted by a brown crystal below the zoned one.

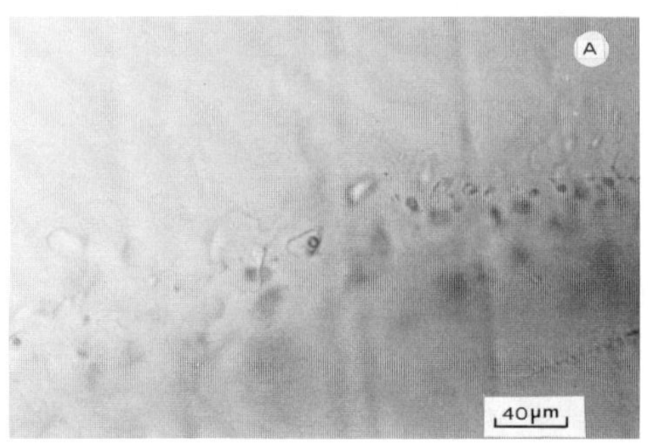

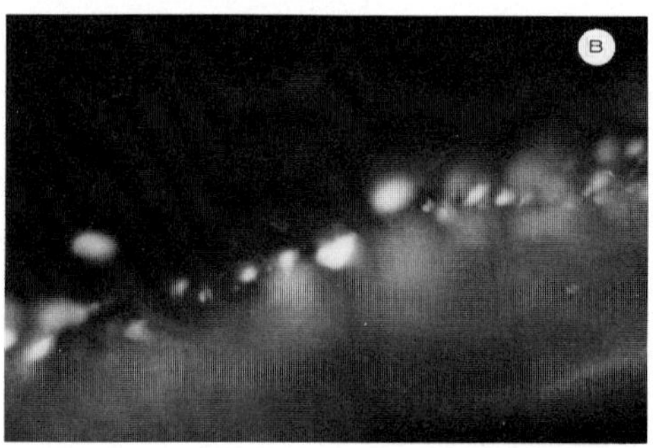

FIG. 11.—A secondary plane of oil inclusions in early cement, Bone Spring Limestone, southern Guadalupe Mountains, Texas. (A) transmitted light and (B) fluorescence.

perature but final melting occurred around −45°C. These characteristics suggest a calcium-chloride-rich brine (Roedder, 1984).

STABLE CARBON AND OXYGEN ISOTOPIC COMPOSITION OF CEMENTS

Stable oxygen and carbon isotopic analyses of early and late cements and host limestone are given in Table 2 and illustrated graphically in Figure 13. Although we were able to distinguish several zones of early cement based on petrographic characteristics, these zones generally varied less than 1‰ with no discernable pattern and are lumped together for convenience. The average oxygen and carbon isotopic composition for early cement is −0.35‰ and 0.59‰, respectively; late cement values are −19‰ and −8.48‰. Late cements exhibited fine cathodoluminescent zonation, usually on a smaller scale than we could sample. In one sample bright and dull cathodoluminescent zones could be sampled separately (Table 2, analyses 1a and 1b) and were similar in isotopic composition.

TABLE 2.—STABLE CARBON AND OXYGEN ISOTOPE ANALYSES OF EARLY AND LATE CALCITE CEMENTS FROM FRACTURE FILLINGS IN THE BONE SPRING LIMESTONE. ALL ISOTOPE MEASUREMENTS IN PPT.

Sample	$\delta^{13}C$	$\delta^{18}O$
Host limestone 2	−0.21	−1.07
Host limestone 4	−0.80	−0.85
Early cement 2c1/2	1.19	−0.29
Early cement 2c3	0.17	0.59
Early cement 2c4	−0.27	−0.76
Early cement 2c5	1.53	−2.19
Early cement 7a1/2	0.77	−0.33
Early cement 7a6	1.79	−1.59
Early cement 2	1.27	0.14
Early cement 3a	0.53	0.12
Early cement 3b	0.31	0.89
Early cement 4a	−0.44	0.03
Early cement 4b	−0.36	−0.49
Late cement 2c6	−0.16	−8.98
Late cement 1a	−0.19	−8.15
Late cement 1b	−0.28	−8.05
Late cement 7	−0.14	−8.72

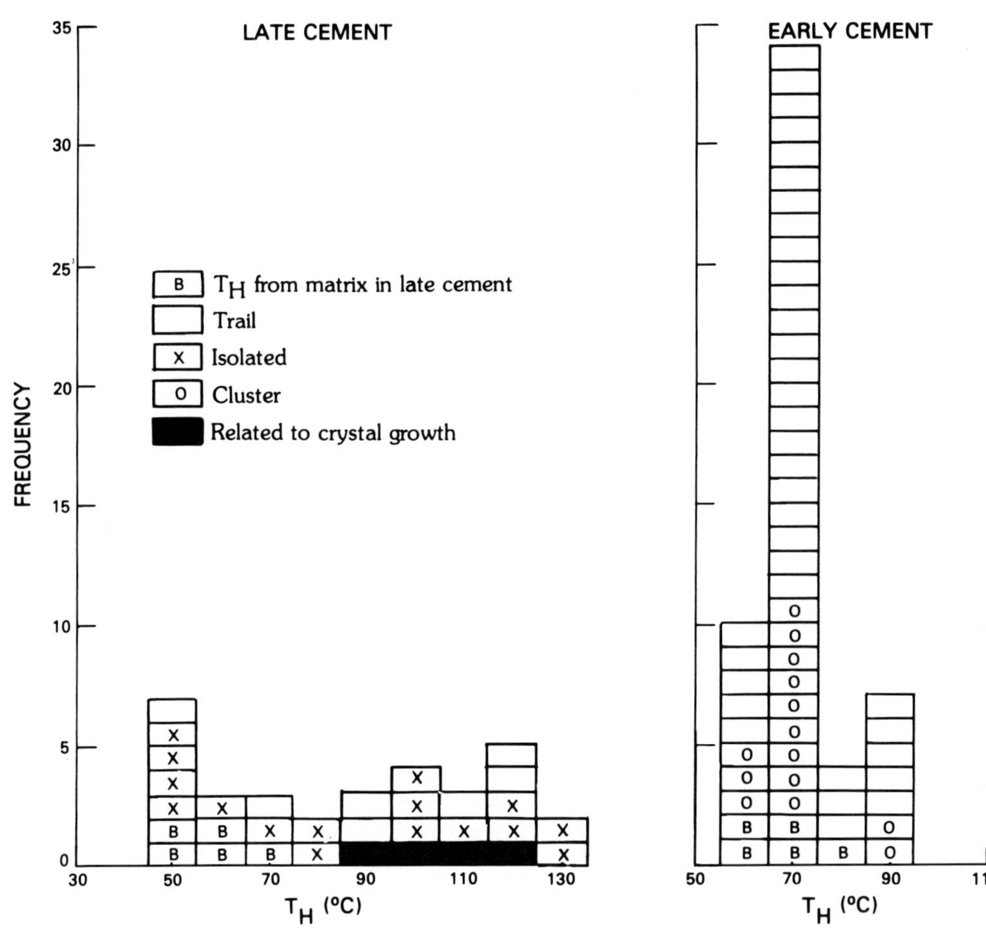

FIG. 12.—Fluid inclusion homogenization temperatures in early and late cements, Bone Spring Limestone, southern Guadalupe Mountains, Texas. We interpret primary fluid inclusions in these samples as: inclusions related to crystal growth; inclusions occurring in isolated randomly arrayed clusters; or as isolated large irregular-shaped inclusions.

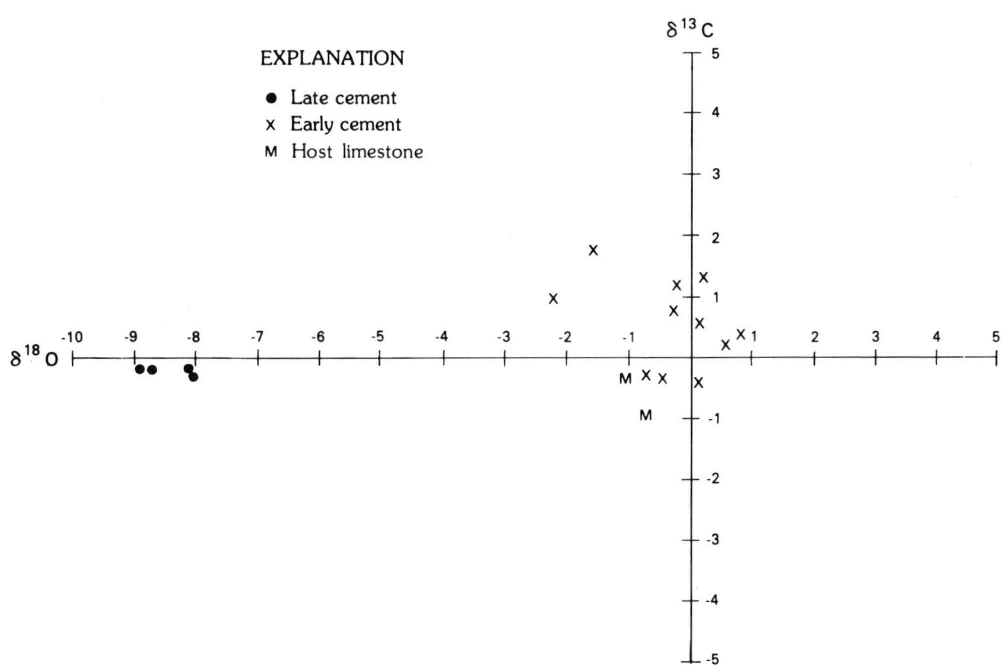

FIG. 13.—Isotope data for early and late calcite cements in fracture fillings, Bone Spring Limestone, southern Guadalupe Mountains, Texas.

The difference in $\delta^{18}O$ between early and late cements is about 8‰. If precipitation took place from water with constant oxygen isotopic composition, it would indicate a temperature difference of about 54°C at $\delta^{18}O$ of water at +4‰ and 68°C at $\delta^{18}O$ of water at +10‰ (Table 3). The observed isotopic change in calcite, however, may be due, in part, to isotopic changes in precipitating water as well as to temperature changes. In succeeding sections we will estimate isotopic composition of the precipitating water using fluid-inclusion homogenization temperature (T_h) to constrain possible temperature ranges.

TABLE 3.—POSSIBLE TEMPERATURE RANGES INDICATED BY 8‰ CHANGE IN $\delta^{18}O$ BETWEEN EARLY AND LATE CALCITE CEMENTS FOR SOME REASONABLE RANGES IN TEMPERATURE AND $\delta^{18}O$ WATER. ALL ISOTOPE MEASUREMENTS IN PPT.

Early Cement			Late Cement		
$\delta^{18}O$(cement)	$\delta^{18}O(H_2O)$	T°C	$\delta^{18}O(CaCO_2)$	$\delta^{18}(H_2O)$	T°C
−0.35	+4.0	38	−8.48	+4.0	92
−0.35	+5.0	44	−8.48	+5.0	100
−0.35	+6.0	50	−8.48	+6.0	108
−0.35	+7.0	56	−8.48	+7.0	117
−0.35	+8.0	62	−8.48	+8.0	125
−0.35	+9.0	69	−8.48	+9.0	134
−0.35	+10.0	76	−8.48	+10.0	144
−0.35	+11.0	83	−8.48	+11.0	153
−0.35	+12.0	91	−8.48	+12.0	163

BURIAL AND THERMAL HISTORY

Deposition of the Leonardian Bone Spring Limestone and its subsequent burial in the southern Guadalupe Mountains area were nearly complete by the end of the Permian (King, 1948; Fig. 14). The Bone Spring Limestone remained buried at depths near 2 km until the Tertiary. Mesozoic sedimentation was relatively thin and offset by subsequent erosional events (McGowen et al., 1983; Adams, 1965). Although maximum burial depth was reached near the close of the Permian, maximum burial did not correspond with the time of maximum temperature. Calcite cement in the early fractures contain fluid inclusions that have homogenization temperatures (T_h) of about 70°C, 50°C lower than those of definite primary fluid inclusions (average T_h = 110°C) in the later calcite cement (Fig. 12). These late vertical fractures strike N 20° W and appear genetically related to the early Tertiary (Laramide) Basin and Range faulting. These faults also have strong north-northwest orientation (Fig. 15). Tertiary tectonism was accompanied by volcanism in the area (Horak, 1975; Calzia and Hiss, 1978), which may account for the higher temperature fluids indicated by our data.

If it is assumed $\delta^{18}O$ water remained constant, an 8‰ shift in $\delta^{18}O$ between the early and late cements indicates formation at a temperature difference of at least 50°C, based on the calcite-water oxygen isotope geothermometer of Epstein et al. (1953). This relative temperature difference is consistent with the observed difference in T_h in the two inclusion groups. Vitrinite reflectance data from the Bone Spring Limestone (R_m = 0.7 percent; Fig. 16), interpreted using a calibration to maximum temperature (Barker and Pawlewicz, in prep.), also argues the maximum temperature reached in this section of the Delaware Basin was about 120°C. The average T_h of 110°C measured in primary fluid inclusions from the late cement is comparable to the maximum temperature indicated by vitrinite refectance geothermometry. These geothermometric data indicate maximum burial temperature for the western Delaware Basin was recorded in the Tertiary.

FLUID HISTORY

Several lines of evidence indicate the source of the oil trapped in the Bone Spring fracture-filling cements is a more deeply buried and thermally mature sediment. Our samples

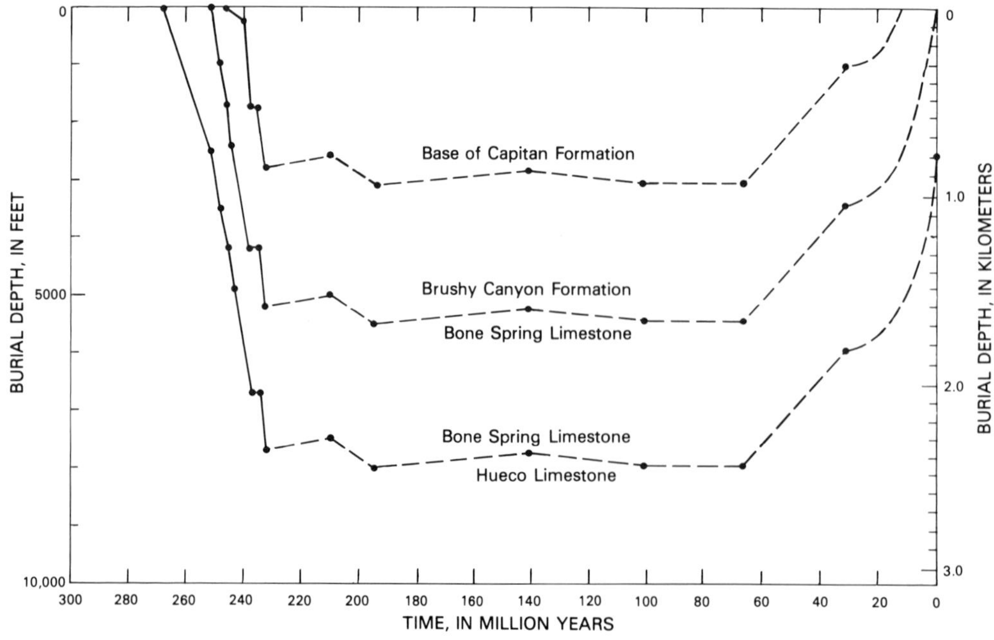

FIG. 14.—Burial history reconstruction, Bone Spring Limestone, southern Guadalupe Mountains, Texas. Based on stratigraphic data, tectonic, and depositional history reconstruction of King (1948); Adams (1965); and Horak (1975).

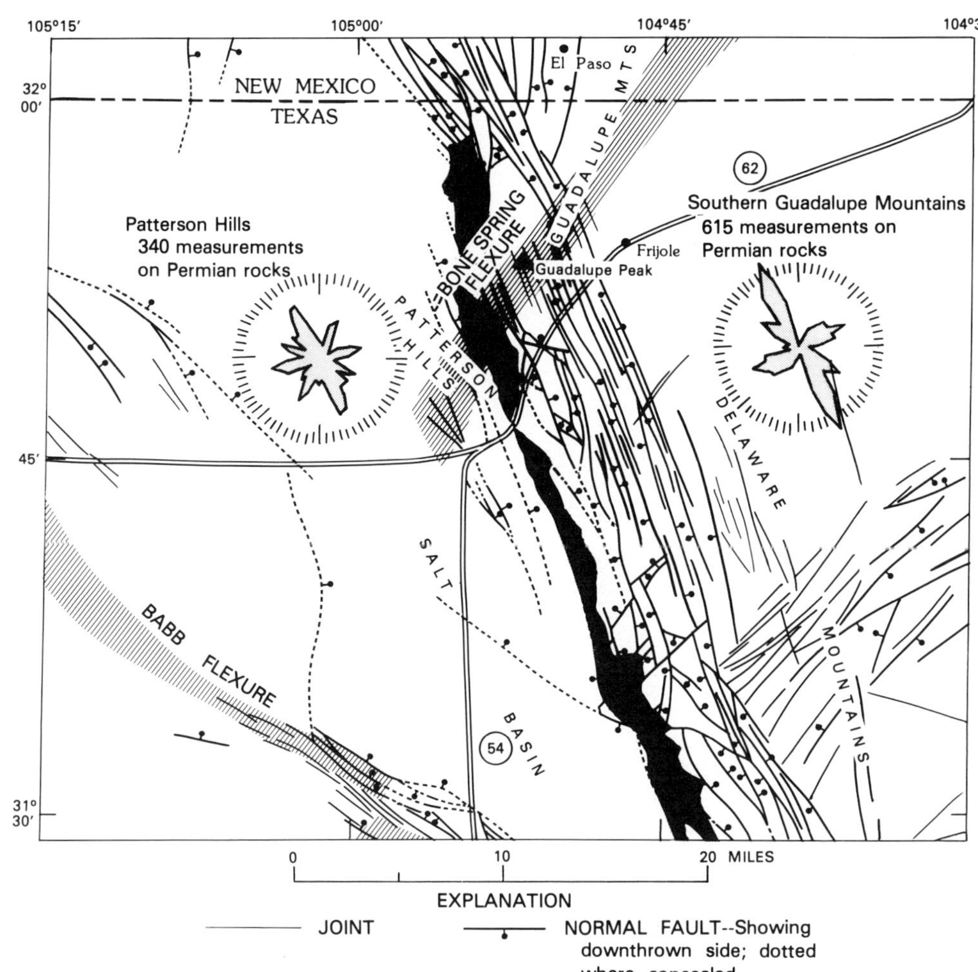

FIG. 15.—Fracture and fault orientation, southern Guadalupe Mountains, Texas. Based on King (1948). Roses are trends of joints grouped as percent of total per 5° of arc. Black color is the outcrop of the Bone Spring Limestone.

of Bone Spring Limestone have a R_m of 0.7% that suggests these rocks have just entered the oil window (Tissot and Welte, 1978). Oil in the primary fluid inclusions fluoresces a color indicating an API gravity of at least 45° which is much more mature than expected for a unit just beginning to generate oil. Oil with this API gravity is similar to oil from more deeply buried Delaware Basin sources (Holmquest, 1966). The aqueous fluid inclusion freezing data suggest the fluid precipitating cement in the Bone Spring fractures was a calcium-chloride-rich brine, a composition that commonly indicates a deep basin source (White, 1965).

The yellow-fluorescent secondary inclusions trapped along late fractures contain a lower API gravity oil that is more consistent with the thermal maturity of the Bone Spring. These inclusions may be evidence of oil migration from a less thermally mature source, possibly the Bone Spring Limestone itself.

The oxygen isotopic composition of the deep basin brine may be estimated by combining fluid inclusion-derived temperatures with the oxygen isotopic composition of enclosing calcite in the relationship of Epstein et al. (1953). We calculate that early cements precipitated from waters having an oxygen isotopic composition of +9.2‰ and late calcite from waters at +7.5‰. Considering the ranges of T_h averaged for these calculations, the 1.7‰ difference between these values is not considered significant. The two cement generations may have precipitated from water with essentially the same oxygen isotopic composition.

These oxygen isotope compositions are not unreasonable for deep basin brines. Fluid from relatively large inclusions in halite from deeply buried Delaware Basin evaporites originally contained water with $\delta^{18}O$ of about +10‰ before mixing with isotopically light meteoric water (O'Neil et al., in prep.: cited by Roedder, 1984, p. 318). Other deep basin brines show a similar signature—water from the Jurassic Smackover Formation has $\delta^{18}O$ of +5‰ (Moore and Druckman, 1981), and water from the Sunniland Formation of south Florida has $\delta^{18}O$ of +10‰ (Halley, unpublished data). Land and Prezbindowski (1981) have published a compendium of analyses of waters from Lower Cretaceous carbonates of south-central Texas and many analyses are in the range of 0 to +10‰. Some waters have isotopic compositions greater than +12‰ and one value reaches 29.6‰. This suggests that both early and late cements were formed at depth, prior to the influx of meteoric waters that accompanied uplift of the Guadalupe Mountains. This interpretation is consistent with the fluid inclusion and vitrinite reflectance data presented earlier.

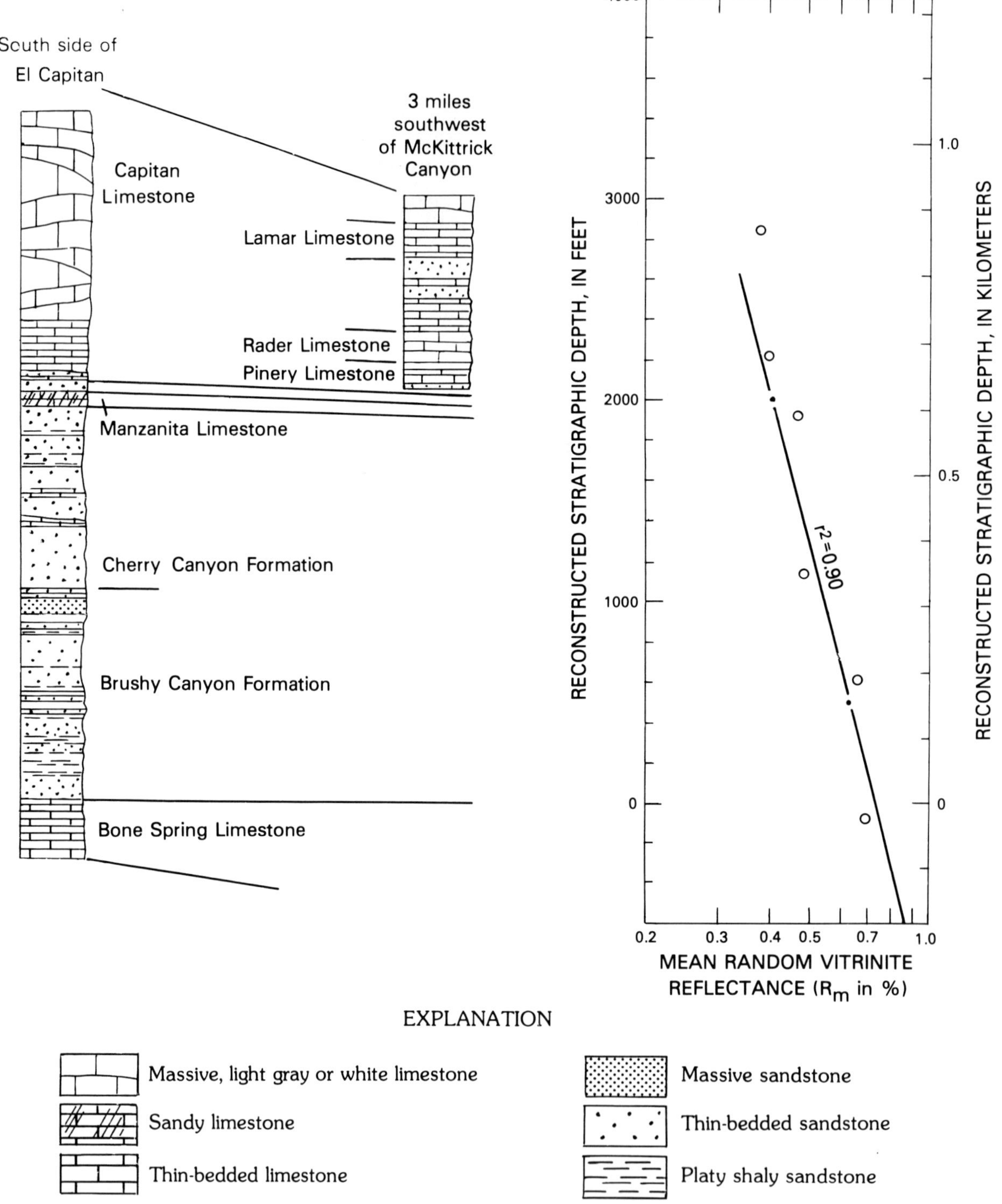

FIG. 16.—Vitrinite reflectance versus stratigraphic depth, selected Permian carbonates of southern Guadalupe Mountains, Texas. Stratigraphic thickness and nomenclature from King (1948).

DISCUSSION

In order to estimate Paleozoic geothermal gradients in the basin, we use the average geothermal gradient of 30°C/km for Paleozoic basins in a stable platform environment given by Tissot and Welte (1978). This paleogeothermal gradient is used to estimate paleodepth in the Delaware Basin during the Paleozoic, and, in conjunction with our burial history reconstruction, the time of certain diagenetic events. A paleogeothermal gradient of 30°C/km during the Paleozoic indicates that oil migrated in the Late Permian after being at a burial depth of about 1–2 km at oil-inclusion homogenization temperatures of 55° to 75°C. The continued trapping of oil fluid inclusions from the early- to late-cement time implies that the petroleum generation and migration recorded in the Bone Spring Limestone began in the Late Permian and minor oil migration or remobilization continued at least into the early Tertiary.

The present-day geothermal gradient in the area is about 20°C/km (Geothermal Survey of North America Subcommittee, 1976). Both fluid inclusion data and vitrinite reflectance data from the area suggest a higher geothermal gradient in the past. T_h data from deep burial, and void-filling, blocky calcite cement in the Capitan Limestone indicate a temperature of about 75°C at a stratigraphic position approximately 0.9 km above the Bone Spring Limestone (B.L. Crysdale, pers. commun., 1984). Assuming these data are comparable to the Bone Spring late-cement average T_h of about 110°C, they suggest a paleogeothermal gradient of about 50°C/km.

Vitrinite reflectance between the Lamar Limestone member of the Bell Canyon Formation (basinal equivalent of the Capitan Limestone) and the deeper Bone Spring Limestone increases from 0.4% to 0.7% (Fig. 16). These R_m values correspond to temperatures of about 75°C and 120°C, respectively, when interpreted in terms of maximum burial temperature (Barker and Pawlewicz, in prep.). Using the same vertical separation (0.9 km), the R_m data convert to a geothermal gradient of about 50°C/km. These high paleogeothermal gradient estimates are in better agreement with observed maximum temperature (about 120°C at the maximum burial depth of 2 km) than is the modern geothermal gradient. We suggest that the higher temperatures in the late cement are due to an increased geothermal gradient during the Laramide orogeny and later development of the Basin and Range Province in the early Tertiary. Paleogeothermal gradient estimates, when compared with the present-day geothermal gradient, indicate that heat flow has decreased markedly on the western edge of Delaware Basin.

CONCLUSIONS

Comparative geothermometric analysis of Bone Spring Limestone exposures on the western margin of the Delaware Basin support the following conclusions.

(1) Oil migration occurred in the Bone Spring during the Late Permian at about 70°C and after burial at depths of 1–2 km.

(2) Maximum burial temperature of about 120°C was recorded in the Tertiary. Some oil migrated at this time.

(3) Fluid inclusion and vitrinite reflectance data from the Bone Spring Limestone suggest the maximum geothermal gradient was about 50°C/km, which is much higher than the present-day geothermal gradient of 20°C/km and higher than the estimated Paleozoic geothermal gradient of 30°C/km. The geothermal gradient has apparently decreased in the western Delaware Basin.

(4) Isotopic and fluid inclusion evidence indicate that the sources of oil-bearing solutions were apparently deep basin brines. The brines had oxygen isotope characteristics typical of highly evaporated water similar to water produced from wells in Upper Permian evaporites in the basin center.

ACKNOWLEDGMENTS

We thank M. J. Pawlewicz for vitrinite reflectance measurements and B. L. Crysdale for kerogen separation. Charles Threldkeld and April Vuletich kindly performed mass spectrometry and sample preparation for stable isotope determinations in these rocks. Robert Burruss, Richard Major, and Nancy Naeser read earlier versions of the manuscript and made comments which greatly improved the final product.

REFERENCES

Adams, J. E., 1965, Stratigraphic-tectonic development of the Delaware Basin: American Association of Petroleum Geologists Bulletin, v. 49, p. 2140–2148.

Anderson, Gene, 1975, Coring and core analysis handbook: Petroleum Publishing Company, Tulsa, p. 69.

Barker, C. E., 1982, A rapid method for concentrating sedimentary organic matter for vitrinite reflectance analysis: Journal of Sedimentary Petrology, v. 52, p. 663–664.

———., and Reynolds, T. J., 1984, Preparing doubly polished sections of temperature sensitive sedimentary rocks: Journal of Sedimentary Petrology, v. 54, p. 635–636.

Baskin, D. K., 1979, A method for preparing phytoclasts for vitrinite reflectance analysis: Journal of Sedimentary Petrology, v. 49, p. 633–635.

Burruss, R. C., 1981, Hydrocarbon fluid inclusions in studies of sedimentary diagenesis: Mineralogical Association of Canada, Short Course Handbook, v. 6, p. 138–156.

Calzia, J. P. and Hiss, W. L., 1978, Igneous rocks in the northern Delaware Basin, New Mexico and Texas: New Mexico Bureau of Mines and Mineral Resources Circular 159, p. 39–46.

Epstein, Samuel, Buchsbaum, Ralph, Lowenstam, H. A., and Urey, H. C., 1953, Revised carbonate-water isotopic temperature scale: Geological Society of America Bulletin, v. 64, p. 1315–1326.

Geothermal Survey of North America Subcommittee of the AAPG Research Committee, 1976, Geothermal gradient map of North America: American Association of Petroleum Geologists and U.S. Geological Survey, scale 1:5,000,000.

Goldstein, T. P., 1983, Geocatalytic reactions in formation and maturation of petroleum: American Association of Petroleum Geologists Bulletin, v. 67, p. 152–159.

Holmquest, H. J., 1966, Stratigraphic analysis of source-bed occurrences and reservoir oil gravities: American Association of Petroleum Geologists Bulletin, v. 50, p. 1478–1486.

Horak, R. E., 1975, Tectonic relationship of the Permian Basin to the Basin and Range Province, *in* Hills, J. S., ed., Exploration from the Mountains to the Basin: American Association of Petroleum Geologists, Southwest Section, Joint meeting, Transactions, El Paso Geological Society, p. 60–94.

King, P. B., 1948, Geology of the southern Guadalupe Mountains, Texas: U.S. Geological Survey Professional Paper 215, 183 p.

Land, L. S., and Prezbindowski, D. R., 1981, The origin and evolution of saline formation water, Lower Cretaceous carbonates, south-central Texas, U.S.A.: Journal of Hydrology, v. 54, p. 51–74.

McGowen, J. H., Granata, G. E., and Seni, S. J., 1983, Depositional setting of the Triassic Dockum Group, Texas panhandle and eastern New Mexico, *in* Reynolds, M. W., and Dolly, E. D., eds., Mesozoic Paleogeography of West-Central United States: Society of Economic Paleontologists and Mineralogists, Rocky Mountain Section, p. 13–38.

Moore, C. H., and Druckman, Yehezkeel, 1981, Burial diagenesis and porosity evolution, upper Jurassic Smackover, Arkansas and Louisiana: American Association of Petroleum Geologists Bulletin, v. 65, p. 597–628.

Newell, N. D., Rigby, J. K., Fischer, A. G., Whiteman, A. J., Hickox, J. E., and Bradley, J. S., 1953, The Permian Reef complex of the

Guadalupe Mountains region, Texas and New Mexico: San Francisco, W. H. Freeman, 236 p.

Pray, L. C., and Stehli, F. G., 1963, Allochthonous origin, Bone Spring "patch reefs," west Texas: Geological Society of America, Special Paper 73, p. 218–219.

Prezbindowski, D. R., 1980, Microsampling technique for stable isotopic analysis of carbonates: Journal of Sedimentary Petrology, v. 50, p. 643–644.

Roedder, Edwin, 1984, Fluid inclusions: Mineralogical Society of America, Reviews in Mineralogy, v. 12, 644 p.

Scholle, P. A., and Halley, R. B., 1980, Upper Paleozoic depositional and diagenetic facies in a mature petroleum province: U.S. Geological Survey, Carbonate Short Course Notes, 195 p.

Sterner, S. M., and Bodnar, R. J., 1984, Synthetic fluid inclusions in natural quartz. I. Compositional types synthesized and applications to experimental geochemistry: Geochimica et Cosmochimica Acta, v. 48, p. 2659–2668.

Tissot, B. P., and Welte, D. H., 1978, Petroleum formation and occurrence: New York, Springer-Verlag, 538 p.

Waples, D. W., 1981, Organic geochemistry for exploration geologists: Minneapolis, Burgess, 151 p.

Werre, R. W., Jr., Bodnar, R. J., Bethke, P. M., and Barton, P. B., Jr., 1979, A novel gas-flow fluid inclusion heating/freezing stage: Geological Society of America Abstracts with Programs, v. 11, p. 539.

White, D. E., 1965, Saline waters of sedimentary rocks, *in* Young, Addison and Galley, J. E., eds., Fluids in Subsurface Environments: American Association of Petroleum Geologists Memoir, v. 4, p. 342–366.

APPENDIX

METHODS

Fluid Inclusions

Stage calibration.—

Phase transition temperatures in fluid inclusions were determined in a heating/freezing stage designed by the U.S. Geological Survey (Werre et al., 1979). The thermocouple was initially calibrated outside the stage by adjusting the digital thermometer response (zero point; positive and negative slope) to the thermocouple temperature while it was immersed in baths at the freezing point of mercury, various water-sodium chloride mixtures, distilled water, zinc, tin, and aluminum over a range of −38.9°C to 660°C. The heating/freezing stage was internally calibrated using the adjusted thermocouple/digital thermometer system to measure artificial fluid inclusions with known homogenization temperature (T_h) and composition (Sterner and Bodnar, 1984), allowing direct measurement of stage gradients and necessary temperature corrections. These standards were used to calibrate the stage in the range of −56.6°C (triple point of CO_2) to +374.5°C (critical temperature of H_2O). These measurements indicate the accuracy of the temperature determinations is +/−1°C in heating mode and +/−0.2°C in freezing mode.

Doubly-polished sections suitable for fluid inclusion analysis were prepared using the method of Barker and Reynolds (1984).

Primary versus secondary fluid inclusions.—

Primary or pseudosecondary fluid inclusions result from fluid trapped in crystal defects during growth. They are the result of random processes affecting crystal growth (Roedder, 1984) and the petrography reflects this origin. Fluid inclusions related to crystal growth typically are large, randomly occurring inclusions. Textural evidence of the processes of primary fluid-inclusion trapping may not be clearly expressed in the rock, however, and their origin may be obscure. Direct evidence of the primary nature of fluid inclusions is rare in sedimentary rocks. Some evidence we used to infer the inclusion origin is shown in Figures 9–11. We also used cathodoluminesence to document that some planes of what appear to be secondary inclusions are actually primary fluid inclusions along geochemical zones in the cement. This cathodoluminescent technique should be used with caution because overheating and stretching of the inclusions can occur from the impinging electron beam. The T_h of the fluid inclusion should be measured first, and then the inclusion should be observed under cathodoluminesence to add evidence about its origin.

Most fluid inclusions are secondary (Roedder, 1984), resulting from fracturing/healing or recrystallization of the host after precipitation. A common assumption, lacking evidence to the contrary, is that an inclusion is secondary. Blind assignment of a fluid inclusion, however, to primary or secondary origin without direct evidence to either origin is equally misleading. In this study, inclusions without any evidence of origin are simply identified as fluid inclusions.

If a minimum burial temperature estimate for a rock is required, then most fluid inclusion types and origins are significant. Several exceptions exist, but the two most common ones are: (1) fluid inclusions that have necked and closed with two phases present (ex: liquid and vapor), resulting in the trapping of variable phase ratios, and (2) trapping of a heterogeneous fluid in an inclusion (ex: noncondensible gas and water).

All measurable inclusion types were analyzed, and, where possible, separated according to textural evidence of origin. The interpretation of the significance of an inclusion's origin may change but the T_h value for this inclusion will not. Arbitrarily excluding certain inclusion types at the microscope can lead to a significant bias in determining the minimum temperature of formation. Therefore, we attempted to measure all fluid inclusions with consistent vapor-to-liquid ratio in a mineral zone or secondary plane. Most changes in a fluid inclusion that make geothermometry invalid result in variable vapor-to-liquid ratio and these inclusions should not be used to estimate minimum temperature of formation. Fluid inclusions with variable vapor/liquid ratio can still be used to measure fluid composition. Using the total scan approach, many assumed or obvious secondary inclusions are found to record conditions similar to those that existed when the detectable primaries formed (Fig. 12), although the exact time of formation is still unknown. Conversely, some inclusions with textural relationships indicative of a primary origin appear to be secondary inclusions formed during neomorphism of the early cement.

Differentiation of aqueous and oil fluid inclusions.—

Although oil inclusions commonly fluoresce and aqueous inclusions do not, identification of aqueous inclusions based solely on the presence of nonfluorescent fluid can be misleading. Condensate-type hydrocarbons composed mainly of alkanes do not fluoresce. The different behavior of oil and aqueous inclusions during cooling or heating in the microscope stage are better operational criteria for differentiating these inclusions. Distinct freezing and melting behavior is a characteristic of aqueous fluids and distinguishes them from oil fluid inclusions which show indistinct freezing behavior or none (Burruss, 1981). Hydrocarbon fluids also have a greater compressibility than brine (Roedder, 1984) and require less undercooling below T_h to nucleate a vapor phase. We have observed oil fluid inclusions nucleating a vapor phase with less than 5°C undercooling versus 30°C to 40°C undercooling required for aqueous inclusions. This operational method of differentiating aqueous and oil inclusions is based on diagnostic physical characteristics of the fluids rather than relying on the sometimes ambiguous fluorescence properties. Using fluoresence, heating and cooling behavior, and vapor phase nucleation allows a clear discrimination of the gross fluid type in an inclusion.

Petroleum metastability problem.—

Petroleum is thermodynamically metastable in crustal temperature and pressure conditions (Tissot and Welte, 1978). Thus, petroleum trapped in an inclusion is susceptible to alteration with increasing burial and/or temperature. This possible change is important because fluid inclusion geothermometry assumes that the fluid in the mineral chamber has not changed molar volume after trapping. Petroleum fluid inclusions tend to react isochemically to form high molecular weight solid bitumen or with increasing temperature may crack to form smaller molecular weight hydrocarbons. Changes of fluid density within inclusions will change T_h by altering the vapor-to-liquid ratio and may invalidate the assumption that T_h indicates the minimum entrapment temperature.

Fortunately, bitumen is petrographically detectable as dark, translucent brown to black opaque solids (Roedder, 1984). Petroleum fluid inclusions in the Bone Spring Limestone do not contain optically detectable bitumens and apparently this problem does not affect our samples. Detecting thermal cracking of originally higher molecular weight oil could be difficult to observe optically because solids would not necessarily be formed. If thermal cracking occurred in all inclusions of a sample, they would tend to have similar vapor/liquid ratios. Petroleum cracking reactions in the absence of catalysts do not occur at geologically significant rates in temperatures below about 150°C (Goldstein, 1983). Thermal maturity of the Bone Spring (discussed in the burial history section) indicates it has not experienced temperatures in excess of 120°C, and apparently thermal cracking of petroleum is not important in our samples.

Pressure correction.—

The most direct use of fluid inclusion homogenization temperature data is to assume the data represent a minimum temperature of entrapment (or host mineral formation) and a pressure correction should not be applied. Several conditions are necessary if T_h is to be used as the inclusion trapping temperature (T_t). Common methods to correct T_h to T_t require knowledge of the phases present in an inclusion, *P-V-T* properties of this system, and depth of inclusion formation (pressure correction method) or similar physico-chemical data from a different composition inclusion (intersecting isochore method). These requirements make pressure correction of fluid inclusion data a difficult, error-prone procedure.

Oil inclusions may not require a pressure correction and T_h may be equal to T_t (Burruss, 1981). In our study, T_h data from oil fluid inclusions were not pressure-corrected, but we only use them to estimate the minimum temperature of host mineral formation.

Light Stable Isotopes

Calcite cement and host rock were sampled from polished rock surfaces using a 0.5- to 1.5-mm-diameter carbide dental burr mounted on a miniature vertical mill (Prezbindowski, 1980). Many petrographically distinct zones in the early cement were wide enough (about 2 mm) for a homogeneous sample to be taken for whole-rock isotopic analysis. Some replicate samples were vacuum-roasted at 200°C to remove volatile organic matter to assure that organic carbon in the calcite was not significantly influencing carbon isotope data. Carbon dioxide gas was produced from 10–50 mg of ground sample by reaction in anhydrous phosphoric acid at 25°C in the U.S. Geological Survey Organic Geochemistry Laboratory. Isotopes were analyzed in a MAT 251 PM ratio mass spectrometer. This system is calibrated through a laboratory standard gas produced from National Bureau of Standards carbonate standard TSNBS-19. Routine replicate analyses of TSNBS-19 in this laboratory have a precision of 0.1°/oo (1 standard deviation) for oxygen and slightly less (0.07) for carbon. Water analyses are reported relative to standard mean ocean water (SMOW).

Vitrinite Reflectance

Samples for vitrinite reflectance analysis were taken from roadcut exposures of the Bone Spring Limestone along U.S. Highway 62, starting about 2.5 km east of its intersection with Texas State Highway 54 and continuing to the Lamar Limestone Member of the Bell Canyon Formation, exposed in another highway 62 roadcut near the McKittrick Canyon turnoff. Stratigraphic thickness between samples was determined from measured sections of King (1948). Sedimentary organic matter (OM) was prepared from these samples by digestion in dilute HCl to remove carbonate. The OM was isolated from the acid insoluble residue using mechanical disaggregation in aqueous zinc bromide solution with a density of 2.2 g/cc (Barker, 1982). The isolated OM was rinsed, freeze-dried, mounted on glass slides with epoxy, and polished (Baskin, 1979). Reflectance was measured using vertical illumination on a Zeiss[1] Universal microscope fitted with a MPM-01 microphotometric system. The photometer was restricted with a pinhole diaphram to read a 3 μm spot on the sample at 500× total magnification using a 40×/0.85 n.a. lens under oil immersion (n_e = 1.518). The system was calibrated by a Zeiss leucosapphire standard that has a reflectance of 0.58% with filtered 546 nm light.

[1]Use of trade and company names is for descriptive purposes only and does not imply endorsement by the U.S. Geological Survey.